Thermal Energy Systems

Design and Analysis

Thermal Energy Systems

Design and Analysis

Steven G. Penoncello

University of Idaho
Moscow, Idaho

CRC Press
Taylor & Francis Group
Boca Raton   London   New York

CRC Press is an imprint of the
Taylor & Francis Group, an **informa** business

CRC Press
Taylor & Francis Group
6000 Broken Sound Parkway NW, Suite 300
Boca Raton, FL 33487-2742

© 2015 by Taylor & Francis Group, LLC
CRC Press is an imprint of Taylor & Francis Group, an Informa business

No claim to original U.S. Government works

Printed on acid-free paper
Version Date: 20150716

International Standard Book Number-13: 978-1-4822-4599-8 (Hardback)

Library of Congress Cataloging-in-Publication Data

Penoncello, S. G. (Steven G.)
    Thermal energy systems : design and analysis / author, Steven G. Penoncello.
        pages cm
    Includes bibliographical references and index.
    ISBN 978-1-4822-4599-8 (hardcover : alk. paper) 1. Heat engineering. 2. Heat--Transmission. 3. Thermodynamics--Data processing. I. Title.

TJ260.P38 2015
621.402--dc23
                                                                    2015007851

Visit the Taylor & Francis Web site at
http://www.taylorandfrancis.com

and the CRC Press Web site at
http://www.crcpress.com

Contents

Preface

Thermal fluid energy system design and analysis has seen a significant transformation since the advent and widespread availability of the digital computer. Before computational power was at the fingertips of engineers, the design of such systems required extensive hand calculations, plotting performance and system curves, looking for intersection points to arrive at a design that might work as expected. Digital computing capability has changed this. Now engineers can model a thermal system, including detailed component performance and system characteristics, to solve for operating points without the laborious hand calculations required in the past. Simulation techniques applied to a system model allow the engineer to confidently predict the performance of the system before components are purchased and the system is built. Optimization techniques allow the engineer to consider changing certain system components or parameters to achieve a system that operates with respect to lowest cost, highest thermal efficiency, highest exergetic efficiency, and so on. Today's engineer can accomplish all of this using computer software.

After a brief introduction to thermal energy system's design and analysis in Chapter 1, Chapter 2 provides an introduction to engineering economics. Topics include the time value of money, discrete interest factors, economic decision making, and depreciation and taxes. The goal of this chapter is to give the reader a very basic understanding of the time value of money and how engineering economics plays a role in system design. Although these concepts can be stand-alone, there are several places in the book where the engineering economy concepts are used.

This book is intended for senior students in mechanical or chemical engineering. However, it is also appropriate for practicing engineers as a reference. The book is written assuming that the reader has an introductory background in engineering thermodynamics, heat transfer (conduction and convection), and fluid mechanics. Chapter 3 provides a brief review of this information, but it is not meant to serve as an in-depth resource for this prerequisite material. This chapter also includes an introduction to the concept of exergy and its role in environmental impact and sustainability.

There are *many* types of thermal fluid energy systems. In fact, there are far too many to include all of them in this book. Therefore, the approach taken in this book is to discuss the basics of fluid and energy transport in thermal energy systems. Chapter 4 covers the basics of fluid transport. In this chapter, the focus is on hydraulic systems including pipe networks, both gravity feed and pump feed. This chapter includes a brief review of the fundamental governing equations for the most common fluid flow scenarios seen in thermal energy systems.

Chapter 5 covers energy transport in thermal energy systems. The discussion centers around heat exchangers. This chapter provides a brief review of heat exchanger analysis methods including the logarithmic mean temperature difference (LMTD) method and the effectiveness-NTU (e-NTU) method. This chapter also goes into considerable detail for four specific types of heat exchangers commonly seen in industrial applications; double pipe (tube-in-tube), shell and tube, plate and frame, and cross flow. A general heat exchanger model is developed for each of these heat exchanger types, including the thermodynamic, heat transfer, and hydraulic performance. Heat exchangers for specialized applications are also discussed, including counterflow regenerative heat exchangers, condensers, evaporators, and boilers.

Chapter 6 is an introduction to simulation, evaluation, and optimization of thermal energy systems. Here, many of the concepts in previous chapters are brought together to show how a complete system can be modeled.

The material in this book can be covered in a typical 3-credit, 15-week semester (45-class meetings). A typical semester schedule is shown below.

Class Periods	Topics Covered
2	Introductory material (Chapter 1)
5	Engineering economics (Chapter 2)
4	Application of conservation and balance laws (Chapter 3)
1	Review of fluid flow fundamentals (Chapter 4)
2	Minor losses (Chapter 4)
1	Series and parallel pipe networks (Chapter 4)
2	Economic pipe diameter (Chapter 4)
2	Pump performance and selection (Chapter 4)
1	Cavitation (Chapter 4)
1	Series and parallel pump systems (Chapter 4)
1	The affinity laws for pumps (Chapter 4)
5	Heat exchangers, LMTD, and e-NTU methods (Chapter 5)
1	Regenerative HX, condensers, evaporators, boilers (Chapter 5)
2	Double-pipe heat exchangers (Chapter 5)
2	Shell and tube heat exchangers (Chapter 5)
1	Plate and frame heat exchangers (Chapter 5)
1	Cross-flow heat exchangers (Chapter 5)
2	Thermal energy system simulation (Chapter 6)
1	Fitting component performance data (Chapter 6)
2	Optimization using Lagrange multipliers (Chapter 6)
1	Optimization using software (Chapter 6)

This schedule will leave about five open class periods that can be used for exams, quizzes, homework discussion, or project work. When the author teaches this course, a project is usually assigned after the engineering economics discussion. Student teams work on their projects throughout the semester. The final exam period (2 hours) is used as a poster session where the student design teams display their results. This type of project work and final presentation can accommodate a fairly large group of students, provided there is enough space available for the poster session.

Steve Penoncello
Moscow, Idaho

Additional material is available from the CRC Website: http://www.crcpress.com/product/isbn/9781482245998

Acknowledgments

This book has been in the making for over 20 years. It represents a collection of topics that the author has consistently taught in a senior-level mechanical engineering course on thermal energy system design. Many years of research notes, course notes, and computer codes have been distilled into this manuscript.

There are many people who helped to contribute to the successful completion of this project. To begin with, the author would like to thank his college professors for excellent educational and learning experiences: Dr. Donald Naismith, Dr. Mason Somerville, Dr. J. Peter Sadler, Dr. Nanak Grewal, Professor Palmer Reiten, and Dr. Gene Kemper from the University of North Dakota and Dr. Richard Stewart, Dr. Richard Jacobsen, Dr. Ronald Gibson, and Dr. Gylfi Arnason from the University of Idaho. These are the individuals who had the most positive influence on the author's intellectual maturity.

Thank you to the University of Idaho Sabbatical Leave Evaluation Committee for recommending that the author be granted a sabbatical leave to focus solely on this book. In the current academic *publish-or-perish* environment where textbooks are not generally viewed as scholarship, it is refreshing to know that there are entities in the university environment that *do* view writing a textbook as scholarship. It is doubtful that this book could have ever been completed without the continuous uninterrupted time that the sabbatical leave provided.

Many ME 435 students at the University of Idaho have used this book in its draft forms in the author's course. The author is appreciative to those students who have come forward and identified even the smallest of errors in previous drafts of the book. The many solicited and unsolicited comments about the book are also greatly appreciated.

The publishing team at Taylor & Francis/CRC Press has been a pleasure to work with during the publication phase of this book: Jonathan Plant, Executive Editor for Mechanical, Aerospace, and Nuclear Engineering; Arlene Kopelhoff, Editorial Assistant; Cynthia Klivecka, Project Editor; Laurie Oknowsky, the final Project Coordinator; and Kate Gallo, the initial Project Coordinator. This team was quick to respond to what seemed like endless questions and guided the author every step of the way.

Much of this book was written in the presence of the peaceful serenity of Big Lake in north central Minnesota. Working on the manuscript of this book while taking an occasional glimpse of the sun sparkling on the waters of the lake provided inspiration beyond all imagination.

Family played a big role in the successful completion of this book. The author would like to thank his parents for supporting him in the many opportunities he has pursued in life. During a discussion about potential career paths as a youth, the author recalls his father asking, "Have you ever thought about engineering?" Thank you, dad, for planting the seed.

It is difficult to find the words to express the author's deepest and sincere gratitude to his children: Gina, Nick, Matthew, and Stacey, and their own families, for their continuing love and support. The pride you feel and exhibit for your father means more than you can ever imagine. You are all very special people and are destined for great things. Thank you for the encouraging words during this project.

The writing of a textbook is not a trivial task. During the development of this book, the author spent nearly every free minute of his time on the development of the manuscript. Unfortunately, the one who suffers most from this is the spouse. The author wishes to express his most heartfelt and sincere thanks to his wife, Jean. Your past and continued love and support through this project made it all worthwhile. Now, let us retire very soon and enjoy what the future has to offer!

Steve Penoncello
Moscow, Idaho

Author

Steven G. Penoncello received his BS and MS in mechanical engineering from the University of North Dakota in 1978 and 1980, respectively. He received his PhD in mechanical engineering from the University of Idaho in 1986. He has been a registered professional engineer (mechanical engineering) in the State of Idaho since 1993.

Dr. Penoncello has been teaching courses and doing research in the thermal sciences since 1980. He has held academic positions at the University of North Dakota (instructor from 1980 to 1983, assistant professor from 1986 to 1988, and associate professor from 1988 to 1990), and the University of Idaho (visiting assistant professor from 1985 to 1986, associate professor from 1990 to 1995, and professor since 1995). He has also served in administrative positions at the University of Idaho (mechanical engineering department chair from 1995 to 1999, associate dean for research and graduate studies in the College of Engineering from 1999 to 2005, and director of the Center for Applied Thermodynamic Studies since 2005).

Dr. Penoncello has taught undergraduate and graduate courses in thermodynamics, heat transfer, fluid mechanics, air conditioning, solar engineering, refrigeration engineering, internal combustion engines, energy technology, and thermal energy systems design. His research involves the determination of standard reference quality formulations for the calculation of the thermophysical properties of fluids and fluid mixtures of scientific and engineering interest. He has coauthored one book, two book chapters, and over 35 technical papers in the area of thermophysical properties.

Dr. Penoncello has been an active member of the American Society of Mechanical Engineers (ASME) since 1978. He has served as a member of the K-7 Committee on Thermophysical Properties in the Heat Transfer Division of the ASME from 1988 to 2013. He has also served as a mechanical engineering program evaluator for the Accreditation Board on Engineering and Technology (ABET), representing the ASME from 1999 to 2007.

Dr. Penoncello's background in thermal energy systems design and analysis started during his master's research at the University of North Dakota. His master's thesis topic was the analytical modeling and experimental verification of an innovative heat pump system designed for cold climates. This work took a full system approach and involved the simultaneous application of thermodynamics, fluid mechanics, and heat transfer.

In 1991, the faculty of the University of Idaho, Department of Mechanical Engineering, undertook the task to critically evaluate and update their undergraduate curriculum. During this process, a conscious decision was made to revise the curriculum to allow the undergraduate students to have a significant design experience in several areas of the discipline including solid mechanics and thermal sciences. This process resulted in several new design-based courses including the senior capstone design experience (two courses), an updated machine design course, and a new course in thermal energy systems design. Dr. Penoncello took the lead in the development of the thermal energy systems design course. This book represents a collection of the topics that he teaches in his course that has been developed over the past 20 years.

1

Introduction

1.1 Thermal Energy Systems Design and Analysis

Thermal energy systems are abundant in the commercial and residential sectors. They are *thermal* because they often involve the transfer of heat. They typically use working fluids to transport *energy*. They are *systems* because there are several components connected together to form some sort of process or cycle. Some examples of *thermal energy systems* are as follows:

- A heating and cooling system for a commercial or residential building
- A network of pipes and pumps used to transport a fluid to and from various processes within an industrial facility
- A vapor power cycle used to deliver electrical power
- A vapor compression refrigeration cycle used for cooling or heating

In the design of such systems, the engineer seeks to select components of the system such that the system performs as required while minimizing cost.

For the engineer to analyze or design thermal energy systems, it is important to understand the fundamental concepts expressed in terms of various conservation laws and balances. In addition, phenomenological rate equations such as Newton's law of cooling used in convective heat transfer are required. To complete the picture, the real performance of various components in the system must be known. This is often obtained from equipment manufacturer's specifications. Once all these equations are determined for a specific system, they form the mathematical model of the system. The challenge is to solve these equations to find the system's operating point.

In the era before modern computing technology became widely available, the solution of the mathematical model was done by constructing many graphs representing the equations and looking for intersection points. As an example, consider two straight lines in the x–y plane on a Cartesian coordinate. If the lines are not parallel, there will be a single intersection that can be easily determined graphically. Now take this example and extrapolate it to multiple equations with multiple unknowns. The task becomes daunting and time-consuming if done graphically.

The advent and availability of modern computing technology has reduced the cumbersome task of graphical solutions to simply solving an $n \times n$ system of equations. Modern equation-solving software makes this task quite easy, even for the most complex set of equations. Solution of $n \times n$ systems of equations can even be done in many spreadsheet applications that include a solver add-in. In thermal energy systems, an added feature that enhances the equation-solving software is the ability to calculate thermophysical properties of the fluids involved in real time within the software. The topic of software is discussed further in Section 1.2.

Having a working mathematical model of a thermal system based on physics and equipment performance information allows the engineer to start thinking about optimization of the system. Optimization techniques allow a system to be designed for minimum cost while still meeting required constraints. Many equation-solving software packages incorporate optimization algorithms to allow studies similar to this.

1.2 Software

There are several excellent software packages that are capable of solving $n \times n$ systems of equations in addition to performing optimization. There are only a few, however, that incorporate state-of-the-art thermophysical property information for many different working fluids. The National Institute of Standards and Technology (NIST) maintains the most current fluid property calculation software known as REFPROP. REFPROP is available from the NIST for a reasonable cost. This software has the capability to produce property tables and property diagrams for a wide variety of fluids and fluid mixtures. REFPROP, by itself, does not solve simultaneous equations or perform optimization; it is merely a property calculator. However, REFPROP does come with a dynamic link library (REFPROP.dll) that allows an interface with Microsoft Excel. Once REFPROP is linked with Excel, fluid properties can be accessed through a series of functions. If the Excel spreadsheet has the solver add-in enabled, it is possible to solve a complex set of equations simultaneously.

There are several programs capable of solving simultaneous equations and performing optimization. Engineering Equation Solver (EES—pronounced *ease*) from F-Chart Software (fchart.com) has the added capability to access thermophysical properties of fluids as the equations are being solved. The fluid database in EES is very similar to REFPROP. In nearly all instances, EES uses the same up-to-date modern formulations used by REFPROP. F-Chart Software also sells an add-in that allows EES to interface directly with the REFPROP.dll. In addition to properties of fluids, EES also has access to ideal gas models for many substances, properties of secondary heat transfer fluids (brines), thermodynamic properties of moist air, properties of incompress-ible solids and liquids, and many others.

The primary software package that will be used to solve complex prob-lems in this book is EES. Most of the problems at the end of chapters will require the use of an equation solver with the capability to access fluid properties. On the basis of the earlier discussions, it is recommended that instructors who adopt this book also consider using EES in their classes.

1.3 Thermal Energy System Topics

There are far too many different types of thermal energy systems to include in a single book. In this book, hydraulic systems are considered in addi-tion to systems utilizing heat exchangers. The last chapter in the book includes a discussion on thermal energy system simulation, evaluation, and optimization. The working fluids considered are mostly liquids, but there will be examples where gas flows and condensing and/or boiling flows are considered.

1.4 Units and Unit Systems

Units are a very important part of any engineering analysis. Far too often, units are not written down because they seem *obvious* in a problem. However, countless errors in design and analysis can be traced to a lack of proper unit analysis. As tedious and mundane as it may seem, careful unit analysis should be carried out in all engineering problems.

There are five dimensions used in most engineering unit systems: mass (M), force (F), length (L), time (T), and temperature (θ). From phys-ics, it is understood that force and mass are related through Newton's sec-ond law of motion. Newton stated that force is proportional to mass and acceleration.

$$F \propto ma \tag{1.1}$$

The proportionality constant in this equation is $1/g_c$. Using this constant, Equation 1.1 can be written as follows:

$$F = \frac{ma}{g_c}$$

(1.2)

The magnitude and unit of g_c depend on the unit system being considered. In most unit systems, either force or mass is considered *fundamental* and the other is *derived* from Newton's second law of motion. Table 1.1 shows four common unit systems used in engineering and scientific analysis. The base unit for each of the dimensions along with the MFLTθ units for each dimension is given.

In the four unit systems shown, the length unit is always designated as L and the time unit as T. However, the mass and force units have different MFLT designations (e.g., mass is not always designated as M and force is not always F). This results from mass being derived from force (British Gravitational System) or force being derived from mass (SI and CGS systems) from Newton's second law. In the inch–pound (IP) unit system, force and mass have fundamental units of F and M, respectively.

Notice that in all systems except the IP system, $g_c = 1$ (dimensionless). In the IP system, g_c has a value and unit: 32.174 lbm·ft/lbf·s². The reason for this strange constant is that both mass and force are considered fundamental in the IP unit system. The value of g_c is used to make 1 lbm = 1 lbf at sea level, where the acceleration due to gravity is 32.174 ft/s². This is problematic in the sense that although the international community has adopted the SI unit system, many U.S. industries still use the IP system. One of the major points of confusion related to unit systems is how to write equations containing force and mass dimensions. In the past, before the SI system became a standard, many engineering textbooks incorporated g_c in equations that required a force–mass conversion. For example, Newton's second law of motion was

TABLE 1.1

Four Common Unit Systems Used in Engineering Analysis and Their Base MFLTθ Units

Unit	International System of Units (SI)	Inch–Pound (IP) System (English System)	British Gravitational System	Centimeter–Gram–Second (CGS) System
Mass	kg (M)	lbm (M)	slug ($FL^{-1}T^2$)	g (M)
Force	N (MLT^{-2})	lbf (F)	lbf (F)	dyne (MLT^{-2})
Length	m (L)	ft (L)	ft (L)	cm (L)
Time	s (T)	s (T)	s (T)	s (T)
Temperature	K or °C (θ)	R or °F (θ)	R or °F (θ)	K or °C (θ)
g_c	1	$32.174\dfrac{\text{lbm} \cdot \text{ft}}{\text{lbf} \cdot \text{s}^2}$	1	1

written as shown in Equation 1.2. In the SI system, $g_c = 1$. Therefore, once the SI system became the international standard, Newton's second law is written as

$$F = ma \tag{1.3}$$

It should be noted that Equations 1.2 and 1.3 are the same in every unit system shown in Table 1.1 *except* the IP system. In this book, equations will be written *without* the constant g_c to be consistent with most modern engineering and science books. However, it is important to understand that when working in the IP system, the constant g_c may need to be incorporated. Careful unit analysis will reveal this, as shown in the following example.

EXAMPLE 1.1

A mass of 10 lbm is at sea level where the acceleration due to gravity is 32.174 ft/s². Determine the force (in lbf) that this mass exerts at sea level.

SOLUTION: The force–mass relationship is governed by Newton's second law of motion. Using Equation 1.3

$$F = mg = (10 \text{ lbm})\left(32.174 \frac{\text{ft}}{\text{s}^2}\right)$$

When the units are multiplied through, it is clear that the units are *not* lbf, as expected. This suggests that the force–mass conversion constant in the IP system, g_c, is needed. Dividing by g_c results in

$$F = mg = \frac{(10 \text{ lbm})\left(32.174 \dfrac{\text{ft}}{\text{s}^2}\right)}{32.174 \dfrac{\text{lbm} \cdot \text{ft}}{\text{lbf} \cdot \text{s}^2}} = \underline{10 \text{ lbf}} \quad \leftarrow$$

Careful unit analysis revealed the need to use g_c in this problem. In fact, when working in the IP system, one should always be vigilant to the possibility of needing g_c for a force–mass conversion.

It is important to understand that g_c is *nothing but a constant*. It is *not* the acceleration due to gravity. It is the reciprocal of the proportionality constant in Newton's second law of motion when working in the IP system. It came about because the IP system declares both force and mass as fundamental units.

Conversion between units is often required. The official publication for conversion factors is published by NIST (Thompson and Taylor 2008). Appendix A lists several common conversion factors based on this publication. The values in Appendix A were computed using EES.

EXAMPLE 1.2

The thermal conductivity of AISI 316 Stainless Steel at 90°C is 14.54 W/m K. Convert this value of thermal conductivity to (Btu/h·ft·°F) and (Btu/s·in·°F).

SOLUTION: Using the conversion factors found in Appendix A

$$k = \left(14.54\,\frac{W}{m\,K} \right) \left| \frac{5.7779 \times 10^{-2}\,\frac{Btu}{h \cdot ft \cdot °F}}{1\,\frac{W}{m\,K}} \right| = 8.401\,\frac{Btu}{h \cdot ft \cdot °F} \quad \leftarrow$$

$$k = \left(14.54\,\frac{W}{m\,K} \right) \left(1.3375 \times 10^{-5}\,\frac{Btu}{s \cdot in \cdot °F}\,\frac{m\,K}{W} \right) = 1.945 \times 10^{-4}\,\frac{Btu}{s \cdot in \cdot °F} \quad \leftarrow$$

This solution shows two different ways to conduct the unit analysis. The first conversion is perhaps the more understandable. From Appendix A, it can be seen that 1 W/m K = 5.7779 × 10⁻² Btu/h·ft·°F. The conversion, directly after the vertical bar in the equation, shows this equivalency in a fraction. The second equation is a *space-saver* conversion. Rather than write 1 W/m K in the denominator, the units are written in a format that allows for easy analysis. Either approach is suitable.

Careful, meticulous analysis of units is important in any engineering calculation. It is good practice to conduct unit analysis as shown earlier, no matter what experience level you feel you have. Engineering students and engineers one day from retirement (and everyone in between) should conduct unit analysis very carefully on all of their work.

1.5 Thermophysical Properties

Thermal energy system design and analysis ultimately requires knowledge of the thermophysical properties of the substances involved in the system. Thermophysical properties encompass the common *thermodynamic* properties such as density or heat capacity in addition to *transport* properties such as thermal conductivity or viscosity.

There are countless handbooks and book appendices that contain tables of thermophysical properties of substances. In addition, there are many software packages available that can be used to calculate thermophysical properties. It is important for the engineer to understand the accuracy of such tables and software calculations. Often, tables of properties or values calculated from software are accepted as fact. The real fact is that behind each table or software calculation there is some sort of mathematical *formulation*. This formulation is a collection of equations used to compute the properties. The development of a property formulation ultimately requires experimental data. For some substances such as water or nitrogen, there is a plethora of experimental data over a wide range of pressure and temperature values. However, for some substances the data are very limited. Moreover, as time marches on experimental measurement techniques improve to the point where experimental data on a fluid measured in 2010 may be orders of magnitude more accurate compared to the same data measured in 1940. Therefore, it is of utmost importance for the engineer to understand the *source* of the thermophysical property data being used. Using the most current information for thermophysical properties removes a level of uncertainty in the design or analysis.

The most current state-of-the-art information on thermophysical properties of substances can be found in scientific and technical journals. In the REFPROP and EES software packages referenced earlier, every effort is made to make sure that the formulations used for thermophysical property calculation represent the current standards. Thermophysical properties of many substances, calculated using EES, are given in Appendix B.

1.5.1 Viscosity

Many thermophysical properties are fairly intuitive. For example, the density of a substance is measured in mass per unit volume. Viscosity is one of those properties that can be expressed with several different units, which may make it a confusing property to understand. From a purely intuitive perspective, the viscosity of a substance is related to how easily it flows. For example, when the oil in your car's engine is cold it takes more pumping power to move it compared to oil that is warmed up. Therefore, the viscosity of the cold oil is higher than the viscosity of the warm oil.

1.5.1.1 Dynamic Viscosity

To understand what viscosity represents, it is important to go back to its fundamental definition. The *absolute* or *dynamic* viscosity, μ, is related to the shearing stress and the velocity gradient (strain rate) in the fluid as it moves.

$$\tau = \mu \frac{du}{dy} \tag{1.4}$$

The units of absolute viscosity can be determined from this expression. In the SI system, dynamic viscosity has units of Pa·s or N·s/m², as demonstrated in the following equation:

$$\mu = \tau \frac{dy}{du} [=] (\text{Pa}) \left(\frac{\text{m} \cdot \text{s}}{\text{m}} \right) = \text{Pa} \cdot \text{s} = \frac{\text{N} \cdot \text{s}}{\text{m}^2} \tag{1.5}$$

In Equation 1.5, the symbol [=] means "is dimensionally equal to." By expanding the definition of the newton, an alternative unit for absolute viscosity can be determined as follows:

$$\mu [=] \frac{\text{N} \cdot \text{s}}{\text{m}^2} = \left(\frac{\text{kg} \cdot \text{m}}{\text{s}^2} \right) \left(\frac{\text{s}}{\text{m}^2} \right) = \frac{\text{kg}}{\text{m} \cdot \text{s}} \tag{1.6}$$

Equation 1.5 has units based on force, whereas Equation 1.6 has units based on mass. Perhaps a more common unit of dynamic viscosity is developed from the CGS system.

$$\mu = \tau \frac{dy}{du} [=] \left(\frac{\text{dyne}}{\text{cm}^2} \right) \left(\frac{\text{cm} \cdot \text{s}}{\text{cm}} \right) = \frac{\text{dyne} \cdot \text{s}}{\text{cm}^2} = \text{P (poise)} \tag{1.7}$$

The poise, P, is named after Jean Louis Poiseuille (1799–1869), who did much work in the area of fluid mechanics.

In the IP unit system, the force-based units of dynamic viscosity can be derived as follows:

$$\mu = \tau \frac{dy}{du} [=] \left(\frac{\text{lbf}}{\text{ft}^2} \right) \left(\frac{\text{ft} \cdot \text{s}}{\text{ft}} \right) = \frac{\text{lbf} \cdot \text{s}}{\text{ft}^2} \tag{1.8}$$

This set of units can be converted to a mass-based unit by

$$\mu [=] \frac{\text{lbf} \cdot \text{s}}{\text{ft}^2} g_c = \frac{\text{lbf} \cdot \text{s}}{\text{ft}^2} \left(32.174 \frac{\text{lbm} \cdot \text{ft}}{\text{lbf} \cdot \text{s}^2} \right) [=] \frac{\text{lbm}}{\text{ft} \cdot \text{s}} \tag{1.9}$$

Notice that to convert the absolute viscosity from a force-based unit to a mass-based unit in the IP unit system the force-based value must be multiplied by g_c.

1.5.1.2 Kinematic Viscosity

Another measure of viscosity is the *kinematic* viscosity. The kinematic viscosity is defined as follows:

$$\nu = \frac{\mu}{\rho} \tag{1.10}$$

Kinematic viscosity is a measure of the resistance of the flow under the influence of gravity. The units of kinematic viscosity can be derived based on the definition shown in Equation 1.10. In the SI system, the kinematic viscosity has units of m²/s as shown in Equation 1.11.

$$v = \frac{\mu}{\rho}[=]\frac{kg}{m \cdot s}\frac{m^3}{kg} = \frac{m^2}{s} \tag{1.11}$$

In the IP unit system, the kinematic viscosity has units of ft²/s.

$$v = \frac{\mu}{\rho}[=]\frac{lbm}{ft \cdot s}\frac{ft^3}{lbm} = \frac{ft^2}{s} \tag{1.12}$$

A more common unit of kinematic viscosity is derived from the CGS unit system. The stoke (St) is defined as cm²/s.

$$v = \frac{\mu}{\rho}[=]\frac{g}{cm \cdot s}\frac{cm^3}{g} = \frac{cm^2}{s} = St \text{ (stoke)} \tag{1.13}$$

1.5.1.3 Newtonian and Non-Newtonian Fluids

If the viscosity of a fluid is constant at a fixed state (e.g., defined by temperature and pressure), then the fluid is known as a Newtonian fluid. Therefore, a plot of the shear stress as a function of the velocity gradient for a Newtonian fluid at a given state is a straight line. The slope of the line is the fluid's viscosity. This is shown in Figure 1.1.

The viscosity of a fluid is truly a function of temperature and pressure. Therefore, changing the temperature and pressure of Newtonian fluid changes the slope of the stress–strain rate (velocity gradient) curve shown in Figure 1.1. Even though the slope changes, the relationship is still linear,

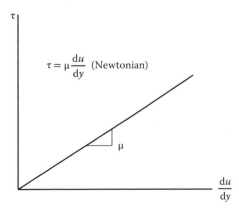

FIGURE 1.1
Newtonian fluid behavior.

which indicates a new value for the viscosity. In liquids and low-pressure vapors, the viscosity of a Newtonian fluid is only weakly dependent on pressure and is often considered a function of temperature only. For gases and supercritical states, the pressure may have a significant effect on the viscosity in addition to the temperature. This is shown in Figure 1.2, which shows a dynamic viscosity versus temperature plot for water. This plot shows the behavior of the dynamic viscosity over a wide temperature range for a range of pressures. Notice that in the liquid and vapor regions the subcritical isobars basically collapse onto one another especially at lower temperatures, making the viscosity a temperature function. As the pressure and temperature increase, the isobars start to separate indicating that pressure, in addition to temperature, influence the dynamic viscosity.

Many common industrial fluids behave as Newtonian fluids including water, toluene, acetone, benzene, air, nitrogen, and oxygen. However, there are many other substances that do not obey Equation 1.4. These are known as non-Newtonian fluids. A non-Newtonian fluid is one in which the viscosity is a function of some mechanical variable (or variables) such as shearing stress or time. It is important for the engineer to have a cursory understanding of how non-Newtonian fluids behave because many of the substances flowing in industrial process are non-Newtonian. The study of non-Newtonian fluids is a complex subject within the broader science of *rheology*. The presentation that follows gives a brief overview of non-Newtonian fluids.

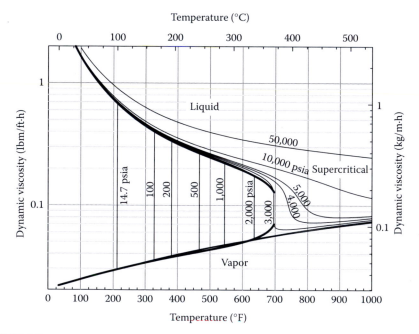

FIGURE 1.2
Dynamic viscosity behavior for water.

Non-Newtonian fluids can be generally classified into one of two categories: *shear thinning* or *shear thickening*. Shear-thinning fluids are substances that experience a decrease in their resting viscosity and flow easier when agitated by a shear stress. In contrast, a shear-thickening fluid behaves just the opposite of shear-thinning fluids. In a shear-thickening fluid, the viscosity of the fluid increases when an external shearing stress is applied.

Shear-thinning fluids can be further classified as *pseudoplastic, thixotropic,* or *Bingham plastic*. A pseudoplastic substance experiences a viscosity decrease with an applied shearing stress. When the applied stress is removed, the fluid reverts back to a more solid-like phase. A thixotropic substance experiences a decrease in viscosity with an applied shear stress, but its viscosity continues to decrease with time. A Bingham plastic behaves like a Newtonian fluid once the initial yield stress of the fluid is exceeded.

Shear-thickening fluids can be classified as *rheopectic* or *dilatant*. A rheopectic substance experiences an increase in viscosity over time. A dilatant substance experiences an increase in viscosity with an applied shear stress, but it does not change over time.

Table 1.2 shows a summary of the different types of non-Newtonian fluids along with some common examples of each. On the basis of the descriptions shown in Table 1.2, the behavior of the time-independent non-Newtonian fluids can be superimposed on Figure 1.2 as shown in Figure 1.3.

For pseudoplastic and dilatant substances, the constants K and n in the equations shown in Figure 1.3 are known as the *consistency index* and flow *behavior index*, respectively. The equation relating the shear stress to the velocity gradient is known as the Ostwald–deWaele equation (also known as the Power Law).

$$\tau = K \left(\frac{du}{dy} \right)^n \tag{1.14}$$

On the basis of the shape of the curves in Figure 1.3, it can be seen that for a pseudoplastic substance, $n < 1$ and for a dilatant material, $n > 1$. For a Newtonian fluid, $n = 1$ and $K = \mu$.

TABLE 1.2

Classification and Examples of Non-Newtonian Fluids

	Shear Thinning	**Shear Thickening**
Time-dependent materials	Thixotropic: yogurt, good quality paint, motor oils	Rheopectic: gypsum pastes, printer inks
Time-independent materials	Pseudoplastic: molten polymers, paper pulp suspensions, mayonnaise	Dilatant: wet beach sand, cornstarch paste, silly putty
With a yield stress	Bingham plastic: toothpaste, blood, molten chocolate, mashed potatoes	

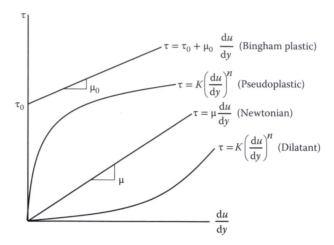

FIGURE 1.3
Behavior of Newtonian and time-independent non-Newtonian fluids.

EXAMPLE 1.3

The following steady-state shear data have been measured and reported for a polymer solution at 298 K.

du/dy (1/s)	0.205	0.280	0.377	0.510	0.692	0.944	1.29	1.582	1.77	2.17	2.43
τ(Pa)		49.69	51.94	54.46	56.44	59.02	61.93	65.48	68.07	70.0	71.6	73.85

Use these data to determine the parameters of the Ostwald–deWaele model and the Bingham plastic model. Comment on the validity of the models. Which model best represents the data? What type of fluid can this polymer be classified as?

SOLUTION: There are several questions to resolve with this problem. To answer these questions, the data need to be plotted and the parameters of each model need to be determined. Many software packages have the capability to curve-fit two-dimensional data to the abovementioned models. Figure E1.3 shows the data and two resulting curve fits using the EES software package. The solid line represents the Bingham plastic model and the dashed line is the Ostwald–deWaele model. The constants of each model were determined from the software.

From this plot, it is easily seen that the Ostwald–deWaele model is a much more accurate representation of the data compared to the Bingham plastic model. Therefore, the data are best represented by the following equation:

$$\tau = 63.3\left(\frac{du}{dy}\right)^{0.159} \quad \text{where} \quad \tau[=]\ \text{Pa} \quad \text{and} \quad \frac{du}{dy}[=]\frac{1}{s}$$

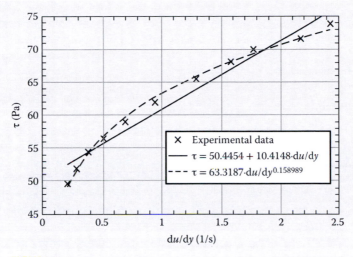

FIGURE E1.3

From this model, it can be seen that the value of $n = 0.159$, which is less than one. Therefore, the polymer solution at 298 K behaves like a non-Newtonian *pseudoplastic* fluid.

1.6 Engineering Design

There are many things in engineering that have definite answers. For example, adding $2 + 2$ gives 4 with complete certainty (at least in Base 10!). *Engineering design*, on the other hand is not as straightforward! *Design* is the *art* of engineering. As engineers, we like to rely on a series of steps to achieve a goal. Unfortunately, design is something that is very difficult to quantify let alone delineate in a list of steps. Engineering design is often thought of as the place where art and experience meet science and technology.

By its very nature, design is open-ended. For example, for any given thermal energy system, there may be *many* designs that accomplish a task. They may look very different and contain different equipment, but they accomplish a required task. These many different design solutions are called *workable designs*. Because of the large variability in these workable designs, their costs may be vastly different. Obviously, cost is a major concern in industry. Some systems may result in more pollution than others. As with cost, pollution is something industries are concerned about, and in many cases, regulations exist that must be followed.

This book will not attempt to cover the detail of the engineering design process. Most readers of this text may be concurrently enrolled in a capstone engineering design course where this subject is discussed in detail.

Other readers may already be practicing engineers who have experience in the engineering design process. The ever-increasing role of the engineer with regard to environmental impact and resource sustainability, however, is an important ethical topic that is discussed in the following section.

1.6.1 Engineering Design and Ethics

Engineers are expected to conduct themselves in an ethical manner. The American Society of Mechanical Engineers (ASME) maintains an ethics website where it is stated, "Ethics is a standard of moral practice that is adhered to for the sake of the industry and which is guided by personal conscience. Ethics addresses fundamental issues of right and wrong, and is an important issue in the field of mechanical engineering, in which proper ethical behavior can avoid legal issues associated with actions such as patent infringement." The ASME Code of Ethics for Engineers (asme.org) is shown below.

The Fundamental Principles

Engineers uphold and advance the integrity, honor, and dignity of the engineering profession by:

   I. using their knowledge and skill for the enhancement of human welfare;

  II. being honest and impartial, and serving with fidelity their clients (including their employers) and the public; and

 III. striving to increase the competence and prestige of the engineering profession.

The Fundamental Canons

1. Engineers shall hold paramount the safety, health, and welfare of the public in the performance of their professional duties.
2. Engineers shall perform services only in the areas of their competence; they shall build their professional reputation on the merit of their services and shall not compete unfairly with others.
3. Engineers shall continue their professional development throughout their careers and shall provide opportunities for the professional and ethical development of those engineers under their supervision.
4. Engineers shall act in professional matters for each employer or client as faithful agents or trustees, and shall avoid conflicts of interest or the appearance of conflicts of interest.
5. Engineers shall respect the proprietary information and intellectual property rights of others, including charitable organizations and professional societies in the engineering field.

6. Engineers shall associate only with reputable persons or organizations.
7. Engineers shall issue public statements only in an objective and truthful manner and shall avoid any conduct which brings discredit upon the profession.
8. Engineers shall consider environmental impact and sustainable development in the performance of their professional duties.
9. Engineers shall not seek ethical sanction against another engineer unless there is good reason to do so under the relevant codes, policies and procedures governing that engineer's ethical conduct.
10. Engineers who are members of the Society shall endeavor to abide by the Constitution, By-Laws and Policies of the Society, and they shall disclose knowledge of any matter involving another member's alleged violation of this Code of Ethics or the Society's Conflicts of Interest Policy in a prompt, complete and truthful manner to the chair of the Committee on Ethical Standards and Review.

These fundamental principles and canons are certainly germane to the field of thermal energy system design; in particular, Canon 8. In this Canon, the role of the engineer with respect to environmental impact is clearly delineated. As thermal energy system designers, mechanical engineers must strive to consider environmental impact and sustainability in every aspect of their work. Considering environmental impact means more than just a passing thought. This Canon implies that the engineer must strive to minimize the environmental impact of his/her design.

Designing a system to minimize environmental impact often is counter to the profit motive of industry. Companies are in business to make money. However, this does not relieve the engineer of his/her ethical duties regarding environmental impact and sustainability. These issues need to be brought to the attention of middle and upper management. It is ultimately their decision on how to proceed, but the engineer has fulfilled his/her ethical responsibility by bringing the matter to their attention.

Problems

1.1   At sea level, a person has a mass of 175 lbm. Determine the weight of this person in newtons (N).
1.2   At sea level on the earth, an astronaut weighs 175 lbf. Determine the astronaut's weight in pounds-force (lbf):
   a.   On the surface of the moon where $g = 1.6$ m/s^2
   b.   On the surface of Mars where $g = 3.7$ m/s^2

1.3   An automobile travels a distance of 1800 m in 1 minute. Determine the speed of the automobile in
   a.   miles/h (mph)
   b.   km/h (kph)

1.4   A pump is being used in a system that is circulating liquid hexane at 80°F. The volumetric flow rate of the hexane through the pump is 125 gallons/min (gpm). Determine the flow rate of the hexane through the pump in
   a.   L/s
   b.   m³/s
   c.   lbm/h
   d.   kg/s

1.5   A 20% propylene glycol solution is circulating through a system at a temperature of 0°F. Determine the dynamic viscosity of the propylene glycol solution in
   a.   lbm/ft·s
   b.   lbf·h/ft²
   c.   centipoise (cP)

1.6   A 20% propylene glycol solution is circulating through a system at a temperature of 0°F. Determine the kinematic viscosity of the propylene glycol solution in
   a.   ft²/h
   b.   m²/s
   c.   centistokes (cSt)

1.7   The shearing stress for a sample of soy milk at 20°C is characterized by the following experimental data:

Strain rate $\gamma = du/dy$ (1/s)	2.1	25.2	45.9	80.1	160.7	325	412.1	644	855
Shear stress τ (Pa)	0.032	0.227	0.458	0.643	1.144	2.075	2.6	3.737	4.655

   Derive an equation that relates the shearing stress and the velocity gradient (strain rate), and identify what type of fluid the soy milk is classified as.

1.8   The shearing stress for a pasteurized carrot juice in the temperature range 35°C–85°C is characterized by the following experimental data:

Strain rate $\gamma = du/dy$ (1/s)	19.8	55.7	183.9	422.4	627.5	852.5	1056	1224	1588
Shear stress τ (Pa)	0.084	0.295	0.543	1.062	1.459	1.768	2.123	2.379	2.766

   Derive an equation that relates the shearing stress and the velocity gradient (strain rate), and identify what type of fluid the carrot juice is classified as.

1.9   Pick a topic in the general area of thermal energy systems of interest to you and research the ethical issues associated with environmental impact and/or sustainability related to subject you selected. Write a short paper summarizing your findings. Cite all of your references. Use reliable references (Wikipedia is not considered a reliable source).

2

Engineering Economics

2.1 Introduction

Engineering economics is often taught as a semester-long course. The presentation in this chapter is not meant to be a substitute for such a course. It is, however, a summary of the salient features of engineering economic analysis. For more in-depth coverage of these topics and others not included in this chapter, the reader is encouraged to consult one of several excellent references on this subject.

In many cases, the optimization of thermal energy systems is done to minimize the total system *cost*. This cost is generally made up of the initial cost of the components, labor costs, operating costs, maintenance costs, replacement costs, and salvage value. In addition, inflation, depreciation, taxes, and insurance also play a role in the system cost. These costs come at different times during the life of the system. Therefore, to design a cost-optimized system we need some way to convert all of the expenditures and incomes to a common point in time. This is the challenge in engineering economic analyses.

The value of money is determined by two factors: the actual face value of the money and the time value of the money. The time value of money is a result of the *cost* of money; that is, the *interest rate* or *rate of return*. The cost of money can be viewed from two perspectives. From the borrower's perspective, the cost of money is the interest paid. From the point of view of the investor, the cost of money is the rate of return required before an investment is viewed as viable.

2.2 Common Engineering Economics Nomenclature

Throughout this chapter, several economic variables are used. Table 2.1 defines some of the common variables used in economic analysis.

TABLE 2.1

Common Variables Used in Engineering Economic Analysis

Variable	Definition
A	Uniform periodic payments or incomes
B or BV	Book value
D	Depreciation costs
F	Money considered at a future time
G	Uniform cost or income gradient
i	Interest rate or rate of return
I or IC	Initial cost of an item or investment
M	Maintenance costs
n	Number of time periods considered in an analysis
O	Operation costs
P	Money considered at the present time
S	Salvage value of an item
t	Tax rate

2.3 Economic Analysis Tool: The Cash Flow Diagram

The *cash flow diagram* is a useful tool for visualizing the problem. Consider the following example:

A heating system for a multitenant dwelling has an initial cost of $20,000. The yearly operation and maintenance charges are $1000. Increased rent (to help defray the cost of the heating system) results in an extra $5000 per year of income. The heating plant has a life of 10 years, at which time it can be sold for $7000.

The cash flow diagram for this scenario is shown in Figure 2.1. Notice the following in Figure 2.1:

- The costs (or disbursements) are pointing down.
- Incomes are pointing up.
- The annual incomes and costs are indicated at the *end* of the year, even though they may be distributed throughout the year. This may be confusing, especially in year 10. Even though the system will be sold for $7000 at the end of year 10, there must be incomes and maintenance costs from year 9 to 10. These cash flows are indicated at the end of year 10.

The incomes and disbursements are additive *at any point in time* on the cash flow diagram. This means that the cash flow diagram shown in Figure 2.1 can be equivalently drawn as in Figure 2.2.

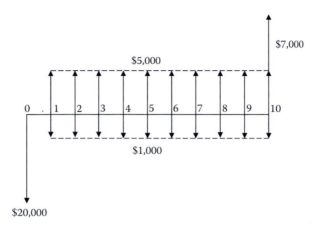

FIGURE 2.1
Economic cash flow diagram.

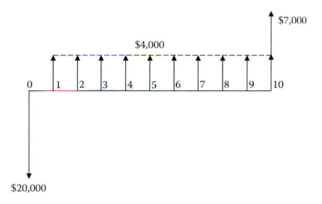

FIGURE 2.2
An equivalent version of the cash flow diagram shown in Figure 2.1.

In most engineering economic analysis problems, the task is to determine how to transform all of the costs on a cash flow diagram to a single point in time. Because $1 today is not equivalent to $1 in the past or the future due to inflation and other economic factors, money has *time value*. Collapsing a cash flow diagram to a single point in time requires that the time value of money is considered.

2.4 Time Value of Money

If there were no such thing as an interest rate, economics would be fairly simple (and banks would be out of business!). For example, if you were to borrow $12,000 from a bank and pay it back over a year in equivalent monthly installments at 0% interest, your monthly payment would be $12,000/12 = $1,000. This is easy.

However, in the real world, money costs money. If you borrow money, you are expected to pay for the borrowed money over time. Likewise, if you invest money you expect a return on your investment over time. The cost of money over time is known as interest. The return on an investment over time is usually called the rate of return. Often, the word *interest* is used for both scenarios.

If the interest were *simple interest*, then the interest payment could be figured into the principal borrowed and divided over the number of payment periods. In our earlier example, if you were to borrow the $12,000 at a simple interest rate of 10% per year, your total interest payment would be (12,000)(0.10) = $1,200. This means you would have to pay the bank a total of $12,000 + $1,200 = $13,200. The monthly payment would then be $13,200/12 = $1,100.

Neither of these examples is common in the economic world. In reality, interest is *compounded*. Compounding means that interest is charged (or paid) on the principal and the previously charged (or accrued) interest. For example, consider an investment of $10,000 at an annual interest rate of 6%, compounded annually. At the end of the first year, the current $10,000 would be worth ($10,000) (1.06) = $10,600. At the end of the second year, the original $10,000 would be worth ($10,600)(1.06) = $11,236. At the end of the third year, it would be ($11,236) (1.06) = $11,910, and so on. Notice in this case, the interest earned gains interest in addition to the principal. This is the idea behind compounding.

2.4.1 Finding the Future Value of a Present Sum: The Single Payment Compound Amount Factor

Using the investment example in Section 2.4, we can develop a mathematical expression that will allow us to determine the value of an investment at any time in the future. The value of the $10,000 investment at the end of the first year is given as follows:

$$(\$10,000)(1.06) = \$10,600$$

At the end of the second year, the value is calculated as follows:

$$(\$10,600)(1.06) = \$11,236$$

$$[(\$10,000)(1.06)](1.06) = \$11,236$$

$$(\$10,000)(1.06)^2 = \$11,236$$

Notice the recursion here. If we are interested in the future value, F, of the $10,000 at any year n, and the interest rate is compounded annually at 6%, this can be found by the following equation:

$$F = (\$10,000)(1.06)^n$$

This thinking can be generalized even further. Notice that the factor 1.06 is equal to (1 + 0.06), where 0.06 is the interest rate, expressed as a decimal rather than a percentage value. Therefore, using this concept, we can calculate the future value, F, of an initial investment made at the present time, P, at any time period, n, for the interest rate, i, as follows:

$$F = P(1+i)^n \tag{2.1}$$

The expression $(1+i)^n$ is known as an *interest factor*. In particular, this interest factor is called the *single payment compound amount factor*. In economic analyses, this factor is written in shorthand notation as follows:

$$F = P\left(\frac{F}{P}, i, n\right) \tag{2.2}$$

Given the expression for this factor, values or tables of values of the single payment compound amount factor for various interest rate and compounding periods can be constructed.

When calculating the interest factor, it is important that the interest rate, i, is consistent with the compounding period. The following example demonstrates this concept.

EXAMPLE 2.1

An investment opportunity that provides an annual rate of return of 6% is available. Determine the future value of a $1000 investment 10 years from now if the interest is compounded (a) annually and (b) monthly.

SOLUTION: For both cases, the future sum can be found using the single payment compound amount factor

$$F = P\left(\frac{F}{P}, i, n\right) = \$1,000\left(\frac{F}{P}, i, n\right)$$

For annual compounding, the interest rate $i = 0.06$ and the number of compounding periods $n = 10$. Therefore, the interest factor is

$$\left(\frac{F}{P}, 0.06, 10\right) = (1+0.06)^{10} = 1.79085$$

The value of the future sum for this annual compounding scenario can be found as follows:

$$F = \$1000 \underset{1.79085}{\left(\frac{F}{P}, 0.06, 10\right)} = \underline{\underline{\$1791}}$$

For monthly compounding, the interest rate is $i = 0.06/12 = 0.005$ and the number of compounding periods is $n = (10)(12) = 120$. Then

$$\left(\frac{F}{P}, 0.005, 120\right) = (1+0.005)^{120} = 1.81940$$

$$F = \$1000 \underset{1.81940}{\left(\frac{F}{P}, 0.005, 120\right)} = \underline{\underline{\$1819}}$$

From an investment perspective, more frequent compounding results in a larger future sum because the interest is gaining interest at a faster rate.

2.4.2 Finding the Present Value of a Future Sum: The Present Worth Factor

The present worth factor is an interest factor that allows the conversion of a future sum to a present value at a given interest rate for a given number of compounding periods. This interest factor is helpful in determining how much money should be invested at a given interest rate to achieve a future goal. Recall that the single payment compound amount factor was used to convert a present sum to a future value at a given interest rate. The present worth factor is used to do just the opposite. Therefore, it is simply the reciprocal of the single payment compound amount factor

$$P = F(1+i)^{-n} = F\left(\frac{P}{F}, i, n\right) \tag{2.3}$$

2.4.3 Nominal and Effective Interest Rates

When a financial institution quotes an interest rate to you, they are quoting what is known as a *nominal* annual rate. Sometimes, they may also quote you a compounding period. For example, 6% compounded monthly means that the nominal annual rate is 6% while the compounding period is monthly.

When the compounding period is more frequent than annual, the nominal interest rate does not reflect the true rate due to the interest itself being compounded. In other words, the compounded interest is subjected to the time value of money. In this case, the interest that more accurately reflects the time value of money is known as the *effective* interest. The effective interest rate for a nominal rate of i for m compounding periods in 1 year can be determined by

$$i_{\text{eff}} = \left(\frac{F-P}{P}\right) = \frac{1}{P}\left[P\left(\frac{F}{P}, \frac{i}{m}, m\right) - P\right] = \left(\frac{F}{P}, \frac{i}{m}, m\right) - 1 \tag{2.4}$$

For example, the effective interest rate for an investment scenario of 6% compounded monthly is

$$i_{\text{eff}} = \left(1 + \frac{i}{12}\right)^{12} - 1 = \left(1 + \frac{0.06}{12}\right)^{12} - 1 = 0.0618 = \underline{\underline{6.18\%}} \quad \Leftarrow \tag{2.5}$$

Notice that in Equation 2.5 the interest rate is expressed as (0.06/12) because the interest is compounded monthly in this example ($m = 12$).

2.4.4 Finding the Future Value of a Uniform Series: The Compound Amount Factor

In many situations, uniform payments or disbursements occur in an economic scenario. For example, when money is borrowed from a bank for a car loan the payments are uniform monthly payments. These payments can be converted to an equivalent future value using the *compound amount factor*. This conversion allows one to compute how much interest is paid over the life of the loan. A similar scenario can be envisioned for an investment fund where uniform disbursements are paid out on a regular basis (e.g., a retirement fund).

To develop the expression for the compound amount factor, envision the loan payback scenario shown in Figure 2.3. Each of the uniform payments, A, can be envisioned as a present value at its respective position on the cash flow diagram. Therefore, the single payment compound amount factor can be used to move each of those individual values to the future. For the scenario shown in Figure 2.3,

$$F = A(1+i)^9 + A(1+i)^8 + A(1+i)^7 + A(1+i)^6 + A(1+i)^5$$
$$+ A(1+i)^4 + A(1+i)^3 + A(1+i)^2 + A(1+i)^1 + A(1+i)^0$$

(2.6)

Because A is uniform, Equation 2.6 can be written as follows:

$$F = A\left[(1+i)^9 + (1+i)^8 + (1+i)^7 + \quad \ldots \quad + (1+i)^1 + 1\right]$$

(2.7)

For any number of compounding periods, n, Equation 2.7 becomes

$$F = A\left[(1+i)^{n-1} + (1+i)^{n-2} + (1+i)^{n-3} + \quad \ldots \quad + (1+i) + 1\right]$$

(2.8)

The term in brackets in Equation 2.8 converts the uniform series A to an equivalent future value. The term in brackets is known as the *compound amount factor*

$$\left(\frac{F}{A}, i, n\right) = (1+i)^{n-1} + (1+i)^{n-2} + (1+i)^{n-3} + \quad \ldots \quad + (1+i) + 1$$

(2.9)

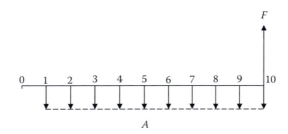

FIGURE 2.3
Cash flow diagram showing the equivalence between a uniform series, A, and a future sum, F.

Equation 2.9 is not a particularly convenient way to write this factor. However, with some algebra the factor can be written in a more useful way. Multiplying both sides of Equation 2.9 by $(1 + i)$ results in

$$\left(\frac{F}{A},i,n\right)(1+i)=(1+i)^n+(1+i)^{n-1}+(1+i)^{n-2}+\ \ldots\ +(1+i)^2+(1+i) \quad (2.10)$$

Subtracting Equation 2.9 from Equation 2.10, we get,

$$\left(\frac{F}{A},i,n\right)[(1+i)-1]=(1+i)^n-1 \quad (2.11)$$

Solving Equation 2.11 for the interest factor, we obtain,

$$\left(\frac{F}{A},i,n\right)=\frac{(1+i)^n-1}{i} \quad (2.12)$$

Therefore, the conversion of a uniform series to an equivalent future value is given by

$$F=A\left[\frac{(1+i)^n-1}{i}\right]=A\left(\frac{F}{A},i,n\right) \quad (2.13)$$

2.4.5 Finding the Equivalent Uniform Series That Represents a Future Value: The Uniform Series Sinking Fund Factor

Given a future value, F, it is possible to find an equivalent uniform series that has the same cash equivalence. This is the reciprocal problem described in Section 2.4.4. Therefore, the conversion F to A can be accomplished by

$$A=F\left[\frac{i}{(1+i)^n-1}\right]=F\left(\frac{A}{F},i,n\right) \quad (2.14)$$

This interest factor is known as the *uniform series sinking fund factor*.

2.4.6 Finding the Present Value of a Uniform Series: The Uniform Series Present Worth Factor

Figure 2.4 shows a cash flow diagram that demonstrates the equivalence of a uniform series to a present value. There are several ways to develop the interest factor required here. One strategy is to use the compound amount factor and convert the uniform series to a future value. Then, the present worth factor can be used to convert to a present value. Using this strategy,

$$P=\left[A\left(\frac{F}{A},i,n\right)\right]\left(\frac{P}{F},i,n\right) \quad (2.15)$$

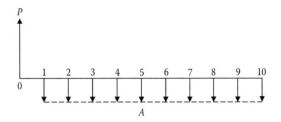

FIGURE 2.4
Cash flow diagram showing the equivalence between a uniform series, *A*, and a present value, *P*.

Substituting the interest factors in Equation 2.15 gives

$$P = A\left[\frac{(1+i)^n - 1}{i}\right](1+i)^{-n} = A\left[\frac{(1+i)^n - 1}{i(1+i)^n}\right] \tag{2.16}$$

The interest factor developed here is known as the *uniform series present worth factor*,

$$\left(\frac{P}{A}, i, n\right) = \frac{(1+i)^n - 1}{i(1+i)^n} \tag{2.17}$$

2.4.7 Finding the Uniform Series That Is Equivalent to a Present Value: The Capital Recovery Factor

Given a present value, an equivalent uniform series can be found using the reciprocal of the uniform series *present worth* factor. This reciprocal is known as the *capital recovery factor*

$$\left(\frac{A}{P}, i, n\right) = \frac{i(1+i)^n}{(1+i)^n - 1} \tag{2.18}$$

Using the capital recovery factor, the uniform series *A*, which is equivalent to a present value *P*, is

$$A = P\left(\frac{A}{P}, i, n\right) = P\left[\frac{i(1+i)^n}{(1+i)^n - 1}\right] \tag{2.19}$$

2.4.8 Finding the Present Value of a Uniform Linearly Increasing Series: The Gradient Present Worth Factor

There are several instances in an industrial scenario where costs are incurred uniformly over time (e.g., every year), but the costs increase each year. Consider the cash flow diagram shown in Figure 2.5. In this figure, the frequency of the series is uniform in time. However, the magnitude of the series increases linearly as time goes on. In addition, notice that the series starts at the end of the *second* period. The value G in this figure is called the *gradient*. It represents the linear increase in the series over time. To determine the equivalent present value of this series, each series value can be treated as a future sum and brought back to the present using the present worth factor

$$P = \frac{G}{(1+i)^2} + \frac{2G}{(1+i)^3} + \frac{3G}{(1+i)^4} + \quad \cdots \quad + \frac{9G}{(1+i)^{10}} \tag{2.20}$$

Writing this equation for any cash flow diagram with n periods,

$$P = \frac{G}{(1+i)^2} + \frac{2G}{(1+i)^3} + \frac{3G}{(1+i)^4} + \quad \cdots \quad + \frac{(n-1)G}{(1+i)^n} \tag{2.21}$$

Factoring out the value of the gradient, G, results in

$$P = G\left[\frac{1}{(1+i)^2} + \frac{2}{(1+i)^3} + \frac{3}{(1+i)^4} + \quad \cdots \quad + \frac{(n-1)}{(1+i)^n}\right] \tag{2.22}$$

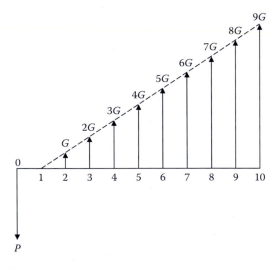

FIGURE 2.5

Cash flow diagram showing the equivalence of a uniform linearly increasing series and a present value.

The term in brackets is the interest factor known as the *gradient present worth factor*. However, this form of the factor is not very convenient to work with. After performing several algebraic manipulations, Equation 2.22 can be written as

$$P = G\left[\frac{(1+i)^n - 1}{i^2(1+i)^n} - \frac{n}{i(1+i)^n}\right] = G\left(\frac{P}{G}, i, n\right) \qquad (2.23)$$

2.4.9 Summary of Interest Factors

The interest factors developed in this section are known as *discrete* interest factors because the interest compounding occurs over discrete time periods. Table 2.2 summarizes the interest factors developed in this section.

2.5 Time Value of Money Examples

In this section, several examples will be presented that demonstrate how the discrete interest factors developed in Section 2.4 can be applied in various analyses.

TABLE 2.2

Interest Factors for Discrete Compounding

Name	Converts	Symbol	Computed by
Single payment compound amount	P to F	$(F/P, i, n)$	$(1+i)^n$
Present worth	F to P	$(P/F, i, n)$	$(1+i)^{-n}$
Uniform series sinking fund	F to A	$(A/F, i, n)$	$\dfrac{i}{(1+i)^n - 1}$
Compound amount	A to F	$(F/A, i, n)$	$\dfrac{(1+i)^n - 1}{i}$
Capital recovery	P to A	$(A/P, i, n)$	$\dfrac{i(1+i)^n}{(1+i)^n - 1}$
Uniform series present worth	A to P	$(P/A, i, n)$	$\dfrac{(1+i)^n - 1}{i(1+i)^n}$
Gradient present worth	G to P	$(P/G, i, n)$	$\dfrac{(1+i)^n - 1}{i^2(1+i)^n} - \dfrac{n}{i(1+i)^n}$

EXAMPLE 2.2

A bank is advertising new car loans at 3.5%. Consider a situation where $20,000 is borrowed from this bank to buy a new car. The loan is to be paid back in equal monthly installments over a 6-year period. Determine the monthly payment and the effective interest rate of the loan.

SOLUTION: In most economic analysis problems, it is a good idea to start with a cash flow diagram to help visualize the problem. The cash flow diagram for this problem is shown in Figure E2.2

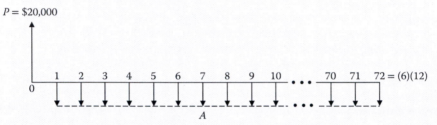

FIGURE E2.2

The $20,000 that the bank lends for purchase of the car is viewed as an income received. The monthly payments are expenses. Notice that there are 72 periods in this cash flow diagram because the life of the loan is 6 years, which is equivalent to 72 months. On the basis of this scenario, the interest rate paid per month is $i = 0.035/12$. The monthly payment can be found by applying the capital recovery factor

$$A = P\left(\frac{A}{P}, i, n\right) = (\$20,000)\left(\underbrace{\frac{A}{P}, \frac{0.035}{12}, 72}_{0.0154184}\right) = \underline{\$308.37}$$

The capital recovery factor was calculated using the expression developed in Section 2.4.7 and shown in Table 2.2. There are many online loan calculators. There are also many loan calculator applications for smart phones and tablets. These calculators use the expressions shown in this example to determine loan payments.

For this loan, the bank advertises an interest rate of 3.5%. This is the nominal annual interest rate. The effective annual rate of this loan is

$$i_{eff} = \left(\frac{F}{P}, \frac{i}{m}, m\right) - 1 = \left(\underbrace{\frac{F}{P}, \frac{0.035}{12}, 12}_{1.03557}\right) - 1 = 0.03557 = \underline{3.56\%}$$

As expected, the interest rate advertised by the bank is slightly lower than the actual interest rate being paid because of the monthly compounding. If the compounding period is yearly, then the effective rate is equal to the nominal rate.

With installment loans, it is interesting to see how the monthly payment is distributed between payment of interest and repayment of the principal. Many banks offer a payment schedule known as an *amortization* schedule. The following example shows how to develop an amortization schedule.

EXAMPLE 2.3

Develop an amortization schedule for the car loan described in Example 2.2.

SOLUTION: From Example 2.2, the monthly payment is $308.37. In the first month, the interest paid is the initial loan amount multiplied by the monthly interest rate,

$$I_1 = P\left(\frac{i}{12}\right) = (\$20,000)\left(\frac{0.035}{12}\right) = \$58.33$$

The balance of the payment is used to pay the principal,

$$P_1 = A - I_1 = \$(308.37 - 58.33) = \$250.04$$

Therefore, at the end of the first month the principal balance on the loan is

$$\text{Bal}_1 = \$20,000 - P_1 = \$(20,000 - 250.04) = \$19,749.96$$

The second payment results in

$$I_2 = P_1\left(\frac{i}{12}\right) = (\$19,749.96)\left(\frac{0.035}{12}\right) = \$57.60$$

$$P_2 = A - I_2 = \$(308.37 - 57.60) = \$250.77$$

$$\text{Bal}_2 = P_1 - P_2 = \$(19749.96 - 250.77) = \$19,499.19$$

This process repeats for subsequent months. For any time period m, during the loan period,

$$I_m = P_{m-1}\left(\frac{i}{12}\right) \qquad P_m = A - I_m \qquad \text{Bal}_m = P_{m-1} - P_m$$

This type of calculation can be easily adapted to a spreadsheet. The various columns can be totaled to determine the total payments made and the interest paid over the life of the loan. Figure E2.3 is an abridged table that shows the resulting amortization schedule for this example.

m	Payment = $308.37		Balance
	Interest	Principal	
0			$ 20,000.00
1	$ 58.33	$ 250.03	$ 19,749.97
2	$ 57.60	$ 250.76	$ 19,499.20
3	$ 56.87	$ 251.50	$ 19,247.71
4	$ 56.14	$ 252.23	$ 18,995.48
5	$ 55.40	$ 252.96	$ 18,742.51
	• • •		
68	$ 4.46	$ 303.91	$ 1,224.53
69	$ 3.57	$ 304.80	$ 919.73
70	$ 2.68	$ 305.69	$ 614.04
71	$ 1.79	$ 306.58	$ 307.47
72	$ 0.90	$ 307.47	$ (0.00)
	$ 2,202.49	$ 20,000.00	

FIGURE E2.3

The interest paid each month decreases while the amount paid to the principal increases. The life of the loan, and thus the total amount of interest paid, can be reduced significantly by making extra payment to the principal *early* in the life of the loan.

The total amount of interest paid for this loan is $2,202.49. The longer the term of the loan, the more the interest paid. However, the longer loan term allows for a smaller payment. The consumer must balance both these issues when deciding what loan term to select.

The following example shows how interest factors are used to manipulate a complex cash flow scenario.

EXAMPLE 2.4

A company is considering investing in some new equipment to enhance one of their production lines. It is estimated that the addition of this new equipment will increase annual profits by $50,000. The annual cost to operate the equipment is $1,000. After the first year, maintenance costs are anticipated to be $1,000, increasing by $1,000 each subsequent year. The equipment is expected to last 10 years, at which time it can be sold for $5,000. The company's board of directors has specified that a minimum rate of return of 20% annually must be realized from the equipment investment. Determine the maximum amount of money that the company can spend now to meet this investment scenario. Assume annual compounding.

SOLUTION: On the basis of the information given in the problem, the cash flow diagram is shown in Figure E2.4.

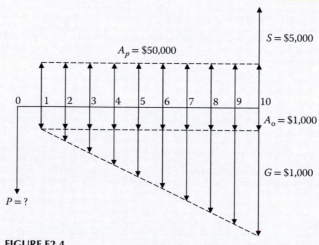

FIGURE E2.4

There are several ways to determine P, which depend on the common point in the cash flow diagram where all incomes and expenses are referred to. In this case, the present value is desired. Therefore, bringing all incomes and expenses to the present (year 0) results in

$$P + A_0\left(\frac{P}{A}, i, n\right) + G\left(\frac{P}{G}, i, n\right) = A_p\left(\frac{P}{A}, i, n\right) + S\left(\frac{P}{F}, i, n\right)$$

The interest rate and number of periods are known; therefore, the interest factors can be found. Solving for P,

$$P = (A_p - A_0)\left(\frac{P}{A}, 0.20, 10\right) - G\left(\frac{P}{G}, 0.20, 10\right) + S\left(\frac{P}{F}, 0.20, 10\right)$$

$$P = (\$50{,}000 - \$1{,}000)\underbrace{\left(\frac{P}{A}, 0.20, 10\right)}_{4.19247} - (\$1{,}000)\underbrace{\left(\frac{P}{G}, 0.20, 10\right)}_{12.88708} + (\$5{,}000)\underbrace{\left(\frac{P}{F}, 0.20, 10\right)}_{0.16151}$$

$$P = \$193{,}352$$

This calculation indicates that an investment in the equipment of $193,352 will meet the company's minimum required rate of return of 20%. If the equipment is less expensive than this, the rate of return would be higher than 20%. Conversely, if the equipment is more expensive than $193,352 then the rate of return drops below 20% and the improvement is viewed as financially unacceptable.

In certain instances, it is desirable to determine the interest rate or the rate of return for a given scenario. This is demonstrated in the following problem.

EXAMPLE 2.5

Suppose the equipment cost for the processing facility in Example 2.4 has an initial cost of $225,000. Keeping all other parameters as defined in Example 2.4, determine the corresponding rate of return on this equipment investment.

SOLUTION: The cash flow diagram for this scenario is the same as that in Example 2.4 with the exception that the initial cost of the equipment is known,

$P = \$225,000$. Therefore, the balancing of incomes and expenses at year 0 results in the same equation as seen in Example 2.4,

$$P + A_0\left(\frac{P}{A},i,n\right) + G\left(\frac{P}{G},i,n\right) = A_p\left(\frac{P}{A},i,n\right) + S\left(\frac{P}{F},i,n\right)$$

$$P + G\left(\frac{P}{G},i,n\right) = \left(A_p - A_0\right)\left(\frac{P}{A},i,n\right) + S\left(\frac{P}{F},i,n\right)$$

Writing out the interest factors,

$$P + G\left[\frac{(1+i)^n - 1}{i^2(1+i)^n} - \frac{n}{i(1+i)^n}\right] = \left(A_p - A_0\right)\left[\frac{(1+i)^n - 1}{i(1+i)^n}\right] + S(1+i)^{-n}$$

The only unknown in this equation is the interest rate, i. However, the difficulty is that this equation cannot be solved explicitly for the interest rate. Therefore, the solution is iterative. This information can be programmed into the Engineering Equation Solver (EES) as follows:

```
"GIVEN: Example 2.4 with an initial investment of $225,000 in the
equipment"

    P   = 225000[$]    "Initial cost of equipment"
    A _ p = 50000[$]    "Annual increase in profit"
    A _ o = 1000[$]     "Annual operating cost"
    G   = 1000[$]       "Maintenance gradient"
    S   = 5000[$]       "Salvage value"
    n   = 10            "Number of years in the scenario"


"FIND: The rate of return on the investment"


"SOLUTION:"
"Equating all incomes and costs to the present,"
    P + G*P\G = (A _ p - A _ o)*P\A + S*P\F

"Interest factors"
    P\A = ((1+i)^n - 1)/(i*(1+i)^n)
    P\G = ((1+i)^n - 1)/(i^2*(1+i)^n) - n/(i*(1 + i)^n)
    P\F = (1 + i)^(-n)
```

Solving this equation set results in $i = 0.1562$. Therefore, the annual rate of return for an equipment cost of $225,000 is 15.62%.

2.6 Using Software to Calculate Interest Factors

Without software, interest factors must be calculated using the expressions shown in Table 2.2. Alternatively, tables of calculated interest factors can be used. However, this often requires interpolation.

Spreadsheet programs are very versatile. Many of the modern programs have built-in interest factor functions. Each spreadsheet program may vary in how the interest factor calculation is accessed within the software, so the user is encouraged to carefully read the help menu to understand if the factors being used are those needed in the analysis. Having interest factors accessible with a function call is very convenient. Combined with solver capability, iterative problems can easily be solved. As an example, the Microsoft Excel solution of Example 2.5 is shown in Figure 2.6. Notice that all of the interest factors were calculated using the PV function (the Help Menu shows how to use this function). The maintenance gradient costs were calculated using the PV function for each individual year since there is no gradient function available.

		Example 2.5 - Using Microsoft Excel		
GIVEN:		A_p =	$ 50,000.00	Annual profit increase due to equipment
		A_o =	$ 1,000.00	Annual operation cost
		S =	$ 5,000.00	Salvage value of equipment at year n
		G =	$ 1,000.00	Maintenance gradient
		n =	10	Number of periods in the analysis (years)
FIND:		i =	15.62%	Rate of return (interest rate)
				Calculated with Goal Seek

$$P + A_o\left(\frac{P}{A},i,n\right) + G\left(\frac{P}{G},i,n\right) = A_p\left(\frac{P}{A},i,n\right) + S\left(\frac{P}{F},i,n\right)$$

COSTS:		P =	$ 225,000.00	Initial cost of equipment
		A_o*(P/A,i,n) =	$4,902.40	Calculated with PV function
	1	G*(P/F,i,1) =	$0.00	Calculated with PV function
	2	G*(P/F,i,2) =	$748.06	Calculated with PV function
	3	G*(P/F,i,3) =	$1,293.99	Calculated with PV function
	4	G*(P/F,i,4) =	$1,678.76	Calculated with PV function
	5	G*(P/F,i,5) =	$1,935.96	Calculated with PV function
	6	G*(P/F,i,6) =	$2,093.02	Calculated with PV function
	7	G*(P/F,i,7) =	$2,172.30	Calculated with PV function
	8	G*(P/F,i,8) =	$2,191.97	Calculated with PV function
	9	G*(P/F,i,9) =	$2,166.67	Calculated with PV function
	10	G*(P/F,i,10) =	$2,108.21	Calculated with PV function
		PV COSTS =	$ 246,291.34	
INCOMES:		A_p*(P/A,i,n) =	$245,120.11	Calculated with PV function
		S*(P/F,i,n) =	$1,171.23	Calculated with PV function
		PV INCOMES =	$246,291.34	
		PV(COST - INCOME) =	$ (0.00)	Set to zero in Goal Seek

FIGURE 2.6
Microsoft Excel solution of Example 2.5.

Equation solving software, such as EES, is another convenient way to calculate interest factors. Example 2.5 was solved using EES with the interest factors entered as equations from Table 2.2.

Most equation solving software allows the user to write functions or subroutines that can be stored in a library. This allows for very easy retrieval of the function value by simply providing a function call within the equation set. As an example, the following EES code shows a function that can be used to calculate the capital recovery factor:

```
"The following function can be used to calculate the capital
recovery factor. Inputs to the function are the interest rate (i)
and the corresponding number of periods (n)"
FUNCTION A\P (i,n)
  A\P = i*(1 + i)**n/((1 + i)**n −1)
END
```

2.7 Economic Decision Making

Interest factors allow the conversion of money in a cash flow diagram to a common point in time. This idea provides the basis for decision making. In the corporate world, decisions to proceed with a project are often made based on which alternative is best from the economic point of view.

Several different methods are used to compute the best economic alternative. Two common methods are the Present Worth Method and the Annual Cost Method.

2.7.1 Present Worth Method

In the present worth method, all expenditures and incomes in the cash flow diagram are converted to an equivalent present worth (the symbol commonly used is PW). When the present worth of each alternative is compared, a decision can be made by selecting the alternative with the numerically highest present worth (e.g., less negative or more positive).

The present worth method cannot be used to compare alternatives that have different economic lives, unless an equipment replacement scheme is devised that makes the lives of all alternatives equal. In the case of alternatives with different annual lives, the *annual cost method*, described in Section 2.7.2, is preferred.

EXAMPLE 2.6

Two machines are being considered for a manufacturing process. Machine A has an initial cost of $800 with operating costs of $600 per year. Machine B costs $1000 with operating costs of $500 per year. Both machines have a 5-year life with no salvage value. If the required rate of return is 15%, compounded annually, which machine should be purchased?

SOLUTION: Assuming that each machine results in the same profit, the only difference between alternatives is the initial and operating costs of each machine. A cash flow diagram for each machine is shown in Figure E2.6.

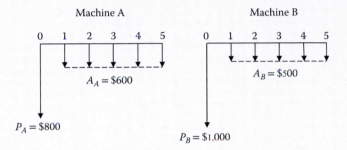

FIGURE E2.6

Assuming annual compounding, the present worth (PW) of each alternative is

$$PW_A = -P_A - A_A\left(\frac{P}{A}, i, n\right) = -\$800 - \$600\underset{3.52155}{\left(\frac{P}{A}, 0.15, 5\right)} = -\$2,811$$

$$PW_B = -P_B - A_B\left(\frac{P}{A}, i, n\right) = -\$1000 - \$500\underset{3.52155}{\left(\frac{P}{A}, 0.15, 5\right)} = -\$2,676 \quad \leftarrow$$

The alternative with the highest PW is machine B; therefore, it should be selected in this case.

2.7.2 Annual Cost Method

In the annual cost method, all expenditures and incomes are converted to an equivalent annual series (the symbol commonly used is AC). When the value of this series is known, a decision can be made based on the alternative with the lowest annual cost.

The annual cost method can be used to compare alternatives with unequal economic lives. If the lives are different, it is assumed that the shorter-life alternative will be replaced with identical equipment (unless otherwise specified).

EXAMPLE 2.7

Two machines are being considered for a manufacturing process. Machine A has an initial cost of $800 with operating costs of $600 per year. Machine B costs $1500 with operating costs of $500 per year. Machine A has a 5-year life, and machine B has a 10-year life. Neither machine has a salvage value. If the required rate of return is 8%, which machine is the best economic alternative?

SOLUTION: Assuming that both machines return the same profit, the only costs that are different in each scenario are the initial and operating costs of each machine. The cash flow diagrams for each machine are shown in Figure E2.7A.

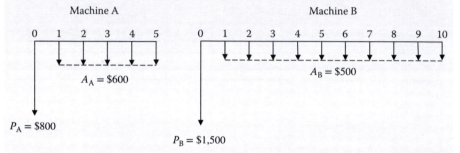

FIGURE E2.7A

For annual compounding, the annual cost of each of these cash flows is as follows:

$$AC_A = -\$800\underbrace{\left(\frac{A}{P}, 8\%, 5\right)}_{0.25046} - \$600 = -\$800$$

$$AC_B = -\$1500\underbrace{\left(\frac{A}{P}, 8\%, 10\right)}_{0.14903} - \$500 = \underline{\underline{-\$724}} \leftarrow$$

This analysis indicates that machine B has the lowest annual cost and should be selected.

The alternatives, A and B, have different economic lives in this example. As stated earlier, the annual cost method can be used in this situation by assuming that the shorter-lived alternative is replaced with identical equipment. To verify that this assumption is valid, consider the replacement of machine A with identical equipment at the end of its 5-year life, as shown in the cash flow diagram in Figure E2.7B.

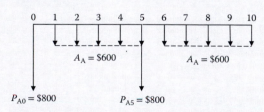

FIGURE E2.7B

The annual cost of this scenario is

$$AC_A = -P_{A0}\left(\frac{A}{P}, i, 10\right) - \left[P_{A5}\left(\frac{P}{F}, i, 5\right)\right]\left(\frac{A}{P}, i, 10\right) - A_A$$

$$AC_A = -\$800\underbrace{\left(\frac{A}{P}, 8\%, 10\right)}_{0.14903} - \left[\$800\underbrace{\left(\frac{P}{F}, 8\%, 5\right)}_{0.68058}\right]\underbrace{\left(\frac{A}{P}, 8\%, 10\right)}_{0.14903} - \$600 = -\$800$$

Notice that the annual cost of this 10-year scenario consisting of an exact replacement at year 5 is identical to the 5-year single machine scheme. This justifies the notion that the lifetime of the alternatives does not need to be the same when using the annual cost method so long as the shorter-lived alternative is replaced with identical equipment.

It may be possible to replace machine A with a different machine at the end of 5 years. For example, consider the replacement scheme in which machine A is replaced with another machine costing $800 at year 5, but due to improvements in the machine its operating costs have dropped to $350. The cash flow diagram for this scenario is shown in Figure E2.7C.

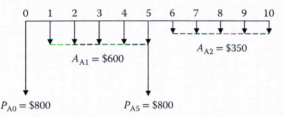

FIGURE E2.7C

The annual cost of this cash flow diagram is

$$AC_A = -\$800\underbrace{\left(\frac{A}{P}, 8\%, 10\right)}_{0.14903} - \left[\$800\underbrace{\left(\frac{P}{F}, 8\%, 5\right)}_{0.68058}\right]\underbrace{\left(\frac{A}{P}, 8\%, 10\right)}_{0.14903}$$

$$- \left[\$600\underbrace{\left(\frac{P}{A}, 8\%, 5\right)}_{3.99271}\right]\underbrace{\left(\frac{A}{P}, 8\%, 10\right)}_{0.14903} - \left[\$350\underbrace{\left(\frac{F}{A}, 8\%, 5\right)}_{5.86660}\right]\underbrace{\left(\frac{A}{F}, 8\%, 10\right)}_{0.069029} = -\$699$$

In this case, it is more economical to purchase machine A and install a new machine at year 5 that has cheaper operating costs.

This example demonstrates that the annual cost method can be used to compare alternatives with unequal lives, assuming that the shorter-lived alternative is replaced with identical equipment. However, if the shorter-lived alternative can be replaced with better equipment in the future, the resulting analysis may change the decision. It is important for the engineer to carefully consider replacement schemes when implementing the annual cost method.

2.7.3 Selection of Alternatives

The methods presented here are based solely on economics. In the selection of alternatives, there are other factors that contribute to the final decision. Some of these factors include environmental concerns, political issues, choice of vendors, and so on. As an example, consider a company that is doing things to gain the public's respect by how it treats the environment. This company may select a more expensive alternative that leads to smaller levels of pollution. This could be an important factor in winning over new customers to purchase the company's products.

2.8 Depreciation and Taxes

Capital outlay (CO) items such as buildings, automobiles, airplanes, equipment, and so on decrease in value over the years due to wear and/or obsolescence. This devaluation can be considered a production cost that reduces a company's earnings. Therefore, considering the devaluation of CO items should reduce taxes and increase profits.

Depreciation is a way to account for the devaluation of CO items and reflects a more accurate picture of earnings. Depending on what is being depreciated, the U.S. Internal Revenue Service (IRS) allows several different methods to calculate the yearly amount allowed.

Taxes are paid by corporations on profits. Corporate rates are generally much higher than consumer tax rates. The corporate tax rate is a composite of federal, state, and local rates. It is common for corporate tax rates to be in the neighborhood of 50% and higher. Since depreciation represents an equivalent cost due to the devaluation of CO items, it reduces profits. The reduction in profit reduces the tax.

2.8.1 After-Tax Cash Flow

Profits realized by a company are taxable. Depreciation reduces the taxes because the depreciation can be deducted from the profits before taxes are calculated. In any year k, the after-tax profit realized is given by the following equation:

$$A_{AT,k} = A_{BT,k} - T_k \tag{2.24}$$

In this equation, $A_{AT,k}$ and $A_{BT,k}$ are the annual profit after taxes and before taxes, respectively, and T_k is the tax imposed in year k. The tax imposed on the annual earnings can be written as follows:

$$T_k = t\left(A_{BT,k} - D_k\right) \tag{2.25}$$

In Equation 2.25, t is the tax rate and D_k is the depreciation allowed in year k. Substituting Equation 2.25 into Equation 2.24 results in the following:

$$A_{AT,k} = A_{BT,k} - t\left(A_{BT,k} - D_k\right) \tag{2.26}$$

This equation shows the tax advantage provided by depreciation.

The U.S. IRS allows depreciation to be determined for CO items based on when the equipment was put into service, what type of equipment it is, and what recovery period is expected for the equipment. For equipment put into service from 1980 to 1986, the Accelerated Cost Recovery System (ACRS) is used. For equipment put into service after 1986, the Modified Accelerated Cost Recovery System (MACRS) is used. The details of how depreciation is calculated using ACRS and MACRS are detailed in IRS Publications 534 (ACRS) and 946 (MACRS). The details of these methods are beyond the scope of this chapter and are often covered in most textbooks on engineering economy. However, the methods are *accelerated* depreciation schemes that give the business a better tax advantage early in the depreciation scenario with decreasing depreciation as time goes on. The following sections show two depreciation schemes: straight line depreciation (SLD) and an accelerated scheme known as the sum of the years' digits (SYD) depreciation.

2.8.2 Straight Line Depreciation

In the SLD model, the amount that can be depreciated annually is constant. The annual depreciation allowed in any year k for an item is given by

$$D_k = \frac{IC - S}{N} \tag{2.27}$$

In this expression, IC is the initial cost of the item being depreciated, S is the salvage value of the equipment at the end of its life, and N is the service life (or tax life) of the item. From this equation, it can be seen that depreciation rate is $1/N$.

2.8.3 Sum of the Years' Digits

This is an accelerated depreciation scheme, allowing a larger deduction in the early years of the life of the item being depreciated. The depreciation allowed in any year k is given by the following equation:

$$D_k = \frac{2(IC - S)(N - k + 1)}{N(N + 1)} \tag{2.28}$$

From this expression, the annual decrease in depreciation is given by the *arithmetic gradient*

$$\frac{2}{N(N + 1)} \tag{2.29}$$

It is interesting to note that the reciprocal of this gradient is the sum of the year's digits. For example, if $N = 5$,

$$\frac{N(N+1)}{2} = \frac{5(5+1)}{2} = 15 = 5+4+3+2+1$$

This is where the depreciation method gets its name.

The following example shows the application of the SLD and SYD depreciation models and their effect on a company's profits, including taxes.

EXAMPLE 2.8

A company is considering the addition of a new machine to one of its processing lines. The initial cost of the machine is $300,000. The addition of the machine is expected to increase the company's profit by $85,000 per year. The service life of the machine is 10 years, and there is no salvage value. The company's corporate tax rate is 51%. Determine the rate of return provided by this machine for the following scenarios:

    a. Before taxes with no depreciation
    b. After taxes with no depreciation
    c. After taxes using SLD
    d. After taxes using SYD

SOLUTION: (a) Figure E2.8A shows the cash flow diagram before taxes without considering any type of depreciation scheme.

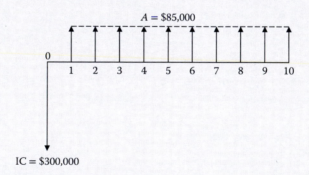

FIGURE E2.8A

The rate of return for this cash flow diagram can be found by solving the following equation:

$$IC = A\left(\frac{P}{A}, i, n\right)$$

$$\$300,000 = (\$85,000)\frac{(1+i)^{10} - 1}{i(1+i)^{10}} \quad \rightarrow \quad i = \underline{\underline{25.38\%}} \quad \Leftarrow$$

(b) Figure E2.8B shows the cash flow diagram after taxes, but without including any depreciation model.

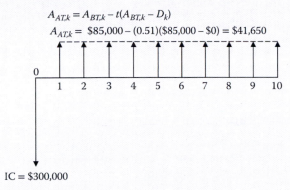

$$A_{AT,k} = A_{BT,k} - t(A_{BT,k} - D_k)$$
$$A_{AT,k} = \$85,000 - (0.51)(\$85,000 - \$0) = \$41,650$$

IC = \$300,000

FIGURE E2.8B

The rate of return can then be found as follows:

$$IC = A_{AT}\left(\frac{P}{A}, i, n\right)$$

$$\$300,000 = (\$41,650)\frac{(1+i)^{10}-1}{i(1+i)^{10}} \quad \rightarrow \quad i = 6.46\% \quad \Leftarrow$$

Notice that the rate of return has dropped substantially due to taxes. Parts (c)
and (d) demonstrate the advantage of using depreciation schemes.

(c) Using SLD, the constant annual amount that can be deducted from the
profits is given by the following equation:

$$D_k = \frac{IC - S}{N} = \frac{\$300,000 - \$0}{10} = \$30,000$$

This amount can be deducted from the before-tax profit. Figure E2.8C shows
the cash flow diagram for this scenario.

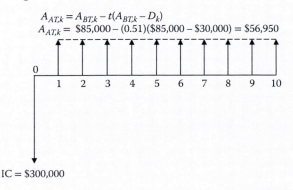

$$A_{AT,k} = A_{BT,k} - t(A_{BT,k} - D_k)$$
$$A_{AT,k} = \$85,000 - (0.51)(\$85,000 - \$30,000) = \$56,950$$

IC = \$300,000

FIGURE E2.8C

Solving for the interest rate,

$$IC = A\left(\frac{P}{A}, i, n\right) \quad \rightarrow \quad \$300,000 = (\$56,950)\frac{(1+i)^{10} - 1}{i(1+i)^{10}}$$

$$i = 13.75\% \quad \Leftarrow$$

(d) With the SYD method, the depreciation changes each year according to the expression

$$D_k = 2\frac{(IC - S)(N - k + 1)}{N(N + 1)}$$

Since the depreciation decreases each year, the after-tax cash flow will also decrease each year. For example, the depreciation for the first 2 years can be found by the following equation:

$$D_1 = 2\frac{(\$300,000 - \$0)(10 - 1 + 1)}{10(10 + 1)} = \$54,545.45$$

$$D_2 = 2\frac{(\$300,000 - \$0)(10 - 2 + 1)}{10(10 + 1)} = \$49,090.91$$

The after-tax cash flow for the first 2 years can then be found:

$$A_{AT,k} = A_{BT,k} - t\left(A_{BT,k} - D_k\right)$$
$$A_{AT,1} = \$85,000 - (0.51)(\$85,000 - \$54,545.45) = \$69,468.18$$
$$A_{AT,2} = \$85,000 - (0.51)(\$85,000 - \$49,090.91) = \$66,686.36$$

The yearly decrease in after-tax cash flow due to the linearly decreasing depreciation is given by

$$G = A_{AT,1} - A_{AT,2} = \$69,468.18 - \$66,686.36 = \$2,781.82$$

This is the linear gradient value that reduces the after-cash flow each year by this amount. The cash flow diagram can be envisioned as shown in Figure E2.8D.

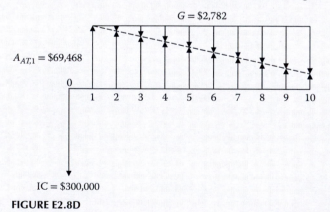

FIGURE E2.8D

This cash flow diagram shows the effect of the linear gradient decreasing the annual after-tax profits each year. This demonstrates the advantage provided by an accelerated depreciation scheme. More after-tax cash is available early in the life of the economic scenario. This provides businesses the opportunity to make more money early. For a new start-up business, this may be very important to the success of the company. For businesses that are established, this provides more money early that can be used for future product development.

This gradient can be moved to the baseline of the cash flow diagram, as shown in Figure E2.8E.

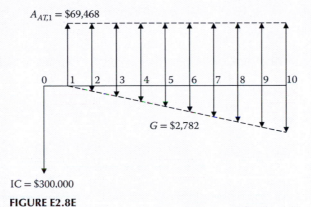

FIGURE E2.8E

Using this cash flow diagram, the rate of return can be found:

$$IC+G\left(\frac{P}{G},i,n\right)=A_{AT,1}\left(\frac{P}{A},i,n\right)$$

$$\$300,000+(\$2,781.82)\left[\frac{(1+i)^{10}-1}{i^2(1+i)^{10}}-\frac{10}{i(1+i)^{10}}\right]=(\$69,461.18)\frac{(1+i)^{10}-1}{i(1+i)^{10}}$$

$$i=15.13\% \quad \Leftarrow$$

Table E2.8 summarizes the results of this problem.

TABLE E2.8

Summary of Results

Tax and Depreciation Scenario	Rate of Return (%)
a. No taxes or depreciation	25.38
b. Taxes with no depreciation	6.46
c. Taxes with SLD	13.75
d. Taxes with SYD depreciation	15.13

The results of Example 2.8 reveal several important concepts:

1. If there were no taxes, the rate of return on the investment of the machine is very attractive! However, in the words of Benjamin Franklin, "In this world nothing can be said to be certain, except death and taxes." Therefore, the result of part (a) is unrealistic. However, it does indicate the maximum possible rate of return.

2. Depreciation provides a significant tax break. In part (b), it is seen that without depreciation the rate of return drops significantly. However, using the different depreciation schemes the rate of return increases by more than a factor of two over the no-depreciation scenario.

3. The depreciation schemes considered in Example 2.8 result in about the same rate of return. The difference in the rates of return is due to the time value of money. In Example 2.8, it appears that the SYD model gives a slight advantage.

Problems

2.1 Design a spreadsheet program that will allow you to compute the interest factors shown in Table 2.2. Use this spreadsheet to calculate the interest factors for a nominal interest rate of 7% compounded quarterly for a period of 6 years.

2.2 Design an interest factor calculator using EES that will allow you to compute the interest factors shown in Table 2.2. Use the EES Diagram Window to input the annual nominal interest rate, compounding periods per year, and number of years. Return the calculated interest factors in the Diagram Window. Use the calculator to determine the interest factors shown in Table 2.2 for a nominal interest rate of 8% compounded monthly for a period of 20 years.

2.3 A credit card advertises a nominal interest rate of 12.5% on any outstanding balance. If the interest is compounded monthly, determine the effective interest rate for this credit card.

2.4 Proud parents wish to establish a college savings fund for their newly born child. Monthly deposits will be made into an investment account that provides an annual rate of return of 4% compounded monthly. Four withdrawals from the savings fund will be made to pay for college expenses. The estimated need is $25,000 when the child turns 18 years old; $28,000 at 19 years; $31,000 at 20 years; and $34,000 at age 21. The last monthly payment to the investment account occurs when the child turns 21. This is also the time that the last withdrawal is made. Determine the monthly deposit required to meet this goal.

2.5 A small-scale silver mine in north Idaho is for sale. A mining engineer estimates that at current production levels, the mine will yield an annual net income of $80,000 for 15 years, after which the silver will be exhausted. If an investor's required rate of return is 15% compounded annually, what is the maximum amount that the investor can bid on this property?

2.6 A company has a short-term storage need. A proposal is brought forth to build a temporary warehouse at a cost of $16,000. The annual maintenance and operating cost of the facility is estimated to be $900 per year. Using the temporary storage space saves the company $3600 per year. If the company's required rate of return is 10%, how many years must the warehouse last?

2.7 For the cash flow diagram shown in Figure P2.7, determine the following:
   a. The equivalent present value at year 0
   b. The equivalent future value at year 10
   c. The equivalent uniform annual value

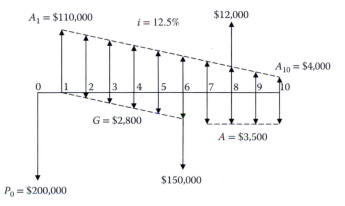

FIGURE P2.7

2.8 The engineering supervisor in a company is recommending the purchase of a new machine for a production line. The machine has an initial cost of $285,000 and is expected to be functional for 20 years. At the end of its life, it can be sold for $80,000. The addition of this machine will increase production, which results in an increase in annual income of $75,000. The machine will require annual maintenance. If the company's minimum rate of return is 25%, how much can be spent on annual maintenance?

2.9 A company has invested in a new machine for its production line. The initial cost of the machine is $9000 and it is expected to last for 5 years with no salvage value at that time. Annual operation costs are anticipated to be relatively constant at $750 per year. However, due to degradation the maintenance costs are expected to increase each year. A best guess for the annual maintenance costs is shown in Table P2.9.

TABLE P2.9

Expected Yearly Maintenance Cost

Year	Expected Maintenance Cost
1	$800
2	$1000
3	$1400
4	$1800
5	$3000

Determine how much money should be invested in a fund that earns 6% annually, compounded monthly, to completely pay for this machine.

2.10   In a chemical processing plant, liquid cyclohexane (ρ = 48.5 lbm/ft³) flows through a piping system (pump, piping, valves, etc.) at a rate of 1200 gpm as it is being transported from one process to the next. The pressure drop through the pipe system is 6.4 psi. The facility is automated and runs 24 h/day, 365 days/year. The cost of energy to operate the pump in the system is $0.12/kWh. Annual maintenance costs for the piping system are 3% of the annual operating cost. The annual income resulting from the piping system is $17,500. The initial cost of the piping system is $60,000. The system is expected to last for 15 years with no salvage value. Determine the rate of return that this piping system provides.

2.11   A company is considering investing in a new machine for its production line. Management has specified that $150,000 is available to invest in the machine. This money will be invested in a fund that earns 8% annually, compounded quarterly, to pay for the machine, its operation cost, and maintenance costs. A machine has been identified that has a uniform annual operation cost of $1250. The maintenance cost of the machine is zero in the first year, $200 in the second year, then increasing by an additional $200 per year after that. The machine has a life of 7 years with a salvage value that is estimated to be 5% of its initial cost. Determine the maximum amount of money that can be spent to purchase this machine.

2.12   The sketch of a proposed heat recovery system in a factory is shown in Figure P2.12. The heat recovery components are the two heat exchangers; the pump, the flow control valve, and the accompanying piping and fittings.

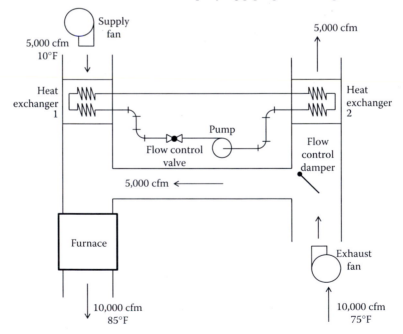

FIGURE P2.12

The initial cost of the proposed heat recovery system is $350,000. The annual operation and maintenance costs are projected to be $10,000. The salvage value of the system at the end of its useful life (projected to be 25 years) is $40,000. The annual savings in fuel costs resulting from this system is estimated to be $45,000 per year.

a. Assuming annual compounding, what is the rate of return for this heat recovery system?

b. If management has set the minimum internal rate of return to be 12% for an energy recovery system like this, what is the maximum initial cost that can be spent on the system (assuming that all other costs and incomes are the same)?

2.13 A company has a need for compressed air in one of its factory operations. A compressor has been selected for this task. The company is now considering three options to drive the compressor. These options are (1) a gasoline engine, (2) a diesel engine, and (3) an electric motor. The required input to the compressor is 400 hp, and the compressor must operate 8 h/day for 270 days each year. The economic and operating data of the engines and motor are shown in Table P2.13.

Any of the three options is expected to have a 15-year life. If the interest rate is 18%, determine which option should be selected.

TABLE P2.13

Compressor Drive Cost Data

Parameter	Gasoline Engine	Diesel Engine
Initial cost	$10,000	$12,000
Salvage value	$3,200	$5,800
Fuel consumption	0.48 lbm/hp-h	0.36 lbm/hp-h
Fuel cost	3.28 $/gal	3.89 $/gal
Fuel density	43.7 lbm/ft³	46.7 lbm/ft³
Parameter	Electric Motor	
Initial cost	$18,000	
Salvage value	$2,400	
Electricity cost	0.11 $/kWh	

2.14 A heat treating process (Process 1) can be added to a processing line for $30,000. The annual operating costs are $12,000 and its life is expected to be 8 years with a $10,000 salvage value at that time. An alternative process (Process 2) can be installed for $21,000. Its annual operating costs are $15,000 and its life is also expected to be 8 years with a $7,000 salvage value. In both cases, the annual increase in revenue due to the addition of the heat treating process is $20,000. The company's tax rate is 52% and the method of depreciation is SYD. Compute the rate of return of each heat treating process after depreciation and taxes. On the basis of this *rate of return* analysis, which alternative should be selected?

2.15   A company is considering the purchase of a computer system for $18,000. The
       operating costs will be $10,000 per year and the useful life is expected to be
       5 years with a salvage value of $5000 at that time. The present annual sales
       volume should increase by $16,000 as a result of acquiring the new computer
       system. The company's tax rate is 50%. Determine the rate of return on this
       investment for the following scenarios:
       a.   Before taxes
       b.   No depreciation
       c.   SLD
       d.   SYD depreciation

3

Analysis of Thermal Energy Systems

3.1 Introduction

Thermal energy systems can be thought of as providing one of two basic functions: (1) deliver power or (2) transport energy using a fluid (or fluids) to provide heating and/or cooling. The thermodynamic analysis of such systems is a critical part of their successful design. This chapter presents some of the fundamental concepts in thermodynamic analysis of thermal energy systems. It is assumed that the reader has had a formal course in engineering thermodynamics. Because of this, much of what is contained in this chapter may be review. This chapter is not meant to be a substitute for a full course in thermodynamics. Where material is considered review, the presentation will be brief.

3.2 Nomenclature

Perhaps one of the more frustrating things experienced by an engineering student is the inconsistent nomenclature seen in the three engineering sciences of fluid mechanics, thermodynamics, and heat transfer. There are certain unavoidable variable duplications that are used to represent different quantities. A few of the more common variable duplications are shown in Table 3.1. This variable duplication is unfortunate because it tends to *compartmentalize* each subject in the student's mind.

The analysis and design of thermal energy systems often requires the simultaneous application of all these engineering sciences. Therefore, a consistent nomenclature is advantageous. As nomenclature is introduced throughout this text, it will be defined and used consistently throughout. However, there will be cases where variable duplication cannot be avoided.

In many instances, variables specified in capital letters are *total* values. For example, W is the *total* work done in a process. Typical units for W are kJ in the SI system and Btu in the IP system. On the other hand, w is the work done per unit mass, W/m. Variables specified on a per unit mass basis are

TABLE 3.1

Some Common Variable Duplication Found in Three Basic Engineering Sciences

Variable	Fluid Mechanics	Thermodynamics	Heat Transfer
h	—	Specific enthalpy	Convective heat transfer coefficient
Q	Volumetric flow rate	Total heat transfer	—
q	—	Specific heat transfer	Heat transfer rate
U	—	Total internal energy	Overall heat transfer coefficient

often termed *specific*. This convention is used for most thermodynamic properties. For example, U is the total internal energy of a substance (an *extensive* property) at a given state, whereas $u = U/m$ is the specific internal energy (an *intensive* property). Of course, there are exceptions to this convention. For example, T is an intensive property, but it is commonly written in an upper-case letter. The nomenclature in this book will strive to follow the total and specific convention used in many engineering thermodynamics textbooks.

3.3 Thermophysical Properties of Substances

Thermophysical properties include the *thermodynamic* and *transport* properties of a substance. Thermodynamic properties are those that can be determined from an *equation of state* for the fluid. Typical thermodynamic properties used in thermal energy systems analysis are pressure, temperature, density, specific volume, internal energy, enthalpy, entropy, isobaric heat capacity, and isochoric heat capacity. Transport properties are properties of a substance, such as thermal conductivity, dynamic viscosity, and kinematic viscosity.

Ultimately, thermodynamic properties of the working fluid(s) in the thermal energy system will be required to complete the analysis. At any given thermodynamic state, there are a myriad of properties. Quite often, the properties required in a thermal energy system analysis are not explicitly known. Therefore, they need to be determined from knowledge of other properties at the state in question. The *State Postulate* is important in helping identify how many properties are needed to fix the thermodynamic state. The State Postulate may be paraphrased as follows:

> Two independent, intensive properties are required to fix the thermodynamic state of a pure substance.

Two properties, x and y, are independent if they are represented by a *single point* on a thermodynamic property diagram. Consider the pressure–volume (P–v) diagram shown in Figure 3.1. This figure shows the liquid and vapor behavior of a typical fluid. There are three *isobars* indicated on this diagram, P_1, P_2, and P_3. In addition, there is a single *isotherm*, T_1. Recall that for a pure

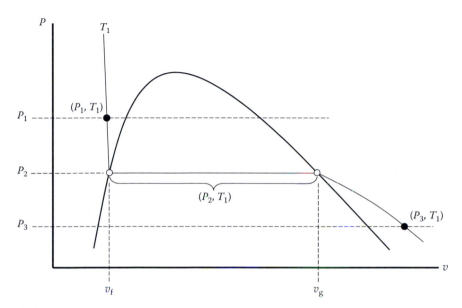

FIGURE 3.1
Pressure–volume diagram for a typical pure fluid.

fluid isotherms on a *P–v* diagram are rather steep in the liquid phase, are parallel to the isobars in the two-phase region, and curve downward in the vapor phase. Two *single-phase* points are identified on this diagram. (P_1,T_1) is a point in the liquid phase and (P_3,T_1) is a point in the vapor phase. These are unique points that identify a thermodynamic state. Therefore, (P_1,T_1) and (P_3,T_1) are *independent* property pairs. They are also intensive because they do not depend on the mass of the substance.

By contrast, (P_2,T_1) are *not* independent because they are coincident in the two-phase region. There are an infinite amount of states that exist at (P_2,T_1). For example, if you were interested in finding the *specific volume* at (P_2,T_1), you would have to conclude that there are an *infinite* number of specific volumes at this pressure and temperature, bounded by the *saturated liquid* specific volume, v_f, and the *saturated vapor* specific volume, v_g. Recall that the pressure P_2 is known as the *vapor pressure* or the *saturation pressure* of the fluid at temperature T_1.

3.3.1 Thermodynamic Properties in the Two-Phase Region

Referring to Figure 3.1 again, it can be seen that in the two-phase region of a pure fluid, (P,v) and (T,v) are independent properties that fix a state somewhere between the saturation-specific volumes, v_f and v_g. In this two-phase region, there is another thermodynamic property that can be used as one of the independent properties required by the State Postulate; the *quality* of the fluid.

The quality of a fluid in the two-phase region is defined as the ratio of the mass of the saturated vapor in the two-phase mixture to the total mass of the two-phase mixture

$$x = \frac{m_g}{m} = \frac{m_g}{m_f + m_g} \tag{3.1}$$

From this definition, it can be seen that the quality is somewhere between 0 and 1. On the saturated liquid line, where there is no saturated vapor present, $x = 0$. Likewise, on the saturated vapor line, $x = 1$.

The quality can be used to determine a variety of thermodynamic properties in the two-phase region. It can be shown that the specific volume, internal energy, enthalpy, and entropy (v, u, h, and s, respectively) in the two-phase region can be found by

$$v = (1-x)v_f + xv_g$$
$$u = (1-x)u_f + xu_g$$
$$h = (1-x)h_f + xh_g \tag{3.2}$$
$$s = (1-x)s_f + xs_g$$

Notice that these equations do not violate the State Postulate because either pressure or temperature is required to fix the saturation values (subscripts "f" and "g"). Therefore, (T,x) and (P,x) are independent, intensive properties that fix the thermodynamic state in the two-phase region.

3.3.2 Important Thermodynamic Properties and Relationships

There are several important thermodynamic relationships that are used to determine fluid properties. Perhaps the most important relationship between properties is the *equation of state*. This equation contains a wealth of thermodynamic information. The equation of state is often expressed by pressure (P) as a function of temperature (T) and specific volume (v)

$$P = P(T,v) \tag{3.3}$$

The functional form of the equation of state depends on the substance being modeled. In the case of the real behavior of the fluid, the equation of state can be quite complex. On the other hand, if the substance is modeled as an ideal gas, the equation of state is simple

$$P(T,v) = \frac{RT}{v} \tag{3.4}$$

In Equation 3.4, R is the gas constant for the fluid. The choice of equation of state depends on how the fluid properties are being modeled. This is discussed in detail in Section 3.3.3.

In addition to pressure, temperature, and specific volume, other properties of interest in thermal energy system analysis are internal energy (u), enthalpy (h), entropy (s), and heat capacities (c_v and c_p). The internal energy and enthalpy are related to the heat capacities of the fluid. The *isochoric heat capacity*, c_v, is defined as

$$c_v = \left(\frac{\partial u}{\partial T}\right)_v \tag{3.5}$$

Partial derivatives in thermodynamics are often written with subscripts to indicate the property that is being held constant during the differentiation. In Equation 3.5, the specific volume, v, is being held constant. This is why c_v is known as the isochoric (constant volume) heat capacity.

In a similar manner, the *isobaric heat capacity*, c_p, is defined as

$$c_p = \left(\frac{\partial h}{\partial T}\right)_P \tag{3.6}$$

In Equation 3.6, the derivative is calculated while the pressure is held constant. Therefore, the heat capacity calculated using Equation 3.6 is known as the isobaric (constant pressure) heat capacity.

Entropy is related to internal energy and enthalpy through the *Gibbs equations* (often referred to as the *Tds equations*). The Gibbs equations can be derived from the first and second laws of thermodynamics. There are several forms of the Gibbs equations. However, the relationships between internal energy, enthalpy, and entropy for a pure fluid are given by the following equations:

$$Tds = du + Pdv \tag{3.7}$$

$$Tds = dh - vdP \tag{3.8}$$

Equations 3.3 through 3.8 will be used in Section 3.3.3 to derive expressions for changes in thermodynamic properties based on different fluid property models.

3.3.3 Evaluation of Thermodynamic Properties

There are three thermodynamic property models that are often used in thermal energy system analyses; (1) the real fluid model, (2) the incompressible substance model, and (3) the ideal gas model. It is important for the engineer to understand how these three models are developed and when they can be used.

3.3.3.1 Real Fluid Model

The real fluid model is the most accurate representation of the thermodynamic properties of a fluid. This model includes an equation of state for the fluid developed from experimental data. Often, the equation of state is of the form $P(T,v)$. However, modern equations of state are typically cast in terms of the *Helmholtz energy* of the fluid as a function of temperature and volume, $A(T,v)$. Independent of the form used for the equation of state, thermodynamic property relationships can be derived, which allow for the calculation of properties at a given state. These calculations can be quite tedious, depending on the complexity of the equation of state. The resulting properties can be compiled in a table of properties or the calculations can be embedded in software, such as REFPROP or Engineering Equation Solver (EES).

Although the real fluid model is meant to represent the real behavior of a fluid, it is entirely possible that an equation of state is not accurate. Consider the case of water. There have been several equations of state published for water over the years. The International Association for the Properties of Water and Steam (IAPWS) is an international association that reviews equations of state for water and recommends which one will be used as the international standard. The current IAPWS equation of state for steam is much more accurate compared to equations of state 40–50 years previous. However, there are still tables of water properties available based on the old formulations. It is very important that the engineer understands the accuracy of the real fluid model being used to determine the thermodynamic properties. The most accurate model should always be used. REFPROP contains formulations that are recommended as standards for property calculation. In nearly all cases, the fluids available in EES are referenced to the same source as used in REFPROP.

3.3.3.2 Incompressible Substance Model

The incompressible substance model is meant to *estimate* the *solid* or *liquid* properties of a substance. Notice that this model is only an estimate of the properties. This model is used quite often in the case where an equation of state is not valid in the liquid phase (some older equations of state were only valid in the vapor phase). It can also be used when the phase of the fluid is known to be liquid, but only its temperature is given. The equation of state of an incompressible substance is given by

$$v = \text{constant} \tag{3.9}$$

In addition, the internal energy of an incompressible substance is only a function of temperature

$$u = u(T) \tag{3.10}$$

Together, Equations 3.9 and 3.10 formulate the incompressible substance model. Properties of incompressible substances can be calculated using Equations 3.5 through 3.8 along with Equations 3.9 and 3.10.

Equation 3.9 implies that the volume of an incompressible substance is independent of temperature and pressure. This is why engineers might consider the density of liquid water to be 1000 kg/m³ or 62.4 lbm/ft³. Notice, no temperature or pressure is specified. However, in reality, it is understood that the liquid density is really a function of temperature and, to a lesser extent, pressure. Appendix B.1 shows saturated liquid properties of several fluids over a range of temperatures. For all the fluids shown in this appendix, the density of the saturated liquid (and thus the specific volume) varies with temperature. This raises an interesting question, "What good is the incompressible substance model?" The answer to this question can be seen by considering Figure 3.2. In Figure 3.2, the specific volume (v), isobaric heat capacity (c_p), dynamic viscosity (μ), thermal conductivity (k), and the Prandtl number (Pr) of liquid water at 50°C and a variety of pressures are compared to the saturated liquid values at the same temperature. The vertical axis of this plot represents the percent deviation between the properties calculated at the saturated liquid state compared to the real-fluid liquid state.

$$\%\Delta\left(\text{Property}\right) = 100\frac{\left[\text{property}\left(P,T\right) - \text{property}\left(T,x=0\right)\right]}{\text{property}\left(P,T\right)} \qquad (3.11)$$

Notice that the percent deviations are all within ±0.2% at pressures up to 2000 kPa. As the pressure gets larger, the deviation from the saturated liquid value becomes greater. However, even at 10,000 kPa, the percent deviations are within 1%. This exercise indicates that the incompressible substance model can be used to estimate liquid properties quite accurately by evaluating the properties on the saturated liquid line *at the given temperature*; no pressure is required. As long as the pressure is not excessive, then the saturated liquid properties are a reasonable estimate of the liquid properties at the given temperature.

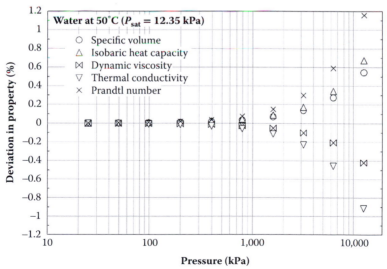

FIGURE 3.2
Comparison of properties of water at 50°C calculated on the saturated liquid line compared to the liquid values defined by (P, $T = 50$°C).

EXAMPLE 3.1

Liquid heptane is flowing through a pipe at 50°F at a volumetric flow rate of 60 gpm. The pipe's inside diameter is 2.067 in. Determine the Reynolds number of this flow.

SOLUTION: The Reynolds number is used in a variety of hydraulic and heat transfer calculations. The Reynolds number for flow in a circular pipe is defined as

$$Re = \frac{\rho VD}{\mu}$$

The volumetric flow rate of the heptane is given. The volumetric flow rate is related to the velocity of the fluid and the pipe diameter.

$$\dot{V} = AV = \left(\frac{\pi D^2}{4}\right) V \quad \rightarrow \quad \therefore \ V = \frac{4\dot{V}}{\pi D^2}$$

Substituting this expression for velocity into the Reynolds number calculation and simplifying results in the following:

$$Re = \frac{4\rho \dot{V}}{\pi D \mu}$$

The pressure of the heptane in the pipe is unknown, but the problem states that it is in the liquid phase. Therefore, the properties required can be estimated using the incompressible substance model. From Appendix B.1, the following properties of heptane at 50°F are determined:

$$\rho = 43.207 \, \frac{\text{lbm}}{\text{ft}^3} \qquad \mu = 1.1837 \, \frac{\text{lbm}}{\text{ft} \cdot \text{h}}$$

The Reynolds number can now be calculated as follows:

$$Re = \frac{4\rho \dot{V}}{\pi D \mu} = \frac{4\left(43.207 \, \frac{\text{lbm}}{\text{ft}^3}\right)(60 \text{ gpm})\left|\left(\frac{8.0208 \text{ ft}^3}{\text{gpm} \cdot \text{h}}\right)\right.}{\pi(2.067 \text{ in.})\left(1.1837 \, \frac{\text{lbm}}{\text{ft} \cdot \text{h}}\right)\left|\left(\frac{\text{ft}}{12 \text{ in.}}\right)\right.} = \underline{\underline{129,847}} \quad \leftarrow$$

The conversion factors were found in Appendix A. This example demonstrates that liquid property values can be estimated as saturated liquid properties at the given temperature.

In the analysis of thermal energy systems, the *difference* in internal energy, enthalpy, or entropy between two thermodynamic states is usually required. When a *single* value of internal energy, enthalpy, or entropy is used (e.g., in combustion calculations), it is important that all formulations used to determine the properties have the same *datum* or *reference* state.

For the incompressible substance model, the internal energy change can be determined by using Equation 3.5. Since the internal energy is only a function of temperature for the incompressible substance model, Equation 3.5 can be rewritten as

$$\tilde{c}_v = \frac{du}{dT} \tag{3.12}$$

In Equation 3.12, a tilde (~) is used to indicate the incompressible substance model's value of the heat capacity. This equation can be integrated between any two states to determine the change in the substance's internal energy.

$$u_2 - u_1 = \int_{T_1}^{T_2} \tilde{c}_v \, dT \tag{3.13}$$

For an incompressible substance, it is often reasonable to assume that the heat capacity is constant at the average temperature of the process. This simplifies Equation 3.13 to the following form:

$$u_2 - u_1 = \tilde{c}_{v,\text{avg}} (T_2 - T_1) \tag{3.14}$$

The enthalpy change of an incompressible substance can be determined by considering Equation 3.6. Since $h = u + Pv$, the partial derivative in Equation 3.6 can be expanded as shown in the following equation:

$$\tilde{c}_p = \left(\frac{\partial h}{\partial T}\right)_P = \left(\frac{\partial u}{\partial T}\right)_P + \left(\frac{\partial (Pv)}{\partial T}\right)_P = \tilde{c}_v + \left(\frac{\partial (Pv)}{\partial T}\right)_P \tag{3.15}$$

The partial derivative of the Pv product in Equation 3.15 can be written as

$$\left(\frac{\partial (Pv)}{\partial T}\right)_P = P\left(\frac{\partial v}{\partial T}\right)_P + v\left(\frac{\partial P}{\partial T}\right)_P = 0 \tag{3.16}$$

This derivative is equal to zero because both partial derivatives on the right-hand side are zero. The partial derivative of v is zero because v is constant for the incompressible substance model. The pressure derivative is also zero because pressure is being held constant during the differentiation. Substituting this result into Equation 3.15 reveals that

$$\tilde{c}_p = \tilde{c}_v \equiv \tilde{c} \tag{3.17}$$

Equation 3.17 implies that the isobaric and isochoric heat capacities are equal for the incompressible substance. Therefore, Equation 3.14 can be rewritten as

$$u_2 - u_1 = \tilde{c}_{avg}(T_2 - T_1) \tag{3.18}$$

The enthalpy change of using the incompressible substance model can be determined by substituting $u = h - Pv$ into Equation 3.18.

$$h_2 - h_1 = \tilde{c}_{avg}(T_2 - T_1) - v(P_2 - P_1) \tag{3.19}$$

Notice that the enthalpy of an incompressible substance is a function of both temperature and pressure. However, if the pressure difference between the two thermodynamic states in question is small, the last term on the right-hand side of Equation 3.19 is very small. In these cases, the enthalpy difference can be estimated by

$$h_2 - h_1 \approx \tilde{c}_{avg}(T_2 - T_1) = u_2 - u_1 \tag{3.20}$$

The entropy change of an incompressible substance between two thermodynamic states can be determined by applying Equation 3.7. For the incompressible substance, $dv = 0$. Therefore, Equation 3.7 can be written as

$$T ds = du \tag{3.21}$$

Substituting Equations 3.12 and 3.17 into Equation 3.21, results in

$$ds = \frac{\tilde{c}}{T} dT \tag{3.22}$$

This equation can be integrated between the two states resulting in

$$s_2 - s_1 = \int_{T_1}^{T_2} \frac{\tilde{c}}{T} dT \tag{3.23}$$

Considering the heat capacity to be constant at the average temperature, Equation 3.23 can be written as

$$s_2 - s_1 = \tilde{c}_{avg} \ln \frac{T_2}{T_1} \tag{3.24}$$

Equation 3.24 indicates that the entropy of an incompressible substance is only a function of temperature. When Equation 3.24 is used to compute entropy changes, the temperatures must be expressed on the *absolute* temperature scale (K or R).

EXAMPLE 3.2

Liquid benzene enters a heat exchanger at 90°C and leaves at 50°C. The pressure drop of the benzene through the heat exchanger is 50 kPa. Use the incompressible substance model to determine the change in the internal energy, enthalpy, and entropy of the benzene as it passes through the heat exchanger.

SOLUTION: The property changes are calculated using Equations 3.18, 3.19, and 3.24. All of these equations require the heat capacity of the benzene. As demonstrated in Figure 3.2, this value can be reasonably estimated as the saturated liquid heat capacity at the average temperature of the benzene. Using Appendix B.1, the average heat capacity of the liquid benzene is found to be

$$T_{avg} = \frac{T_1 + T_2}{2} = \frac{(90+50)°C}{2} = 70°C \quad \rightarrow \quad \tilde{c}_{avg} \approx 1.8712\frac{kJ}{kg \cdot K}$$

Therefore, the property changes of the benzene through the heat exchanger are

$$u_2 - u_1 = \tilde{c}_{avg}(T_2 - T_1) = \left(1.8712\frac{kJ}{kg \cdot K}\right)(50-90)K = -74.85\frac{kJ}{kg} \quad \Leftarrow$$

$$h_2 - h_1 = \tilde{c}_{avg}(T_2 - T_1) - v(P_2 - P_1) = \tilde{c}_{avg}(T_2 - T_1) - \frac{(P_2 - P_1)}{\rho}$$

$$= \left(1.8712\frac{kJ}{kg \cdot K}\right)(50-90)K - \left(\frac{m^3}{824.66\ kg}\right)(-50\ kPa)\left(\frac{kN}{kPa \cdot m^2}\right)\left(\frac{kJ}{kN \cdot m}\right)$$

$$= -74.79\frac{kJ}{kg} \quad \Leftarrow$$

$$s_2 - s_1 = \tilde{c}_{avg}\ln\frac{T_2}{T_1} = \left(1.8712\frac{kJ}{kg \cdot K}\right)\ln\left[\frac{(50+273.15)K}{(90+273.15)K}\right] = -0.218\frac{kJ}{kg \cdot K} \quad \Leftarrow$$

Notice that the enthalpy difference is very close to the internal energy difference. Therefore, if the pressure drop was unknown, but assumed to be small, then the enthalpy difference could be estimated to be the same as the internal energy difference, as shown in Equation 3.20. In the enthalpy difference calculation, the density of the benzene was found in Appendix B.1 at the average temperature.

3.3.3.3 Ideal Gas Model

The ideal gas model is meant to provide an *estimate* for the properties of gases and vapors. The equation of state of the ideal gas can be written in its most universal sense as

$$P\bar{v} = \bar{R}T \tag{3.25}$$

The overbars are used to indicate *molar* quantities. For example, the units of the molar-specific volume can be written in different unit systems as

$$\bar{v}[=]\frac{cm^3}{gmol} \text{ (CGS)} \qquad \bar{v}[=]\frac{m^3}{kmol} \text{ (SI)} \qquad \bar{v}[=]\frac{ft^3}{lbmol} \text{ (IP)} \qquad (3.26)$$

The molar value of $\bar{R}$ is the *universal gas constant*. Table 3.2 shows the value and units of the universal gas constant in several different unit sets. The equation of state can also be expressed on a *mass* basis by introducing the molecular mass of the gas.

$$Pv = \left(\frac{\bar{R}}{M}\right)T = RT \qquad (3.27)$$

In Equation 3.27, R is not universal; it depends on the molecular mass of the gas being analyzed.

In addition to the equation of state, the internal energy of an ideal gas is only a function of temperature. Therefore, the complete ideal gas model can be formulated by

$$Pv = RT \quad \text{and} \quad u = u(T) \qquad (3.28)$$

Figure 3.3 shows a pressure–internal energy diagram for cyclohexane. In Figure 3.3, it can be seen that the isotherms are nearly vertical for a good share of the liquid and vapor regions. The nearly vertical behavior of the isotherms in these regions indicate that the internal energy is only a function of temperature, $u = u(T)$. This plot demonstrates the validity of using the incompressible substance model in the liquid phase and the ideal gas model in the vapor phase.

The internal energy change of an ideal gas can be determined using Equation 3.5. Since the internal energy of an ideal gas is only a function of temperature, the ideal gas isochoric heat capacity can be written as

$$c_v^0 = \frac{du}{dT} \qquad (3.29)$$

TABLE 3.2

Values of the Universal Gas Constant in Different Units

Unit System	$\bar{R}$
SI	$8.314\dfrac{kJ}{kmol \cdot K} = 8314\dfrac{J}{kmol \cdot K}$
IP	$1545\dfrac{ft \cdot lbf}{lbmol \cdot R} = 1.986\dfrac{Btu}{lbmol \cdot R}$
CGS	$8.314 \times 10^7 \dfrac{dyne \cdot cm}{gmol \cdot K}$

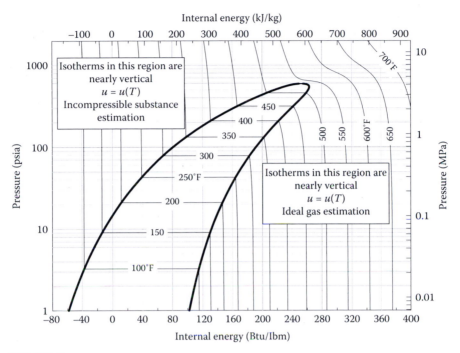

FIGURE 3.3
Pressure–internal energy diagram for cyclohexane.

A superscript "0" is used to identify the heat capacity as an ideal gas value. This equation can be integrated to determine the internal energy change of the ideal gas:

$$u_2 - u_1 = \int_{T_1}^{T_2} c_v^0 \, dT \tag{3.30}$$

The enthalpy change of an ideal gas can be determined from Equation 3.6. For an ideal gas, the enthalpy can be expressed by

$$h = u + Pv = u + RT \tag{3.31}$$

According to Equation 3.31, the enthalpy of an ideal gas is only a function of temperature. Therefore, Equation 3.6 can be written as

$$c_p^0 = \frac{dh}{dT} \tag{3.32}$$

The enthalpy change of the ideal gas can be determined by integrating Equation 3.32 between the temperature limits of the process.

$$h_2 - h_1 = \int_{T_1}^{T_2} c_p^0 \, dT \tag{3.33}$$

The entropy change of an ideal gas can be determined by either of the Gibbs equations; Equation 3.7 or 3.8. Rearranging Equation 3.7 to solve for ds gives

$$ds = \frac{du}{T} + \frac{P}{T}dv \tag{3.34}$$

Substituting Equation 3.29 for du and the ideal gas equation of state for P/T results in

$$ds = \frac{c_v^0}{T}dT + \frac{R}{v}dv \tag{3.35}$$

The entropy change of an ideal gas between two thermodynamic states can be found by integrating Equation 3.35.

$$s_2 - s_1 = \int_{T_1}^{T_2} \frac{c_v^0}{T}dT + R\ln\frac{v_2}{v_1} \tag{3.36}$$

Manipulating the second Gibbs equation, Equation 3.8, in a similar manner, gives an alternate expression for the entropy change of an ideal gas.

$$s_2 - s_1 = \int_{T_1}^{T_2} \frac{c_p^0}{T}dT - R\ln\frac{P_2}{P_1} \tag{3.37}$$

Notice that for an ideal gas, the internal energy and enthalpy are functions of temperature only. However, the entropy of an ideal gas is a function of *both* temperature and pressure.

Equations 3.30, 3.33, 3.36, and 3.37 are the expressions required to compute internal energy, enthalpy, and entropy differences of an ideal gas between two thermodynamic states. All of these equations contain an integral involving the ideal gas heat capacity.

The heat capacity of ideal gases can vary substantially over a given temperature range. The exception to this behavior is seen in the noble gases; the last column of the periodic chart of the elements. For the noble gases (He, Ne, Ar, Kr, Xe, and Rn) the ideal gas isobaric heat capacity is constant and given by

$$c_p^0 = \frac{5}{2}R \qquad \text{(noble gases only)} \tag{3.38}$$

If the heat capacity dependence on temperature is known, then the equations for internal energy, enthalpy, and entropy changes can be integrated. Most thermodynamics textbooks contain tables that show this temperature dependence. However, for many thermal energy system calculations, it is often sufficient to assume that the heat capacity is constant at *the average temperature* between the two states being analyzed. Therefore, the ideal gas property changes can be written as

$$u_2 - u_1 = c_{v,avg}^0 \left(T_2 - T_1\right)$$

$$h_2 - h_1 = c_{p,avg}^0 \left(T_2 - T_1\right)$$

$$s_2 - s_1 = c_{v,avg}^0 \ln\frac{T_2}{T_1} + R\ln\frac{v_2}{v_1} \qquad (3.39)$$

$$s_2 - s_1 = c_{p,avg}^0 \ln\frac{T_2}{T_1} - R\ln\frac{P_2}{P_1}$$

In the equations representing the entropy change of an ideal gas in Equation set 3.39, the temperatures must be on the absolute scale. For the internal energy and enthalpy changes, the temperatures can be either relative or absolute since the equations contain a temperature difference.

The ideal gas heat capacities are related to each other. This can be demonstrated by considering the differential of Equation 3.31.

$$dh = du + RdT + TdR = du + RdT \qquad (3.40)$$

Substituting Equations 3.29 and 3.32 into Equation 3.40 gives

$$c_p^0 dT = c_v^0 dT + RdT \qquad (3.41)$$

This equation can be simplified to

$$c_p^0 - c_v^0 = R \qquad (3.42)$$

EXAMPLE 3.3

A sample of krypton ($M = 83.8$ kg/kmol) gas is initially at 300°C, 400 kPa. The gas undergoes a process and ends up at a pressure of 200 kPa. Determine the change in enthalpy of the gas if the process is

    a. Isothermal (constant temperature)
    b. Isentropic (constant entropy)

SOLUTION:

    a. The enthalpy of an ideal gas is a function of temperature only. Therefore, if the process is isothermal, the enthalpy change of the substance must be

$$h_2 - h_1 = 0\frac{\text{kJ}}{\text{kg}} \quad \Leftarrow$$

b. Krypton is a noble gas. Therefore, its heat capacities are constant. The isobaric heat capacity of the gas can be found from Equation 3.38.

$$c_p^0 = \frac{5}{2}R = \frac{5}{2}\left(\frac{\bar{R}}{M}\right) = \frac{5}{2}\frac{\left(8.314\dfrac{kJ}{kmol \cdot K}\right)}{\left(83.8\dfrac{kg}{kmol}\right)} = 0.248\frac{kJ}{kg \cdot K}$$

The enthalpy change of the gas can be determined from the enthalpy difference equation in Equation set 3.39.

$$h_2 - h_1 = c_p^0\left(T_2 - T_1\right)$$

Since the entropy is constant, the last equation in Equation set 3.39 can be written as

$$c_p^0 \ln\frac{T_2}{T_1} = R \ln\frac{P_2}{P_1}$$

Solving this equation for the final temperature, T_2

$$T_2 = T_1\left(\frac{P_2}{P_1}\right)^{R/c_p^0}$$

The exponent on the pressure ratio can be rewritten as

$$\frac{R}{c_p^0} = \frac{R}{5R/2} = \frac{2}{5}$$

The final temperature in the isentropic process can now be found by

$$T_2 = T_1\left(\frac{P_2}{P_1}\right)^{R/c_p^0} = (300 + 273.15)\,K\left(\frac{200\ kPa}{400\ kPa}\right)^{2/5} = 434.37\ K = 161.2°C$$

Therefore, the enthalpy change of the krypton gas during the isentropic process is

$$h_2 - h_1 = c_p^0\left(T_2 - T_1\right) = \left(0.248\frac{kJ}{kg \cdot K}\right)(161.2 - 300)\,K = -34.4\frac{kJ}{kg} \quad \Leftarrow$$

3.4 Suggested Thermal Energy Systems Analysis Procedure

Thermal energy systems range from very simple to quite complex. Independent of the complexity of the system being studied, the procedure to analyze thermal systems is fairly consistent from system to system. A suggested procedure for thermal energy system analysis is outlined below.

1. Draw a sketch of the thermal energy system being analyzed. This sketch does not have to be terribly artistic, but it should contain common symbols used in system sketches. Some of the more common engineering sketches of equipment found in thermal energy systems are shown in Figure 3.4. Sketches found in Figure 3.4 will be used throughout this book.

2. Label the sketch with the known information. This will help when it comes time to begin the actual analysis of the system.

3. Draw a *system boundary* to identify what is being analyzed. This is perhaps one of the most important aspects of thermal energy system analysis, and one that is often overlooked. Conservation and balance laws that are needed to conduct the system analysis are *meaningless* unless a system boundary is drawn. The location of the system boundary is completely up to the engineer. Wise selection of the system boundary can make the analysis quite simple, even for a

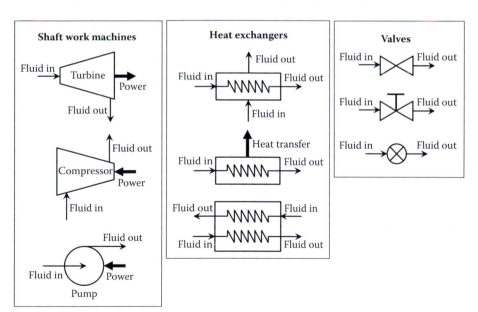

FIGURE 3.4
Common sketches used to identify thermal energy system components.

complex thermal system. On the other hand, some system boundaries may complicate the analysis to the point where it cannot be completed because information about fluid properties, flow rates, or energy transfers is unknown.

4. Once the system boundary is identified, then the proper conservation and/or balance laws can be applied to solve the problem. Conservation and balance laws are discussed in Section 3.5 of this chapter.

A good, structured procedure is important in thermal energy system analysis. The procedure outlined above is only a suggestion. Each individual may approach the analysis a bit differently, but the items above will most likely be included in the analysis.

3.5 Conserved and Balanced Quantities

The laws of fluid mechanics, thermodynamics, and heat transfer required to conduct a system analysis are formulated to define some sort of quantity. For example, the first law of thermodynamics quantifies *energy*, and the second law of thermodynamics quantifies *entropy*. A physical quantity can be classified as *conserved* if it cannot be created or destroyed. There are four physical quantities that are conserved; mass (in a nonnuclear reaction), energy, momentum, and electrical charge. Quantities that are not conserved are *balanced*. Examples of balanced quantities pertinent to thermal energy systems are entropy and *exergy*.

3.5.1 Generalized Balance Law

Figure 3.5 shows a picture of a system and a generalized balanced quantity, Ω at an instant in time. Dots are used above the quantity Ω to indicate a time rate. The quantities $\dot{\Omega}_{in}$ and $\dot{\Omega}_{out}$ represent the rate of flow of Ω into and out of the system, respectively. The flow may be the result of Ω being carried into and out of the system with mass, or it may be the result of Ω being directly transferred into the system (e.g., heat or work).

Since the quantity Ω is balanced, it must be true that

$$\begin{bmatrix} \text{Time rate of } \Omega \text{ crossing} \\ \text{the system boundary and} \\ \text{passing into the system} \end{bmatrix} - \begin{bmatrix} \text{Time rate of } \Omega \text{ crossing} \\ \text{the system boundary and} \\ \text{passing out of the} \\ \text{system} \end{bmatrix}$$

$$+ \begin{bmatrix} \text{Time rate of } \Omega \text{ being} \\ \text{generated within the} \\ \text{system} \end{bmatrix} = \begin{bmatrix} \text{Time rate of storage} \\ \text{of } \Omega \text{ in the} \\ \text{system} \end{bmatrix}$$

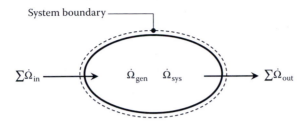

FIGURE 3.5
Generalized system.

The first two terms represent the *net* rate of Ω entering the system. If the *generated* term on the left-hand side is positive, the quantity Ω is literally generated within the system. On the other hand, if the generated term is negative, the quantity Ω is actually *destroyed* inside the system. If the *storage* term on the right-hand side is positive, the quantity Ω is being built up inside the system. Conversely, if the storage term is negative, Ω is being depleted inside the system. In equation form, this expression can be written as

$$\sum \dot{\Omega}_{in} - \sum \dot{\Omega}_{out} + \dot{\Omega}_{gen} = \dot{\Omega}_{sys} \qquad (3.43)$$

A more common way to express Equation 3.43 is to write the storage term on the right-hand side as a time derivative. The result is the *generalized balance law.*

$$\sum \dot{\Omega}_{in} - \sum \dot{\Omega}_{out} + \dot{\Omega}_{gen} = \frac{d\Omega_{sys}}{dt} \qquad (3.44)$$

Equation 3.44 is the starting point for the development of any of the equations that are commonly used in thermal energy systems analysis and design.

3.6 Conservation of Mass

Conservation of mass equation can be developed by making the substitution, $m = \Omega$ in Equation 3.44. Mass is a conserved quantity. Therefore, the generation term is zero. This results in the following expression:

$$\sum \dot{m}_{in} - \sum \dot{m}_{out} = \frac{dm_{sys}}{dt} \qquad (3.45)$$

The mass flow rates in the above equations can be expressed in terms of the fluid density, ρ; the cross-sectional area of the flow passage, A; and the mean velocity of the fluid flow, V. In addition, the mass stored in the system is a function of the density of the fluid and the volume enclosed by the system boundary. Incorporating these ideas into Equation 3.45 gives

$$\sum (\rho A V)_{in} - \sum (\rho A V)_{out} = \frac{d(\rho \mathcal{V})_{sys}}{dt} \tag{3.46}$$

If the mass flow rate entering the system is the same as the mass flow rate leaving, then there can be no mass stored inside the system. A system operating under this condition is said to be operating at a *steady-flow* condition. For a steady-flow condition, Equation 3.45 can be written as

$$\sum \dot{m}_{in} = \sum \dot{m}_{out} \tag{3.47}$$

To this point, the equations developed for the system are *rate* equations. They are describing the condition of the system *at an instant in time*. If the behavior of the system over a specified time is required, then any of these equations can be integrated over time. For example, to investigate how a system's mass might change during a process, Equation 3.45 can be integrated over time, resulting in

$$\sum m_{in} - \sum m_{out} = (m_2 - m_1)_{sys} \tag{3.48}$$

EXAMPLE 3.4

Consider the tank and water supply system as shown in Figure E3.4A. The diameter of the supply pipe is $D_1 = 20$ mm, and the average velocity leaving the supply pipe is $V_1 = 0.595$ m/s. A shut-off valve is located at $z = 0.1$ m in the exit pipe, which has a diameter of $D_2 = 10$ mm. The tank diameter is $D_t = 0.3$ m. The density of the water is uniform at 998 kg/m³.

   a. Find the time to fill the tank to $H_0 = 1$ m, assuming the tank is initially empty and the shut-off valve is closed. Neglect the volume associated with the short pipe connecting the tank to the shut-off valve.
   b. At the instant the water level reaches $H_0 = 1$ m, the shut-off valve is opened. The instantaneous average velocity of the outflow depends on the water depth above z and can be expressed by

$$V_2 = 0.85 \sqrt{g[H(t) - z]}$$

   where g is the acceleration due to gravity. Determine whether the tank continues to fill or begins to empty immediately after the valve is opened.
   c. Determine the steady-flow value of the water depth.

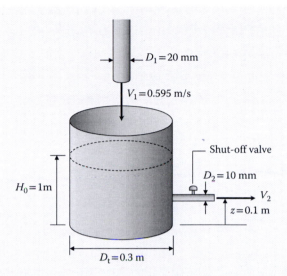

FIGURE E3.4A

SOLUTION: Part (a) of this problem asks how long it takes to fill the empty tank to a height of $H_0 = 1$ m. An appropriate system boundary for this analysis is one that encompasses the inside of the tank up to a height of 1 m as shown in Figure E3.4B. During the time required to fill this system, a total mass, m_{in} enters and the mass inside the system changes by an amount, $m_2 - m_1$. From this analysis, it is clear that we are investigating what is happening inside the system over a finite period. Therefore, the appropriate form of the conservation of mass is given by Equation 3.48.

$$\sum m_{in} - \sum m_{out} = \left(m_2 - m_1\right)_{sys}$$

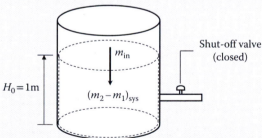

FIGURE E3.4B

For the filling process, this equation reduces to

$$m_{in} = \left(m_2 - m_1\right)_{sys}$$

Each of the terms in this equation can be expanded as follows:

$$(\rho A_1 V_1)t = (m_2 - m_1)_{sys} = m_2$$

$$(\rho A_1 V_1)t = \rho V_2$$

$$\frac{\pi D_i^2}{4} V_1 t = \frac{\pi D_i^2}{4} H_0$$

The only unknown in this equation is the time required to fill, t. Solving for t gives

$$t = \frac{H_0}{V_1}\left(\frac{D_i^2}{D_i^2}\right)$$

$$t = \frac{1\text{ m}}{0.595\text{ m/s}}\frac{(0.3\text{ m})^2}{(0.020\text{ m})^2} = 378.2\text{ s} = 6.3\text{ min} \quad \leftarrow$$

Part (b) asks whether the water level rises or falls when the shut-off valve is opened. To determine this, the system must be analyzed at the instant in time when the valve is opened. When the valve is opened, water will flow across the system boundary at the outlet as indicated in Figure E3.4C. At that instant, we are interested in what happens to $H(t)$, the instantaneous height of the water in the tank.

This analysis is not conducted over time, but rather at an instant in time. Therefore, Equation 3.45 is appropriate. Since there is one mass flow entering and one mass flow leaving the system, Equation 3.45 is written as

$$\dot{m}_{in} - \dot{m}_{out} = \frac{dm_{sys}}{dt}$$

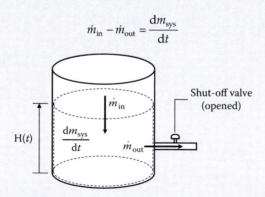

FIGURE E3.4C

If the sign of the storage term dm_{sys}/dt is positive, then the tank continues to fill. On the other hand, if the storage term is negative, the tank begins to drain. The mass flow rates on the left-hand side of this equation can be rewritten in terms of the density of the fluid, cross-sectional areas, and velocities.

$$\rho A_1 V_1 - \rho A_2 V_2 = \frac{dm_{sys}}{dt}$$

The velocity, V_2, is given in the problem statement in terms of the height of the water in the tank. Substituting this into the above equation and simplifying

$$\rho A_1 V_1 - \rho A_2 \left(0.85\sqrt{g[H(t)-z]}\right) = \frac{dm_{sys}}{dt}$$

$$\frac{dm_{sys}}{dt} = \rho\left(\frac{\pi}{4}\right)\left[D_1^2 V_1 - 0.85 D_2^2 \sqrt{g(H_0 - z)}\right]$$

The value of H_0 is used for $H(t)$, because this is the height of the water in the tank at the instant the valve is opened. Performing the calculations

$$\frac{dm_{sys}}{dt} = \left(998\ \frac{kg}{m^3}\right)\left(\frac{\pi}{4}\right)\left[(0.020\ m)^2\left(0.595\ \frac{m}{s}\right)\right.$$

$$\left. -(0.85)(0.010\ m)^2\sqrt{\left(9.81\frac{m}{s^2}\right)(1\ m - 0.1\ m)}\right]$$

$$\frac{dm_{sys}}{dt} = -0.0114\ \frac{kg}{s} \quad \leftarrow$$

Since the storage term is negative, the tank will begin to empty when the shut-off valve is opened.

Part (c) of this problem asks for the water depth resulting in a steady-flow scenario. It is important to understand that the value of the derivative computed in part (b) is instantaneous. Since the velocity of the water leaving the tank through the bottom pipe is a function of the depth of the water, this derivative will change with time. For a steady-flow condition, $dm_{sys}/dt = 0$. This allows for the solution of the height, H. From the analysis in part (b)

$$\frac{dm_{sys}}{dt} = 0 = \rho\left(\frac{\pi}{4}\right)\left[D_1^2 V_1 - 0.85 D_2^2 \sqrt{g(H-z)}\right]$$

$$0.85 D_2^2 \sqrt{g(H-z)} = D_1^2 V_1$$

$$g(H-z) = \left(\frac{D_1^2 V_1}{0.85 D_2^2}\right)^2$$

$$H = z + \frac{1}{g}\left(\frac{D_1^2 V_1}{0.85 D_2^2}\right)^2 = (0.1\ m) + \frac{1}{(9.81\ m/s^2)}\left[\frac{(0.020\ m)^2(0.595\ m/s)}{0.85(0.010\ m)^2}\right]^2$$

$$H = 0.90\ m \quad \leftarrow$$

When the valve is opened, the water depth will decrease from 1 to 0.9 m, at which point the incoming flow balances the outgoing flow and a steady-flow scenario is achieved.

3.7 Conservation of Energy (The First Law of Thermodynamics)

Conservation of energy is also known as the first law of thermodynamics. To develop an equation for the conservation of energy, $\Omega = E$ is substituted into Equation 3.44, where E represents the total energy of a system. For non-reacting systems, the total energy (extensive form) of a substance is made up of *internal energy* due to molecular activity, *kinetic energy* due to motion, and *potential energy* due to elevation. Therefore,

$$E = U + \frac{mV^2}{2} + mgz \tag{3.49}$$

The energy of a substance per unit mass, the intensive form, is given by

$$e = \frac{E}{m} = u + \frac{V^2}{2} + gz \tag{3.50}$$

Making the substitution $\Omega = E$ into Equation 3.44 results in the rate form of the conservation of energy

$$\sum \dot{E}_{in} - \sum \dot{E}_{out} = \frac{dE_{sys}}{dt} \tag{3.51}$$

For flow systems, it is convenient to separate the energy inputs and outputs due to heat and work transfer from the energy inputs and outputs carried by mass flow of the fluids involved.

$$\sum \left(\dot{E}_{in} - \dot{E}_{out} \right)_{HW} + \sum \left(\dot{E}_{in} - \dot{E}_{out} \right)_{flow} = \frac{dE_{sys}}{dt} \tag{3.52}$$

The first term in Equation 3.52 represents the net energy transfer rate associated with heat and/or work transfer to/from the system. This term can be written as

$$\sum \left(\dot{E}_{in} - \dot{E}_{out} \right)_{HW} = \sum \left(\dot{Q} + \dot{W} \right)_{in} - \sum \left(\dot{Q} + \dot{W} \right)_{out} \tag{3.53}$$

The second term on the left-hand side of Equation 3.52 represents the net energy transfer rate resulting from the fluid entering and leaving the system. This term can be rewritten as

$$\sum \left(\dot{E}_{in} - \dot{E}_{out} \right)_{flow} = \sum \dot{m}_{in} \left(e + Pv \right)_{in} - \sum \dot{m}_{out} \left(e + Pv \right)_{out} \tag{3.54}$$

In this equation, *e* is the specific energy of the fluid as given in Equation 3.50 and *Pv* is the specific work required to move the fluid across the system boundary. The quantity *Pv* is often referred to as *flow work*. Substituting Equation 3.50 into Equation 3.54 gives

$$\left(\dot{E}_{in} - \dot{E}_{out}\right)_{flow} = \sum \dot{m}_{in}\left(u + \frac{V^2}{2} + gz + Pv\right)_{in} - \sum \dot{m}_{out}\left(u + \frac{V^2}{2} + gz + Pv\right)_{out} \quad (3.55)$$

This equation can be rewritten by realizing that $u + Pv$ is the enthalpy, *h*, of the fluid. Therefore,

$$\left(\dot{E}_{in} - \dot{E}_{out}\right)_{flow} = \sum \dot{m}_{in}\left(h + \frac{V^2}{2} + gz\right)_{in} - \sum \dot{m}_{out}\left(h + \frac{V^2}{2} + gz\right)_{out} \quad (3.56)$$

Substituting Equations 3.53 and 3.56 into Equation 3.52 gives

$$\left[\sum\left(\dot{Q} + \dot{W}\right)_{in} - \sum\left(\dot{Q} + \dot{W}\right)_{out}\right] + \sum \dot{m}_{in}\left(h + \frac{V^2}{2} + gz\right)_{in}$$
$$- \sum \dot{m}_{out}\left(h + \frac{V^2}{2} + gz\right)_{out} = \frac{dE_{sys}}{dt} \quad (3.57)$$

This is the general conservation of energy equation (the first law of thermodynamics). Although this may seem fairly complex, it can be simplified for many common engineering applications.

A confusing issue in many thermodynamics textbooks is the sign convention for heat and work. Most textbooks adopt the convention that positive energy transfers are heat into a system and work out of a system, whereas negative energy transfers are heat out of a system and work into a system. In the generalized approach taken here, the sign of heat and work transfers are always considered positive in sign, independent of direction. In cases where a heat transfer or work transfer term is unknown, it can be assumed to be either into or out of the system. If the algebraic sign of the term turns out to be negative, this means that the assumed direction is opposite of the actual direction; the magnitude of the result is correct.

This generalized approach takes the sign mystery out of the calculation. It allows one to focus on the result without worrying if a sign error was made somewhere in the analysis. Several examples of application of the conservation of energy equation follow.

EXAMPLE 3.5

Steam at 1.6 MPa and 350°C enters a steam turbine at a flow rate of 16 kg/s. The steam leaves the turbine as a saturated vapor at 30°C. The turbine delivers 9 MW of power. A schematic of the turbine and its operating conditions are given in Figure E3.5. Determine the heat transfer rate from this turbine.

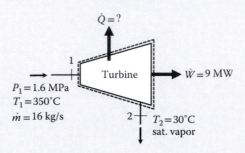

FIGURE E3.5

SOLUTION: In many analyses, turbines are considered to be adiabatic (no heat transfer). However, this problem asks for the heat transfer rate. This means the turbine is losing heat, perhaps a result of the insulation around the turbine casing degrading.

   The system boundary to be analyzed is shown in the sketch. Since the flow rate is given as a constant value, the turbine must be operating in a steady-flow mode. For turbines, it is common to consider the kinetic and potential energy changes to be negligible. Under these conditions, Equation 3.57 reduces to

$$\left[0 - \left(\dot{Q} + \dot{W} \right)_{out} \right] + \dot{m}\left(h_1 - h_2 \right) = 0$$

Solving this equation for the heat transfer rate

$$\dot{Q} = \dot{m}\left(h_1 - h_2 \right) - \dot{W}$$

Given the input and exhaust conditions specified in the problem, the enthalpy values* of the steam are as follows: h_1 = 3146.0 kJ/kg and h_2 = 2555.6 kJ/kg. Therefore, the heat transfer rate from the turbine is

$$\dot{Q} = \left(16\,\frac{kg}{s} \right)\left(3146.0 - 2555.6 \right)\frac{kJ}{kg} - \left(9\ MW \right)\left| \frac{1000\ kW}{MW} \right. = \underline{446.4\ kW} \quad \leftarrow$$

The sign of the heat transfer rate (+) is consistent with the assumed direction (out of the system). Notice that the magnitude of this heat loss represents about 5% of the power output of the turbine. Better insulation on the turbine casing would increase the power output.

---

* Throughout this book, thermodynamic properties are computed using EES from F-Chart Software, http://www.fchart.com

EXAMPLE 3.6

Water flows through a shower head steadily and exits at a volumetric flow rate of 2.6 gpm. An electric resistance heater placed in the water pipe heats the water from 61°F to 110°F as shown in Figure E3.6A. Determine the electrical power required by the heater.

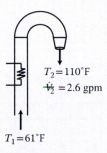

$T_2 = 110°F$
$\dot{V}_2 = 2.6$ gpm

$T_1 = 61°F$

FIGURE E3.6A

SOLUTION: The heater is a device that has electrical current flowing through it. The flow of electrical current is analogous to power. Therefore, the electrical resistance heater can be replaced by a power transfer that crosses the system boundary as shown in Figure E3.6B.

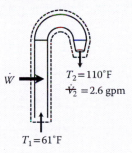

$\dot{W}$

$T_2 = 110°F$
$\dot{V}_2 = 2.6$ gpm

$T_1 = 61°F$

FIGURE E3.6B

With negligible kinetic and potential changes between the inlet and outlet of the system, the conservation of energy equation 3.57 becomes

$$\dot{W} + \dot{m}\left(h_1 - h_2\right) = 0$$

Since the water remains in the liquid phase and the pressure drop through the system is relatively small, a reasonable approximation to the enthalpy change is $\Delta h \approx \tilde{c}_{p,avg}\Delta T$. Substituting this approximation into the above equation and solving for the power input gives

$$\dot{W} = \dot{m}\tilde{c}_{p,avg}\left(T_2 - T_1\right)$$

The average isobaric heat capacity of the water can be estimated at the average temperature on the saturated liquid curve as follows:

$$T_{avg} = \frac{T_1 + T_2}{2} = \frac{61°F + 110°F}{2} = 85.5°F$$

$$\tilde{c}_{avg}\left(T_{avg} = 85.5°F, x = 0\right) = 0.9984 \text{ Btu/lbm} \cdot \text{R}$$

The mass flow rate of the water needs to be determined. In the problem statement, the volume flow rate is given at the outlet of the shower. This can be converted to a mass flow rate since the temperature of the water is known at that point.

$$\dot{m} = \rho_2 \dot{V}_2$$

Using the incompressible substance model for a liquid, the density of the water at the outlet can be estimated as the density of the saturated liquid at T_2.

$$\rho_2\left(T_2 = 110°F, x = 0\right) = 61.86 \text{ lbm/ft}^3$$

Now, the mass flow rate of the water can be found by

$$\dot{m} = \left(61.86 \ \frac{\text{lbm}}{\text{ft}^3}\right)\left(2.6 \ \frac{\text{gal}}{\text{min}}\right)\left|\frac{0.133680556 \ \text{ft}^3}{\text{gal}}\right|\frac{60 \ \text{min}}{\text{h}} = 1290 \ \frac{\text{lbm}}{\text{h}}$$

The power required can then be found by

$$\dot{W} = \left(1290 \ \frac{\text{lbm}}{\text{h}}\right)\left(0.9984 \ \frac{\text{Btu}}{\text{lbm} \cdot \text{R}}\right)(110 - 61) \text{R}\left|\frac{\text{kW} \cdot \text{h}}{3412.1415 \ \text{Btu}}\right|$$

$$\dot{W} = 18.5 \text{ kW} \quad \leftarrow$$

EXAMPLE 3.7

A 4-ft³ rigid tank contains saturated R-134a at 100 psia as shown in Figure E3.7A. Initially, 20% of the volume is occupied by liquid and the rest by vapor. A valve at the top of the tank is now opened, and saturated vapor is allowed to escape slowly from the tank. Heat is transferred to the refrigerant such that the pressure inside the tank remains constant. The valve is closed when the last drop of liquid in the tank is vaporized. Determine the total heat transfer for this process.

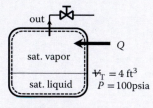

FIGURE E3.7A

SOLUTION: In this problem, it is helpful to envision the process on a thermo-dynamic diagram. Figure E3.7B shows a temperature–volume (*T–v*) diagram for this process. The end state of the process can easily be identified as the saturated vapor state at 100 psia, labeled "2" on the diagram. State 1 is not known at this point. However, since 20% of the volume of the tank is occupied by liquid with the rest being vapor, state 1 must lie somewhere in the two-phase region on the 100 psia isobar.

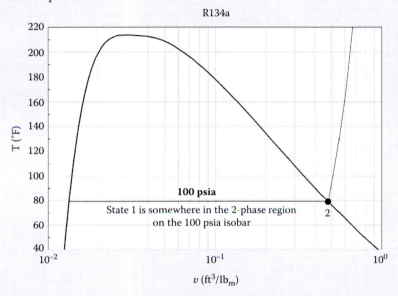

R134a

FIGURE E3.7B

The energy entering the system is the heat transfer, and there is only one mass flow (out of the system). The mass flow rate leaving the tank is not constant in this example. Therefore, this is not a steady-flow process. Assuming that the kinetic and potential effects are negligible, the conservation of energy equation 3.57 reduces to

$$\dot{Q} - \dot{m}_{\text{out}} h_{\text{out}} = \frac{dE_{\text{sys}}}{dt}$$

This equation can be integrated between the initial state defined in the problem, and the end state (when the last drop of liquid vaporizes).

$$\int_{t_1}^{t_2} \dot{Q} \, dt - \int_{t_1}^{t_2} \dot{m}_{\text{out}} h_{\text{out}} \, dt = \int_{E_1}^{E_2} dE_{\text{sys}}$$

The first integral on the left-hand side of this equation represents the total amount of heat transferred to the R-134a during the process.

$$\int_{t_1}^{t_2} \dot{Q} \, dt = Q$$

The R-134a in the vessel is maintained at a constant pressure of 100 psia due to the heat transfer. This means that the enthalpy of the fluid leaving the system is constant. Therefore, the conservation of energy equation can be written as

$$Q - h_{out} \int_{t_1}^{t_2} \dot{m}_{out}\, dt = \int_{E_1}^{E_2} dE_{sys}$$

The integral containing the mass flow rate can be modified using the conservation of mass equation (Equation 3.45) applied to the system.

$$-\dot{m}_{out} = \frac{dm_{sys}}{dt}$$

Substituting this expression into the conservation of energy equation results in

$$Q - h_{out} \int_{t_1}^{t_2} \left(-\frac{dm_{sys}}{dt} \right) dt = \int_{E_1}^{E_2} dE_{sys}$$

$$Q + h_{out} \int_{m_1}^{m_2} dm_{sys} = \int_{E_1}^{E_2} dE_{sys}$$

Integrating both sides results in

$$Q + h_{out}(m_2 - m_1) = E_2 - E_1$$

The energy terms on the right-hand side represent the energy change of the R-134a inside of the system. Neglecting differences in kinetic and potential energies,

$$E_2 - E_1 = m_2 e_2 - m_1 e_1 = m_2 u_2 - m_1 u_1$$

Substitution of this expression into the previous equation and solving for the heat transfer gives

$$Q + h_{out}(m_2 - m_1) = m_2 u_2 - m_1 u_1$$
$$Q = m_2(u_2 - h_{out}) - m_1(u_1 - h_{out})$$

During the process, the R-134a is in a saturation condition inside the tank. At any time, the mass of the R-134a in the tank is

$$m = m_g + m_f$$

In this equation the subscript "g" represents the saturated vapor, and "f" is the saturated liquid. These masses are related to the volume occupied by each phase.

$$m_g = \rho_g V_g \quad \text{and} \quad m_f = \rho_f V_f$$

Therefore, the total mass of R-134a in the vessel at any time is

$$m = \rho_g V_g + \rho_f V_f$$

The initial mass in the tank (state 1) is

$$m_1 = \rho_g V_{g,1} + \rho_f V_{f,1}$$

At the end of the process, there is no saturated liquid left. Therefore, the mass at that point is

$$m_2 = \rho_g V_{g,2} = \rho_g V_T$$

The saturated liquid and saturated vapor densities can be found in Appendix B.2 (interpolation is required). Therefore, the initial and final masses are

$$m_1 = \left(2.094 \frac{\text{lbm}}{\text{ft}^3}\right) \left[(0.80)(4 \text{ ft}^3)\right] + \left(75.05 \frac{\text{lbm}}{\text{ft}^3}\right) \left[(0.20)(4 \text{ ft}^3)\right]$$

$$m_1 = \underset{\text{sat vapor in the tank}}{6.70 \text{ lbm}} + \underset{\text{sat liquid in the tank}}{60.04 \text{ lbm}} = \underset{\text{total initial mass in the tank}}{66.74 \text{ lbm}}$$

$$m_2 = \left(2.094 \frac{\text{lbm}}{\text{ft}^3}\right)(4 \text{ ft}^3) = \underset{\text{final mass in the tank}}{8.375 \text{ lbm}}$$

The enthalpy of the R-134a leaving the vessel is the saturated vapor enthalpy at 100 psia; $h_{\text{out}} = 113.83$ Btu/lbm.

The internal energies, u_1 and u_2, are the specific internal energies of the R-134a in the vessel at the initial and end states, respectively. The internal energy in the final state is the saturated vapor internal energy at 100 psia and can be found to be $u_2 = 105.0$ Btu/lbm

The R-134a in the vessel at the initial state is a mixture of saturated liquid and saturated vapor. Therefore, the internal energy at that state can be found using the quality relationship.

$$u_1 = u_f + x_1 u_{fg} = u_f + x_1 \left(u_g - u_f\right)$$

The quality in the initial state is

$$x_1 = \frac{m_{g,1}}{m_{\text{total}}} = \frac{6.70 \text{ lbm}}{66.74 \text{ lbm}} = 0.10$$

Then, the internal energy at the initial state can be found by

$$u_1 = 37.62 \frac{\text{Btu}}{\text{lbm}} + (0.10)(105.0 - 37.62)\frac{\text{Btu}}{\text{lbm}} = 44.39 \frac{\text{Btu}}{\text{lbm}}$$

Now, the required heat transfer can be computed.

$$Q = (8.375 \text{ lbm})(105.0 - 113.83)\frac{\text{Btu}}{\text{lbm}} - (66.74 \text{ lbm})(44.39 - 113.83)\frac{\text{Btu}}{\text{lbm}}$$

$$Q = \underline{4561 \text{ Btu}} \quad \leftarrow$$

Even though only 20% of the vessel is occupied by liquid in the initial state, a significant amount of energy in the form of heat is required to vaporize it.

3.8 Entropy Balance (The Second Law of Thermodynamics)

The entropy balance is an expression of the second law of thermodynamics that is particularly useful in analysis of thermal equipment and processes. The second law of thermodynamics helps to identify irreversibilities in a process or system. Entropy is not a conserved quantity since it can be *generated* by a process. *Entropy generation* occurs as a result of irreversibility. There are many things that cause irreversibility in a process. Common causes of irreversibility in thermal equipment are mechanical friction (e.g., bearings supporting a rotating shaft), fluid friction due to flow, heat transfer through a finite temperature difference, abrupt expansion or contraction, and mixing.

Making the substitution $\Omega = S$ into Equation 3.44 results in the *entropy balance*.

$$\sum \dot{S}_{in} - \sum \dot{S}_{out} + \dot{S}_{gen} = \frac{dS_{sys}}{dt} \tag{3.58}$$

Entropy can be carried into and out of a system by heat and by virtue of the mass flow crossing the system boundary. Therefore, it is convenient to separate the entropy rates due to heat transfer from the entropy rates due to flow.

$$\sum \left(\dot{S}_{in} - \dot{S}_{out} \right)_Q + \sum \left(\dot{S}_{in} - \dot{S}_{out} \right)_{flow} + \dot{S}_{gen} = \frac{dS_{sys}}{dt} \tag{3.59}$$

The term $\dot{S}_{gen}$ is the *entropy generation rate* or *entropy production rate*. The magnitude of the entropy generation gives an indication of the irreversibility associated with a thermal component or a process. The sign of the entropy generation is an indication of the *type* of process. By its very nature, entropy generation can never be negative. For real-world, irreversible processes, the entropy generation is always positive. If the component or process is thermodynamically reversible (an ideal process), then the entropy generation is zero. This can be quantified by Equation 3.60 shown below:

$$\dot{S}_{gen} \begin{cases} > 0 \text{ for all real-world processes} \\ = 0 \text{ for reversible processes} \\ < 0 \text{ is impossible} \end{cases} \tag{3.60}$$

The reversible process is of particular interest to engineers. Even though a reversible process does not exist in the real world, it serves as a performance benchmark that can be used in thermal system design. It is also useful in defining *isentropic efficiencies* of various devices (e.g., pumps or turbines). The isentropic efficiency is reviewed in Section 3.8.2.

The entropy transfer rate due to heat transfer across the system boundary, the first term on the left-hand side of Equation 3.59, can be written as

$$\sum \left(\dot{S}_{in} - \dot{S}_{out} \right)_Q = \sum \left(\frac{\dot{Q}}{T_b} \right)_{in} - \sum \left(\frac{\dot{Q}}{T_b} \right)_{out} \qquad (3.61)$$

In Equation 3.61, the temperature, T_b is the boundary temperature on the system boundary where the heat transfer is taking place. This temperature must be expressed on the *absolute scale* (K or R). Notice that Equation 3.61 is written to accommodate different boundary temperatures for a given system corresponding to where the heat is crossing. Substituting Equation 3.61 into Equation 3.59 gives

$$\left[\sum \left(\frac{\dot{Q}}{T_b} \right)_{in} - \sum \left(\frac{\dot{Q}}{T_b} \right)_{out} \right] + \sum \left(\dot{S}_{in} - \dot{S}_{out} \right)_{flow} + \dot{S}_{gen} = \frac{dS_{sys}}{dt} \qquad (3.62)$$

Writing the entropy transfer rates due to mass flow as a function of the mass flow rate, a general expression for the entropy balance can be written as

$$\left[\sum \left(\frac{\dot{Q}}{T_b} \right)_{in} - \sum \left(\frac{\dot{Q}}{T_b} \right)_{out} \right] + \left[\sum (\dot{m}s)_{in} - \sum (\dot{m}s)_{out} \right] + \dot{S}_{gen} = \frac{dS_{sys}}{dt} \qquad (3.63)$$

Entropy is a thermodynamic property. Therefore, it is a function of state. On the other hand, entropy generation is *not* a property. It depends on the process. If the path taken by the process is reversible, then the entropy generation is zero. If the process is irreversible, then there are several possible paths that can be taken, depending on the severity of the irreversibility. Because of this, entropy generation is considered a *path* function, similar to heat and work.

EXAMPLE 3.8

Steam at 1.6 MPa and 350°C enters a steam turbine at a flow rate of 16 kg/s. The steam leaves the turbine as a saturated vapor at 30°C. The turbine delivers 9 MW of power. The turbine is insulated, but not perfectly. The average temperature of the insulation on the outer surface of the turbine is 70°C. Figure E3.8 shows this turbine and its operating conditions. Determine the entropy generation rate for the turbine.

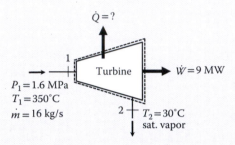

FIGURE E3.8

SOLUTION: This is the same turbine that was analyzed in Example 3.5. For the conditions given, the entropy balance equation (Equation 3.63) can be written as

$$-\frac{\dot{Q}}{T_s} + \dot{m}\left(s_1 - s_2\right) + \dot{S}_{gen} = 0$$

$$\therefore \ \dot{S}_{gen} = \dot{m}\left(s_2 - s_1\right) + \frac{\dot{Q}}{T_s}$$

The heat transfer rate in Example 3.5 was found to be 446.4 kW. This is a heat loss from the turbine. The inlet and exhaust entropy values can be found since the states are fully identified; $s_1 = 7.0713 \text{ kJ/kg·K}$ and $s_2 = 8.4520 \text{ kJ/kg·K}$. With these values,

$$\dot{S}_{gen} = \dot{m}\left(s_2 - s_1\right) + \frac{\dot{Q}}{T_s}$$

$$\dot{S}_{gen} = \left(16\frac{\text{kg}}{\text{s}}\right)(8.4520 - 7.0713)\frac{\text{kJ}}{\text{kg·K}} + \frac{446.4 \text{ kW}}{(70 + 273.15)\text{K}}$$

$$\dot{S}_{gen} = 22.09 \text{ kW/K} + 1.30 \text{ kW/K} = 23.39 \text{ kW/K} \quad \leftarrow$$

Notice that the entropy generation due to the fluid flow is 22.09/23.39 = 0.94 = 94% of the total. The entropy generation due to the heat transfer is only 6%. Therefore, any engineering efforts to eliminate irreversibility should be focused on the flow process within the turbine.

3.8.1 Reversible and Adiabatic Process

A thermodynamic process is considered reversible if the system can be returned back to its original state without any changes in the system *or the surroundings* that the system interacts with. In reality, reversible process cannot occur. However, they serve as a benchmark in engineering design. If the flow process through the component is reversible, there is no entropy generation. Therefore, Equation 3.63 becomes

$$\left[\sum \left(\frac{\dot{Q}}{T_b} \right)_{in} - \sum \left(\frac{\dot{Q}}{T_b} \right)_{out} \right] + \left[\sum (\dot{m}s)_{in} - \sum (\dot{m}s)_{in} \right] = \frac{dS_{sys}}{dt} \tag{3.64}$$

Now consider a reversible process that has one flow entering, one flow leaving, and is operating at steady-flow conditions. For this special case, Equation 3.64 becomes

$$\left[\sum \left(\frac{\dot{Q}}{T_b} \right)_{in} - \sum \left(\frac{\dot{Q}}{T_b} \right)_{out} \right] + \dot{m} \left(s_{in} - s_{out} \right) = 0 \tag{3.65}$$

Consider the case where the process is also adiabatic. For this situation, Equation 3.65 simplifies to

$$s_{in} = s_{out} \tag{3.66}$$

Equation 3.66 implies that for a steady-flow, single-flow device, a reversible *and* adiabatic process is *isentropic*. This is an important discovery in the thermodynamic analysis of many flow devices.

3.8.2 Isentropic Efficiencies of Flow Devices

The concept of an isentropic process is particularly helpful in defining a performance parameter of a flow device known as the isentropic efficiency. The definition of the isentropic efficiency is dependent on what purpose the device is serving.

3.8.2.1 Turbines

The purpose of a turbine is to deliver power. The maximum power that can be delivered from a turbine occurs when the flow process through the turbine is isentropic. In the real world, we expect that the turbine delivers less power due to irreversibilities. Therefore, the isentropic efficiency of a turbine can be expressed as

$$\eta_t = \frac{w_t}{w_{t,s}} \tag{3.67}$$

In Equation 3.67, w_t is the actual work produced by the turbine and $w_{t,s}$ is the work produced by an isentropic turbine. Applying the conservation of energy to a steady state, adiabatic turbine, the work production is simply the difference in enthalpy between the inlet and outlet. Therefore, the isentropic efficiency can be expressed as

$$\eta_t = \frac{w_t}{w_{t,s}} = \frac{h_{in} - h_{out}}{h_{in} - h_{out,s}} \tag{3.68}$$

Application of the entropy balance to a steady-flow, adiabatic turbine results in

$$\frac{\dot{S}_{gen}}{\dot{m}} = s_{gen} = s_{out} - s_{in} \tag{3.69}$$

The *h–s* diagram, commonly known as the Mollier diagram, is particularly useful in visualizing the isentropic efficiency and entropy generation in a turbine as shown in Figure 3.6. This figure shows that the exhaust state of the isentropic expansion and the actual expansion through the turbine are the same *pressure, P_{out}*.

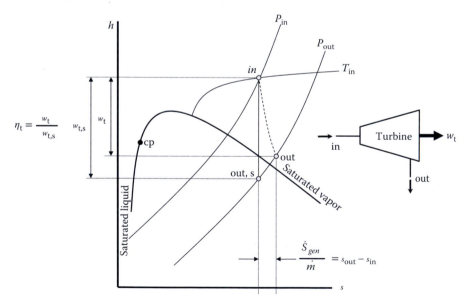

FIGURE 3.6
Mollier diagram showing the isentropic efficiency and entropy generation for an adiabatic turbine.

3.8.2.2 Compressors, Pumps, and Fans

Compressors, pumps, and fans are designed to move fluids. Compressors and fans move gases, whereas pumps move liquids. These devices accomplish fluid movement by utilizing an energy input. If the device is reversible and adiabatic, then all of the input energy can be focused on moving the fluid. In the case of a real-world device, the major sources of irreversibility are mechanical friction and fluid friction. Therefore, to move the fluid at the same rate, more energy is required in the real case. Using this reasoning, the isentropic efficiency of a compressor, fan, or pump can be expressed by

$$\eta_m = \frac{w_s}{w} = \frac{h_{out,s} - h_{in}}{h_{out} - h_{in}} \qquad (3.70)$$

The subscript "m" is used to indicate a device that *moves* a fluid. Application of the entropy balance to a steady state, adiabatic compressor, fan, or pump results in

$$\frac{\dot{S}_{gen}}{\dot{m}} = s_{gen} = s_{out} - s_{in} \qquad (3.71)$$

Figure 3.7 shows the *h–s* diagram for a compressor. As with the turbine, the isentropic efficiency and entropy generation can be seen graphically.

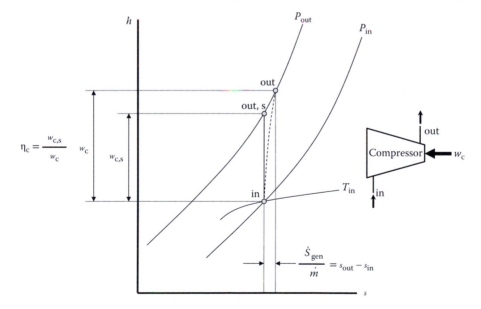

FIGURE 3.7

h–s diagram showing the isentropic efficiency and entropy generation for an adiabatic compressor.

h–s diagrams for the fan and pump are similar. For the pump, the process occurs in the liquid phase. Figure 3.7 indicates that the exhaust state of the isentropic compression and actual compression are at the same pressure.

3.8.2.3 Nozzles

Nozzles are utilized in many different systems. Perhaps the most well-known large-scale application is in rocketry. However, nozzles also play an important role on the small scale. For example, nozzles are used in turbines to increase the kinetic energy of the fluid before impinging on the turbine blades, causing a transfer of momentum resulting in a rotating shaft. From these two examples, it becomes clear that the purpose of a nozzle is to increase the fluid's kinetic energy.

Nozzles are generally considered to be adiabatic. If the flow is frictionless, then the velocity at the exit of the nozzle will be at a maximum. Since the purpose of a nozzle is to increase the kinetic energy of the fluid passing through it, the isentropic efficiency of a nozzle can be expressed as

$$\eta_n = \frac{ke}{ke_s} = \frac{V^2/2g}{V_s^2/2g} = \frac{V^2}{V_s^2} \tag{3.72}$$

In the evaluation of Equation 3.72, the isentropic exhaust of the nozzle is at the same pressure as the actual exhaust.

3.8.2.4 Diffusers

The diffuser uses a decrease in the fluid's velocity to affect a pressure increase. Like nozzles, diffusers are considered to be adiabatic since the fluid passes through the device quickly leaving very little time to transfer heat. Since the purpose of a nozzle is to increase the fluid's pressure, its isentropic efficiency is defined as

$$\eta_d = \frac{\Delta P_d}{\Delta P_{d,s}} = \frac{P_{out} - P_{in}}{P_{out,s} - P_{in}} \tag{3.73}$$

In Equation 3.73, P_{out} is the actual outlet pressure of the diffuser and $P_{out,s}$ is the exit pressure from the isentropic diffuser evaluated at the same exit velocity as the actual diffuser.

EXAMPLE 3.9

GIVEN: Air at 25°C, 100 kPa enters the compressor of a gas turbine power plant and compressed to 460°C, 1600 kPa. After compression, the air is cooled to 300°C with no drop in pressure as it passes through a heat exchanger as shown in Figure E3.9A. Cold water flowing at 2 kg/s cools the air in the heat exchanger. The water enters at 25°C, 200 kPa, and leaves at 110°C with no drop in pressure. Determine the following:

   a. Isentropic efficiency of the compressor
   b. Compressor's power draw
   c. Entropy generation in the compressor and heat exchanger

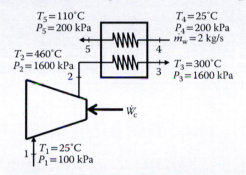

FIGURE E3.9A

SOLUTION: The thermodynamic properties of the air and water for the states given in this problem can all be found since two independent, intensive properties are known at each state. The properties, calculated using EES, are shown in Table E3.9. The two independent properties that fix the thermodynamic state are identified in **bold** type. Notice in the property tables above that state 2s has been calculated using pressure and entropy as known input variables.

TABLE E3.9

Working Fluid Properties

		Thermodynamic Properties of Air			
State	P (kPa)	T (°C)	v (m³/kg)	h (kJ/kg)	s (kJ/kg·K)
1	**100**	**25**	0.85575	298.45	6.86412
2	**1600**	**460**	0.13246	750.13	6.99523
2s	**1600**	375.9	0.11728	659.59	**6.86412**
3	**1600**	**300**	0.10356	579.22	6.73249
		Thermodynamic Properties of Water			
State	P (kPa)	T (°C)	v (m³/kg)	h (kJ/kg)	s (kJ/kg·K)
4	**200**	**25**	0.001	105.01	0.36718
5	**200**	**110**	0.00105	461.46	1.41878

a. The isentropic efficiency of the compressor can be found using Equation 3.70 as follows:

$$\eta_c = \frac{h_{2s} - h_1}{h_2 - h_1} = \frac{(659.59 - 298.45) \text{ kJ/kg}}{(750.13 - 298.45) \text{ kJ/kg}} = 0.7996 = \underline{80.0\%} \quad \leftarrow$$

b. To compute the power draw of the compressor, a system boundary is drawn around the compressor as shown in Figure E3.9B and the conservation of energy equation is applied to the resulting system.

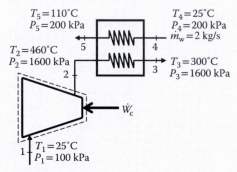

FIGURE E3.9B

For steady flow, the conservation of energy equation can be written as

$$\dot{W}_c + \dot{m}_a \left(h_1 - h_2 \right) = 0$$
$$\therefore \quad \dot{W}_c = \dot{m}_a \left(h_2 - h_1 \right)$$

The enthalpy values can be found since the thermodynamic states are identified at all states, but the mass flow rate of the air is unknown. However, the mass flow rate of the air can be found because it is related to the heat transfer rate in the heat exchanger. Figure E3.9C shows a system boundary surrounding the heat exchanger.

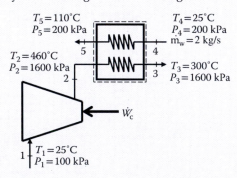

FIGURE E3.9C

For this system boundary, the conservation of energy is written as

$$\left(\dot{m}_a h_2 + \dot{m}_w h_4\right) - \left(\dot{m}_a h_3 + \dot{m}_w h_5\right) = 0$$

$$\dot{m}_a \left(h_2 - h_3\right) = \dot{m}_w \left(h_5 - h_4\right)$$

$$\therefore \quad \dot{m}_a = \dot{m}_w \left(\frac{h_5 - h_4}{h_2 - h_3}\right) = \left(2 \ \frac{\text{kg}}{\text{s}}\right) \left[\frac{(461.46 - 105.01) \ \text{kJ/kg}}{(750.13 - 579.22) \ \text{kJ/kg}}\right] = 4.17 \ \frac{\text{kg}}{\text{s}}$$

Now, the power requirement for the compressor can be found by

$$\dot{W}_c = \dot{m}_a \left(h_2 - h_1\right) = \left(4.17 \ \frac{\text{kg}}{\text{s}}\right)(750.13 - 298.45)\frac{\text{kJ}}{\text{kg}}\left|\left(\frac{\text{kW} \cdot \text{s}}{\text{kJ}}\right)\right| = \underline{\underline{1884 \ \text{kW}}} \quad \leftarrow$$

c. The entropy generation in each of the devices can be found by applying the entropy balance equation to the system boundary surrounding the device. For the compressor, the entropy balance results in

$$\dot{S}_{\text{gen},c} = \dot{m}_a \left(s_2 - s_1\right) = \left(4.17 \ \frac{\text{kg}}{\text{s}}\right)(6.99523 - 6.86412)\frac{\text{kJ}}{\text{kg} \cdot \text{K}}\left|\left(\frac{\text{kW} \cdot \text{s}}{\text{kJ}}\right)\right|$$

$$= \underline{\underline{0.5469 \ \frac{\text{kW}}{\text{K}}}} \quad \leftarrow$$

For the heat exchanger, there are multiple flows in and out. Applying the entropy balance equation to the heat exchanger results in

$$\left(\dot{m}_a s_2 + \dot{m}_w s_4\right) - \left(\dot{m}_a s_3 + \dot{m}_w s_5\right) + \dot{S}_{\text{gen,HX}} = 0$$

Solving for the entropy generation,

$$\dot{S}_{\text{gen,HX}} = \left(\dot{m}_a s_3 + \dot{m}_w s_5\right) - \left(\dot{m}_a s_2 + \dot{m}_w s_4\right) = \dot{m}_a \left(s_3 - s_2\right) + \dot{m}_w \left(s_5 - s_4\right)$$

$$\dot{S}_{\text{gen,HX}} = \left[\left(4.17 \ \frac{\text{kg}}{\text{s}}\right)(6.73249 - 6.99523)\frac{\text{kJ}}{\text{kg} \cdot \text{K}}\right.$$

$$\left. + \left(2 \ \frac{\text{kg}}{\text{s}}\right)(1.41878 - 0.36718)\frac{\text{kJ}}{\text{kg} \cdot \text{K}}\left|\left(\frac{\text{kW} \cdot \text{s}}{\text{kJ}}\right)\right|\right]$$

$$\dot{S}_{\text{gen,HX}} = \underline{\underline{1.007 \ \text{kW/K}}} \quad \leftarrow$$

This calculation reveals that the heat exchanger has nearly twice the entropy generation as the compressor. The magnitude of the entropy generation is a result of irreversibility. In the compressor, the main source of irreversibility is primarily mechanical and fluid friction. In the heat exchanger, the primary source is the heat transfer through a finite temperature difference.

3.9 Exergy Balance: The Combined Law

Envisioning thermodynamic quantities requires abstract thought. For example, we can physically sense heat, but how can it be quantified? From your previous course(s) in thermodynamics, you learned that heat is a form of energy that is transferred due to a temperature difference. However, heat was not always considered energy. From 1697 to 1703, German chemist Georg Stahl proposed the *phlogiston theory*. Stahl theorized that heat was a fire-like element called *phlogiston*. Phlogiston was contained within substances and was released during combustion which produced the sensation of heat. The phlogiston theory was superseded by the *caloric theory*, developed by French chemist Antoine Lavoisier with a series of papers published over the years 1768–1787. Lavoisier proposed that heat was actually a fluid called *caloric* and it flowed into a substance. As caloric entered a substance, it would expand. The caloric theory eventually gave way to the concept of heat as energy.

In 1845, James P. Joule published a paper entitled "On the Mechanical Equivalent of Heat." Joule's work became the foundation of the first law of thermodynamics. During the years 1850–1865 several individuals, including Sadi Carnot, William Thomson (Lord Kelvin), William Rankine, and Rudolf Clausius, developed the ideas that eventually became the second law of thermodynamics. In 1865, Clausius proposed a new *transformational property*, entropy, and its relationship to heat as energy. These famous works are the basis for how we understand heat (as energy) in the first and second laws of thermodynamics today.

In 1953, Slovene mechanical engineer Zoran Rant introduced the word *exergy* in his PhD dissertation, "Exergy: A Useful Concept." Rant's work was based on the earlier works of American engineer J. Willard Gibbs on *available energy* in 1873, the concept of *free energy* developed by German physicist Hermann Helmholtz in 1882, and the idea of the *possibility of doing work* developed by French physicist Louis Gouy in 1889.

This section expands on the concept of exergy—its definition and practical application to thermal energy systems analysis and design. As demonstrated by the brief history given above, exergy is a relatively new idea. It is an abstraction, just like energy. However, it is perhaps one of the more practical abstractions that has risen from thermodynamics.

3.9.1 What Is Exergy?

A paraphrase of the first and second laws of thermodynamics is given below.

> Energy cannot be created or destroyed [The First Law]. Energy can only be transformed. The transformation of energy always happens in such a way that the useful energy content is diminished [The Second Law].

Exergy (also called *available energy* or *availability*) is the *useful energy* content of a substance. Like energy, exergy is a commodity that is present in all substances. However, the difference between energy and exergy is that exergy can be destroyed in a process. This idea can be demonstrated by considering a natural energy resource, such as coal, oil, or natural gas. Figure 3.8 shows the energy and exergy content of such an energy reserve as a function of time. Notice that the energy content of the resource never changes, even as the reserve is utilized. This is a demonstration of the first law of thermodynamics; energy is conserved. The exergy content of the resource, however, decreases with time as a result of using the resource. Figure 3.8 shows that the exergy content of the resource will reach zero at some time. This must be true because exergy is destroyed in all real-world processes. Figure 3.8 also shows a second exergy curve that reaches the zero axis further in the future. This curve represents the exergy content of the resource with improved designs meant to reduce the exergy destruction. By improving system designs to reduce the exergy destruction, the life of the resource can be extended significantly. This is an *ethical responsibility* of engineers as referenced in Chapter 1 (Section 1.6.1) of this book. Improving designs to reduce exergy destruction prolongs the life (sustainability) of the natural resource being used. This has a direct relationship to *environmental impact*.

From this discussion, it can be seen that exergy is a commodity. However, it is a commodity that is also destroyed in all real processes. Therefore, it is the responsibility of the engineer to design systems resulting in *exergy conservation*. Notice that the often-heard phrase *energy conservation* is a misnomer. Energy is *automatically* conserved according to the first law of thermodynamics. Engineers strive to conserve exergy.

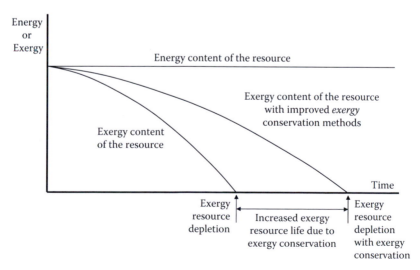

FIGURE 3.8
Energy and exergy content of an energy resource as a function of time.

3.9.1.1 Definition of Exergy

Exergy is defined as the maximum theoretical work that can be done as a system interacts and achieves equilibrium with its surroundings. The surroundings are often called the *dead state*. At the dead state (specified by P_0, T_0), the substance has no exergy content. From this definition, it can be seen that exergy is analogous to work. Work is the form of energy that we seek in our engineering systems. Work is pure exergy.

3.9.2 Exergy Balance

Exergy is a quantity that is destroyed, not generated, in a process by irreversibility. Therefore, an equivalent way to write the general balance equation (Equation 3.44) is

$$\sum \dot{\Omega}_{in} - \sum \dot{\Omega}_{out} - \dot{\Omega}_{des} = \frac{d\Omega_{sys}}{dt} \tag{3.74}$$

The subscript "des" signifies a destroyed quantity (like exergy). Notice that this term is opposite in sign to a quantity that is generated (like entropy).

Making the substitution $\Omega = X$ into Equation 3.74 results in the *exergy balance*.

$$\sum \dot{X}_{in} - \sum \dot{X}_{out} - \dot{X}_{des} = \frac{dX_{sys}}{dt} \tag{3.75}$$

The variables X or x will be used in this book to specify total or specific exergy, respectively. Subscripts will be used, particularly in the case of specific exergy, x, so it is not confused with quality.

Exergy is carried in and out of a system by energy (heat and work), and by mass flow. Rewriting Equation 3.75 to reflect this results in

$$\sum \left(\dot{X}_{in} - \dot{X}_{out} \right)_Q + \sum \left(\dot{X}_{in} - \dot{X}_{out} \right)_W + \sum \left(\dot{m}_{in} x_{f,in} - \dot{m}_{out} x_{f,out} \right) - \dot{X}_{des} = \frac{dX_{sys}}{dt} \tag{3.76}$$

To apply Equation 3.76 to a system, the exergy terms need to be quantified. The equations that follow in the remainder of this section are presented without derivation. The interested reader is encouraged to consult an undergraduate-level textbook in mechanical engineering thermodynamics for the details of these derivations, such as the works by Moran (2014), Cengel (2011), and Klein (2012).

Exergy transport rate due to heat transfer, the first term on the left-hand side of Equation 3.76, can be written as

$$\sum \left(\dot{X}_{in} - \dot{X}_{out} \right)_Q = \sum \left[\dot{Q}_{in} \left(1 - \frac{T_0}{T_b} \right)_{in} - \dot{Q}_{out} \left(1 - \frac{T_0}{T_b} \right)_{out} \right] \tag{3.77}$$

In Equation 3.77, T_b is the system boundary temperature where the heat is being transferred and T_0 is the dead-state temperature, both expressed as absolute temperatures.

The exergy transferred to and from the system by work, the second term on the left-hand side of Equation 3.76, is given by

$$\sum \left(\dot{X}_{in} - \dot{X}_{out} \right)_W = \sum \left(\dot{W}_{in} - \dot{W}_{out} \right) + P_0 \frac{d\mathcal{V}}{dt} \tag{3.78}$$

In Equation 3.78, the term $P_0 \left(d\mathcal{V}/dt \right)$ represents the rate of work that is done on or by the surroundings due to a change in volume of the system over time. If the time derivative of the system volume is negative, then the system is being compressed by the surroundings. This is a *free* source of exergy that can be capitalized on. If the system is expanding, then the volume derivative is positive. This represents work that is done on the environment and is not useful. Therefore, it ends up reducing the exergy output in the form of work. In a steady-flow scenario or a case where the system volume does not change over time, the volume derivative vanishes.

The third term on the left-hand side of Equation 3.76 is the net exergy transported by mass as it crosses the system boundary. In this equation, x_f, is known as *flow exergy* and it is defined as

$$x_f = \left(h - h_0 \right) - T_0 \left(s - s_0 \right) + \frac{V^2}{2} + gz \tag{3.79}$$

The subscripts "0" in Equation 3.79 represent the properties at the dead state, which is defined by P_0 and T_0.

The fourth term on the left-hand side of Equation 3.76 is the exergy destruction rate. Exergy destruction is related to entropy production. Exergy destruction is also called *irreversibility*. The exergy destruction rate term is given by

$$\dot{X}_{des} = T_0 \dot{S}_{gen} \tag{3.80}$$

The magnitude of the exergy destruction in a process is of great interest to engineers. Equation 3.80 shows how the abstract concept of entropy generation is used in a very practical sense to design and analyze systems.

If the flow of exergy into and out of a system is unsteady, then there will be storage of exergy in the system. This is represented by the derivative on the right-hand side of Equation 3.76. The exergy of a system is given by

$$X_{sys} = \left(U - U_0 \right) - T_0 \left(S - S_0 \right) + P_0 \left(\mathcal{V} - \mathcal{V}_0 \right) + \frac{V^2}{2} + gz \tag{3.81}$$

Using Equation 3.81, the rate of exergy storage inside the system can be written as

$$\frac{dX_{sys}}{dt} = \frac{d}{dt} \left[\left(U - U_0 \right) - T_0 \left(S - S_0 \right) + P_0 \left(\mathcal{V} - \mathcal{V}_0 \right) + \frac{V^2}{2} + gz \right] \tag{3.82}$$

As with previous equations, the subscript "0" refers to the dead-state properties.

Substituting Equations 3.77 through 3.82 into Equation 3.76 would make a very complex equation! Therefore, in exergy analysis, it is easier to start with Equation 3.76 and substitute Equations 3.77 through 3.82 as needed. Fortunately, for most exergy analyses, the resulting exergy balance equation is quite manageable.

In the development of the exergy balance, notice that there are terms which are representative of the first and second laws of thermodynamics. The equations above contain enthalpy and internal energy (first-law properties) and entropy (a second-law property). This happens because the equations for the exergy balance were derived using a combination of the first and second laws. Therefore, the exergy balance is sometimes known as the *combined law.*

EXAMPLE 3.10

Steam at 1.6 MPa and 350°C enters a steam turbine at a flow rate of 16 kg/s. The steam leaves the turbine as a saturated vapor at 30°C. The turbine delivers 9 MW of power. The turbine is insulated, but not perfectly. The average temperature of the insulation on the outer surface of the turbine is 70°C. The environment surrounding the turbine can be considered to be the dead state at 25°C, 100 kPa. Figure E3.10 shows this turbine. Determine the exergy destruction rate in the turbine.

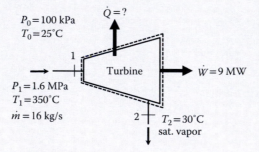

FIGURE E3.10

SOLUTION: This is the same turbine that was considered in Examples 3.5 and 3.8. In Example 3.5, the turbine was analyzed from a first-law point of view. Example 3.8 considered the turbine from a second-law perspective. In this example, the performance of the turbine will be considered relative to exergy.

There are several ways to determine the exergy destruction. Perhaps the simplest is to consider Equation 3.80. The entropy generation rate has previously been determined in Example 3.8. Therefore,

$$\dot{X}_{des} = T_0 \dot{S}_{gen} = (25 + 273.15)\,\text{K}\left(23.39\,\frac{\text{kW}}{\text{K}}\right) = \underline{6974\ \text{kW}} \quad \leftarrow$$

Notice that the temperature used in this calculation must be on the *absolute* scale. Although this provides the required answer, it really does not reveal much about the turbine. More can be learned about the turbine by considering application of Equation 3.76 to the turbine.

$$-\dot{X}_{out,Q} - \dot{X}_{out,W} + \dot{m}(x_{f1} - x_{f2}) - \dot{X}_{des} = 0$$
$$\therefore \quad \dot{X}_{des} = \dot{m}(x_{f1} - x_{f2}) - \dot{X}_{out,Q} - \dot{X}_{out,W}$$

This formulation is particularly informative, because each term has significance to how the turbine is performing as shown below:

$$\dot{X}_{des} = \underbrace{\dot{m}(x_{f1} - x_{f2})}_{\text{Net exergy flow input}} - \underbrace{\dot{X}_{out,Q}}_{\text{Exergy loss due to heat}} - \underbrace{\dot{X}_{out,W}}_{\text{Exergy output}}$$

This equation is helpful because it shows the magnitude of each of the exergy components of the turbine: the net incoming exergy, the exergy loss due to heat transfer, the exergy output (power), and the exergy destruction.

Neglecting kinetic and potential energy differences, the net exergy transfer rate due to mass flow can be written as

$$\dot{m}(x_{f1} - x_{f2}) = \dot{m}\left\{\left[(h_1 - h_0) - T_0(s_1 - s_0)\right] - \left[(h_2 - h_0) - T_0(s_2 - s_0)\right]\right\}$$
$$\dot{m}(x_{f1} - x_{f2}) = \dot{m}\left[(h_1 - h_2) - T_0(s_1 - s_2)\right]$$

Using the property values determined in Examples 3.5 and 3.8, the net exergy transfer rate into the turbine is

$$\dot{m}(x_{f1} - x_{f2}) = \left(16\frac{\text{kg}}{\text{s}}\right)\left[(3146.0 - 2555.6)\frac{\text{kJ}}{\text{kg}} - (298.15)\text{K}(7.0713 - 8.4520)\frac{\text{kJ}}{\text{kg}\cdot\text{K}}\right]$$
$$\dot{m}(x_{f1} - x_{f2}) = 16,033 \text{ kW}$$

The rate that exergy is lost from the turbine due to heat transfer is given by

$$\dot{X}_{out,Q} = \dot{Q}\left(1 - \frac{T_0}{T_b}\right) = (446.4 \text{ kW})\left[1 - \frac{(25+273.15)\text{K}}{(70+273.15)\text{K}}\right] = 59 \text{ kW}$$

The heat transfer rate from the turbine was determined in Example 3.5. The exergy transfer rate from the turbine is the power delivered

$$\dot{X}_{out,W} = \dot{W} = 9000 \text{ kW}$$

Therefore, the exergy destruction rate in the turbine is

$$\dot{X}_{des} = \dot{m}(x_{f1} - x_{f2}) - \dot{X}_{out,Q} - \dot{X}_{out,W} = (16,033 - 59 - 9,000)\text{kW} = \underline{\underline{6,974 \text{ kW}}} \quad \leftarrow$$

This is the same result that was found using the simpler equation, $\dot{X}_{des} = T_0\dot{S}_{gen}$. However, this result is more meaningful because the relationship between the net exergy input and the exergy destruction is clearly seen. Because of the steam flow entering and leaving, the turbine is provided with 16,033 kW of flow exergy. This exergy rate is converted to power (9000 kW), an exergy transfer due to heat (59 kW), and the remainder is destroyed (6974 kW).

As demonstrated in Example 3.10, the exergy destruction rate is more physically meaningful than the entropy generation rate. For example, the entropy generation rate of the turbine in Example 3.10 is 23.39 kW/K. But what does this really mean? Unfortunately, the only thing this indicates is that there are irreversibilities in the turbine (the entropy generation rate is positive). However, the number, by itself, does not tell us how bad the irreversibility is. On the other hand, the full exergy analysis of the turbine clearly reveals how much exergy is input to the turbine, and how it is distributed. The exergy analysis gives a much clearer and meaningful picture to how the turbine is operating.

3.9.3 Exergy Accounting and Exergy Flow Diagrams

Performing a full exergy analysis of a device reveals much information about how the device is utilizing its exergy *resource* to complete its exergy *task*. Exergy accounting is a way to summarize the results of a full exergy analysis. Consider the turbine that was analyzed in Example 3.10. A full exergy accounting of the turbine is given in Table 3.3.

This table indicates that 56.1% of the incoming exergy to the turbine is ultimately delivered as power. Only 0.4% of the incoming exergy is lost due to the heat transfer. The remaining 43.5% of the exergy is destroyed.

Another useful way to visualize the flow of exergy through the turbine is to construct an *exergy flow diagram* based on the results of a full exergy analysis. An exergy flow diagram for the turbine analyzed in Example 3.10 is shown in Figure 3.9. The block arrows in the exergy flow diagram are drawn somewhat to scale. When this is done, a depiction of what happens to the incoming exergy can be easily visualized.

TABLE 3.3

Exergy Accounting for the Turbine of Example 3.10

Exergy Inputs	Value	% of Total
Net exergy flow rate provided by the steam	16,033 kW	100
Total exergy input	16,033 kW	

Exergy Outputs	Value	% of Total Input
Power	9,000 kW	56.1
Heat loss	59 kW	0.4
Total exergy output	9,059 kW	
Exergy destruction (input–output)	**6,974 kW**	**43.5**

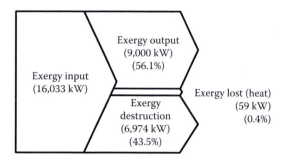

FIGURE 3.9
Exergy flow diagram for the turbine analyzed in Example 3.10.

3.9.4 Exergetic Efficiencies of Flow Devices

A full exergy accounting of a flow device allows for the determination of the *exergetic efficiency* (sometimes called the *second law efficiency*) of the device. The exergetic efficiency of a flow device is given by

$$\eta_x = \frac{\text{exergy output (exergy task)}}{\text{exergy input (exergy resource)}} \tag{3.83}$$

This indicates that the exergetic efficiency is a task/resource index. It indicates how well the device is utilizing its exergy resource to perform the required task.

3.9.4.1 Turbines

As seen in Example 3.10, the exergy output of a turbine is the power delivered. The exergy input to the turbine is the *net* exergy flow rate. Therefore, the exergetic efficiency of a turbine can be expressed as

$$\eta_{x,t} = \frac{\dot{W}_{out}}{\dot{m}\left(x_{f,in} - x_{f,out}\right)} \tag{3.84}$$

For the turbine analyzed in Example 3.10, the exergetic efficiency is 56.1%.

3.9.4.2 Compressors, Pumps, and Fans

The exergy output of a device that moves a fluid (compressor, pump, or fan) is the increase of flow exergy of the fluid passing through it. The exergy input to the device is power. Therefore, the exergetic efficiency of a compressor, pump, or fan can be expressed as

$$\eta_{x,m} = \frac{\dot{m}\left(x_{f,out} - x_{f,in}\right)}{\dot{W}_{in}} \tag{3.85}$$

The subscript "m" is used to indicate a device that moves a fluid.

3.9.4.3 Heat Exchangers

Heat exchanger performance can also be quantified with an exergetic efficiency. Consider the case where the hot and cold fluids do not mix in the heat exchanger. As the hot fluid passes through the heat exchanger, it gives up its exergy to the cold fluid. Therefore, the exergy input to the heat exchanger can be considered to be the exergy decrease of the hot fluid. The exergy output can be thought of as the exergy increase of the cold fluid. Therefore, a suitable expression for the exergetic efficiency of a heat exchanger, where the fluids are unmixed, can be written as

$$\eta_{x,\text{HX}} = \frac{\dot{m}_c \left(x_{f,\text{out}} - x_{f,\text{in}} \right)_c}{\dot{m}_h \left(x_{f,\text{in}} - x_{f,\text{out}} \right)_h} \tag{3.86}$$

Recall that there is no expression for the isentropic efficiency of a heat exchanger. This is because the fluids in a heat exchanger always exist at different temperatures. Even if one could envision frictionless flow, there is still a temperature difference that makes the heat exchanger irreversible. This shows the very practical nature of the exergy analysis compared to second law (entropy) analysis. Given a device, one can always devise its exergy *task* and identify its exergy *resource*.

EXAMPLE 3.11

Water enters the tubes of a boiler in a power plant at 1200 psia and 120°F, as shown in Figure E3.11. Steam exits the tubes at 1400°F with virtually no drop in pressure. The water is boiled by combustion gases entering the boiler at 2300°F, 1 atm. The combustion gases flow through the boiler with a negligible pressure drop. The mass flow rates of the steam and combustion gases are 1.4×10^6 and 1.13×10^7 lbm/h, respectively. The combustion gases can be modeled as air. The flows are steady and there is no stray heat transfer from the boiler. The dead-state temperature can be taken as 70°F. Determine the following:

    a. Exergy destruction rate in the boiler
    b. Exergetic efficiency of the boiler

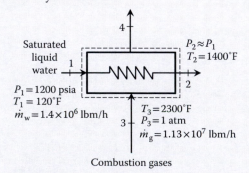

FIGURE E3.11

SOLUTION: Figure E3.11 shows a system boundary drawn around the boiler. A full exergy analysis will be done on the boiler, which will facilitate the calculation of the exergy destruction rate and the exergetic efficiency. Since there are several property lookups required for this problem, EES will be used to solve the problem. The following statements set up the given information for the problem:

```
"GIVEN: A steam boiler in a power plant"
"Water parameters"
   w$ = 'steam_iapws'  "real fluid model for water"
   P_1 = 1200 [psia]
   T_1 = 120 [F]
   P_2 = P_1
   T_2 = 1400 [F]
   m_dot_w = 1.4E6 [lbm/hr]


"Combustion gases (modeled as air)"
   g$ = 'air_ha'  "real fluid model for air"
   T_3 = 2300 [F]
   P_3 = 1 [atm]*convert(atm,psia)
   P_4 = P_3
   m_dot_g = 1.13E7 [lbm/hr]


"Dead state temperature"
   T_0 = converttemp(F,R,70[F])


"FIND: (a) Exergy destruction rate in the boiler
       (b) Exergetic efficiency of the boiler"
```

The exergy balance equation (Equation 3.76) applied to the boiler results in the following equation:

$$\left(\dot{m}_w x_{f1} + \dot{m}_g x_{f3}\right) - \left(\dot{m}_w x_{f2} + \dot{m}_g x_{f4}\right) - \dot{X}_{des} = 0$$

$$\dot{X}_{des} = \dot{m}_g\left(x_{f3} - x_{f4}\right) - \dot{m}_w\left(x_{f2} - x_{f1}\right)$$

Expanding the above equation for the flow exergy values results in

$$\dot{X}_{des} = \dot{m}_g\left[\left(h_3 - h_4\right) - T_0\left(s_3 - s_4\right)\right] - \dot{m}_w\left[\left(h_2 - h_1\right) - T_0\left(s_2 - s_1\right)\right]$$

```
"SOLUTION:"
"The Exergy Balance for the boiler is given by,"
X_dot_des = m_dot_g*((h_3 - h_4) - T_0*(s_3 - s_4)) &
            - m_dot_w*((h_2 - h_1) - T_0*(s_2 - s_1))
```

At state 4, the combustion gases leaving the boiler, the pressure is known. Application of the conservation of energy equation to the boiler allows the enthalpy at state 4 to be found, thus fixing the state.

$$\left(\dot{m}_w h_1 + \dot{m}_g h_3\right) - \left(\dot{m}_w h_2 + \dot{m}_g h_4\right) = 0$$

$$\dot{m}_g\left(h_3 - h_4\right) = \dot{m}_w\left(h_2 - h_1\right)$$

```
"Conservation of Energy is used to determine the outlet enthalpy
of the gases,"
  m_dot_g*(h_3 - h_4) = m_dot_w*(h_2 - h_1)
```

The exergetic efficiency of the boiler is determined from Equation 3.86.

$$\eta_{x,HX} = \frac{\dot{m}_w \left(x_{f2} - x_{f1} \right)}{\dot{m}_g \left(x_{f3} - x_{f4} \right)} = \frac{\dot{X}_c}{\dot{X}_h}$$

The flow exergy rates of the hot and cold fluids can be written as

$$\dot{X}_c = \dot{m}_w \left[\left(h_2 - h_1 \right) - T_0 \left(s_2 - s_1 \right) \right]$$
$$\dot{X}_h = \dot{m}_g \left[\left(h_3 - h_4 \right) - T_0 \left(s_3 - s_4 \right) \right]$$

```
"The exergetic efficiency of the boiler is,"
  eta_x_HX = X_dot_c/X_dot_h
  X_dot_c = m_dot_w*((h_2 - h_1) - T_0*(s_2 - s_1))
  X_dot_h = m_dot_g*((h_3 - h_4) - T_0*(s_3 - s_4))
```

All that remains is to find the properties.

```
"Properties"
  h_1 = enthalpy(w$,P = P_1,T = T_1)
  s_1 = entropy(w$,P = P_1,T = T_1)
  h_2 = enthalpy(w$,P = P_2,T = T_2)
  s_2 = entropy(w$,P = P_2,T = T_2)
  h_3 = enthalpy(g$,P = P_3,T = T_3)
  s_3 = entropy(g$,P = P_3,T = T_3)
  s_4 = entropy(g$,P = P_4,h = h_4)
```

The resulting exergy destruction in the boiler is $\dot{X}_{des} = 6.817 \times 10^8$ Btu/h, and the corresponding exergetic efficiency of the boiler is $\eta_x = 0.6183$ (61.83%). This result indicates that the boiler utilizes nearly 62% of the incoming exergy from the combustion gases to boil the water. The remaining 38% of the exergy is destroyed in the process. This exergy destruction is related to the large temperature difference in the boiler.

3.10 Energy and Exergy Analysis of Thermal Energy Cycles

Components and devices are often connected together, forming a cycle. There are many different cycles. However, they can be generally categorized as cycles that (1) deliver power (e.g., a steam power cycle) or (2) transport energy in the form of heat (e.g., a heat pump cycle). The individual components of the cycle can be analyzed using any of the conservation or balance laws discussed in Sections 3.6 through 3.9. In addition to component performance, the engineers are also interested in how the *cycle* performs.

Figure 3.10 shows a sketch of power and refrigeration cycles. The actual cycle is inside of the circle labeled either "P" (for power cycle) or "R" (for refrigeration cycle). The cycle can take on many different configurations. Figure 3.10 also shows that the cycle must communicate thermally with energy reservoirs. In the case of a power cycle, the energy reservoirs are required by the second law of thermodynamics. In the case of a refrigeration cycle, the second law requires an energy input to move heat from a low temperature to a high temperature.

Figure 3.10 also shows a system boundary drawn around the cycle. These systems allow for the application of the conservation and balance laws without knowing anything about the details of the cycle. For example, the conservation of energy equation can be used to determine the work output of the power cycle and the work input to the refrigeration cycle as a function of the heat transfer interactions with the thermal energy reservoirs, as shown in Figure 3.10.

3.10.1 Cycle Energy Performance Parameters

There are many different ways to quantify the performance of a thermodynamic cycle. The most common cycle performance index is the *thermal efficiency*. The thermal efficiency of any cycle is defined as

$$\eta_{th} = \frac{\text{energy sought}}{\text{energy that costs}} \tag{3.87}$$

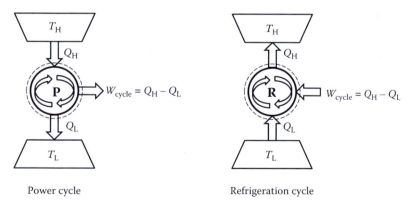

Power cycle   Refrigeration cycle

FIGURE 3.10
Sketches of a power cycle and a refrigeration cycle.

As seen from Equation 3.87, the thermal efficiency is an energy-based efficiency. Applying this equation to the power cycle shown in Figure 3.10 results in

$$\eta_{th,power} = \frac{W_{cycle}}{Q_H} \tag{3.88}$$

Power delivery cycles, particularly vapor power cycles, often use the *heat rate* as a performance index. The heat rate of a power cycle is defined as

$$HR = \frac{\text{heat transfer rate into the cycle}}{\text{net power delivery from the cycle (kW)}} = \frac{\dot{Q}_H}{\dot{W}_{cycle}} [=] \frac{\text{energy}/h}{kW} = \frac{\text{energy}}{kWh}$$

$$\tag{3.89}$$

At first glance, the heat rate appears to be the reciprocal of the thermal efficiency of the power cycle. This is true, except that it is *dimensional*. Notice that it carries units of energy/kWh. The energy can be expressed in terms of Btu or an SI equivalent (e.g., kJ). Power plant engineers prefer to use the heat rate because it gives them a better sense of how much fuel is required (energy) per kWh of energy delivered by the plant.

EXAMPLE 3.12

The thermal efficiency of a steam power cycle is determined to be 36%. Determine the heat rate of the cycle.

SOLUTION: The heat rate is simply the reciprocal of the thermal efficiency, modified by the proper unit conversion to result in energy/kWh. In the SI system, the heat rate of this cycle is

$$HR_{SI} = \frac{1}{0.36} \left| \left(\frac{kJ}{kW \cdot s} \right) \left(\frac{3,600 \ s}{h} \right) \right. = 10,000 \frac{kJ}{kWh} \quad \leftarrow$$

In the English (IP) unit system,

$$HR_{IP} = \frac{1}{0.36} \left| \left(\frac{3412.1 \ Btu}{kW \cdot h} \right) \right. = 9478 \frac{Btu}{kWh} \quad \leftarrow$$

As shown, the calculated heat rates reveal how much energy (kJ or Btu) must be used to generate a single kWh of energy.

For the refrigeration cycle shown in Figure 3.10, the *energy that costs* is the work input to the cycle. However, the *energy sought* depends on what the refrigeration cycle is being used for. If the goal is to keep a space cool, then the energy sought is the low temperature heat transfer, Q_L. Then the thermal efficiency of the cycle is

$$\eta_{th,refrig} = \frac{Q_L}{W_{cycle}} \equiv COP_C \tag{3.90}$$

The thermal efficiency of a refrigeration cycle is more commonly expressed in decimal form, rather than a percent. In this case, it is called the *cooling coefficient of performance, COP_C*. The reason for this is that the thermal efficiency of a refrigeration cycle is often well in excess of 100%. It is not uncommon to see refrigeration systems operating with thermal efficiencies of 200%–400% (which translates to a COP_C between 2.0 and 4.0).

An alternative performance parameter used by the refrigeration industry is the *energy efficiency ratio*. The energy efficiency ratio of a refrigeration cycle is defined as

$$EER = \frac{\text{refrigeration rate (capacity)}}{\text{power input}} = \frac{\dot{Q}_L}{\dot{W}_{cycle}} [=] \frac{\text{energy/h}}{W} = \frac{\text{energy}}{W \cdot h} \tag{3.91}$$

Comparing Equations 3.90 and 3.91 reveals that the only difference between the COP_C and the EER is that the COP_C is dimensionless, whereas the EER has units. Most often, the refrigeration rate (also known as the *capacity*) of the cycle is expressed in Btu/h. However, it is possible to express the capacity in kW. Higher values of EER correspond to less operating cost.

EXAMPLE 3.13

A small window air conditioner has an Energy Star label that indicates it has an EER of 11.6 with a power input of 800 W. Determine the cooling capacity of the air conditioner (Btu/h) and its COP_C.

SOLUTION: The cooling capacity of the air conditioner can be calculated using Equation 3.91.

$$EER = \frac{\dot{Q}_L}{\dot{W}_{cycle}} \quad \therefore \quad \dot{Q}_L = \dot{W}_{cycle}(EER) = (800\ W)\left(11.6\frac{Btu/h}{W}\right) = 9280\frac{Btu}{h} \quad \leftarrow$$

The COP_C can be found by using the proper conversion factor.

$$COP_C = \left(11.6\frac{Btu}{W \cdot h}\right)\left(\frac{W \cdot h}{3.4121\ Btu}\right) = 3.40 \quad \leftarrow$$

Although a misnomer, this air conditioner would probably be marketed as a 9000 Btu unit.

It is possible to use a refrigeration cycle to provide heating. In this case, the energy sought is Q_H (refer to Figure 3.10). This type of cycle is called a *heat pump* cycle. The heat pump cycle operates in the same cycle as the refrigeration cycle. The only difference is that the heat rejected to the high-temperature reservoir is the energy sought. Therefore, the thermal efficiency of a heat pump cycle is given by

$$\eta_{th,hp} = \frac{Q_H}{W_{cycle}} \equiv COP_H \tag{3.92}$$

As with the refrigeration cycle, the thermal efficiency of the heat pump cycle is often in excess of 100%. Therefore, the *heating coefficient of performance, COP_H*, is commonly used rather than the thermal efficiency. As with the refrigeration cycle, the COP_H is the decimal form of the thermal efficiency.

3.10.1.1 Maximum Thermal Efficiency of a Cycle

Refer, once again to the power cycle shown in Figure 3.10. The work delivered by the cycle is given by

$$W_{cycle} = Q_H - Q_L \tag{3.93}$$

Substituting this into the power cycle thermal efficiency equation (Equation 3.88) and simplifying results in

$$\eta_{th,power} = 1 - \frac{Q_L}{Q_H} \tag{3.94}$$

Equation 3.94 shows that the thermal efficiency of a power cycle can *never* reach 100%, even in an ideal scenario. This is because Q_L must always be present due to the second law of thermodynamics. When efficiency is expressed in terms of a percent, we tend to automatically assume that 100% is the best possible efficiency. However, as demonstrated, this is not true for the power cycle.

A similar conclusion can be drawn for the refrigeration cycle shown in Figure 3.10. Substituting the expression for the cycle work into the COP_C equation and simplifying gives

$$COP_C = \frac{Q_L}{Q_H - Q_L} = \frac{1}{Q_H / Q_L - 1} \tag{3.95}$$

Likewise, the heating coefficient performance of the cycle can be expressed as

$$COP_H = \frac{Q_H}{Q_H - Q_L} = \frac{1}{1 - Q_L / Q_H} \tag{3.96}$$

Equations 3.95 and 3.96 indicate that there is really *no bound* on the coefficients of performance. They depend on the ratio of the heat transferred to/from each thermal reservoir. In fact, they can be (and often are) in excess of 1.0.

The second law pioneers, Sadi Carnot, William Thomson (Lord Kelvin), and William Rankine helped to solve the problem of the maximum thermal efficiency of a cycle. Carnot proposed that the maximum thermal efficiency of a cycle occurs when all process are *reversible*. Concurrently, Kelvin and Rankine stated that for a reversible cycle, the ratio of the heat transfers is equivalent to the ratio of the *absolute temperatures* of the thermal reservoirs.

$$\left(\frac{Q_H}{Q_L}\right)_{rev} = \frac{T_H}{T_L} \tag{3.97}$$

Expressions for the maximum thermal efficiency of a cycle can now be developed by substituting Equation 3.97 into Equations 3.94 through 3.96. The resulting equations are known as *Carnot efficiency* expressions. For a power cycle, the Carnot efficiency is given by

$$\eta_{th,Carnot} = 1 - \frac{T_L}{T_H} \tag{3.98}$$

Likewise, the Carnot COP_C can be written as

$$COP_{C,Carnot} = \frac{1}{T_H / T_L - 1} \tag{3.99}$$

For the heat pump cycle, the Carnot COP_H is given by

$$COP_{H,Carnot} = \frac{1}{1 - T_L / T_H} \tag{3.100}$$

EXAMPLE 3.14

Ocean Thermal Energy Conversion (OTEC) technology exploits the tempera-ture gradient in the ocean to drive a power cycle. Consider a case where the warm surface waters are at a temperature of 28°C. The low-temperature reser-voir for the power cycle is located at a depth of 500 ft, where the water tempera-ture is 10°C. During normal operation, the power cycle is delivering 50 MW of power, while transferring 1600 MW of heat to the low-temperature reservoir. Determine the thermal efficiency of this OTEC power cycle. How does this compare with the maximum thermal efficiency of the cycle?

SOLUTION: A sketch of the cycle is shown in Figure E3.14.

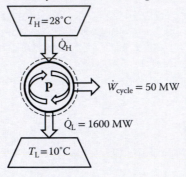

FIGURE E3.14

The thermal efficiency of the power cycle is given by Equation 3.88.

$$\eta_{th,power} = \frac{W_{cycle}}{Q_H}$$

The conservation of energy equation applied to the system boundary sur-rounding the cycle allows for the calculation of the heat transfer rate from the high-temperature reservoir.

$$\dot{Q}_H = \dot{W}_{cycle} + \dot{Q}_L = (50 + 1600)\,MW = 1650\ MW$$

Therefore, the thermal efficiency of the cycle is

$$\eta_{th,power} = \frac{50\ MW}{1650\ MW} = 0.0303 = \underline{\underline{3.03\%}} \quad \leftarrow$$

The maximum thermal efficiency of the power cycle is the Carnot efficiency given by Equation 3.98.

$$\eta_{th,Carnot} = 1 - \frac{T_L}{T_H} = 1 - \frac{(10 + 273.15)\,K}{(28 + 273.15)\,K} = 0.0598 = \underline{\underline{5.98\%}} \quad \leftarrow$$

This example demonstrates the problem with the "100% is best" thinking. The calculated thermal efficiency of 3.03% seems dismal until it is realized that the best possible efficiency for a cycle of this type under the conditions given is 5.98%.

3.10.2 Exergetic Cycle Efficiency

The use of the cycle thermal efficiency is well understood. However, the previous section demonstrated that the maximum value of the thermal efficiency is not 100%, but rather the Carnot efficiency. The concept of exergy and the *exergetic efficiency* provides an alternate framework for determining the performance of a cycle.

The exergetic efficiency of a cycle is defined in the same way that it was for a device, Equation 3.83, which is repeated below:

$$\eta_x = \frac{\text{exergy output (exergy task)}}{\text{exergy input (exergy resource)}} \tag{3.101}$$

3.10.2.1 Power Cycles

Consider a power cycle operating between two thermal reservoirs as shown in Figure 3.11. This figure indicates that the low-temperature reservoir is at the dead-state temperature. If the dead state is considered as the atmosphere, then this depiction of the power cycle is accurate; the ultimate low-temperature sink is the atmosphere.

An alternative expression for the exergetic efficiency can be developed based on Figure 3.11. For a power cycle, the exergy task is the delivery of the cycle work. The exergy input is the exergy associated with the heat transfer from the high-temperature reservoir. Therefore, the exergetic efficiency of a power cycle can be written as

$$\eta_{x,power} = \frac{W_{cycle}}{X_{Q_H}} \tag{3.102}$$

Applying the exergy balance equation (Equation 3.76) to the system boundary that encloses the complete power cycle results in

$$X_{Q_H} - X_{Q_0} - X_{W_{cycle}} - X_{des} = 0 \tag{3.103}$$

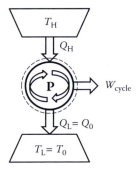

FIGURE 3.11
Power cycle operating between T_H and the dead state.

There is no exergy associated with the heat transferred to the dead state, because it is at the dead-state temperature T_0. This can be proven with the following equation:

$$X_{Q_0} = Q_0 \left(1 - \frac{T_0}{T_0}\right) = 0 \qquad (3.104)$$

Substituting Equation 3.104 into Equation 3.103 and solving for the cycle work gives

$$X_{W_{cycle}} = W_{cycle} = X_{Q_H} - X_{des} \qquad (3.105)$$

Substitution of Equation 3.105 into Equation 3.102 results in

$$\eta_{x,power} = 1 - \frac{X_{des}}{X_{Q_H}} \qquad (3.106)$$

Equation 3.106 is an important equation, because it demonstrates that the maximum exergetic efficiency of a power cycle is 100% for the case of a reversible cycle with no exergy destruction. Unlike the thermal efficiency that is bounded by the Carnot efficiency at the upper limit, the exergetic efficiency does reach a maximum of 100% in the case of an ideal cycle with no irreversibility.

The exergy content of the heat transfer from the high-temperature reservoir relative to the dead state at T_0 is

$$X_{Q_H} = Q_H \left(1 - \frac{T_0}{T_H}\right) \qquad (3.107)$$

Substitution of Equation 3.107 into Equation 3.102 results in

$$\eta_{x,power} = \frac{W_{cycle}}{Q_H \left(1 - \dfrac{T_0}{T_H}\right)} = \frac{\eta_{th,power}}{\left(1 - \dfrac{T_0}{T_H}\right)} = \frac{\eta_{th,power}}{\eta_{th,Carnot}} \qquad (3.108)$$

Equation 3.108 indicates that the exergetic efficiency of a power cycle operating between T_H and T_0 is directly proportional to the thermal efficiency and inversely proportional to the Carnot efficiency. In addition, since the Carnot efficiency of a power cycle is less than 1, the exergetic efficiency of a power cycle must be larger than the thermal efficiency.

3.10.2.2 Refrigeration and Heat Pump Cycles

Consider a refrigeration cycle operating between a low-temperature source, T_L, and the dead state, T_0, as the sink. This cycle is shown in Figure 3.12. The exergetic efficiency of this cycle can be determined by applying Equation 3.101.

$$\eta_{x,refrig} = \frac{-X_{Q_L}}{W_{cycle}} \qquad (3.109)$$

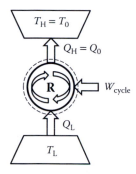

FIGURE 3.12
Refrigeration cycle operating between T_L and the dead state.

A negative sign must be used in the numerator of Equation 3.109, because the exergy *output* defined in Equation 3.101 is really an input to the refrigeration cycle. The exergy balance applied to the system boundary surrounding the cycle shown in Figure 3.12 results in

$$X_{Q_L} + X_{W_{cycle}} - X_{des} = 0 \qquad (3.110)$$

Solving Equation 3.110 for the exergy transferred due to Q_L and substituting into Equation 3.109 reveals the following:

$$\eta_{x,refrig} = 1 - \frac{X_{des}}{W_{cycle}} \qquad (3.111)$$

Similar to what was seen for the power cycle, Equation 3.111 indicates that the exergetic efficiency of a refrigeration cycle has an upper limit of 100% in the case of a completely reversible cycle with no exergy destruction.

The exergy associated with the heat transfer from the low-temperature source is given by

$$X_{Q_L} = Q_L \left(1 - \frac{T_0}{T_L} \right) \qquad (3.112)$$

Substituting Equation 3.112 into Equation 3.109 results in

$$\eta_{x,refrig} = \frac{-Q_L \left(1 - \dfrac{T_0}{T_L} \right)}{W_{cycle}} = \frac{Q_L}{W_{cycle}} \left(\frac{T_0}{T_L} - 1 \right) = \frac{COP_C}{COP_{C,Carnot}} \qquad (3.113)$$

Following similar thinking, the exergetic efficiency of a heat pump cycle can also be developed. The result, shown in Equation 3.114, is left as an exercise for the reader.

$$\eta_{x,hp} = \frac{Q_H}{W_{cycle}} \left(1 - \frac{T_0}{T_H} \right) = \frac{COP_H}{COP_{H,Carnot}} \qquad (3.114)$$

3.10.2.3 Significance of the Exergetic Cycle Efficiency

As demonstrated in Sections 3.10.2.1 and 3.10.2.2, the exergetic efficiency of any cycle is always bounded by 0 and 1 (0%–100%). This makes the exergetic cycle efficiency a much more intuitive index for cycle performance compared to the thermal efficiency. This fact, coupled with the definition of the exergetic cycle efficiency, Equation 3.101, demonstrates the task/resource nature of the exergetic efficiency. High values of exergetic efficiency indicate that the exergy resource is well matched for the exergy task the cycle is providing.

EXAMPLE 3.15

Two refrigeration systems are being considered for a cold storage application: a standard vapor compression refrigeration system and an absorption refrigeration system. In both cases, the cold space must be maintained at –5°C. To maintain this temperature, heat must be removed from the cold space at a rate of 35 kW. The vapor compression system requires an electrical input of 10 kW to accomplish the refrigeration. The absorption refrigeration system utilizes a 58 kW heat input at 50°C to drive the cycle instead of electricity. Both cycles sink to the atmosphere, which can be considered the dead state, at 22°C. Determine the COP_C and exergetic efficiency of each refrigeration cycle.

SOLUTION: A sketch of the two refrigeration cycles is shown in Figure E3.15. For the vapor compression system, the COP_C is given by

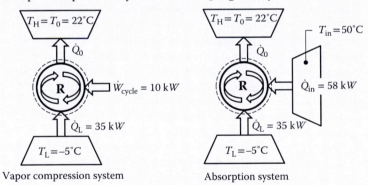

Vapor compression system   Absorption system

FIGURE E3.15

$$COP_{C,vap} = \frac{\dot{Q}_L}{\dot{W}_{cycle}} = \frac{35 \text{ kW}}{10 \text{ kW}} = 3.50$$

The energy input to the absorption cycle is heat. Therefore, the COP_C is determined by

$$COP_{C,abs} = \frac{\dot{Q}_L}{\dot{Q}_{in}} = \frac{35 \text{ kW}}{58 \text{ kW}} = 0.60$$

The calculated values of the COP_C for each cycle indicate that the vapor compression cycle is performing better, relative to energy. The higher COP_C of the vapor compression cycle indicates that more refrigeration can be accomplished per kW of energy input compared to the absorption cycle.

The exergetic efficiency of the vapor compression cycle can be found using Equation 3.113.

$$\eta_{x,vap} = \frac{Q_L}{W_{cycle}}\left(\frac{T_0}{T_L}-1\right) = COP_C\left(\frac{T_0}{T_L}-1\right) = (3.5)\left[\frac{(22+273.15)\,\text{K}}{(-5+273.15)\ \text{K}}-1\right] = 0.352$$
$$= 35.2\%$$

The exergetic efficiency of the heat-driven absorption cycle can be derived from Equation 3.101 and is given by

$$\eta_{x,abs} = \frac{-\dot{X}_{\dot{Q}_L}}{\dot{X}_{\dot{Q}_{in}}} = \frac{\dot{Q}_L\left(\frac{T_0}{T_L}-1\right)}{\dot{Q}_{in}\left(1-\frac{T_0}{T_{in}}\right)} = COP_{C,abs}\frac{\left(\frac{T_0}{T_L}-1\right)}{\left(1-\frac{T_0}{T_{in}}\right)} = 0.60\frac{\left[\frac{(22+273.15)\,\text{K}}{(-5+273.15)\,\text{K}}-1\right]}{\left[1-\frac{(22+273.15)\,\text{K}}{(50+273.15)\,\text{K}}\right]}$$
$$= 69.7\%$$

Although the vapor compression cycle is the better cycle at transporting and utilizing energy, it is the absorption cycle that is almost twice as efficient at utilizing the exergy input to achieve the refrigeration task. In this section, the exergetic efficiency was referred to several times as a task/resource index. The task of the refrigeration cycle (independent of what type it is) is to move heat, Q_L. With the vapor compression system, this is accomplished with work, W_{cycle}. With the absorption system, the cooling is accomplished with a heat input, Q_{in}. The low value of exergetic efficiency for the vapor compression cycle indicates that the exergy input (work) is not well matched to the exergy output (heat). On the other hand, the exergetic efficiency of the absorption system is much higher because the exergy input is *heat* and the exergy task is movement of *heat*. Using heat to move heat is a much better exergetic solution than using work to move heat.

3.10.2.4 Energy/Exergy Conundrum

Example 3.15 reveals an engineering trade-off between thermal efficiency and exergetic efficiency. Recall that thermal efficiency is the ratio of energy sought to *energy that costs*. Therefore, systems with high thermal efficiency are less expensive to operate. On the other hand, systems with high exergetic efficiency are making the best use of the exergy resource to perform a task. Therefore, designing a thermal system to have a high thermal efficiency is good from the economic point of view, whereas designing a system to have high exergetic efficiency is good from the environmental (exergy resource) perspective. As demonstrated in Example 3.15, what is good for high thermal efficiency (minimum cost) is not necessarily good for the environment (exergy resource). This is the conundrum consumers and industry face.

For example, businesses are in business to make money. Therefore, it seems logical for a business to design systems that maximize thermal efficiency, thereby reducing operating costs. However, what impact does this practice have on the environment? Even if exergy is not a consideration in system designs, it is *still being destroyed* in the system! The problem that industry faces is that designing to maximize exergetic efficiency may not result in a good profit stream to keep the company in business. However, for an industry to gain various certifications, minimum levels of performance must be met. For example, the Energy Star program was created by the United States Environmental Protection Agency in 1992 under authority of the Clean Air Act. The Energy Star program is voluntary, but minimum standards have been established that industries must meet for various types of systems to have a product worthy of the Energy Star certification. These minimum requirements are meant to minimize environmental impact by reducing air pollution. Consumers then have a choice between products that have achieved the Energy Star certification or products that do not carry the certification. Industries that certify their products through the Energy Star program are practicing good *exergy conservation*.

Consumers are faced with many energy/exergy decisions. For example, should you heat your home with electric resistance heating or air-cooled solar collectors? Electric resistance heating is 100% efficient in that all the electricity (work) entering the heater leaves as heat. However, this is a device that uses work to move heat. As seen in Example 3.15, this is not a very exergetically efficient method of heating. Of course, this also depends on where the electricity originates. If you live in a region where electrical energy is delivered from a hydroelectric plant, then you are relying on the hydrological cycle, which can be considered renewable. However, if you live in a region where electrical energy is delivered from a fossil fuel–fired power plant, then as the electricity is used to run the electric heater, the exergy resource (the fossil fuel) is being depleted. From an environmental point of view, you would be much better heating your home with the air-cooled solar collectors because of their high exergetic efficiency (heat moving heat). Of course, the conundrum is that air-cooled solar collectors cost quite a bit more than electric resistance heaters.

The purpose of this presentation is not to solve the energy/exergy conundrum. Instead, it is meant to provoke some high-level exergy-based thinking in the engineers who will shape the future of the world.

3.11 Detailed Analysis of Thermal Energy Cycles

In this section, some common thermal energy cycles for power delivery and refrigeration are analyzed on the component level. This is not meant to be an exhaustive presentation. The purpose is to consider the application of the equations developed thus far in this chapter to analyze cycles from the energy and exergy points of view. The methods shown here can easily be extrapolated to much more complex cycles.

3.11.1 Solution Strategy for Cycle Analysis

Consider the steam power cycle with reheat shown in Figure 3.13. The pressures and temperatures at several states are identified in the accompanying table. In addition, the isentropic efficiencies of the turbines and the pump are known. A full energy analysis of this cycle results in values for the energy transfer rates indicated on the diagram. Knowing the energy transfer rates, the thermal efficiency of the cycle can also be found.

To determine the energy transfer rates and thermal efficiency, the following solution strategy is suggested:

1. Construct a complete table of thermodynamic properties at each state in the cycle.

2. Apply balance and conservation laws to determine the required information.

When constructing the table of properties (step 1), it is important to always be mindful of the State Postulate for a pure fluid, as discussed in Section 3.3.

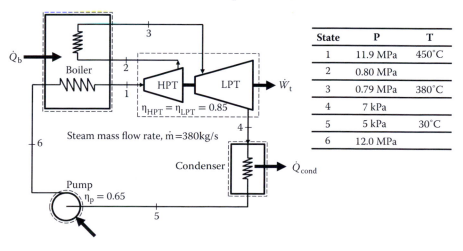

State	P	T
1	11.9 MPa	450°C
2	0.80 MPa	
3	0.79 MPa	380°C
4	7 kPa	
5	5 kPa	30°C
6	12.0 MPa	

FIGURE 3.13
Steam power cycle with reheat with known properties in the accompanying table.

Recall that the State Postulate for a pure fluid specifies that two independent, intensive properties are required to fix the thermodynamic state. Therefore, to construct the table of properties, two independent, intensive properties need to be identified at each state. Referring to the cycle shown in Figure 3.13, three of the states (1, 3, and 5) are defined by pressure and temperature. The P,T combination at these three states are independent. This can be verified by a quick check of the saturation temperatures for the given pressures. Therefore, all of the properties at states 1, 3, and 5 can be found.

At states 2 and 4, only the pressure is known. However, these states are the exhaust of the turbines and a key performance indicator of the turbines is known; the isentropic efficiency. Writing the isentropic efficiency expression for the high-pressure turbine (HPT) gives

$$\eta_{\text{HPT}} = \frac{h_1 - h_2}{h_1 - h_{2s}} \tag{3.115}$$

The unknown in Equation 3.115 is the enthalpy leaving the HPT, h_2. The enthalpy leaving the isentropic turbine, h_{2s} can be found because the pressure and entropy are known at that state.

$$h_{2s} = h(P_2, s_{2s})$$
$$s_{2s} = s_1 \tag{3.116}$$

The entropy at state 1 is known because the pressure and temperature are known. The three equations summarized in Equations 3.115 and 3.116 allow for the calculation of the enthalpy at state 2. State 2 is now identified with pressure and enthalpy. Therefore, the rest of the properties can be found. Similar calculations allow for the determination of the properties at state 4, the exhaust of the low-pressure turbine.

State 6 is the exhaust of the pump. The only property known at this state is the pressure. However, the isentropic efficiency of the pump is known.

$$\eta_p = \frac{h_6 - h_{5s}}{h_6 - h_5} \tag{3.117}$$

The unknown in Equation 3.117 is the enthalpy leaving the pump, h_6. The isentropic enthalpy leaving the pump, h_{6s}, is determined from the pump exhaust pressure and inlet entropy.

$$h_{6s} = h(P_6, s_{6s})$$
$$s_{6s} = s_5 \tag{3.118}$$

The entropy at state 5 is known because pressure and temperature are known at that state. Equations 3.117 and 3.118 can then be solved for the enthalpy, h_5. Then, state 5 is fixed with pressure and enthalpy.

Table 3.4 shows the property table determined using the procedure outlined above for the cycle shown in Figure 3.13. The properties in Table 3.4

were determined using EES. The values shown in the **bold** font represent the independent, intensive properties that fix each state.

Notice that the quality, x, column in Table 3.4 contains values of 100 and –100. When using EES to calculate properties, a quality value of 100 means that the state is in the vapor or supercritical phase. If the value of x is –100, this implies that the state is in the liquid phase.

Once a table of properties is constructed, it is often helpful to superimpose the states on a thermodynamic diagram. This helps visualize how the cycle is operating. EES is capable of drawing several different property plots. Furthermore, if the properties are stored in arrays, then the EES arrays table can be overlaid on the property plot. Figure 3.14 shows a temperature–entropy diagram for the cycle shown in Figure 3.13.

The second step of the solution strategy is to perform the required thermodynamic analyses. This often includes the application of balance or

TABLE 3.4

Calculated Properties for the Cycle Shown in Figure 3.13

State	P (MPa)	T (°C)	h (kJ/kg)	s (kJ/kg·K)	x
1	**11.9**	**450.0**	3211.50	6.30846	100
2	**0.80**	170.4	**2701.64**	6.51132	0.9674
3	**0.79**	**380.0**	3225.57	7.51573	100
4	**0.007**	39.0	**2468.49**	7.94374	0.9571
5	**0.005**	**30.0**	125.74	0.43676	–100
6	**12.0**	30.8	**139.87**	0.44374	–100

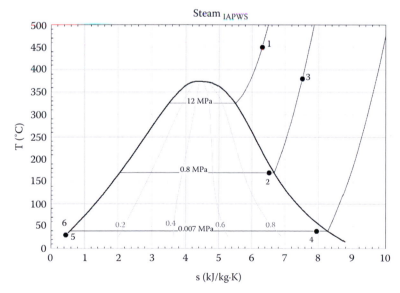

FIGURE 3.14
Temperature–entropy diagram for the cycle shown in Figure 3.13.

conservation laws to the cycle. Figure 3.13 shows four different system boundaries surrounding various components in the cycle.

The power delivered by the turbines can be determined by considering a system boundary surrounding both the high-pressure and low-pressure turbines. If the individual power delivered by each turbine is needed, then two system boundaries need to be analyzed: one around each turbine. For the combined turbines, the conservation of energy equation reduces to

$$\dot{W}_t = \dot{m}\left[(h_1 + h_3) - (h_2 + h_4)\right] = 481.4 \text{ MW} \tag{3.119}$$

The power required by the pump can be found by applying the conservation of energy equation to a system boundary surrounding the pump.

$$\dot{W}_p = \dot{m}(h_6 - h_5) = 5.37 \text{ MW} \tag{3.120}$$

The net power delivered from the cycle can now be determined by

$$\dot{W}_{net} = \dot{W}_t - \dot{W}_p = 476.1 \text{ MW} \tag{3.121}$$

The heat transfer rate into the boiler can be found by applying the conservation of energy equation to the system boundary surrounding the boiler.

$$\dot{Q}_b = \dot{m}\left[(h_1 + h_3) - (h_6 + h_2)\right] = 1167 \text{ MW} \tag{3.122}$$

The heat transfer rate from the condenser can be found in one of two ways. The conservation of energy equation can be applied to a system boundary that surrounds the condenser (Figure 3.13). An alternative way is to apply the conservation of energy equation to a system boundary that encompasses the complete cycle.

$$\dot{Q}_c = \dot{m}(h_4 - h_5) = 890.2 \text{ MW}$$

$$\dot{Q}_c = \dot{Q}_b + \dot{W}_p - \dot{W}_t = 890.2 \text{ MW} \tag{3.123}$$

The thermal efficiency of the cycle can now be calculated.

$$\eta_{th} = \frac{\dot{W}_{net}}{\dot{Q}_b} = 0.348 = 34.8\% \tag{3.124}$$

Many other parameters can be calculated for this cycle including entropy generation rates for the turbines and pump, exergy destruction rates for the turbines and pump, and exergetic efficiencies of the components in the cycle.

Building the table of properties first has a type of *domino effect* on the problem. Once the table is fully constructed, the application of the conservation and balance laws is fairly straightforward. As mentioned earlier, some thermodynamic analysis may be required to complete the property table.

3.11.2 Thermal Energy Cycle Examples

In this section, examples of thermal energy cycles will be presented. The solution strategy suggested in Section 3.11.1 will be used to develop the solution.

EXAMPLE 3.16

Under normal operation, the turbine exhaust of a gas turbine power cycle is very hot and therefore contains a significant amount of exergy. One scheme that utilizes this exergy is called a *combined cycle*, as shown in Figure E3.16. In a combined cycle, the hot exhaust from the gas turbine is used to boil steam in an interstage heat exchanger. The heat exchanger serves as a boiler for a vapor power cycle.

Consider a combined cycle power plant that has a net power delivery of 50 MW. Air enters the compressor of the gas turbine cycle at 14.7 psia, 70°F. The compressor has a pressure ratio of 12 and an isentropic efficiency of 80%. After heat addition in the combustor, the air enters the gas turbine at 2000°F with no appreciable pressure drop through the combustor. The air then expands through the gas turbine, which has an isentropic efficiency of 85%, and leaves the turbine at a pressure of 16 psia. The hot air passes through the interstage heat exchanger and leaves at 14.7 psia, 260°F. In the steam cycle, steam enters the turbine at 1100 psia, 880°F. The steam expands through the steam turbine, which has an isentropic efficiency of 78%, to the condenser pressure, 2 psia. The condenser fully condenses the steam to a saturated liquid condition at 2 psia. The exhaust of the pump, which has an isentropic efficiency of 60%, is 1100 psia. The condenser is cooled with liquid water that enters at 60°F and exits at 95°F.

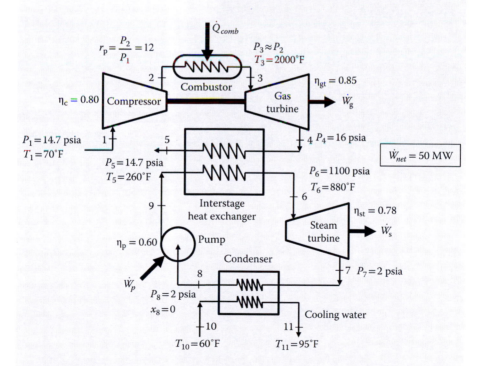

FIGURE E3.16

Determine the following information:

a. Mass flow rate of the air, steam, and cooling water in the cycle (lbm/h).
b. Net power delivered by the gas turbine cycle (MW).
c. Net power delivered by the steam cycle (MW).
d. Thermal efficiency of the combined cycle.
e. A full exergy accounting of the net rate of flow exergy increase of the air as it passes through the combustor of the gas turbine cycle (the dead state can be taken as 14.7 psia, 70°F).

SOLUTION: The first step in the solution to this problem is to build the table of properties. In addition to the properties, it is convenient to also compute the flow exergy at each state for the full accounting requested in part (e). Using methods developed in Section 3.11.1, the full property table for the cycle can be developed (including flow exergy values) using the independent, intensive properties for each state shown in Table E3.16A.

States 2, 4, 7, and 9 require an additional isentropic efficiency analysis of the compressor, gas turbine, steam turbine, and pump to determine the enthalpy at each of those states. States 10 and 11 are the inlet and outlet conditions of the cooling water passing through the condenser. The only information given about the cooling water is the temperature at each state. Therefore, the incompressible substance model can be used to estimate the liquid water properties at the given temperature on the saturated liquid line ($x = 0$). Table E3.16B shows the property table for this cycle (including flow exergies). The properties were calculated using EES.

TABLE E3.16A

Independent Properties Used to Define the Thermodynamic State Points in the Cycle Shown in Figure E3.16

State	Independent Properties	Additional Analysis Required
1	P,T	none
2	P,h	$\eta_c = (h_{2s} - h_1)/(h_2 - h_1)$
3	P,T	none
4	P,h	$\eta_{gt} = (h_3 - h_4)/(h_3 - h_{4s})$
5	P,T	none
6	P,T	none
7	P,h	$\eta_{st} = (h_6 - h_7)/(h_6 - h_{7s})$
8	P,x	none
9	P,h	$\eta_p = (h_{9s} - h_8)/(h_9 - h_8)$
10	T,x	Incompressible substance model
11	T,x	Incompressible substance model

TABLE E3.16B

Table of Thermodynamic Properties for the Cycle Shown in Figure E3.16

State	P (psia)	T (°F)	h (Btu/lbm)	s (Btu/lbm-R)	x_f (Btu/lbm)	x
1	14.7	70.0	126.63	1.63535	0.00	100
2	176.4	736.7	290.71	1.66437	148.71	100
3	176.4	2000.0	635.00	1.85942	389.70	100
4	16	1061.1	374.93	1.89147	112.65	100
5	14.7	260.0	172.45	1.70926	6.6734	100
6	1100	880.0	1432.9	1.59111	591.52	100
7	2	126.0	1035.5	1.78247	92.808	0.9215
8	2	126.0	94.02	0.17499	2.7285	0
9	1100	128.7	99.51	0.17872	6.2364	−100
10		60.0	28.08	0.05554	0.05248	0
11		95.0	63.04	0.12065	0.53047	0

Now that the properties and flow exergy values are identified, the problem can be solved. The net power developed (50 MW) is the sum of the net power delivered from each of the cycles in the combined cycle.

$$\dot{W}_{net} = \dot{W}_{net,gt} + \dot{W}_{net,st}$$

The net power delivered from the gas turbine cycle is the total turbine power delivered less the power required to run the compressor.

$$\dot{W}_{net,gt} = \dot{W}_{gt} - \dot{W}_c = \dot{m}_a \left(h_3 - h_4\right) - \dot{m}_a \left(h_2 - h_1\right)$$

In this equation, the subscript "a" represents the air flowing through the gas turbine cycle. Similarly, the net power developed by the steam cycle is the total turbine power delivered less the power required to run the pump.

$$\dot{W}_{net,st} = \dot{W}_{st} - \dot{W}_p = \dot{m}_s \left(h_6 - h_7\right) - \dot{m}_s \left(h_9 - h_8\right)$$

Here, the subscript "s" is used to designate the steam flow in the vapor power cycle. The above set of equations is three equations with four unknowns, $\dot{W}_{net,gt}$, $\dot{W}_{net,st}$, $\dot{m}_a$, and $\dot{m}_s$. Therefore, another equation is required. The only component in the combined cycle that contains air and steam is the interstage heat exchanger. The conservation of energy applied to a system boundary that surrounds the heat exchanger results in

$$\dot{m}_a h_4 + \dot{m}_s h_9 = \dot{m}_a h_5 + \dot{m}_s h_6$$

This equation introduces no new unknowns to the problem. This equation set can be solved to reveal the mass flow rates of the air and steam, and the net power deliver from the gas turbine and vapor power cycles.

$$\dot{W}_{net,gt} = 30.86 \text{ MW}, \ \dot{W}_{net,st} = 19.14 \text{ MW}, \ \dot{m}_a = 1.097 \times 10^6 \ \frac{\text{lbm}}{\text{h}}, \ \dot{m}_s = 166{,}616 \frac{\text{lbm}}{\text{h}}$$

The mass flow rate of the cooling water flowing through the condenser can be found by applying the conservation of energy equation to a system boundary surrounding the condenser.

$$\dot{m}_s h_7 + \dot{m}_w h_{10} = \dot{m}_s h_8 + \dot{m}_w h_{11}$$

$$\dot{m}_w = \dot{m}_s \left(\frac{h_7 - h_8}{h_{11} - h_{10}} \right) = \underline{4.487 \times 10^6 \ \frac{\text{lbm}}{\text{h}}}$$

The thermal efficiency of the combined cycle can be written as

$$\eta_{th} = \frac{\dot{W}_{net}}{\dot{Q}_{comb}} = \frac{\dot{W}_{net}}{\dot{m}_a \left(h_3 - h_2 \right)} = \underline{0.452 = 45.2\%}$$

For the exergy accounting, it is convenient to calculate the flow exergy values at each state. This is an easy thing to do since the enthalpy and entropy values are already being calculated. The only additional information needed for the flow exergy calculation is the dead-state condition.

The problem asks for the disposition of the net flow exergy increase of the air through the combustor. Using the calculated flow exergy values and mass flow rate of the air through the gas turbine cycle, the input exergy flow rate to the cycle is

$$\dot{X}_{in} = \dot{m}_a \left(x_{f3} - x_{f2} \right) = \underline{77.49 \text{ MW}}$$

This exergy is distributed through the cycle in three ways: (1) exergy output (power delivered), (2) exergy destruction in the cycle components, and (3) exergy lost. The exergy rate output from the combined cycle is known (50 MW). This can be further broken down into the exergy rate output from each of the individual cycles.

$$\dot{X}_g = \dot{W}_{net,gt} = \underline{30.86 \text{ MW}}$$

$$\dot{X}_s = \dot{W}_{net,st} = \underline{19.14 \text{ MW}}$$

The exergy destruction rate in each component can be calculated using the exergy balance equation (Equation 3.76). Applying the exergy balance to the components in the cycle results in

Compressor: $\dot{X}_{des,c} = \dot{W}_c + \dot{m}_a \left(x_{f1} - x_{f2} \right) = \underline{4.942 \text{ MW}}$

Gas turbine: $\dot{X}_{des,gt} = -\dot{W}_{gt} + \dot{m}_a \left(x_{f3} - x_{f4} \right) = \underline{5.459 \text{ MW}}$

Heat exchanger: $\dot{X}_{des,HX} = \left(\dot{m}_a x_{f4} + \dot{m}_s x_{f9} \right) - \left(\dot{m}_a x_{f5} + \dot{m}_s x_{f6} \right) = \underline{5.496 \text{ MW}}$

Steam turbine: $\dot{X}_{des,st} = -\dot{W}_{st} + \dot{m}_s \left(x_{f6} - x_{f7} \right) = \underline{4.949 \text{ MW}}$

Condenser: $\dot{X}_{des,cond} = \left(\dot{m}_s x_{f7} + \dot{m}_w x_{f10} \right) - \left(\dot{m}_s x_{f8} + \dot{m}_w x_{f11} \right) = \underline{3.770 \text{ MW}}$

Pump: $\dot{X}_{des,p} = \dot{W}_p + \dot{m}_s \left(x_{f8} - x_{f9} \right) = \underline{0.097 \text{ MW}}$

Exergy is lost in the cycle in two places; the interstage heat exchanger and the condenser. For the heat exchanger, there is exergy leaving with the air that is simply discarded. This is *lost* exergy in the sense that it is not being used and cannot be recovered once it is dumped to the environment. This lost exergy is

$$\dot{X}_{lost,HX} = \dot{m}_a \left(x_{f5} - x_{f1} \right) = \underline{2.146 \text{ MW}}$$

There is an exergy rise of the cooling water as it passes through the condenser. This exergy is carried away by the cooling water and discarded through the water cooling process. This exergy loss can be calculated as

$$\dot{X}_{lost,w} = \dot{m}_w \left(x_{f11} - x_{f10} \right) = \underline{0.629 \text{ MW}}$$

These exergy calculations can be summarized in tabular form for an easy view of the exergy accounting, as shown in Table E3.16C.

TABLE E3.16C

Exergy Accounting of the Cycle Shown in Figure E3.16

Exergy Input	MW	% of Total
Combustor	77.49	100
Exergy Output		
Gas turbine cycle	30.86	39.8
Vapor power cycle	19.14	24.7
Total combined cycle	50.00	64.5
Exergy Destruction		
Compressor	4.942	6.4
Gas turbine	5.459	7.0
Interstage heat exchanger	5.496	7.1
Steam turbine	4.949	6.4
Condenser	3.770	4.9
Pump	0.097	0.1
Total exergy destruction	24.713	31.9
Exergy Lost		
Air leaving the heat exchanger	2.146	2.8
Cooling water	0.629	0.8
Total exergy lost	2.775	3.6
Total (output + destruction + lost)	**77.49**	**100.0**

This exergy accounting reveals that the exergetic efficiency of the cycle is

$$\eta_{x,cycle} = \frac{\dot{X}_{out}}{\dot{X}_{in}} = \underline{64.5\%}$$

In other words, 64.5% of the net exergy increase of the air passing through the combustion chamber is converted to power, 31.9% of the input exergy is destroyed in the cycle components due to irreversibility, and 3.6% of the input exergy is lost.

Notice that the combustor was not analyzed in the exergy analysis of Example 3.16. Instead, the exergy rate increase of the air passing through the combustor formed the basis of the exergy input to the cycle. A more accurate depiction of what is happening in the combustor is shown in Figure 3.15. Notice that this figure does not indicate a heat transfer. Instead, there are three mass flows that cross the system boundary of the combustor: air from the compressor, fuel into the combustor, and products of combustion leaving the combustor. The heat is released inside of the combustor as a result of the combustion of the fuel and air.

The fuel entering the combustor has exergy content due to its chemistry. This is known as *chemical exergy*. The flow exergy, x_f, that is used in the exergy balance is known as *thermomechanical exergy*. In reality, all fluids have both thermomechanical and chemical exergy. The chemical exergy of many fluids is negligible. However, the chemical exergy of *fuels* is significant. A true exergy accounting would consider the fuel's chemical exergy as the exergy input to the cycle. When this is done, the exergy destruction in the combustor can be calculated. The determination of the chemical exergy of fuels is a complex subject beyond the scope of this book. A good treatment of chemical exergy can be found in the work of Bejan (2006). On the basis of this discussion, it can be concluded that Example 3.16 is not a complete exergy picture of what is happening in the combined cycle. A better exergy accounting would be based on the chemical exergy of the fuel. However, Example 3.16 does serve as an example of how exergy accounting is done in a more complex cycle.

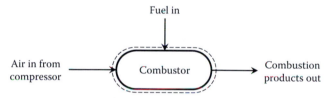

FIGURE 3.15

Flows entering and leaving a combustor in a gas turbine cycle.

EXAMPLE 3.17

Multistage vapor compression refrigeration is utilized to increase the coefficient of performance compared to a single-stage vapor compression cycle operating between the same evaporating and condensing conditions. There are many different refrigeration cycles that utilize multistage compression. One such multistage vapor compression refrigeration cycle that utilizes *flash gas removal* is shown in Figure E3.17A.

Consider a R-22 multistage vapor compression cycle that has a saturated evaporating temperature (SET) of 10°F and a saturated condensing temperature (SCT) of 140°F. The R-22 leaves the condenser as a saturated liquid at the SCT. As the R-22 passes through the first expansion valve, it flashes into saturated liquid and saturated vapor at 110 psia. The flash evaporator separates these two phases: saturated vapor exits at state 3, and saturated liquid exits the flash evaporator at state 8. The saturated liquid leaving the flash evaporator passes through a second expansion valve, causing the R-22 to flash again before it enters the evaporator. The R-22 completely boils in the evaporator and leaves as a saturated vapor, where it enters a compressor with an isentropic efficiency of 80%. After compression, the R-22 vapor recombines with the saturated vapor from the flash evaporator after it has been compressed with a

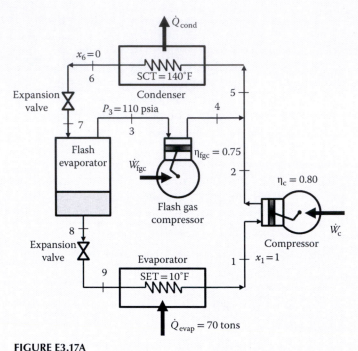

FIGURE E3.17A

flash gas compressor that has an isentropic efficiency of 75%. The refrigeration capacity of the system is 70 tons. Determine the following:

a. The power draw, COP_C, and horsepower per ton (HPT) for the multistage cycle described above.
b. The power draw, COP_C, and HPT for a single-stage vapor compression refrigeration cycle with the same capacity and operating between the same SET and SCT.
c. Investigate the effect of the flash evaporator pressure, P_3, by plotting the COP_C as a function of P_3.

SOLUTION: There are several refrigeration terms involved in this problem statement that are more easily understood by considering a pressure–enthalpy diagram of the cycle as shown in Figure E3.17B. As seen in this figure, the SCT is the temperature in the condenser where the refrigerant changes its phase from vapor to liquid. Likewise, the SET is the temperature in the evaporator where the refrigerant boils.

In this problem, the capacity of the refrigeration system is given as 70 tons. The *capacity* of a refrigeration system is the heat transfer rate occurring at the evaporator. The capacity is often expressed in *tons*. A ton of refrigeration is defined as the steady-state heat transfer rate required to melt 1 ton of ice at 32°F in a 24-hour period. It can be shown that this heat transfer rate is 12,000 Btu/h.

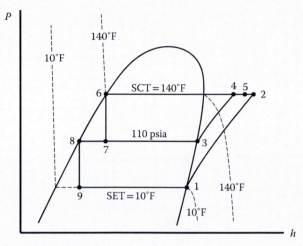

FIGURE E3.17B

Following the solution strategy developed in Section 3.11.1, the first step is to construct the table of thermodynamic properties in the cycle. In this cycle, there is a mass flow split that occurs at the flash evaporator. Therefore, it is helpful to include the mass flow rates at each state in the property table.

According to the *P–h* diagram shown in Figure E317.B, there are three pressures in the cycle. The intermediate pressure (the pressure in the flash evaporator) is given as 110 psia. The high and low pressures can be found from the knowledge of the SCT and the SET. Following this thinking, the pressures at the nine state points in the cycle are given by

$$P_1 = P_9 = P_{sat} \text{ (SET)}$$

$$P_2 = P_4 = P_5 = P_6 = P_{sat} \text{ (SCT)}$$

$$P_3 = P_7 = P_8 = 110 \text{ psia}$$

Notice that the high and low pressures in the cycle are evaluated as the saturation pressure at the SCT and SET, respectively.

Using the techniques developed in Section 3.11.1, the independent properties required to determine the properties at each state are summarized in Table E3.17A. This table indicates that state 5 is defined by pressure and enthalpy. The pressure, as mentioned earlier, is equal to the saturation pressure at the SCT. The enthalpy can be found by applying the conservation of energy and conservation of mass equations to a system boundary that surrounds the point where the flows from the two compressors converge.

$$\dot{m}_4 h_4 + \dot{m}_2 h_2 = \dot{m}_5 h_5$$
$$\dot{m}_4 + \dot{m}_2 = \dot{m}_5$$

TABLE E3.17A

Independent Properties Used to Define the Thermodynamic State Points in the Cycle Shown in Figure E3.17A

State	Independent Properties	Additional Analysis Required
1	P,x or T,x	none
2	P,h	$\eta_c = (h_{2s} - h_1)/(h_2 - h_1)$
3	P,x	none
4	P,h	$\eta_{fgc} = (h_{4s} - h_3)/(h_4 - h_3)$
5	P,h	mass and energy conservation where states 2 and 4 combine
6	P,x or T,x	none
7	P,h	$h_7 = h_6$ (valve)
8	P,x	none
9	P,h	$h_9 = h_8$ (valve)

This is a set of two equations with four unknowns: the mass flow rates and h_5. The capacity of the refrigeration system is known; therefore, the mass flow rate at states 1, 2, 8, and 9 can be found by applying the conservation of energy equation to a system boundary surrounding the evaporator.

$$\dot{Q}_{evap} = \dot{m}_1 \left(h_1 - h_9 \right)$$

This is a third equation with no new unknowns. The mass flow rate split can be determined by analyzing the flash evaporator using conservation of energy and conservation of mass.

$$\dot{m}_7 h_7 = \dot{m}_3 h_3 + \dot{m}_8 h_8$$

$$\dot{m}_7 = \dot{m}_3 + \dot{m}_8$$

To complete this analysis, mass flow rate equivalencies are specified as follows:

$$\dot{m}_7 = \dot{m}_5 = \dot{m}_6$$

$$\dot{m}_8 = \dot{m}_9 = \dot{m}_1 = \dot{m}_2$$

$$\dot{m}_3 = \dot{m}_4$$

The only unknown in this set of equations is the enthalpy at state 5, h_5. This completes the property table. In addition, all of the mass flow rates have also been determined in this analysis! The resulting table of properties and flow rates is shown in Table E3.17B. The properties and flow rates in this table were computed using EES.

Now that all the properties and flow rates are known, the analysis of the cycle can be completed.

a. Analysis of the multistage cycle

The power draw for the cycle is the sum of the power draw by each compressor.

$$\dot{W}_m = \dot{W}_c + \dot{W}_{fgc} = \dot{m}_1 \left(h_2 - h_1 \right) + \dot{m}_3 \left(h_4 - h_3 \right) = \underline{148.1 \text{ hp}}$$

TABLE E3.17B

Table of Thermodynamic Properties and Flow Rates for the Cycle Shown in Figure E3.17A

State	P (psia)	T (°F)	h (Btu/lbm)	s (Btu/lbm·R)	x	$\dot{m}$ (lbm/h)
1	47.55	10	172.05	0.42250	1	10,625.7
2	352.2	222.6	199.81	0.43079	100	10,625.7
3	110	56.51	176.14	0.41365	1	4,950.52
4	352.2	191.0	192.68	0.42010	100	4,950.52
5	352.2	212.4	197.54	0.42745	100	15,576.2
6	352.2	140	119.42	0.29879	0	15,576.2
7	110	56.51	119.42	0.30378	0.3178	15,576.2
8	110	56.51	92.998	0.25258	0	10,625.7
9	47.55	10	92.998	0.25419	0.1424	10,625.7

A parameter of interest to refrigeration engineers is the HPT. This parameter is analogous to the heat rate for a vapor power cycle. It is the dimensional reciprocal of the COP_C. This parameter tells the engineer how much power (in horsepower) is required to affect 1 ton of refrigeration in the cycle.

$$HPT_m = \frac{\dot{W}_m \text{ [hp]}}{\dot{Q}_{evap} \text{ [ton]}} = 2.12 \frac{hp}{ton}$$

b. Analysis of a single-stage vapor compression refrigeration cycle

In this part of the problem, the same parameters calculated in part (a) will be determined for a single-stage vapor compression refrigeration cycle with the same capacity (70 ton) operating between the same SCT and SET. This cycle is shown in Figure E3.17C.

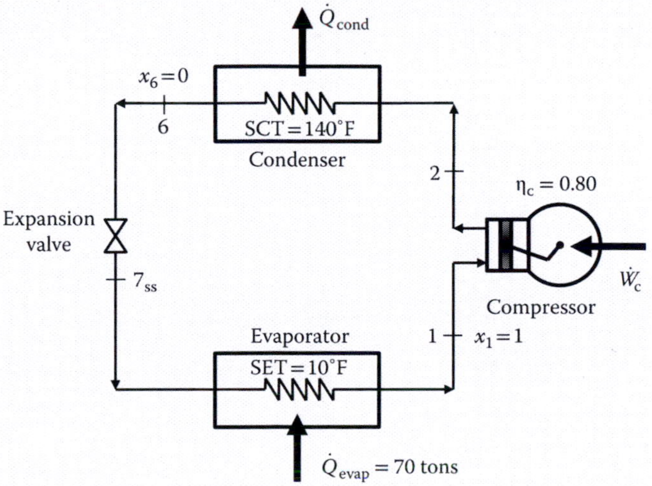

FIGURE E3.17C

The *P–h* diagram shown in Figure E3.17D shows the single-stage cycle superimposed on the same *P–h* diagram that was used for the multistage cycle. In this diagram, the multistage cycle states and processes are grayed out.

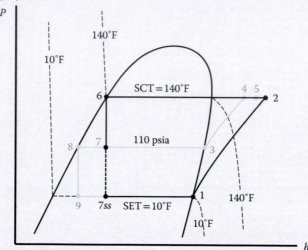

FIGURE E3.17D

The properties at states 1, 2, and 6 are the same that were determined for the multistage cycle. The only new state point is labeled "7ss" (ss stands for single stage). Since process 6-7ss is occurring through the expansion valve, it must be true that

$$h_{7ss} = h_6$$

The mass flow rate in the single-stage cycle is the same in all components, and must be recalculated. This can be done by analyzing the evaporator for the single-stage cycle.

$$\dot{Q}_{evap} = \dot{m}_{ss}\left(h_1 - h_{7ss}\right) \quad \rightarrow \quad \dot{m}_{ss} = \frac{\dot{Q}_{evap}}{\left(h_1 - h_{7ss}\right)} = 15{,}961 \ \frac{\text{lbm}}{\text{h}}$$

Now, the power draw of the compressor in this cycle can be determined by

$$\dot{W}_{c,ss} = \dot{m}_{ss}\left(h_2 - h_1\right) = 174.1 \ \text{hp}$$

The COP_C and HPT of the single-stage cycle can also be found by

$$COP_{c,ss} = \frac{\dot{Q}_{evap}}{\dot{W}_{c,ss}} = 1.90 \qquad HPT_{ss} = \frac{\dot{W}_{c,ss} \ [\text{hp}]}{\dot{Q}_{evap} \ [\text{ton}]} = 2.49 \ \frac{\text{hp}}{\text{ton}}$$

Table E3.17C shows a direct comparison of the single-stage cycle and the multistage cycle.

TABLE E3.17C

Comparison of a Single-Stage Cycle to the Multistage Flash Gas Removal Cycle

Parameter	Single-Stage Cycle	Multistage Cycle
Power draw	174.1 hp	148.1 hp
Cooling coefficient of performance	1.90	2.23
Horsepower per ton	2.49 hp/ton	2.12 hp/ton

Recall that in both cases, the capacity of the cycle is 70 tons. In addition, both cycles are operating between the same evaporating and condensing conditions. Even though the multistage cycle is more complex, and requires more equipment, its operating cost (directly related to the power draw) and thermal performance as indicated by the COP_C and HPT are superior.

c. Parametric study—effect of the flash evaporator pressure

The intermediate pressure in the cycle, which is the pressure in the flash evaporator, can be controlled by the expansion valve placed before the flash evaporator. To determine the effect of this pressure on the system performance, a parametric study can be conducted that varies the flash evaporator pressure over a specified range. Referring to Table E3.17B, it can be seen that the low and high pressures in the cycle are 47.55 and 352.2 psia, respectively. The flash evaporator pressure must be within these bounds. Therefore, a parametric study can be conducted using EES for a range of flash evaporator pressures ranging from 50 to 350 psia. The result of this parametric study is shown in Figure E3.17E.

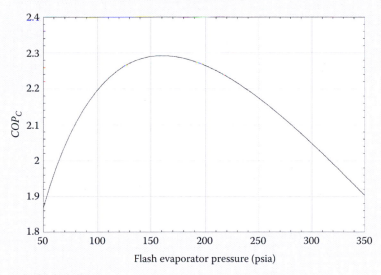

FIGURE E3.17E

This study reveals that there is an *optimum* flash evaporator pressure that maximizes the COP_C of the cycle.

The examples shown in this section follow the suggested solution strategy laid out in Section 3.11.1. There are many possible configurations for thermal energy cycles. These examples were selected to show how a generalized solution strategy can make the analysis manageable, even though the cycle may be quite complex. In many cases, additional analysis may be required to complete the table of properties. However, once the table is complete, the thermal analysis of the cycle is quite straightforward.

Care must be taken with cycles where the mass flow rate splits. Example 3.17 demonstrates that additional analysis is required to determine the flow split in the cycle. In the case of the multistage refrigeration cycle shown in this example, the mass flow split was determined by analyzing the flash evaporator using the conservation of energy and conservation of mass laws. This is true for any other cycle with mass flow splits. Somewhere in the cycle, a component can be identified that allows for the calculation of how the mass flow splits in the cycle. As demonstrated, this often requires simultaneous application of conservation of energy and conservation of mass.

Problems

3.1   The open water tank shown in Figure P3.1 is filled through pipe 1 with a velocity of 7 ft/s and through pipe 3 at a volumetric flow rate of 0.2 ft³/s. The density of the water can be taken as 62.4 lbm/ft³. Determine the velocity through pipe 2 required to maintain a constant water level, H.

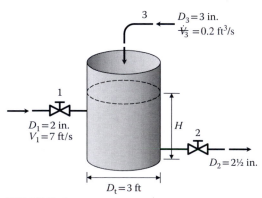

FIGURE P3.1

3.2 If the valve in pipe 1 of Problem 3.1 is suddenly opened to allow a velocity of 12 ft/s, find the instantaneous rate of change of the depth in the tank with time, dH/dt.

3.3 Water enters the covered tank shown in Figure P3.3 with a volumetric flow rate of 5 gpm. The water line has an inside diameter of 0.2423 ft. The air vent on the tank has an inside diameter of 0.1633 ft. The water is at a temperature of 80°F and the air is at 14.5 psia, 80°F. Determine the air velocity leaving the vent at the instant shown in the figure.

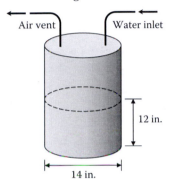

Air vent

Water inlet

12 in.

14 in.

FIGURE P3.3

3.4 Refrigerant 134a enters a pipe with an inside diameter of 3.88 cm at 40 kPa, a quality of 0.15, and a velocity of 0.5 m/s. The refrigerant gains heat as it flows through the pipe and exits as a saturated vapor with a drop in pressure of 0.5 kPa. Determine the following:
   a. Volumetric flow rate at the inlet of the pipe (L/s)
   b. Mass flow rate of the refrigerant through the pipe (g/s)
   c. Volumetric flow rate at the exit of the pipe (L/s)
   d. Velocity of the refrigerant at the exit of the pipe (m/s)
   e. Heat transfer rate to the refrigerant (kW) as it flows through the pipe

3.5 The American Society of Heating, Refrigerating, and Air-conditioning Engineers (ASHRAE) specify that the minimum fresh air ventilation require-ment for a residential building is 0.35 air changes per hour. This means that in 1 hour, 35% of the air contained in a residence should be replaced by fresh, outside air. Consider a residence that has 2000 ft² of living area with 8-ft ceil-ings. The fresh air ventilation requirements of this residence are to be met by a single fan. Determine the following:
   a. Volumetric flow capacity (in cfm) of the fresh air ventilation fan
   b. Circular duct diameter (in in.) required if the fresh ventilation air velocity is not to exceed 780 ft/min

3.6 The hot water needs of a residence are met by a 20-gal hot water heater rated at 6000 Btu/h. The tank is initially filled with hot water at 175°F. As the hot water supply is drawn from the tank, it is replaced by cold water at 50°F at a flow rate sufficient to keep the tank full at all times. During this replenishing process in the tank, the water is heated at the water heater's full rated capacity. Water this hot is great for washing clothes or dishes, but it is far too hot for other purposes, like taking a shower. To use the hot water for a shower, it must be mixed with cold water. Consider a case where the hot water from the tank is mixed with cold water at 50°F as shown in Figure P3.6. The cold water enters the mixing process at a flow rate of 0.95 gpm. After an 8-minute shower, the temperature in the water tank has dropped to 140°F.

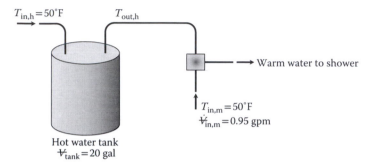

FIGURE P3.6

Determine the following:
a.   Volumetric flow rate of the hot water from the tank during the shower (gpm)
b.   Average temperature of the water delivered to the shower (°F)
c.   Number of gallons of water used during the shower
     Note: When solving this problem, the instantaneous water temperature in the tank can be approximated as a constant value at the average temperature in the tank during the shower.

3.7   Ammonia enters a valve as a saturated liquid at 12 bar with a mass flow rate of 6 kg/min and is steadily throttled to a pressure of 1 bar. Determine the rate of entropy production for this process (in kW/K). If the valve were replaced by a power-recovery turbine operating at steady state, determine the maximum theoretical power that could be developed (in kW).

3.8   Steam leaves the boiler of a power plant at 7 MPa, 500°C as shown in Figure P3.8. As the steam passes to the turbine, there is a small pressure drop due to friction in the pipe and the temperature drops due to a heat loss through the pipe's insulation. The surface temperature of the pipe insulation is 45°C. Because of these effects, the steam enters the turbine at 6.93 MPa, 493°C. The

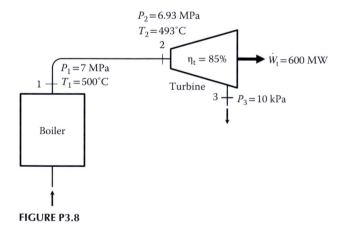

FIGURE P3.8

steam then passes through an adiabatic turbine and exits at 10 kPa. The turbine has an isentropic efficiency of 85% and is delivering 600 MW of power. Determine the following:

a. Heat transfer rate from the pipe connecting the boiler to the turbine (in MW)

b. Entropy generation rate associated with the heat loss from the pipe (kW/K)

3.9 A two-stage compressor is being used to transport methane as shown in Figure P3.9. The methane enters the first compressor stage at 1 bar, 25°C, a flow rate of 0.4 kg/s, and is compressed to 5 bar. The methane then passes through an intercooler and is cooled back to 25°C with no appreciable drop in pressure, after which, it is compressed in the second stage to 12 bar. The isentropic efficiency of both compression stages is 79%. The intercooler is a counterflow heat exchanger, where the methane is being cooled by water. The water enters the intercooler at 20°C and exits at 80°C.

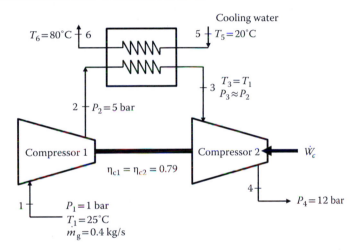

FIGURE P3.9

Determine the following:

a. Mass flow rate of the cooling water in the intercooler (kg/s).

b. Total power draw of both compressors (kW).

c. Investigate the effect of the intermediate pressure P_2 by constructing a plot that shows the total power draw as a function of P_2 for the range 2 bar $\leq P_2 \leq$ 6 bar.

d. One reason for multistage compression with intercooling is to reduce the total power requirement in an overall compression process. Verify that this idea is correct by calculating the power draw for a single compressor used to compress 0.4 kg/s of methane from 1 to 12 bar with an isentropic efficiency of 79%.

3.10 Consider the two-stage methane compressor with intercooling described in Problem 3.9 with the dead state of 1 bar, 25°C. For this thermal system determine the following:

a. Exergy destruction rate in each component (kW)

b. Exergetic efficiency of each component

3.11   Air enters an adiabatic nozzle at 230 kPa, 600°C with a velocity of 60 m/s
        as shown in Figure P3.11. At the nozzle exit, the air is at 70 kPa, 450°C. The
        dead state can be considered to be 100 kPa, 20°C. Determine the following
        information:
        a.   Exit velocity of the air (m/s)
        b.   Isentropic efficiency of the nozzle
        c.   Exergy destruction per kg of air flowing through the nozzle (kJ/kg)
        d.   Devise and calculate an exergetic efficiency for the nozzle

$P_1 = 230$ kPa
$T_1 = 600°C$
$V_1 = 60$ m/s

$P_2 = 70$ kPa
$T_2 = 450°C$

1

2

FIGURE P3.11

3.12   A household refrigerator sold in the United States has a refrigeration capacity
        of 0.14 tons. The Energy Star label on the refrigerator indicates that it has an
        EER of 10.8. Determine
        a.   Power draw of the refrigerator (W)
        b. COP_C of the refrigerator
3.13   A steam power plant's instantaneous operating conditions are summarized in
        Table P3.13.

TABLE P3.13

Instantaneous Steam Power Plant Operating Conditions

Parameter	Value
Net power delivery	462 MW
Heat rate	11,827 Btu/kWh

        For this operating condition determine
        a.   Boiler heat transfer rate (MBtu/h).
        b.   Thermal efficiency of the cycle.
        c.   The heating value of the coal used in this power plant is 28,262 kJ/kg. If
              the plant operates continuously (24 h/day) at the rate defined above, deter-
              mine how many tons of coal are required per day.
3.14   Show that the COP_C and the COP_H of a refrigeration cycle are related by the
        following expression:

$$COP_H = COP_C + 1$$

        Start with a sketch of the refrigeration cycle as shown in Figure 3.10 to develop
        your solution.
3.15   A refrigeration cycle is operating as a heat pump. The source temperature is
        at the dead state, T_0, and the sink is at T_H. Show that the exergetic efficiency of
        this heat pump cycle can be written as

$$\eta_{x,hp} = \frac{Q_H}{W_{cycle}}\left(1 - \frac{T_0}{T_H}\right) = \frac{COP_H}{COP_{H,Carnot}}$$

3.16   In Example 3.15, it was demonstrated that refrigeration cycles utilizing a
       heat input to accomplish refrigeration tend to have a low COP$_C$, but a higher
       exergetic efficiency compared to vapor compression refrigeration cycles using
       electricity to run a compressor. Consider the combined cycle shown in Figure
       P3.16 that utilizes a heat input to accomplish refrigeration.

       The refrigerant used in the cycle is R-143m (trifluoromethyl methyl ether), an
       environmentally friendly refrigerant with low ozone depletion potential (ODP)
       and low global warming potential (GWP). (R-143m is one of the fluids available in
       EES.) The R-143m enters the turbine at 1500 kPa, 100°C, and expands to 500 kPa.
       The isentropic turbine efficiency is 86%. After expansion, the refrigerant is fully
       condensed in the condenser. A pump with an isentropic efficiency of 64% is used
       to raise the R-143m pressure back to the boiler pressure. At the condenser exit,
       the refrigerant flow splits; some going through the power loop just described,
       and the rest going through the refrigeration loop, where the SET is −10°C. The
       R-143m exits the evaporator as a saturated vapor, after which it is compressed
       to the condensing pressure through a compressor with an isentropic efficiency
       of 75%. The turbine delivers just enough power to run the compressor; there is
       no net power delivery from the cycle. The refrigeration capacity of the cycle is
       80 tons. Pressure drops through connecting piping can be considered negligible.

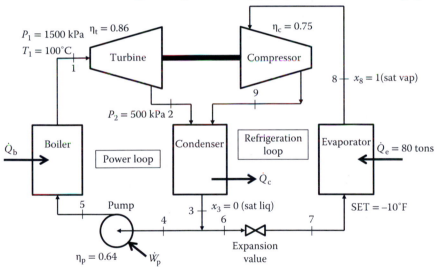

FIGURE P3.16

       Determine the following parameters for this innovative heat-driven cycle:
       a.   Mass flow rate of the refrigerant passing through the refrigeration loop
            (kg/s).
       b.   Power required by the compressor (kW).
       c.   Mass flow rate of the refrigerant passing through the power loop (kg/s).
       d.   Heat transfer rate at the condenser (kW).
       e.   Power draw of the pump (kW).
       f.   Heat transfer rate required at the boiler (kW).
       g.   COP$_C$ of the combined cycle.
       h.   Draw a pressure–enthalpy diagram for R-143m and overlay the 9-cycle
            state points on the diagram.

3.17 Consider the cycle described in Problem 3.16. The heat transfers in the cycle are occurring at boundary temperatures defined as follows:

$$T_{b,b} - T_1 = 5\,\text{K}\left(T_{b,b} = \text{boiler boundary temperature}\right)$$

$$T_3 - T_{b,c} = 2\,\text{K}\left(T_{b,c} = \text{condenser boundary temperature}\right)$$

$$T_{b,e} - T_7 = 2\,\text{K}\left(T_{b,e} = \text{evaporator boundary temperature}\right)$$

For these conditions, determine the following:
a. Exergy destruction rates in each of the components in the cycle (kW)
b. Exergetic efficiency of the cycle

4

Fluid Transport in Thermal Energy Systems

4.1 Introduction

In thermal energy system analysis and design, the primary calculations involving fluid transport include the application of the conservation of mass, conservation of energy, and the determination of head losses in pipes, tubes, ducts, and fittings. The conservation laws are presented in Chapter 3 (Sections 3.6 and 3.7). The determination of pressure losses in a fluid transport system is based on energy conservation principles and is presented in subsequent sections of this chapter. The flow of a fluid in a thermal energy system can be affected in one of two ways: gravity or forced by a pump. This chapter covers both types of fluid flow systems.

Fluid transport systems can be very simple or quite complex. Independent of the complexity of the design, the engineer strives to maintain a cost-optimized system. Piping systems that have large diameters may result in very reasonable operating cost (e.g., small power requirements for pumps), but the capital cost of the system may be prohibitive because of the large diameter. On the other hand, a pipe system with a small diameter may have very reasonable capital cost, but the pumping costs may be overwhelming. This chapter develops equations that can be used to optimize a piping system with respect to cost.

4.2 Piping and Tubing Standards

Pipes are generally designed for high-pressure service. They are made of a variety of materials, including steels, cast iron, wrought iron, clays, and plastics. Independent of the material, pipe sizes are standardized and are referenced based on their *nominal diameter*. Thickness of the pipe wall is designated by the *schedule* number. The steel pipe schedule is usually expressed as a number (e.g., schedule 40). Stainless steel pipe schedules are usually expressed with an "S" after the schedule (e.g., schedule 10S). For iron pipes, the pipe schedule is usually expressed as *std* (standard), *xs* (extra strong), or *xxs* (double extra strong). Pipes are specified based on their nominal diameter and schedule. For example, a pipe that is specified as *2-nom sch 40* has a nominal 2-in. diameter, and the thickness of the pipe wall is consistent with the standard specified for schedule 40. Appendix C lists the standardized pipe dimensions.

EXAMPLE 4.1

Determine the flow area and wall thickness of the following pipes: 3-nom sch 10S (stainless steel), 3-nom sch 40 (plain carbon steel), and 3-nom xxs (cast iron).

SOLUTION: The flow area of the pipe is its cross-sectional area based on the inside diameter (ID) of the pipe.

$$A = \frac{\pi(\text{ID})^2}{4}$$

The thickness of the wall of the pipe can be determined from the outside and inside diameters of the pipe as follows:

$$\text{th} = \frac{\text{OD} - \text{ID}}{2}$$

Appendix C lists the standard outside and inside diameters of pipes. Table E4.1 shows the resulting cross-sectional area and wall thickness of the three 3-nom pipes calculated using the diameters found in Appendix C.

TABLE E4.1

Flow Area (*A*) and Wall Thickness (th)

Pipe	OD (in.)	ID (in.)	*A* (in.²)	th (in.)
3-nom sch 10S	3.5	3.260	8.347	0.120
3-nom sch 40	3.5	3.068	7.393	0.216
3-nom xxs	3.5	2.300	4.155	0.600

This result indicates that all 3-nom pipes are not the same! They have different flow areas and wall thicknesses depending on the schedule.

TABLE 4.1

Common Uses of Different Types of Copper Tubing

Application	K	L	M
Underground water distribution systems		X	X
Above-ground water distribution systems			X
Hydronic heating		X	X
Fuel oil and natural gas distribution systems		X	
Nonflammable medical gases	X	X	
Air conditioning and refrigeration systems		X	
Ground source heat pump systems		X	
Fire sprinkler systems	X	X	X
Compressed air	X	X	X

Pipes can be connected by a variety of joints including threaded, flanged, welded, or bell and spigot. Joints for pipes that are 2-nom or smaller are often threaded. For pipes larger than 2-nom, the joints are often flanged or welded. In the bell and spigot joint, one end of the pipe is enlarged just enough to accept another pipe. In this type of joint, a gasket is used to keep a fluid-tight seal. Bell and spigot joints are usually used in large-diameter iron and PVC pipes.

Tubing has thinner walls compared to an equivalent diameter pipe and is meant for low-pressure service. Tubing can be made of a variety of materials. The most common are copper, steel, and plastic. Tubing is specified by its nominal diameter and *type*. The type specification is analogous to the schedule specification in pipes. There are many tubing types available. The most common copper tubing types are K, L, and M. A type K tube has a thicker wall than a type L tube. Type L tubes are thicker walled compared to a type M tube. Appendix D lists the dimensions of standard copper tubing. The various uses of the different types of copper tubes are listed in Table 4.1.

Tubing can be joined and connected to fittings in a variety of ways, including soldering and brazing. Other tubing fittings include compression fittings and flare fittings.

4.3 Fluid Flow Fundamentals

Consider a pipe transporting a fluid as shown in Figure 4.1. In this figure, a system boundary has been identified, which encompasses the pipe and cuts it at two points identified as 1 and 2. In a general case where friction is considered, there is a possibility of heat being transferred from the system because of friction effects.

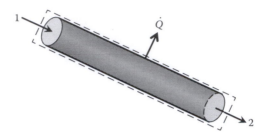

FIGURE 4.1
A pipe transporting a fluid.

For the case of steady state flow, the conservation of energy applied to this system is

$$\frac{-\dot{Q}}{\dot{m}} + (h_1 - h_2) + \frac{1}{2}(V_1^2 - V_2^2) + g(z_1 - z_2) = 0 \tag{4.1}$$

The enthalpy terms can be written in terms of $h = u + Pv$. Making this substitution in Equation 4.1 results in

$$\frac{-\dot{Q}}{\dot{m}} + \left[(u_1 + P_1 v_1) - (u_2 + P_2 v_2) \right] + \frac{1}{2}(V_1^2 - V_2^2) + g(z_1 - z_2) = 0 \tag{4.2}$$

The specific volume is the reciprocal of density. Therefore, the term Pv can be rewritten as P/ρ. Making this substitution and rearranging the equation results in

$$\frac{P_1}{\rho_1} + \frac{V_1^2}{2} + gz_1 = \frac{P_2}{\rho_2} + \frac{V_2^2}{2} + gz_2 + \left[(u_2 - u_1) + \frac{\dot{Q}}{\dot{m}} \right] \tag{4.3}$$

The term in brackets in Equation 4.3 represents the change in the fluid's internal energy and any accompanying heat transfer rate because of friction. This can be expressed as the product of the acceleration due to gravity, g, and the *head loss* due to friction, l_f.

$$\left[(u_2 - u_1) + \frac{\dot{Q}}{\dot{m}} \right] = gl_f \tag{4.4}$$

Notice that the units of the head loss are length, and the product gl_f has units of specific energy. For example, in the SI unit system,

$$gl_f \; [=] \frac{m}{s^2} \cdot m = \frac{m^2}{s^2} \left| \frac{N \cdot s^2}{kg \cdot m} \right. = \frac{N \cdot m}{kg} = \frac{J}{kg}$$

If the fluid is considered incompressible, then its density is constant. Combining this idea along with the head loss represented by Equation 4.4, Equation 4.3 can be rewritten as follows:

$$\frac{P_1}{\rho} + \frac{V_1^2}{2} + gz_1 = \frac{P_2}{\rho} + \frac{V_2^2}{2} + gz_2 + gl_f \qquad (4.5)$$

This form is known as the *specific energy* form of the conservation of energy because each term has units of energy per unit mass. If Equation 4.5 is divided through by g, the conservation of energy equation can be written in *length form* as follows:

$$\frac{P_1}{\rho g} + \frac{V_1^2}{2g} + z_1 = \frac{P_2}{\rho g} + \frac{V_2^2}{2g} + z_2 + l_f \qquad (4.6)$$

It is common to rewrite ρg as γ, the specific weight of the fluid.

$$\frac{P_1}{\gamma} + \frac{V_1^2}{2g} + z_1 = \frac{P_2}{\gamma} + \frac{V_2^2}{2g} + z_2 + l_f \qquad (4.7)$$

Each of the terms in Equation 4.7 represents a *head* value: pressure head, velocity head, elevation head, and head loss due to friction.

Multiplying Equation 4.5 by the fluid density gives the *pressure form* of the conservation of energy equation.

$$P_1 + \frac{\rho V_1^2}{2} + \gamma z_1 = P_2 + \frac{\rho V_2^2}{2} + \gamma z_2 + \gamma l_f \qquad (4.8)$$

If the flow is assumed to be frictionless, then the head loss term is zero and Equation 4.7 can be written as follows:

$$\frac{P_1}{\gamma} + \frac{V_1^2}{2g} + z_1 = \frac{P_2}{\gamma} + \frac{V_2^2}{2g} + z_2 \qquad (4.9)$$

Equation 4.9 is known as *Bernoulli's equation*. This equation represents the ideal condition of an incompressible fluid flowing without friction. Although this situation does not exist in the real world, application of this equation can be useful in providing a benchmark for engineering design.

4.3.1 Head Loss due to Friction in Pipes and Tubes

For fluid flow in circular pipes, the head loss due to friction is expressed as

$$l_f = f \frac{L}{D} \frac{\bar{V}^2}{2g} \tag{4.10}$$

In this equation,

f = Moody friction factor (sometimes called the Darcy–Weisbach factor)
L = length of pipe
D = inside diameter of the pipe
$\bar{V}$ = average velocity of the fluid in the pipe
g = local acceleration due to gravity

If a noncircular pipe is used, the diameter in Equation 4.10 can be replaced with the hydraulic diameter.

$$D_h = \frac{4 \text{ (cross-sectional area)}}{\text{wetted perimeter}} = \frac{4A_c}{P} \tag{4.11}$$

The friction factor, f, can be determined graphically using the Moody diagram, shown in Figure 4.2, or it can also be computed using empirical expressions.

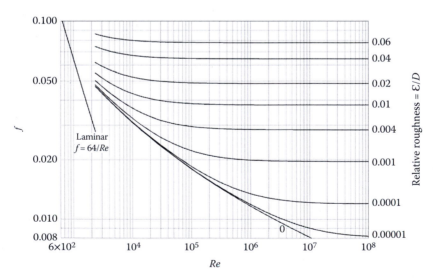

FIGURE 4.2
Moody diagram.

In the laminar flow regime ($Re_D < 2300$), the friction factor is given by

$$f = \frac{64}{Re_D},$$

(4.12)

where the *Reynolds number* is given by

$$Re_D = \frac{\rho \bar{V} D}{\mu} = \frac{\bar{V} D}{v}$$

(4.13)

In this equation, μ is the *dynamic viscosity* of the fluid and v is the *kinematic viscosity*.

In the turbulent region, the friction factor can be computed using one of several correlations shown in Table 4.2. Any of these correlations are sufficient for engineering calculations. In this book, the *Swamee–Jain correlation* is used for friction calculations because of its relative simplicity compared to the other correlations.

In the correlations shown in Table 4.2, ε is the *absolute surface roughness* of the pipe and ε/D is known as the *relative roughness*. Table 4.3 lists typical values of the absolute surface roughness for different types of pipes. The values in this table were taken from Moody (1944).

TABLE 4.2

Turbulent Friction Factor Correlations

Correlation	Equation
Chen	$f = \left[-2.0 \log \left\{ \frac{\varepsilon/D}{3.7065} - \frac{5.0452}{Re} \log \left[\frac{1}{2.8257} \left(\frac{\varepsilon}{D} \right)^{1.1098} + \frac{5.8506}{Re^{0.8981}} \right] \right\} \right]^{-2}$
Churchill	$f = 8 \left[\left(\frac{8}{Re} \right)^{12} + \frac{1}{(B+C)^{1.5}} \right]^{1/12}$
	$B = \left[2.457 \ln \frac{1}{(7/Re)^{0.9} + (0.27\varepsilon/D)} \right]^{16} \qquad C = \left(\frac{37530}{Re} \right)^{16}$
Colebrook	$\frac{1}{\sqrt{f}} = -2 \log \left(\frac{\varepsilon/D}{3.7} + \frac{2.51}{Re\sqrt{f}} \right)$
Haaland	$f = \left\{ -0.782 \ln \left[\frac{6.9}{Re} + \left(\frac{\varepsilon/D}{3.7} \right)^{1.11} \right] \right\}^{-2}$
Swamee–Jain	$f = \frac{0.25}{\left[\log \left(\frac{\varepsilon/D}{3.7} + \frac{5.74}{Re^{0.9}} \right) \right]^2}$

In any fluid friction calculation, there are several variables that must be resolved including the fluid properties, the pipe material, the pipe length, the pipe diameter, the fluid flow rate, and the pressure drop. Consider a design problem where the fluid is known (therefore its properties can be determined), the type of pipe is specified, and the pipe length is known. This leaves three parameters: the pipe diameter, the fluid flow rate, and the pressure drop through the length of pipe. Knowing any two of these three parameters allows for a solution of the conservation of energy equation to find the third. This leads to three possible friction calculation problems. These three problem types depend on which parameters are known. The three problem types can be classified as "Type 1," "Type 2," and "Type 3" problems. Table 4.4 shows these types of problems and indicates which variables are known and unknown in the friction calculation.

The following examples show the procedure for these three types of friction calculations.

TABLE 4.3

Typical Absolute Roughness Values (ε) for Pipes

Pipe Material	foot	centimeter
Drawn tubing	0.000005	0.00015
Commercial steel or wrought iron	0.00015	0.0046
Asphalted cast iron	0.004	0.12
Galvanized iron	0.005	0.15
Cast iron	0.0085	0.26
Wood stave	0.0006–0.003	0.18–0.09
Concrete	0.001–0.01	0.031–0.31
Riveted steel	0.003–0.03	0.091–0.91

Source: Moody, L. F., *Trans. ASME*, 66, 8, 671–684, 1944.

TABLE 4.4

Typical Fluid Friction Calculations

Problem Type	Pressure Drop	Fluid Flow Rate	Pipe Diameter
Type 1	Unknown	Known	Known
Type 2	Known	Unknown	Known
Type 3	Known	Known	Unknown

EXAMPLE 4.2: A Type 1 Problem

Liquid heptane is being transported at a volumetric flow rate of 500 gpm through an 8-nom sch 120 commercial steel pipe. The heptane is flowing in the pipe at an average temperature of 70°F. The pipe is 0.5 mile long. There is no elevation difference from the inlet to the outlet of the pipe. Determine the pressure drop through the pipe.

SOLUTION: A sketch of the horizontal pipe and the system boundary to be analyzed is shown in Figure E4.2.

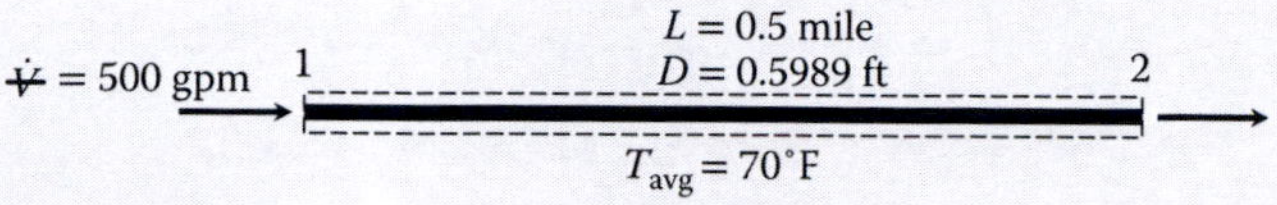

FIGURE E4.2

The average temperature in the pipeline is given. At these conditions, the density and dynamic viscosity of the heptane can be found from Appendix B.1.

$$\rho = 42.625\,\frac{lbm}{ft^3} \qquad \mu = 1.0391\,\frac{lbm}{ft \cdot h}\left(\frac{h}{3600\ s}\right) = 0.0002886\ lbm/ft \cdot s$$

It is more convenient to use the specific weight rather than the density when using Equation 4.8 to calculate the pressure drop. Given the density value, the specific weight of the heptane can be found.

$$\gamma = \rho\frac{g}{g_c} = \left(49.625\,\frac{lbm}{ft^3}\right)\left|\frac{32.174\,\dfrac{ft}{s^2}}{32.174\,\dfrac{lbm \cdot ft}{lbf \cdot s^2}}\right| = 49.625\,\frac{lbf}{ft^3}$$

The inside diameter of the pipe can be found from Appendix C.

$$D = 0.59892\ ft$$

Applying the length form of the conservation of energy equation between the inlet and the exit of the pipe,

$$\frac{P_1}{\gamma} + \frac{V_1^2}{2g} + z_1 = \frac{P_2}{\gamma} + \frac{V_2^2}{2g} + z_2 + l_f$$

The heptane is flowing in a steady state and the pipe diameter is constant. Therefore, the velocity of the heptane does not change as it passes through the pipe. In addition, there is no elevation change from inlet to outlet. Therefore, the above equation reduces to

$$P_1 - P_2 = \gamma l_f$$

Substituting the head loss due to friction into this equation gives

$$\Delta P = P_1 - P_2 = \gamma\left(f\frac{L}{D}\frac{V^2}{2g}\right)$$

The velocity of the heptane in the pipe can be found since the volumetric flow rate is given.

$$V = \frac{\dot{V}}{A} = \frac{4\dot{V}}{\pi D^2} = \frac{4\left(500\,\dfrac{\text{gal}}{\text{min}}\right)\left(\dfrac{2.228\times10^{-3}\ \text{ft}^3\cdot\text{min}}{\text{s}\cdot\text{gal}}\right)}{\pi(0.59892\ \text{ft})^2} = 3.954\,\frac{\text{ft}}{\text{s}}$$

The friction factor can be obtained from any of the correlations listed in Table 4.2. Using the Swamee–Jain correlation,

$$f = \frac{0.25}{\left[\log\left(\dfrac{\varepsilon/D}{3.7} + \dfrac{5.74}{Re^{0.9}}\right)\right]^2}$$

The Reynolds number and relative roughness need to be found.

$$Re = \frac{\rho V D}{\mu} = \frac{\left(42.625\,\dfrac{\text{lbm}}{\text{ft}^3}\right)\left(3.954\,\dfrac{\text{ft}}{\text{s}}\right)(0.59892\ \text{ft})}{0.0002886\,\dfrac{\text{lbm}}{\text{ft}\cdot\text{s}}} = 349{,}731$$

$$\overset{\text{(Table 4.3)}}{\frac{\varepsilon}{D}} = \frac{0.00015\ \text{ft}}{0.59892\ \text{ft}} = 2.505\times10^{-4}$$

Substituting these values into the Swamee–Jain correlation allows for the calculation of the friction factor.

$$f = \frac{0.25}{\left[\log\left(\dfrac{\varepsilon/D}{3.7} + \dfrac{5.74}{Re^{0.9}}\right)\right]^2} = \frac{0.25}{\left[\log\left(\dfrac{2.505\times10^{-4}}{3.7} + \dfrac{5.74}{349{,}731^{0.9}}\right)\right]^2} = 0.01645$$

Therefore, the pressure drop in the pipeline is

$$\Delta P = \gamma f \frac{L}{D}\frac{V^2}{2g}$$

$$\Delta P = \left(42.625\,\frac{\text{lbf}}{\text{ft}^3}\right)(0.01645)\frac{0.5\ \text{mile}\left(\dfrac{5280\ \text{ft}}{\text{mile}}\right)\left(3.954\,\dfrac{\text{ft}}{\text{s}}\right)^2}{0.59892\ \text{ft}\ \ 2\left(32.174\,\dfrac{\text{ft}}{\text{s}^2}\right)}\left(\frac{\text{ft}^2}{144\ \text{in.}^2}\right)$$

$$\Delta P = 5.22\,\frac{\text{lbf}}{\text{in.}^2} = 5.22\ \text{psi} \quad \leftarrow$$

From a design point of view, this type of calculation allows the engineer to start thinking about selecting a pump to overcome the calculated pressure drop. If the fluid flow rate is to be met, then the pump must be able to offset this entire pressure drop.

EXAMPLE 4.3: A Type 2 Problem

Liquid water is flowing through a 1 std type M copper tube at 30°C. The tube is 50 m long. The inlet to the tube is 20 m higher than the outlet. As the water passes through this tube, it experiences a pressure drop of 130 kPa. Determine the volumetric flow rate of water through the tube.

SOLUTION: A sketch of the tube is shown in Figure E4.3 along with the system boundary to be analyzed.

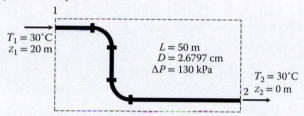

FIGURE E4.3

The properties of the water flowing through the tube can be found in Appendix B.1.

$$\rho = 995.6 \frac{kg}{m^3} \qquad \mu = 2.8705 \frac{kg}{m \cdot h} \left(\frac{h}{3600\ s} \right) = 0.0007974 \frac{kg}{m \cdot s}$$

The specific weight of the water can be calculated once the density is known.

$$\gamma = \rho g = \left(995.6 \frac{kg}{m^3} \right) \left(9.807 \frac{m}{s^2} \right) \left(\frac{N \cdot s^2}{kg \cdot m} \right) = 9763.8 \frac{N}{m^3} = 9.7638 \frac{kN}{m^3}$$

The inside diameter of the tube can be found in Appendix D.

$$D = 2.6797\ cm = 0.026797\ m$$

Applying the conservation of energy in length form to the system boundary results in

$$\frac{P_1}{\gamma} + z_1 = \frac{P_2}{\gamma} + z_2 + l_f$$

Rearranging this equation and substituting the head loss due to friction expression gives

$$\frac{(P_1 - P_2)}{\gamma} = (z_2 - z_1) + f \frac{L}{D} \frac{V^2}{2g}$$

The unknowns in this equation are the friction factor and the velocity. However, the friction factor is related to the velocity through the Reynolds number. Therefore, this solution is iterative. The iterative procedure used to solve a Type 2 problem is listed as follows:

1. Guess the friction factor (usually, $f = 0.02$ is a good starting value).
2. Calculate the velocity of the fluid from the conservation of energy equation.
3. Calculate the Reynolds number using the velocity found in step 2.
4. Use the Reynolds number and the relative roughness to determine the friction factor using the Moody diagram or one of the correlations found in Table 4.2.
5. Compare the friction factor from step 4 to the previous value of the friction factor.
6. If the two values of the friction factor do not match, use the newest value of the friction factor and repeat to step 2 until closure.

Fortunately, this iteration usually closes quickly (within 1 to 2 iterations). Following this process, start by guessing the friction factor.

$$f = 0.02$$

Solving the conservation of energy equation for the velocity,

$$V = \sqrt{\frac{2gD}{Lf}\left[\frac{\Delta P}{\gamma} - (z_2 - z_1)\right]}$$

With the guessed value of f, the velocity is

$$V = \sqrt{\frac{2\left(9.807\,\frac{m}{s^2}\right)(0.026797\ m)}{(50\ m)(0.02)}\left[\frac{130\ kPa\left(\frac{kN}{kPa \cdot m^2}\right)}{\left(9.7638\,\frac{kN}{m^3}\right)} - (0-20)m\right]}$$

$$V = 4.184\,\frac{m}{s}$$

Now, the Reynolds number corresponding to this velocity can be found.

$$Re = \frac{\rho VD}{\mu} = \frac{\left(995.6\,\frac{kg}{m^3}\right)\left(4.184\,\frac{m}{s}\right)(0.026797\ m)}{\left(0.0007974\,\frac{kg}{m \cdot s}\right)} = 140{,}011$$

The relative roughness of the copper pipe is

$$\frac{\varepsilon}{D} = \frac{\overset{\text{(Table 4.3 - drawn tubing)}}{0.00015\ cm}}{2.6797\ cm} = 5.598 \times 10^{-5}$$

Calculating the friction factor,

$$f = \frac{0.25}{\left[\log\left(\frac{\varepsilon/D}{3.7} + \frac{5.74}{Re^{0.9}}\right)\right]^2} = \frac{0.25}{\left[\log\left(\frac{5.598 \times 10^{-5}}{3.7} + \frac{5.74}{140,011^{0.9}}\right)\right]^2} = 0.01708$$

This value of the friction factor is not the same as the guessed value (0.02). Therefore, the process needs to be repeated with the new value of the friction factor. The results are as follows:

$$f = 0.01708$$

$$V = \sqrt{\frac{2gD}{Lf}\left[\frac{\Delta P}{\gamma} - (z_2 - z_1)\right]} = 4.528 \frac{m}{s}$$

$$Re = \frac{\rho VD}{\mu} = 151,507$$

$$\frac{\varepsilon}{D} = 5.598 \times 10^{-5}$$

$$\therefore f = 0.01683$$

The friction factors are close, but not close enough to declare convergence. Another iteration results in the following;

$$f = 0.01683$$

$$V = \sqrt{\frac{2gD}{Lf}\left[\frac{\Delta P}{\gamma} - (z_2 - z_1)\right]} = 4.564 \frac{m}{s}$$

$$Re = \frac{\rho VD}{\mu} = 152,628$$

$$\frac{\varepsilon}{D} = 5.598 \times 10^{-5}$$

$$\therefore f = 0.01681$$

These two friction factors are very close. Therefore, the iteration can be declared complete and the volumetric flow rate of the water can be determined.

$$\dot{V} = AV = \left(\frac{\pi D^2}{4}\right)V = \left[\frac{\pi\left(0.026797 \text{ m}^2\right)}{4}\right]\left(4.564 \frac{m}{s}\right)\left(\frac{1000 \text{ L}}{m^3}\right) = 2.574 \frac{L}{s} \quad \leftarrow$$

This example demonstrates how the iterative procedure can be accomplished with a calculator. This problem can be easily solved using an equation solver as well. The Engineering Equation Solver (EES) code to solve this problem is as follows:

```
"GIVEN: Water flowing through a copper tube"

"Fluid data"
  f$ = 'steam_iapws'
  T = 30[C]
  rho = density(f$,T = T,x = 0)
  gamma = rho*g
  mu = viscosity(f$,T = T,x = 0)

"Pipe Data from Appendix D and Table 4.3"
  L = 50[m]
  D = 0.026787[m]
  epsilon = 0.00015[cm]*convert(cm,m)

"Known system parameters"
  DELTAP = 130[kPa]
  z_1 = 20[m]
  z_2 = 0[m]

"Constants"
  g = 9.807[m/s^2]

"FIND: Volumetric flow through the tube"

"SOLUTION:"
"Conservation of Energy from 1 to 2,"
  DELTAP*convert(kPa,Pa)/gamma = (z_2 - z_1) &
      + f*(L/D)*(V^2/(2*g))

"Determination of the friction factor from the Moody diagram,"
  Re = rho*V*D/mu
  Vol_dot*convert(L/s,m^3/s) = A*V
  A = pi*D^2/4
  f = MoodyChart(Re,epsilon/D)
```

The calculated volumetric flow rate through this tube is 2.571 L/s. This compares very well with the value calculated using the iterative procedure.

The last line in the EES code accesses an internal routine (MoodyChart) that calculates the friction factor as a function of the Reynolds number and the relative roughness using the Churchill correlation shown in Table 4.2.

The calculation procedure for a Type 2 problem can be laborious if done by hand. However, the use of modern equation-solving software, such as EES, makes this problem very straightforward.

EXAMPLE 4.4: A Type 3 Problem

Liquid ethanol is flowing at a rate of 800 gpm in a horizontal, galvanized iron pipe over a distance of 2.5 miles. The average temperature of the ethanol in the pipeline is 60°F. The pump selected for this system can overcome a pressure drop of 45 psi. Determine the nominal diameter required if sch 40 pipe is to be used.

SOLUTION: A sketch of the pipeline and the system boundary to be analyzed is shown in Figure E4.4.

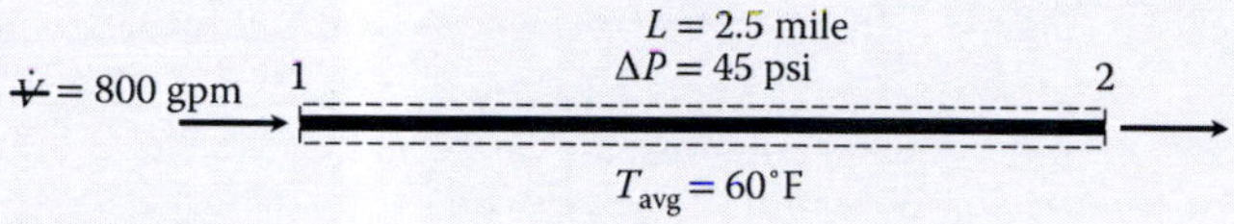

FIGURE E4.4

The ethanol properties can be found in Appendix B.1.

$$\rho = 49.524 \frac{lbm}{ft^3} \qquad \mu = 3.0650 \frac{lbm}{ft \cdot h} \left(\frac{h}{3600\ s} \right) = 0.0008514 \frac{lbm}{ft \cdot s}$$

Given the density, the specific weight of the ethanol can be found.

$$\gamma = \rho \frac{g}{g_c} = \left(49.524 \frac{lbm}{ft^3} \right) \left| \frac{32.174 \frac{ft}{s^2}}{32.174 \frac{lbm \cdot ft}{lbf \cdot s^2}} \right| = 49.524 \frac{lbf}{ft^3}$$

Applying the conservation of energy in length form to the system identified previously results in the following:

$$\frac{P_1}{\gamma} = \frac{P_2}{\gamma} + l_f$$

Rearranging this equation and substituting the expression for the head loss due to friction gives

$$\frac{P_1 - P_2}{\gamma} = \frac{\Delta P}{\gamma} = f \frac{L}{D} \frac{V^2}{2g}$$

The velocity of the fluid is related to the volumetric flow rate.

$$V = \frac{\dot{V}}{A} = \frac{4\dot{V}}{\pi D^2}$$

Substituting this expression into the conservation of energy equation and simplifying gives

$$\frac{\Delta P}{\gamma} = f\,\frac{8L\dot{V}^2}{\pi^2 D^5 g}$$

The unknowns in this equation are the friction factor and the diameter. However, the friction factor is related to the diameter through the Reynolds number. Therefore, this solution is iterative. The iterative procedure for a Type 3 problem is listed as follows:

1. Guess the friction factor (usually, $f = 0.02$ is a good starting value).
2. Calculate the diameter of the pipe from the conservation of energy equation.
3. Calculate the Reynolds number, using the diameter found in step 2.
4. Calculate the relative roughness, using the diameter found in step 2.
5. Use the Reynolds number and the relative roughness to determine the friction factor using one of the correlations found in Table 4.2.
6. Compare the friction factor from step 5 to the previous value of the friction factor.
7. If the two values of the friction factor do not match, use the newest value of the friction factor and repeat to step 2 until closure.

This iterative procedure usually closes quickly (within 1 to 2 iterations). Using this procedure, start by guessing the friction factor:

$$f = 0.02$$

Solve the conservation of energy equation for the diameter.

$$D = \left(\frac{8 f \gamma L\dot{V}^2}{\pi^2 \Delta P g}\right)^{1/5}$$

Using the guessed value for the friction factor, the diameter is calculated as follows:

$$D = \left[\frac{8(0.02)\left(49.524\,\frac{\text{lbf}}{\text{ft}^3}\right)(2.5\ \text{mile})\left(800\,\frac{\text{gal}}{\text{min}}\right)^2\left(\frac{5280\ \text{ft}}{\text{mile}}\right)\left(\frac{2.228\times10^{-3}\ \text{ft}^3\cdot\text{min}}{\text{s}\cdot\text{gal}}\right)^2}{\pi^2\left(45\,\frac{\text{lbf}}{\text{in.}^2}\right)\left(32.174\,\frac{\text{ft}}{\text{s}^2}\right)\left(\frac{144\ \text{in.}^2}{\text{ft}^2}\right)}\right]^{1/5}$$

$$D = 0.6944\ \text{ft}$$

The Reynolds number can be rewritten in terms of the volumetric flow rate as follows:

$$Re = \frac{\rho VD}{\mu} = \frac{\rho D}{\mu}\left(\frac{4\dot{V}}{\pi D^2}\right) = \frac{4\rho\dot{V}}{\pi D\mu}$$

$$Re = \frac{4\left(49.524\dfrac{\text{lbm}}{\text{ft}^3}\right)\left(800\dfrac{\text{gal}}{\text{min}}\right)\left(\dfrac{2.228\times10^{-3}\ \text{ft}^3\cdot\text{min}}{\text{s}\cdot\text{gal}}\right)}{\pi(0.6944\ \text{ft})\left(0.0008514\dfrac{\text{lbm}}{\text{ft}\cdot\text{s}}\right)} = 190{,}097$$

The relative roughness of the pipe is

$$\frac{\varepsilon}{D} = \frac{\overset{\text{(Table 4.3)}}{0.005\ \text{ft}}}{0.6944\ \text{ft}} = 0.0072$$

Using the calculated values of the Reynolds number and the relative roughness, the friction factor can be found using any of the correlations listed in Table 4.1.

$$f = \frac{0.25}{\left[\log\left(\dfrac{\varepsilon/D}{3.7} + \dfrac{5.74}{Re^{0.9}}\right)\right]^2} = \frac{0.25}{\left[\log\left(\dfrac{0.0072}{3.7} + \dfrac{5.74}{190{,}097^{0.9}}\right)\right]^2} = 0.03458$$

This value of the friction factor is not the same as the guessed value (0.02). Therefore, the process must be repeated with the new value of the friction factor. The results are as follows:

$$f = 0.03458$$

$$D = \left(\frac{8f\gamma L\dot{V}^2}{\pi^2\Delta Pg}\right)^{1/5} = 0.7748\ \text{ft}$$

$$Re = \frac{4\rho\dot{V}}{\pi D\mu} = 170{,}379$$

$$\frac{\varepsilon}{D} = 0.006453$$

$$\therefore f = 0.03351$$

Another iteration results in

$$f = 0.03351$$

$$D = \left(\frac{8f\gamma L\dot{V}^2}{\pi^2\Delta Pg}\right)^{1/5} = 0.7699\ \text{ft}$$

$$Re = \frac{4\rho\dot{V}}{\pi D\mu} = 171{,}453$$

$$\frac{\varepsilon}{D} = 0.006494$$

$$\therefore f = 0.03357$$

For iterative calculation purposes, these two friction factor values are very close, and the iteration can be considered closed. This last iteration indicates that the pipe diameter required is

$$D = \underline{0.7699 \text{ ft}} \quad \leftarrow$$

This value must now be translated to a nominal size. Since the pipe is schedule 40, Appendix C shows that this pipe diameter lies between 8-nom and 10-nom.

8-nom sch 40 (std): $D = 0.66508$ ft

10-nom sch 40 (std): $D = 0.83500$ ft

If the smaller pipe diameter is selected, the pressure drop would be higher than 45 psi for a flow rate of 800 gpm. This means that the pump selected for this application would not be able to deliver the required 800 gpm through an 8-nom sch 40 pipe. Therefore, for this application, select the 10-nom sch 40 pipe.

This problem can be easily solved with an equation solver. The EES code for this problem is as follows:

```
"GIVEN: Ethanol flowing in a galvanized pipe"

"Fluid flow rate"
   Vol_dot = 800[gpm]

"Pump pressure increase"
   DELTAP = 45[psi]

"Fluid property data"
   f$ = 'ethanol'
   T = 60[F]
   rho = density(f$,T = T,x = 0)
   gamma = rho*g/g_c
   mu = viscosity(f$,T = T,x = 0)*convert(lbm/ft-hr,lbm/ft-s)

"Pipe Data"
   L = 2.5[mile]*convert(mile,ft)
   epsilon = 0.005[ft]      "Table 4.3"

"Constants"
   g = 32.174[ft/s^2]
   g_c = 32.174[lbm-ft-/lbf-s^2]
"FIND: The pipe diameter required"

"SOLUTION:"
"Conservation of Energy Equation for the system identified"
   DELTAP*convert(psi,lbf/ft^2)/gamma = f*(L/D)*(V^2/(2*g))
```

```
"Determination of the friction factor from the Churchill
    Correlation"
    Re = rho*V*D/mu
    Vol_dot*convert(gpm,ft^3/s) = A*V
    A = pi*D^2/4
    f = MoodyChart(Re,epsilon/D)
```

The solution to this set of equations reveals that $D = 0.7701$ ft, which is very nearly the same answer that was obtained after three iterations. The resulting pipe selection is 10-nom sch 40.

Examples 4.3 and 4.4 show the versatility of an equation solver such as EES to solve iterative pipe flow problems. When equations are entered into the Equation Window of EES, normal structured programming rules do not apply. The equations can be entered in free format. Therefore, there is no need for algebraic rearrangement of equations as was done in the manual solutions shown in Examples 4.3 and 4.4. This is another advantage of a free-format equation solver: the governing equations can be written without need for algebraic rearrangement. This tends to make the code much more understandable.

4.4 Valves and Fittings

Valves and fittings are an integral part of a thermal energy system involving fluid transport. Valves can be used for flow control or to completely shut down the flow. Fittings are used to connect straight runs of pipe together. Each of these components introduces additional head loss due to friction. The head loss due to valves and fittings is often called a *minor loss*. When determining total head loss in a fluid network, it is important to include these effects. Much of the information presented in this section is from Crane Technical Paper No. 410 (The Crane Company 2013). Permission has been granted by the Crane Company to include this information here. This publication has become an industry standard in the area of pipes, valves, and fittings.

To accommodate the friction losses in valves and fittings, the conservation of energy equation for an incompressible fluid flow can be modified to include a minor-loss term.

$$\frac{P_1}{\gamma} + \frac{V_1^2}{2g} + z_1 = \frac{P_2}{\gamma} + \frac{V_2^2}{2g} + z_2 + l_f + l_m \tag{4.14}$$

The minor loss, l_m, is expressed as follows:

$$l_m = \sum_i \left(K_i \frac{V_i^2}{2g} \right) \qquad (4.15)$$

In this equation, the K_i values are known as the *resistance* or *loss* coefficients for the valve or fitting being considered. The velocity head in Equation 4.15 is the velocity head associated with the valve or fitting. Combining Equation 4.15 with Equations 4.10 and 4.14 gives the conservation of energy equation between two points in a pipe network containing fittings and valves.

$$\frac{P_1}{\gamma} + \frac{V_1^2}{2g} + z_1 = \frac{P_2}{\gamma} + \frac{V_2^2}{2g} + z_2 + \sum_i \left(f_i \frac{L_i}{D_i} \frac{V_i^2}{2g} \right) + \sum_i \left(K_i \frac{V_i^2}{2g} \right) \qquad (4.16)$$

Equation 4.16 is often known as the *Modified Bernoulli equation*. Notice that the head loss term due to pipe friction is written to accommodate several different diameter pipes in the system between points 1 and 2. Values of K are determined experimentally and tabulated for easy use. Appendix E shows how K values are calculated for many common valves and fittings.

4.4.1 Valves

Valves are an important part of any fluid transport system. Table 4.5 shows the standard drawing symbols and various uses of common valves used in industrial applications. A more comprehensive guide for valve selection can be found in the *Valve Selection Handbook* (Smith and Zappe 2004).

TABLE 4.5

Common Uses for Valves in Industrial Applications

Type of Valve	Drawing Symbol	Recommended Uses
Gate		Fully open or fully closed service, minimal line pressure drop, infrequent operation
Globe		Flow control (throttling), frequent operation; service with some line resistance
Check		Control the direction of flow and quick automatic reaction to flow change; should be used in conjunction with a gate valve
Ball		On or off service, flow control (throttling), when positive shut-off is necessary, when low profile is necessary
Butterfly		When positive shut-off is necessary, fully open or fully closed service

The great number of valve designs available to an engineer makes it difficult to thoroughly categorize them. However, it is possible to think of valves as either *low resistance* or *high resistance*. Low-resistance valves have a straight-through flow path. Valves of this type include gate valves, ball valves, plug valves, and butterfly valves. These types of valves are usually used for fully open or fully closed service. High-resistance valves are used to control the flow of the fluid (e.g., a globe valve) or change the flow direction of a fluid (e.g., an angle valve).

The American Society of Mechanical Engineers (ASME) has developed a valve class system. The suggested pairing of valve classes to pipe schedules is listed in Table 4.6. The determination of the loss coefficients in Appendix E was developed using these pairings.

The following example shows how to use the data in Appendix E to determine the pressure drop through a valve.

TABLE 4.6

Valve Class Pairings with Pipe Schedules

Valve Class	Paired Pipe
Class 300 and lower	Schedule 40
Class 400 and 600	Schedule 80
Class 900	Schedule 120
Class 1500	Schedule 160
Class 2500 (½ to 6 in.)	XXS
Class 2500 (8 in. and up)	Schedule 160

EXAMPLE 4.5

Water at 60°F is flowing at 250 gpm through a valve connected to a 3-nom sch 40 commercial steel pipe. Determine the pressure drop through the valve if it is a fully opened (a) globe valve and (b) gate valve.

SOLUTION: The pressure drop through the valve can be determined from the head loss through the valve. Considering a system boundary that surrounds only the valve, the conservation of energy equation is

$$\frac{\Delta P_{valve}}{\gamma} = K\frac{V^2}{2g}$$

The specific weight of the water can be found from the density. Using the fluid property data from Appendix B.1,

$$\gamma = \rho\frac{g}{g_c} = 62.363\frac{lbf}{ft^3}$$

The velocity of the water can be found from the volumetric flow rate and the inside diameter of the pipe. Using the pipe data from Appendix C,

$$V = \frac{\dot{V}}{A} = \frac{4\dot{V}}{\pi D^2} = \frac{4\left(250\,\frac{\text{gal}}{\text{min}}\right)\left(2.228 \times 10^{-3}\,\frac{\text{ft}^3 \cdot \text{min}}{\text{s} \cdot \text{gal}}\right)}{\pi\,(0.25567\ \text{ft})^2} = 10.85\,\frac{\text{ft}}{\text{s}}$$

From Appendix E, the K value for a globe valve is

$$K_{\text{globe}} = 340\,f_T$$

The value of the friction factor, f_T, in this expression is calculated from a modified Colebrook correlation.

$$f_T = 0.25\left[\log\left(\frac{\varepsilon/D}{3.7}\right)\right]^{-2}$$

For clean commercial steel pipes, this value is also tabulated on the first page of Appendix E. For a 3-nom commercial steel valve,

$$f_T = 0.017$$

Therefore, the loss coefficient for the globe valve is,

$$K_{\text{globe}} = 340\,f_T = 340(0.017) = 5.78$$

Now, the pressure drop through the globe valve can be found.

$$\Delta P_{\text{globe}} = \gamma K_{\text{globe}}\frac{V^2}{2g} = \left(62.363\,\frac{\text{lbf}}{\text{ft}^3}\right)(5.78)\frac{\left(10.85\,\frac{\text{ft}}{\text{s}}\right)^2}{2\left(32.174\,\frac{\text{ft}}{\text{s}^2}\right)}\left(\frac{\text{ft}^2}{144\ \text{in.}^2}\right)$$

$$= 4.58\,\frac{\text{lbf}}{\text{in.}^2} = 4.58\ \text{psi} \quad \leftarrow$$

For a normal gate valve, the loss coefficient from Appendix E is

$$K_{\text{gate}} = 8\,f_T$$

Therefore, the pressure drop through a 3-nom gate valve is

$$\Delta P_{\text{gate}} = \Delta P_{\text{globe}}\frac{K_{\text{gate}}}{K_{\text{globe}}} = (4.58\ \text{psi})\left(\frac{8\,f_T}{340\,f_T}\right) = 0.11\ \text{psi} \quad \leftarrow$$

This example shows that the globe valve has a significantly higher pressure drop compared to the gate valve. According to Table 4.5, a globe valve is used for flow control, whereas a gate valve is used for on or off service. Therefore, if flow control is desired in the system, there is a pressure-drop penalty to pay. However, if the valve is only being used for on or off purposes, there is only a small pressure drop.

It should also be pointed out that in this example, both valves are fully opened. If the globe valve is used for flow control, then the pressure drop will be higher than the 4.58 psi calculated here.

Swing check valve   Lift check valve

FIGURE 4.3
Swing and lift check valves. (From The Crane company, *Flow of Fluids Through Valves, Fittings and Pipe*, Stamford, Connecticut, 2013. Reprinted with permission.)

4.4.1.1 Check Valves

Check valves are used in fluid transport systems to provide flow in only one direction in the pipe network. If the fluid flow is reversed from its normal operation, the check valve will close, thus preventing backflow. This is particularly useful when a system is used intermittently and backflow is not desired. A good example of this is pumping fluid to an elevated tank. A check valve in the discharge line of the pump will stop backflow from occurring when the pump is shut down. Appendix E shows how the loss coefficient for several different types of check valves can be determined.

Two common types of check valves, the swing check valve and lift check valve, are shown in Figure 4.3. A check valve stops backflow by virtue of a seated disc inside the valve. When the flow is in the proper direction, the disc is moved off its seat.

If the velocity of the flow is too small, the disc will lift, but it will not reach its stop. This results in noisy operation and premature wear of the moving parts in the valve. To alleviate this, experiments have been conducted to determine the minimum velocity required to completely move the disc to its stop, thereby eliminating noisy operation and rapid wear. Appendix E shows empirical expression developed based on these experimental results. The general form of this empiricism is

$$V_{min} = C\sqrt{v} \qquad (4.17)$$

In this equation, the velocity (V) is in ft/s and the specific volume (v) is in ft³/lbm. The constant, C, is given in the expressions found in Appendix E. Notice that in Appendix E, the specific volume is given by the symbol $\bar{V}$. The following example shows how to specify the proper-sized check valve.

EXAMPLE 4.6

A globe-type lift check valve is required in a 2½-nom sch 40 pipe transporting 70°F benzene at 40 gpm. Specify the size of the lift check valve required.

SOLUTION: The velocity of the benzene flowing in the pipe can be determined knowing the volumetric flow rate and the pipe diameter.

$$V = \frac{\dot{V}}{A} = \frac{4\dot{V}}{\pi D^2} = \frac{4(40\,\text{gal}\,/\,\text{min})\big|(2.228\times 10^{-3}\,\text{ft}^3 \cdot \text{min}\,/\,\text{s}\cdot\text{gal})}{\pi\underbrace{(0.20575\,\text{ft})^2}_{\text{Appendix C}}} = 2.68\,\text{ft}\,/\,\text{s}$$

From Appendix E, the minimum velocity required to fully lift the disc is

$$V_{min}\,[\text{ft/s}] = 40\beta^2\sqrt{v\,[\text{ft}^3/\text{lbm}]} = \frac{40\beta^2}{\sqrt{\rho\,[\text{lbm/ft}^3]}}$$

$$\big[\text{Note: } \bar{V} \text{ in Appendix E is the } \textit{specific volume}\big]$$

For a 2½-in. check valve, $\beta = 1$. Therefore,

$$V_{min,2.5\text{-in.}} = \frac{40\beta^2}{\sqrt{\rho}} = \frac{40}{\underbrace{\sqrt{54.786}}_{\text{Appendix B}}} = 5.40\,\frac{\text{ft}}{\text{s}}$$

The pipe velocity is far too low to meet this minimum requirement. A 2½-in. valve will be too large. Stepping down in size to a 2-in. valve,

$$\beta = \frac{D_{valve}}{D_{pipe}} = \frac{\overbrace{0.17225\,\text{ft}}^{\text{Appendix C}}}{0.20575\,\text{ft}} = 0.837$$

$$V_{min,2\text{-in.}} = \frac{40\beta^2}{\sqrt{\rho}} = \frac{40(0.837)^2}{\sqrt{54.786}} = 3.78\,\frac{\text{ft}}{\text{s}}$$

This valve is still too large! For a 1½-in. valve,

$$\beta = \frac{D_{valve}}{D_{pipe}} = \overset{\text{Appendix C}}{\frac{0.13417 \text{ ft}}{0.20575 \text{ ft}}} = 0.652$$

$$V_{min, 2\text{-in.}} = \frac{40\beta^2}{\sqrt{\rho}} = \frac{40(0.652)^2}{\sqrt{54.786}} = 2.29 \frac{\text{ft}}{\text{s}}$$

The velocity in the 2½-in. pipe is sufficient to fully lift the disc in a 1½-in. valve. Therefore, select the 1½-in. check valve and specify reducers to connect the valve to the pipeline.

Most problems encountered with check valves results from oversizing. It may seem feasible to select a check valve that has the same nominal diameter as the pipe. However, as shown in Example 4.6, this may result in a situation where the valve disc is not fully lifted. As pointed out earlier, this leads to noisy operation and premature wear of the moving parts.

In Example 4.6, the check valve is smaller than the pipe size. This means that reducers are needed to connect it to the pipe system. For short-length concentric reducers, the resistance coefficients can be determined from Appendix E, or estimated as follows:

$$K_{d,in} = 0.5\left(1 - \beta^2\right)^2 \tag{4.18}$$

$$K_{d,out} = \left(1 - \beta^2\right)^2 \tag{4.19}$$

In these equations, β is the ratio of the smaller diameter (d) to the larger diameter (D). The subscripts "in" and "out" refer to the flow *into* and *out of* the check valve, respectively. The head losses calculated in Equations 4.18 and 4.19 give the resistance coefficient associated with the smaller diameter (d) in the fitting. To convert these resistances to be consistent with the larger pipe diameter (D), the following expression is used:

$$K_D = \frac{K_d}{\beta^4} \tag{4.20}$$

Using Equation 4.20 allows for the calculation of the head loss due to the reducer using the velocity head in the larger pipe. This is convenient because it eliminates the need to keep track of two velocity heads in the friction calculations.

4.4.1.2 Control Valves

Control valves are used to adjust the flow rate of a fluid. Oftentimes it is convenient to express the flow rate and resulting head loss from a control valve in terms of a *flow coefficient*, C_v. The flow coefficient of a control valve is defined as the volumetric flow of water at 60°F, in gallons per minute, at a pressure drop of 1 psi across the valve. The flow coefficient is used in the United States. Outside the United States, the *flow factor*, K_v, is normally used. The flow factor is defined as the volumetric flow of water at 20°C, in cubic meters per hour, at a pressure drop of 1 bar across the valve. Crane Technical Paper No. 410 (The Crane Company 2013) has an extensive section on control valve sizing and selection.

4.4.2 Fittings

Fittings may be classified as *branching, reducing, expanding,* or *deflecting.* Branching fittings are fittings that split or collect flows such as tees or wyes. Reducing or expanding fittings change the cross-sectional area of the flow such as reducers or bushings. Deflecting fittings change the direction of the flow such as elbows or return bends.

Appendix E gives expressions for the determination of the loss coefficient for several reducing, expanding, and deflecting fittings. The common branch fittings, tees and wyes, require a more detailed treatment. Tees and wyes are used to either split flow (diverging) or combine flow (converging), as shown in Figure 4.4.

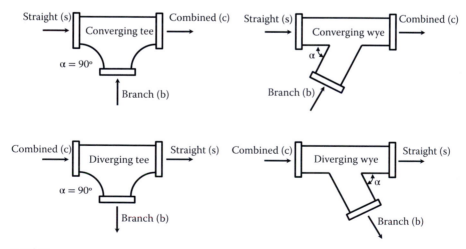

FIGURE 4.4
Converging and diverging tee and wye fittings.

4.4.2.1 Loss Coefficients for Tees and Wyes

There are two loss coefficients for tees and wyes. One of these loss coefficients is assigned to the flow that runs straight through the fitting, K_{run}. The other loss coefficient is for the flow that runs through the branch, K_{branch}. The value of the loss coefficient depends on the geometry of the fitting and the volumetric flow rate through each part of the fitting. Empirical expressions have been developed for these loss coefficients.

For converging flow fittings, the following expression has been developed:

$$K_{run} \text{ or } K_{branch} = C\left[1 + D\left(\frac{y_b}{\beta_b^2}\right)^2 - E(1 - y_b)^2 - F\left(\frac{y_b}{\beta_b}\right)^2\right] \qquad (4.21)$$

In this equation, C, D, E, and F are defined in Table 4.7A, y_b is the fraction of the volumetric flow passing through the branch of the fitting, and β_b is the branch diameter ratio.

$$y_b = \frac{\dot{V}_b}{\dot{V}_c} \qquad (4.22)$$

$$\beta_b = \frac{D_b}{D_c} \qquad (4.23)$$

In Equations 4.22 and 4.23, the subscript "c" is used to designate the combined flow, defined in Figure 4.4. To calculate the head loss across a branch or run, use the velocity in the combined leg of the fitting.

For diverging fittings, the expressions used to calculate the loss coefficients for the branch and run are as follows:

$$K_{branch} = G\left[1 + H\left(\frac{y_b}{\beta_b^2}\right)^2 - J\left(\frac{y_b}{\beta_b^2}\right)\cos\alpha\right] \qquad (4.24)$$

$$K_{run} = My_b^2 \qquad (4.25)$$

The values of G, H, J, and M are summarized in Tables 4.8A and 4.9. When calculating the head loss from a diverging fitting, the velocity in the combined leg is used.

TABLE 4.7A

Loss Coefficient Constants for Converging Fittings–Equation 4.21

	K_{branch}				K_{run}			
α	C	D	E	F	C	D	E	F
30°	See Table 4.7B	1	2	1.74	1	0	1	1.74
45°	See Table 4.7B	1	2	1.41	1	0	1	1.41
60°	See Table 4.7B	1	2	1	1	0	1	1
90°	See Table 4.7B	1	2	0	$K_{run} = 1.55y_b - y_b^2$			

Source: The Crane Company, *Flow of Fluids Through Valves, Fittings and Pipe*, Stamford, Connecticut, 2013. With permission.

TABLE 4.7B

Values for C in Equation 4.21 for K_{branch}

	y_b	
β^2_{branch}	≤ 0.35	>0.35
≤ 0.35	$C=1$	$C=1$
>0.35	$C=0.9(1-y_b)$	$C=0.55$

Source: The Crane Company, *Flow of Fluids Through Valves, Fittings and Pipe*, Stamford, Connecticut, 2013. With permission.

TABLE 4.8A

Constants for Branch Loss Coefficients in Diverging Fittings—Equation 4.24

α	G	H	J
0° to 60°	See Table 4.8B	1	2
90° with $\beta_{branch} \leq 2/3$	1	1	2
90° with $\beta_{branch} = 1$	$1+0.3y^2_b$	0.3	0

Source: The Crane Company, *Flow of Fluids Through Valves, Fittings and Pipe*, Stamford, Connecticut, 2013. With permission.

TABLE 4.8B

Values for G in Equation 4.24 for K_{branch}

β^2_{branch}		
≤ 0.35	$G=1.1-0.7y_b$ for $y_b \leq 0.6$	$G=0.85$ for $y_b > 0.6$
>0.35	$G=1.0-0.6y_b$ for $y_b \leq 0.4$	$G=0.60$ for $y_b > 0.4$

Source: The Crane Company, *Flow of Fluids Through Valves, Fittings and Pipe*, Stamford, Connecticut, 2013. With permission.

TABLE 4.9

Values for M for the Run in Diverging Fittings—Equation 4.25

	y_b	
β^2_{branch}	≤ 0.5	>0.5
≤ 0.40	$M=0.4$	$M=0.4$
>0.40	$M=2(2y_b-1)$	$M=0.3(2y_b-1)$

Source: The Crane Company, *Flow of Fluids Through Valves, Fittings and Pipe*, Stamford, Connecticut, 2013. With permission.

In a diverging or converging fitting, it is possible for the head loss in either the branch or the run to be negative. If this occurs, there is actually a head *gain* through that portion of the fitting. This can occur when the slower moving fluid in the fitting is accelerated to the velocity of the combined flow. Even though there may be head gain in a branch or run, the gain is offset by a higher head loss through the other leg of the fitting resulting in an overall head loss for the entire fitting.

EXAMPLE 4.7

A 4-nom class-300 tee fitting with equal leg diameters has 300 gpm of water at 60°F flowing into the straight run and 100 gpm of 60°F water converging in from the 90° branch. Determine the loss coefficients for the branch and run and the head loss across each flow path.

SOLUTION: According to Table 4.6, a class-300 fitting should be paired with a schedule 40 pipe. Therefore, the inside diameter of the pipe connected to the fitting can be found in Appendix C.

$$D = 0.33550 \text{ ft}$$

The loss coefficients for the branch and run of the tee can be found using Equation 4.21 and Table 4.7A. Since the fitting has equal leg diameters,

$$\beta_{\text{branch}} = 1$$

The temperature of the water is the same for the straight and branch runs. Therefore, the volumetric flow rates are additive. The combined flow through the fitting is

$$\dot{V_c} = \dot{V_b} + \dot{V_s} = (100 + 300)\text{gpm} = 400 \text{ gpm}$$

The volumetric flow fraction through the branch is

$$y_b = \frac{\dot{V_b}}{\dot{V_c}} = \frac{100 \text{ gpm}}{400 \text{ gpm}} = 0.25$$

Using Table 4.7A with the parameters calculated above, the values of the branch coefficients, C, D, E, and F, can be found.

$$C = 0.9(1 - y_b) = 0.9(1 - 0.25) = 0.675$$
$$D = 1$$
$$E = 2$$
$$F = 0$$

Using these values, the loss coefficient for the branch can be found.

$$K_{branch} = C\left[1 + D\left(\frac{y_b}{\beta_b^2}\right)^2 - E(1 - y_b)^2 - F\left(\frac{y_b}{\beta_b}\right)^2\right]$$

$$K_{branch} = (0.675)\left[1 + 1\left(\frac{0.25}{1}\right)^2 - 2(1 - 0.25)^2 - 0\right] = \underline{-0.0422} \quad \leftarrow$$

This indicates that there is actually a head *gain* in the branch because of the acceleration of the fluid to the velocity of the combined flow. The head loss associated with the branch flow can be found once the velocity in the combined leg is known:

$$V_c = \frac{4\dot{V}_c}{\pi D^2} = \frac{4\left(400\,\frac{gal}{min}\right)\left(2.228 \times 10^{-3}\,\frac{ft^3 \cdot min}{s \cdot gal}\right)}{\pi(0.33550\,ft)^2} = 10.08\,\frac{ft}{s}$$

Therefore, the head loss associated with the branch of the tee is

$$l_{branch} = K_{branch}\frac{V_c^2}{2g} = (-0.0422)\frac{\left(10.08\,\frac{ft}{s}\right)^2}{2\left(32.174\,\frac{ft}{s^2}\right)} = \underline{-0.067\,ft} \quad \leftarrow$$

According to Table 4.7A, the loss coefficient for the straight run part of the tee is

$$K_{run} = 1.55y_b - y_b^2 = 1.55(0.25) - (0.25)^2 = \underline{0.325} \quad \leftarrow$$

Therefore, the head loss associated with the run of the tee is

$$l_{run} = K_{run}\frac{V_c^2}{2g} = (0.325)\frac{\left(10.08\,\frac{ft}{s}\right)^2}{2\left(32.174\,\frac{ft}{s^2}\right)} = \underline{0.51\,ft} \quad \leftarrow$$

As expected, there is a net head loss associated with the fitting even though there is a gain in the branch fluid.

4.5 Design and Analysis of Pipe Networks

Engineering drawings of a pipe network tell the engineer many things in addition to the layout of the network. The drawings also indicate the type of joints to be used. Figure 4.5 shows how several common joints used in putting pipe networks together are indicated on an engineering drawing.

In the design and analysis of fluid networks, the same three problem types that were discussed in Section 4.3 exist.

Type 1: Knowing the construction of the pipe network (i.e., the layout and all dimensions) and the required fluid flow rate, determine the pressure drop in the pipe network. This type of calculation is important in the selection of pumps to move the fluid.

Type 2: Knowing the construction of the pipe network and the pressure drop that a pump can overcome, determine the flow rate of the fluid through the network. This type of calculation allows the engineer to determine if a given pump has the capability to move the required amount of fluid through the network.

Type 3: Knowing the volumetric flow rate requirement and the pressure drop that the pump can overcome, determine the proper pipe diameter for the network.

In Section 4.3, these three types of problems have been applied to straight pipe lengths. The problems are no different in a complex pipe network. In addition to straight runs of pipe, there are valves and fittings as well.

| Threaded | Flanged | Welded | Soldered | Bell & spigot |

FIGURE 4.5
Symbols used to indicate joining methods in pipe network drawings.

EXAMPLE 4.8

Liquid cyclohexane at 35°C is pumped at a volumetric flow rate of 10 L/s through a pipe network to a large storage tank as shown in Figure E4.8. Determine the pressure drop (kPa) from the pump discharge (A) to the pipe exit (B).

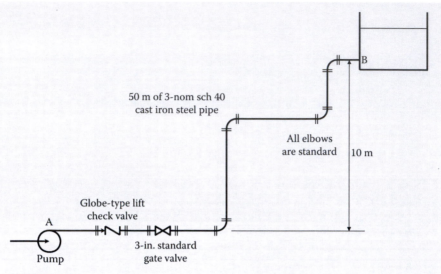

FIGURE E4.8

SOLUTION: The pressure drop from A to B is equal to the pressure loss in the pipe network including the head loss due to friction and the minor losses due to the valves and fittings. Applying the conservation of energy equation between points A and B results in

$$\frac{P_A}{\gamma} + z_A = \frac{P_B}{\gamma} + z_B + \left(f\frac{L}{D} + \sum K \right)\frac{V^2}{2g}$$

Rearranging this equation to solve for the pressure drop that the pump must overcome results in

$$\frac{P_A - P_B}{\gamma} = \frac{\Delta P}{\gamma} = (z_B - z_A) + \left(f\frac{L}{D} + \sum K \right)\frac{V^2}{2g}$$

The properties of the cyclohexane can be found from Appendix B.1

$$\gamma = \rho g = \left(764.30\,\frac{kg}{m^3} \right)\left(9.807\,\frac{m}{s^2} \right)\left(\frac{N \cdot s^2}{kg \cdot m} \right) = 7495.5\,\frac{N}{m^3}$$

$$\mu = 2.7007\,\frac{kg}{m \cdot h}\left(\frac{h}{3600\ s} \right) = 0.0007502\,\frac{kg}{m \cdot s}$$

The inside diameter of the pipe can be found in Appendix C and the roughness of the pipe can be found in Table 4.3.

$$D = 7.7927\ cm = 0.077927\ m$$

$$\varepsilon = 0.26\ cm = 0.0026\ m$$

From the drawing, the elevations of points A and B as well as the length of straight pipe in the network are given as follows:

$$z_A = 0 \text{ m} \qquad z_B = 10 \text{ m} \qquad L = 50 \text{ m}$$

To determine the friction factor, the Reynolds number and relative roughness are required. To determine the Reynolds number, the velocity of the cyclohexane in the pipe is needed. This velocity can be found by,

$$V = \frac{\dot{V}}{A} = \frac{4\dot{V}}{\pi D^2} = \frac{4\left(10\frac{L}{s}\right)\left(\frac{m^3}{1000 \text{ L}}\right)}{\pi (0.077927 \text{ m})^2} = 2.097 \frac{m}{s}$$

The Reynolds number can now be found.

$$Re = \frac{\rho V D}{\mu} = \frac{\left(764.30 \frac{kg}{m^3}\right)\left(2.097 \frac{m}{s}\right)(0.077927 \text{ m})}{0.0007502 \frac{kg}{m \cdot s}} = 166,461$$

The relative roughness of the pipe can also be found.

$$\frac{\varepsilon}{D} = \frac{0.0026 \text{ m}}{0.077927 \text{ m}} = 0.03336$$

Using the Reynolds number and the relative roughness, the friction factor can be found using any of the correlations in Table 4.2. Using the Swamee–Jain correlation,

$$f = \frac{0.25}{\left[\log\left(\frac{\varepsilon/D}{3.7} + \frac{5.74}{Re^{0.9}}\right)\right]^2} = \frac{0.25}{\left[\log\left(\frac{0.03336}{3.7} + \frac{5.74}{166,461^{0.9}}\right)\right]^2} = 0.06011$$

The only task left is to determine the total effect of the minor losses due to the valves and fittings. The check valve size is not given. However, the size can be specified by ensuring that the cyclohexane velocity in the pipe completely lifts the disc from its seat inside the valve. The minimum velocity required to fully lift the disc (Appendix E) is

$$V_{min} = \frac{40 \beta^2}{\sqrt{\rho}}$$

For a 3-in. check valve, $\beta = 1$. Therefore,

$$V_{min,3\text{-in.}} = \frac{40\beta^2}{\sqrt{\rho}} = \underbrace{\frac{40}{\sqrt{47.714}}}_{\text{Appendix B.1}} = 5.79\frac{\text{ft}}{\text{s}}\left|\left(\frac{0.3048 \text{ m}}{\text{ft}}\right)\right| = 1.77\frac{\text{m}}{\text{s}}$$

Notice that this calculation had to be done in English units before converting the velocity to the SI system. This is because expression used in Appendix E to calculate the minimum velocity is empirical. It was developed to use density in lbm/ft³ and the velocity in ft/s. This result indicates that the pipe velocity is higher than the minimum velocity required to completely lift the disc in the check valve. Therefore, a 3-in. check valve will work in this application, and no reducers are needed to connect the check valve to the pipe. Therefore, the loss coefficients required are

$$\sum K = K_{check} + K_{gate} + 4K_{elbow} + K_{exit}$$

The loss coefficients for the individual valves and fittings can be found in Appendix E. For the check valve,

$$K_{check} = 600 f_T$$

The pipe is made of cast iron. Therefore, the value of f_T is computed using the modified Colebrook correlation shown in Appendix E (the table of f_T values in Appendix E, is for commercial steel pipe).

$$f_T = 0.25\left[\log\left(\frac{\varepsilon/D}{3.7}\right)\right]^{-2} = 0.25\left[\log\left(\frac{0.03336}{3.7}\right)\right]^{-2} = 0.05978$$

Using the calculated value of f_T, the loss coefficient for the check valve can be found.

$$K_{check} = 600 f_T = 600(0.05978) = 35.87$$

The loss coefficients for the gate valve, elbows, and pipe exit can be determined from the appropriate loss coefficient expressions and the calculated value of f_T.

$$K_{gate} = 8 f_T = 8(0.05978) = 0.48$$
$$K_{elbow} = 30 f_T = 30(0.05978) = 1.79$$
$$K_{exit} = 1.00$$

These values can be used to determine the total loss coefficient for the minor losses in the system.

$$\sum K = 35.87 + 0.48 + 4(1.79) + 1.00 = 44.51$$

Now, everything is known except the pressure drop from point A to B. Using the conservation of energy equation developed above,

$$\Delta P = \gamma \left[(z_B - z_A) + \left(f \frac{L}{D} + \sum K \right) \frac{V^2}{2g} \right]$$

$$\Delta P = \left(7495.5 \frac{N}{m^3} \right) \left\{ (10-0)m + \left[(0.06011) \left(\frac{50 \text{ m}}{0.077927 \text{ m}} \right) + 44.51 \right] \frac{\left(2.097 \frac{m}{s} \right)^2}{2 \left(9.807 \frac{m}{s^2} \right)} \right\}$$

$$\Delta P = 214,566 \frac{N}{m^2} \left| \left(\frac{Pa \cdot m^2}{N} \right) \left(\frac{kPa}{1000 \text{ Pa}} \right) = \underline{214.6 \text{ kPa}} \quad \leftarrow$$

It is interesting to determine the magnitude of the head loss due to the straight pipe compared to the head loss due to the valves and fittings. For the system analyzed in this problem,

$$l_f = f \frac{L}{D} \frac{V^2}{2g} = 8.64 \text{ m}$$

$$l_m = \left(\sum K \right) \frac{V^2}{2g} = 9.98 \text{ m}$$

In this case, the minor losses are larger than the head loss due to friction in the straight pipe! This is due to the check valve. Notice that its K value is much larger than the other gate valve, elbows, and exit.

Often pipe or tube networks have parallel runs. Figure 4.6 shows a sketch of a pipe system with two parallel flow paths, labeled 1 and 2. The label "c" indicates combined flow. To analyze a system like this, a system boundary is sketched around the individual pipes in the network. Then the conservation of mass and conservation of energy equations are used. In Figure 4.6, if a system boundary was sketched around pipe 1 from A to B and a second system boundary around pipe 2 from A to B, it must be true that the pressure drop from A to B is the *same* through each parallel leg.

$$(P_A - P_B)_1 = (P_A - P_B)_2 \tag{4.26}$$

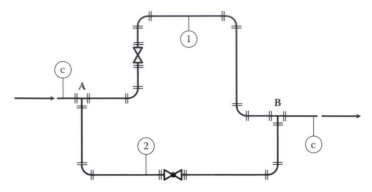

FIGURE 4.6
A parallel pipe network (plan view).

If the diameters of each leg are constant, the conservation of energy between points A and B for each leg of the network can be written as follows:

$$\frac{P_A - P_B}{\gamma} = (z_B - z_A) + \left(f_1 \frac{L_1}{D_1} + \sum K_1 \right) \frac{V_1^2}{2g} + \left(K_{run,A} + K_{run,B} \right) \frac{V_c^2}{2g} \qquad (4.27)$$

$$\frac{P_A - P_B}{\gamma} = (z_B - z_A) + \left(f_2 \frac{L_2}{D_2} + \sum K_2 \right) \frac{V_2^2}{2g} + \left(K_{branch,A} + K_{branch,B} \right) \frac{V_c^2}{2g} \qquad (4.28)$$

Notice that the K values for the tees are separated out from the other fittings. This is because the minor losses of the branch and straight runs of the tee are associated with the combined velocity of the flow entering (or leaving) the tee. In Equations 4.27 and 4.28, V_c is the combined flow velocity. Since the pressure drop through each parallel leg is the same,

$$\begin{aligned}
&\left(f_1 \frac{L_1}{D_1} + \sum K_1 \right) \frac{V_1^2}{2g} + \left(K_{run,A} + K_{run,B} \right) \frac{V_c^2}{2g} \\
&= \left(f_2 \frac{L_2}{D_2} + \sum K_2 \right) \frac{V_2^2}{2g} + \left(K_{branch,A} + K_{branch,B} \right) \frac{V_c^2}{2g}
\end{aligned} \qquad (4.29)$$

The following example illustrates how to apply these equations to a network with parallel flow paths.

EXAMPLE 4.9

Figure E4.9 shows a parallel flow copper tube network. Water at 70°F is flowing through the network such that the pressure drop from A to B is 0.2 psi. Determine the volumetric flow rate (gpm) through each leg of the parallel network and also determine the total volumetric flow (gpm) through the network.

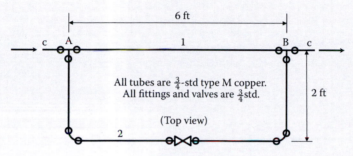

FIGURE E4.9

SOLUTION: The following data pertinent to the system can be found immediately:

$$\gamma = \rho \frac{g}{g_c} = \left(62.298 \frac{\text{lbm}}{\text{ft}^3}\right) \frac{g}{g_c} = 62.298 \frac{\text{lbf}}{\text{ft}^3}$$
Appendix B.1

$$\mu = \left(2.3586 \frac{\text{lbm}}{\text{ft}\cdot\text{h}}\right)\left(\frac{\text{h}}{3600\ \text{s}}\right) = 0.0006552 \frac{\text{lbm}}{\text{ft}\cdot\text{s}}$$
Appendix B.1

$$D = 0.06758\ \text{ft} \qquad \varepsilon = 0.000005\ \text{ft}$$
Appendix D   Table 4.3

$$L_1 = 6\ \text{ft} \qquad L_2 = [6 + 2(2)]\ \text{ft} = 10\ \text{ft}$$

The conservation of energy equations for each tube in the system are shown as follows:

$$\frac{P_A - P_B}{\gamma} = f_1 \frac{L_1}{D} \frac{V_1^2}{2g} + \left(K_{\text{run,A}} + K_{\text{run,B}}\right)\frac{V_c^2}{2g} \quad \text{(Parallel run 1)}$$

$$\frac{P_A - P_B}{\gamma} = \left(f_2 \frac{L_2}{D} + \sum K_2\right)\frac{V_2^2}{2g} + \left(K_{\text{branch,A}} + K_{\text{branch,B}}\right)\frac{V_c^2}{2g} \quad \text{(Parallel run 2)}$$

The friction factors in each tube require the calculation of the Reynolds number and the relative roughness. Since the volumetric flow rate in each tube is not known, this becomes a type 2 iterative solution.

In tube 1, there are two minor losses associated with the tees. The tee at A is a diverging tee and the tee at B is converging. The losses are those associated with the value of K_{run} for each tee. In tube 2, the minor losses include a diverging tee branch (A), two elbows, a gate valve, and a converging tee branch (B). The K values for the elbows and gate valve can be easily determined from Appendix E. However, when calculating the K values for the tees, the value of y_b is unknown since the volumetric flow through each leg is to be determined. This compounds the iterative nature of this calculation.

Because of the iterative nature of this problem, it is most convenient to solve it using an equation solver such as EES. The EES code for this problem will be developed step by step. The preliminary information given in the problem is stated first, along with the calculated properties of the water.

```
"GIVEN: A parallel flow system as shown"
    f$ = 'steam_iapws'
    T = 70[F]
    rho = density(f$,T=T,x=0)
    gamma = rho*g/g_c
    mu = viscosity(f$,T=T,x=0)*convert(lbm/ft-hr,lbm/ft-s)

$IFNOT Parametric Study
    DELTAP_AB = 0.2[psi]
$ENDIF

"Copper tubing"
    D = 0.06758[ft]              "Appendix D"
    epsilon = 0.000005[ft]       "Table 4.3"
    L_1 = 6[ft]
    L_2 = 2*2[ft] + 6[ft]

"Constants"
    g = 32.174[ft/s^2]
    g_c = 32.174[lbm-ft/lbf-s^2]

"FIND: Volume flow rate through each leg"
```

Notice that the pressure drop (DELTAP_AB) is inside of a parametric study *directive*. There are several directive commands that can be used in EES. The parametric study directive allows the user to build a Parametric Table and vary the variable(s) included inside the parametric study directive. When a normal "Solve" command is issued, the solver uses the value within the directive. However, when the "Solve Table" command is executed, the solver ignores the value within the directive and uses the value(s) listed in the Parametric Table. Use of this directive eliminates the need to comment out variables used in a parametric study.

The conservation of energy equations for each leg of the parallel network are shown previously. In EES format, these equations are written as follows:

```
SOLUTION:"
"The pressure drop from A to B is the same for both circuits"
"Tube 1"
   DELTAP_AB*convert(psi,lbf/ft^2)/gamma = f_1*(L_1/D)*(V_1^2/   &
       (2*g))+ (K_run_A + K_run_B)*(V_c^2/(2*g))

"Tube 2"
   DELTAP_AB*convert(psi,lbf/ft^2)/gamma = (f_2*(L_2/D) + K_2)   &
       (V_2^2/(2*g))+(K_branch_A + K_branch_B)*(V_c^2/(2*g))
```

The friction factors for each leg of the network can be calculated using the built-in Moody Chart function in EES, or any of the formulations shown in Table 4.2. Here, the Swamee–Jain correlation is used.

```
"Friction factor in each tube,"
   f_1 = 0.25*log10((epsilon/D)/3.7 + 5.74/Re_1^0.9)^(-2)
   Re_1 = rho*V_1*D/mu
   Vol_dot_1*convert(gpm,ft^3/s) = A*V_1
   f_2 = 0.25*log10((epsilon/D)/3.7 + 5.74/Re_2^0.9)^(-2)
   Re_2 = rho*V_2*D/mu
   Vol_dot_2*convert(gpm,ft^3/s) = A*V_2
   A = pi*D^2/4
```

The loss coefficients need to be found. The most complex calculation here is associated with the tees. The data for the tees are as follows:

```
"Loss coefficieints"
"Tee data"
   beta = 1
   alpha = 90 [deg]
   y_b = Vol_dot_2/Vol_dot_c
```

In the straight tube, the loss coefficients can be found by the following:

```
"Straight leg (subscript 1)"
   K_run_A = BigM*y_b^2
   BigM = IF(y_b,0.5,2,2,0.3)*(2*y_b - 1)
   K_run_B = 1.55*y_b - y_b^2
```

Notice in this set of EES equations, M in Equation 4.25 is assigned the EES variable BigM. This is done because EES variables are *not* case sensitive and *m* is often used to represent a mass (although it is not in this particular problem). Tricks such as this help avoid unintentional variable overlapping, which can make debugging difficult. Also used in this set of equation is the IF statement. The format of the IF statement is

$$Q = IF(A,B,x,y,z)$$

EES compares A and B within the IF statement and assigns the following actions: if A < B, then "x" is executed, if A = B, then "y" is executed, or if A > B, then "z" is executed. In the IF statement, A,B,x,y,z do not need to be constant values. They can also be expressions.

The minor losses in tube 2 of the parallel system can be calculated as follows:

```
"Tube with elbows and valve (Tube 2)"
   K _ 2 = 2*K _ el + K _ gv
   K _ el = 30*f _ T
   K _ gv = 8*f _ T
   f _ T = 0.25*(log10((epsilon/D)/3.7))^(-2)

"In leg 2, the tee flow is a branch flow,"
   K _ branch _ A = BigG*(1 + BigH*(y _ b/beta^2)^2 -          &
     BigJ*(y _ b/beta^2))*cos(alpha)
   BigG = 1 + 0.3*y _ b^2
   BigH = 0.3
   BigJ = 0
   K _ branch _ B = BigC*(1 + BigD*(y _ b/beta^2)^2 -          &
     BigE*(1 - y _ b)^2 - BigF*(y _ b/beta)^2)
   BigC = 0.55
   BigD = 1
   BigE = 1
   BigF = 0
```

Two more equations are needed: one that allows for the solution of the combined flow rate and one that allows for the calculation of the combined flow velocity. Both of these values are needed in the calculation of the loss coefficients for the tees.

```
"The combined flow and combined velocity are"
   Vol_dot_c = Vol_dot_1 + Vol_dot_2
   Vol_dot_c*convert(gpm,ft^3/s) = A*V_c
```

The solution to this iterative set of equations is $\dot{V}_1 = 4.747$ gpm, $\dot{V}_2 = 3.129$ gpm, and $\dot{V}_c = 7.876$ gpm. This result indicates that tube 1, the straight run, carries more of the flow. This is to be expected because tube 1 has a lower head loss due to friction and fittings. In addition, the length of straight tube in tube 1 is smaller than in tube 2, resulting in a lower friction loss.

Having a program such as EES to solve Example 4.9 eliminates the tedious manual iterative calculations. However, another feature of most equation solvers is their ability to conduct and plot the results of parametric studies. Consider a case where you are interested in how the flow splits between the two tubes of the parallel system shown in Example 4.9 as the pressure drop between points A and B varies. This study can be done by setting up a Parametric Table in EES. The results can then be plotted for a visual display of the behavior of the system, as shown in Figure 4.7.

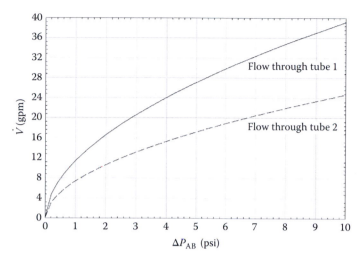

FIGURE 4.7
Result of a parametric study for example problem 4.9.

4.6 Economic Pipe Diameter

Many possible pipe diameters can be used to transport a fixed flow rate through a pipe. A small pipe has a low initial cost, but the pressure drop may be quite high resulting in a significant pumping cost to move the fluid through the pipe. On the other hand, a large pipe will have a significant first cost, but the pressure drop is much lower resulting in a low pumping cost. This thinking leads one to believe that there must be a pipe diameter that minimizes the total cost of the pipe system. This diameter is known as the *economic pipe diameter* or *optimum pipe diameter*.

The expressions developed in this section are based on the work of Darby and Melson (1982) and further refined by Janna (2015). Even though many simplifying assumptions are made as the method is developed, the resulting expressions provide a very reasonable estimate to the economic diameter.

4.6.1 Cost of a Pipe System

There are many costs that are incurred in a pipe system. However, these costs can be grouped into two categories: capital (i.e., initial) and annual costs. The capital cost of a pipe system includes the cost of the pipe itself and costs associated with installation, valves, fittings, pumps, insulation, hangers, and supports. The costs incurred annually are the cost of energy required to move the fluid through the pipe system and any maintenance costs required.

Since capital and annual costs occur at different times, economic interest factors must be considered to align all costs to a common point on the cash flow diagram for the pipe system. In the method presented here, costs are converted to annual costs. Interest is assumed to be compounded annually.

4.6.2 Determination of the Economic Diameter

The strategy used to determine the economic diameter is to develop a *cost function* that represents the total annual costs of the pipe system as a function of the pipe diameter.

$$AC_T = f(D) \tag{4.30}$$

The diameter that results in the lowest annual cost is the economic diameter. This diameter can be found by taking the derivative of Equation 4.30, setting it equal to zero, and solving for D. To ensure that this value is indeed a minimum, the second derivative can be used.

The total annual cost of the pipe system is defined as

$$AC_T = AC_P + AC_H + AC_M + AC_E \tag{4.31}$$

In this equation, AC_P is the annual cost of the pipe in the system, including installation; AC_H the annual cost of the hardware including valves, fittings, pumps, insulation, supports, and so on; AC_M the annual cost of maintenance for the system; and AC_E the annual cost of the energy required to move the fluid through the system.

The initial cost of installed pipe per foot is determined from cost data collected in 1980. These data can be fit to the following equation:

$$\frac{IC_P}{L} = (1+e)^{(yr-1980)} C_1 D^s \tag{4.32}$$

In this equation, C_1 and s are constants. C_1 represents the installed cost per foot of a 12-nom pipe based on 1980 dollars. The exponent s corresponds to various pipe classes. The values of C_1 and s are given in Table 4.10. The term in parenthesis involving the variables, e and yr, is a way to convert the constant C_1 from 1980 dollars to the current dollars. The variable e is an average yearly escalation rate (expressed in decimal form) from 1980 to present, which takes into account inflation. The variable, yr, is the year of installation of the fluid transport system. Equation 4.32 is valid for pipe diameters up to 36 in.

TABLE 4.10

Constants C_1 and s in Equation 4.32

Pipe Class (max working pressure)	$C_1(\$/ft^{s+1})$	s
300 (720 psig)	22.50	1.14
400 (960 psig)	23.14	1.20
600 (1440 psig)	28.28	1.29
900 (2160 psig)	36.10	1.32
1500 (3600 psig)	53.39	1.35

Source: Janna, W. S., *Design of Fluid Thermal Systems*, Stamford, CT, Cengage Learning, 2015.

EXAMPLE 4.10

Determine the installed cost per foot of an 8-nom sch 40 (class 300) pipe in 1980 and 2014 assuming a yearly escalation rate of 2.5%.

SOLUTION: The inside diameter of a 8-nom sch 40 pipe can be found in Appendix C.

$$D = 0.66508 \text{ ft}$$

Also, from Table 4.10,

$$s = 1.14 \qquad C_1 = 22.50 \frac{\$}{\text{ft}^{2.14}}$$

Using Equation 4.32, in 1980, the installed pipe cost was,

$$\frac{IC_P}{L} = C_1 D^s = \left(22.50 \frac{\$}{\text{ft}^{2.14}}\right)(0.66508 \text{ ft})^{1.14} = 14.13 \frac{\$}{\text{ft}} \quad \leftarrow$$

Using the estimated escalation rate of 2.5%, the installed pipe cost in 2014 is,

$$\frac{IC_P}{L} = (1+e)^{(\text{yr}-1980)} C_1 D^s$$

$$\frac{IC_P}{L} = (1+0.025)^{(2014-1980)}\left(22.50 \frac{\$}{\text{ft}^{2.14}}\right)(0.66508 \text{ ft})^{1.14} = 34.65 \frac{\$}{\text{ft}} \quad \leftarrow$$

This shows the estimated effect of inflation on the price of the installed pipe.

Knowing the initial cost per unit length of the pipe and installation, the annual cost can be determined by using the capital recovery interest factor from Chapter 2.

$$\frac{AC_P}{L} = \frac{IC_P}{L}\left(\frac{A}{P}, i, n\right) \tag{4.33}$$

For shorthand purposes, the variable crf will be used for the capital recovery factor. Using this notation, Equation 4.33 is written as

$$\frac{AC_P}{L} = (\text{crf})\frac{IC_P}{L} = (\text{crf})(1+e)^{(\text{yr}-1980)} C_1 D^s \tag{4.34}$$

The fittings, valves, pumps, and other hardware used in a piping system are different for each design. For the economic diameter model, it is sufficient to estimate these costs as a multiple of the initial cost per unit length of the pipe and installation. This multiple, F, typically ranges from 6 to 7. Using this idea, the initial cost of the hardware per foot is given by

$$\frac{IC_H}{L} = F\frac{IC_P}{L} = F(1+e)^{(yr-1980)}C_1 D^s \tag{4.35}$$

The annual cost per unit length of the hardware can be determined using the capital recovery factor.

$$\frac{AC_H}{L} = \frac{IC_H}{L}(crf) = (crf)F(1+e)^{(yr-1980)}C_1 D^s \tag{4.36}$$

Maintenance costs occur annually and are estimated to be a fraction of the initial cost per foot of the pipe and the hardware.

$$\frac{AC_M}{L} = b\left(\frac{IC_P}{L} + \frac{IC_H}{L}\right) \tag{4.37}$$

In this equation, b is the estimated percent (expressed as a decimal) of the initial costs that will be spent on annual maintenance. Typical values for b are 0.01–0.02 (1%–2%).

   Adding Equations 4.34, 4.36, and 4.37 together and algebraically rearranging results in

$$\frac{AC_P}{L} + \frac{AC_H}{L} + \frac{AC_M}{L} = (crf + b)(1 + F)(1 + e)^{(yr-1980)}C_1 D^s \tag{4.38}$$

Multiplying both sides through by length, L, gives

$$AC_P + AC_H + AC_M = L(crf + b)(1 + F)(1 + e)^{(yr-1980)}C_1 D^s \tag{4.39}$$

These are the first three terms of Equation 4.31. The last term in Equation 4.31 is the annual cost of the energy required to move the fluid through the pipe system by the pumps. The annual cost of the energy can be written as follows:

$$AC_E = C_2 \dot{W}_p t \tag{4.40}$$

In this equation, C_2 is the cost of energy, $\dot{W}_P$ is the electrical power input into the pump, and t is the annual operation time of the system. Because of inefficiencies, the power transmitted to the fluid is less than the power input to the pump. The efficiency of the pump can be defined as

$$\eta_P = \frac{\dot{W}}{\dot{W}_P} \tag{4.41}$$

Solving Equation 4.41 for the power input to the pump and substituting into Equation 4.40 gives

$$AC_E = \frac{C_2 \dot{W} t}{\eta_P} \tag{4.42}$$

The power delivered to the fluid can be determined by analyzing a general system boundary that represents a pipe system as shown in Figure 4.8. Assuming the pipe diameter is same throughout the network and the fluid is incompressible, the conservation of energy equation applied to the system identified in Figure 4.8 results in

$$\frac{\dot{W}}{\dot{m}} = (h_2 - h_1) + g(z_2 - z_1) \tag{4.43}$$

Assuming that the minor losses in the pipe system are negligible, Equation 4.43 can be rewritten as follows:

$$\frac{\dot{W}}{\dot{m}} = \frac{P_2 - P_1}{\rho} + g(z_2 - z_1) + f \frac{L}{D} \frac{V^2}{2} \tag{4.44}$$

The velocity of the fluid is related to the mass flow rate by

$$V = \frac{\dot{m}}{\rho A} = \frac{4\dot{m}}{\pi \rho D^2} \tag{4.45}$$

Substituting Equation 4.45 into Equation 4.44 and solving for the power input to the fluid, results in

$$\dot{W} = \dot{m} \left[\frac{P_2 - P_1}{\rho} + g(z_2 - z_1) \right] + \frac{8 f L \dot{m}^3}{\pi^2 \rho^2 D^5} \tag{4.46}$$

Substituting this result into Equation 4.42 gives the annual cost of the energy required to move the fluid through the system.

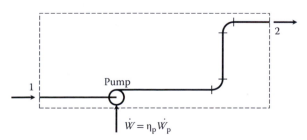

FIGURE 4.8
A pump and pipe network.

$$AC_E = \frac{\dot{m}C_2 t}{\eta_p}\left[\frac{P_2 - P_1}{\rho} + g(z_2 - z_1)\right] + \frac{8C_2 t f L \dot{m}^3}{\eta_p \pi^2 \rho^2 D^5} \tag{4.47}$$

The total annual cost of the pipe network can now be determined by substituting Equations 4.47 and 4.39 into Equation 4.31.

$$AC_T = L(\text{crf} + b)(1 + F)(1 + e)^{(\text{yr}-1980)} C_1 D^s$$
$$+ \frac{\dot{m}C_2 t}{\eta_p}\left[\frac{P_2 - P_1}{\rho} + g(z_2 - z_1)\right] + \frac{8C_2 t f L \dot{m}^3}{\eta_p \pi^2 \rho^2 D^5} \tag{4.48}$$

The derivative of this equation with respect to the pipe diameter is

$$\frac{d(AC_T)}{dD} = sL(\text{crf} + b)(1 + F)(1 + e)^{(\text{yr}-1980)} C_1 D^{s-1} - 5\left(\frac{8C_2 t f L \dot{m}^3}{\eta_p \pi^2 \rho^2 D^6}\right) \tag{4.49}$$

Setting Equation 4.49 to zero and solving for the D gives

$$D_{\text{econ}} = \left[\frac{40 C_2 t f \dot{m}^3}{s(\text{crf} + b)(1 + F)(1 + e)^{(\text{yr}-1980)} C_1 \eta_p \pi^2 \rho^2}\right]^{\frac{1}{s+5}} \tag{4.50}$$

This is the economic diameter that minimizes the cost of the pipe system. Several observations can be made about the economic diameter:

- The derivative of the total annual cost function was taken with respect to the diameter, holding all other variables constant.
- The pipe length is not part of the economic diameter calculation.
- The only fluid property that is included directly is the density. However, the fluid's viscosity is also part of the problem through the friction factor.
- The pressure and elevation head losses through the pipe do not appear in the economic diameter calculation.
- The solution to the economic diameter equation is iterative because the friction factor is unknown.

The velocity of the fluid flowing in a pipe that has a diameter equal to the economic diameter is known as the *economic velocity*. This can be easily computed by,

$$V_{\text{econ}} = \frac{\dot{\forall}}{A} = \frac{4\dot{\forall}}{\pi D_{\text{econ}}^2} = \frac{4\dot{m}}{\pi \rho D_{\text{econ}}^2} \tag{4.51}$$

The following example demonstrates the calculation procedure used to determine the economic diameter using Equation 4.50.

EXAMPLE 4.11

A commercial steel schedule 40 pipe is being used to transport liquid cyclo-hexane at 70°F to a heat exchanger at a flow rate of 1000 gpm. The fittings, valves, supports, and pump are estimated to cost 6.5 times the installed pipe cost per foot. Annual maintenance is estimated to be 2% of the installed pipe and hardware costs per foot. For this service, the pipe classification is determined to be 300. The pump in the system runs 16 h/day, 260 days/ year. The efficiency of the pump is estimated to be 78%. The cost of energy is $0.09/kW·h. The funds to build the pipeline were borrowed at an interest rate of 5% compounded annually for 15 years. Inflation is estimated to be 2.5% per year since 1980. The pipeline was put into service in 2013. Determine the economic diameter of the pipeline (ft) and the economic velocity of the cyclohexane (ft/s).

SOLUTION: The two major complexities with a problem such as this are (1) the iterative nature of the problem and (2) keeping track of the units to ensure that D_{econ} is in feet. The iterative nature of the problem is easily dealt with by using an equation solver such as EES. EES is also able to convert units to a consistent set. However, it is helpful to go through the details of the unit analy-sis here to understand the conversions required. In Equation 4.50, the term in brackets must have units of $[ft^{(s+5)}]$.
 The constants C_1 and s are determined from Table 4.10.

$$s = 1.14 \qquad C_1 = 22.50 \frac{\$}{ft^{2.14}}$$

A time basis must be selected. It makes no difference what the basis is, but once a selection is made, careful conversion of all time-based parameters must be considered. In this example, the time basis selected is seconds.

$$t = \left(16 \frac{h}{day}\right)(260 \text{ day})\left|\left(\frac{3600 \text{ s}}{h}\right)\right| = 1.498 \times 10^7 \text{ s}$$

Also, when computing the mass flow rate, it must have units of lbm/s. This means that the volumetric flow rate and velocity must use seconds as a time basis as well. The volumetric flow rate is given in gpm. This must be converted to ft³/s.

$$\dot{V} = 1000 \text{ gpm} \left|\left(2.228 \times 10^{-3} \frac{ft^3}{s \cdot gpm}\right)\right| = 2.228 \frac{ft^3}{s}$$
Appendix A

Now, the mass flow rate can be found, along with its proper units.

$$\dot{m} = \rho \dot{V} = \left(48.538 \frac{lbm}{ft^3}\right)\left(2.228 \frac{ft^3}{s}\right) = 108.1 \frac{lbm}{s}$$
Appendix B

The value of C_2 is given in $/kW·h. These units will need to be converted to $/ft·lbf.

$$C_2 = \left(0.09 \frac{\$}{kW \cdot h}\right)\left(\frac{kW \cdot h}{2.6552 \times 10^6 \ ft \cdot lbf}\right) = 3.390 \times 10^{-8} \ \frac{\$}{ft \cdot lbf}$$

Appendix A

This will introduce the proper length unit (ft), but it also unveils a new unit problem: the unit, lbf, has crept into the analysis. This means that both lbm and lbf are in the expression. To resolve this issue, g_c will have to be used.

The rest of the constants in Equation 4.50 are dimensionless. Using the units determined above, a unit analysis on the term in brackets in Equation 4.50 reveals the following:

$$\frac{40C_2 t f \dot{m}^3}{s(crf + b)(1 + F)(1 + e)^{(yr - 1980)} C_1 \eta_p \pi^2 \rho^2} \ [=] \ \frac{\left(\dfrac{\$}{ft \cdot lbf}\right)(s)\left(\dfrac{lbm^3}{s^3}\right)}{\left(\dfrac{\$}{ft^{s+1}}\right)\left(\dfrac{lbm^2}{ft^6}\right)\left(g_c \dfrac{lbm \cdot ft}{lbf \cdot s^2}\right)} = ft^{s+5}$$

This analysis indicates that the units have worked out correctly and the force–mass conversion factor g_c must be included in the denominator of Equation 4.50 if the IP unit system is being used.

The friction factor is unknown at this point because the pipe diameter is unknown. Therefore, this is a type 3 iterative solution. Using EES, the given information in the problem can be specified:

```
"GIVEN:  Transporting cyclohexane in a pipe system"
  C_1 = 22.50[$/ft^2.14]
  s = 1.14
  C_2 = 0.09[$/kWh]*convert($/kWh,$/lbf-ft)
  t = 16[hr/day]*260[day]*3600[s/hr]
  Fac = 6.5
  b = 0.02
  eta = 0.78
  e = 0.025
  yr = 2013
  n = 15
  i = 0.05
  crf = i*(1+i)^n/((1+i)^n - 1)

"The pipe is commercial steel class 300# sch 40"
  epsilon = 0.00015[ft]    "Table 4.3"

"Fluid data"
  f$ = 'cyclohexane'
  T_f = 70[F] ; x_f=0
  rho = density(f$,T=T_f,x=x_f)
  mu = viscosity(f$,T=T_f,x=x_f)*convert(lbm/ft-hr,lbm/ft-s)
  V_dot = 1000[gpm]
```

```
"Constants"
  g = 32.174 [ft/s^2]
  g_c = 32.174 [lbm-ft/lbf-s^2]
```

```
"FIND:  The economic diameter of the pipe and the economic
velocity of the cyclohexane"
```

The economic diameter is given by Equation 4.50.

```
"SOLUTION:"
"The optimum diameter is given by,"
  D_econ^(s+5) = (40*C_2*t*f*m_dot^3)/(s*(crf+b)*(1+Fac)*   &
    (1 + e)^(yr-1980)*C_1*eta*pi^2*rho^2*g_c)
```

Notice the inclusion of g_c in the denominator that is used for the force–mass conversion. The friction factor can be determined using any of the correlations in Table 4.2. Using the Swamee–Jain correlation,

$$f = \frac{0.25}{\left[\log \left(\dfrac{\varepsilon/D_{econ}}{3.7} + \dfrac{5.74}{Re^{0.9}} \right) \right]^2}$$

The Reynolds number can be found by

$$Re = \frac{\rho V D_{econ}}{\mu} = \frac{\rho D_{econ}}{\mu} \left(\frac{4\dot{V}}{\pi D_{econ}^2} \right) = \frac{4\dot{m}}{\pi D_{econ}\mu}$$

The mass flow rate of the cyclohexane can be found knowing the volumetric flow rate.

$$\dot{m} = \rho \dot{V}$$

The economic velocity can be computed once the economic diameter is known.

$$\dot{m} = \rho A V_{econ} = \rho \left(\frac{\pi D_{econ}^2}{4} \right) V_{econ}$$

In EES format, the above equations are written as follows:

```
"Friction factor calcluation"
  f = 0.25*log10((epsilon/D_econ)/3.7 + 5.74/Re^0.9)^(-2)

  Re = 4*m_dot/(pi*D_econ*mu)
  m_dot = rho*V_dot*convert(gpm, ft^3/s)
```

```
"Calculation of the economic velocity"
   m_dot = rho*(pi*D_econ^2/4)*V_econ
```

This completes the equation set required to iteratively solve the problem. The resulting economic diameter is $D_{econ} = 0.5012$ ft. The economic velocity corresponding to this diameter is $V_{econ} = 11.29$ ft/s.

The results of Example 4.11 indicate an economic diameter of 0.5012 ft. Referring to Appendix C, it is clear that this is not a standard pipe diameter. For schedule 40 pipe, it falls between 5-nom ($D = 0.42058$ ft) and 6-nom ($D = 0.50542$ ft). This selection is different compared to a pure type 3 problem. Here, *both* pipes will satisfy the volumetric flow rate. The next challenge is to determine which pipe size to specify. To do this, it is helpful to construct the *cost curves*.

4.6.3 Cost Curves

Once an economic diameter is calculated, a standard pipe needs to be selected. The selection of the proper size pipe can be accomplished by analysis of the cost curves for the system. The total annual cost of the system is given by Equation 4.48. This equation can be rearranged as follows:

$$
\begin{aligned}
AC_T &- \frac{\dot{m}C_2 t}{\eta_p}\left[\frac{P_2 - P_1}{\rho} + g(z_2 - z_1)\right] \\
&= L(\text{crf} + b)(1+F)(1+e)^{(\text{yr}-1980)}C_1 D^s + \frac{8C_2 t f L \dot{m}^3}{\eta_p \pi^2 \rho^2 D^5}
\end{aligned}
\tag{4.52}
$$

The left-hand side of Equation 4.52 is independent of the diameter of the pipe. Therefore, it can be replaced by a single variable, Ω, and rewritten as

$$
\Omega_T = L(\text{crf} + b)(1+F)(1+e)^{(\text{yr}-1980)}C_1 D^s + \frac{8C_2 t f L \dot{m}^3}{\eta_p \pi^2 \rho^2 D^5}
\tag{4.53}
$$

This is a total annual cost that includes the effect of the head losses due to pressure and elevation. Dividing this equation through by the length of the pipe results in

$$
\omega_T = \frac{\Omega_T}{L} = (\text{crf} + b)(1+F)(1+e)^{(\text{yr}-1980)}C_1 D^s + \frac{8C_2 t f \dot{m}^3}{\eta_p \pi^2 \rho^2 D^5}
\tag{4.54}
$$

This is a representative cost per foot of the installed pipe, including the head loss effects. This value is *not* the installed cost per foot; it includes the effect of the pressure and elevation head loss through the pipe. The first term on the right-hand side of Equation 4.54 is the annual cost per foot of the pipe, hardware (including pumps, valves, and fittings), and maintenance. The second term is representative of the annual cost of the energy required to move the fluid through the system. Each of these terms is defined as follows:

$$\omega_H = (crf + b)(1 + F)(1 + e)^{(yr-1980)} C_1 D^s \tag{4.55}$$

and

$$\omega_E = \frac{8 C_2 t f \dot{m}^3}{\eta_p \pi^2 \rho^2 D^5} \tag{4.56}$$

Substituting Equations 4.55 and 4.56 into Equation 4.54 results in

$$\omega_T = \omega_H + \omega_E \tag{4.57}$$

Each of the terms in Equation 4.57 can be calculated as a function of the pipe diameter. The resulting curves are known as the cost curves for the system. Figure 4.9 shows the cost curves for Example 4.11.

The curves are drawn assuming that the friction factor is constant and equal to the value determined from the solution of Equation 4.50. The cost curves clearly show the economic trade-off between the pipe size and the cost of energy required to transport the fluid. As the pipe diameter becomes larger, the capital cost of the pipeline increases (ω_H), whereas the energy costs

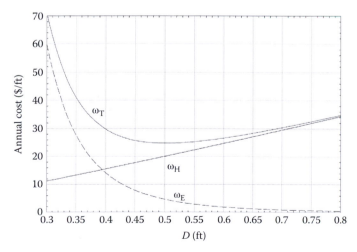

FIGURE 4.9
Cost curves for the pipe system described in Example 4.11.

decrease (ω_E). The trend is exactly opposite for small-diameter pipes. This trade-off results in an optimum point in the total cost curve. This optimum point occurs at the economic diameter.

The cost curves are helpful in visualizing how the pipe diameter influences the economics. In the case of Example 4.11, it was discovered that the economic diameter (0.5012 ft) falls between a 5-nom ($D = 0.42058$ ft) and 6-nom ($D = 0.50542$ ft) pipe. Using Figure 4.8, it can be seen that the annual cost per foot of a 5-nom pipe is larger than a 6-nom pipe. One may argue that both pipes are very close to the economic diameter and either pipe could be selected. Although this argument has some validity, it may be wise to select the larger diameter to accommodate future expansion of the pipeline (e.g., increased volumetric flow rate). Therefore, the proper pipe to select for Example 4.11 is a 6-nom sch 40 commercial steel pipe.

Equation 4.54 provides an alternative way to find the economic diameter. For example, Figure 4.8 indicates that the economic diameter is slightly larger than 0.5 ft, which is consistent with the result calculated from Equation 4.50. Another possible way to find the economic diameter is to use computer software to find the minimum point of Equation 4.54. Most equation-solving software packages have the capability to perform optimization or min/max studies. Appendix F discusses how EES can be used to perform such min/max studies. Using EES to minimize Equation 4.54 with respect to the diameter results in the following solution for Example 4.11: $D_{econ} = 0.5023$ ft. The economic velocity corresponding to this diameter is $V_{econ} = 11.24$ ft/s. Notice that this solution is nearly identical to the solution of Example 4.11, but it is not exactly the same. At first glance, this may seem strange. It would seem that the calculation of the economic diameter using Equation 4.50 should return the same result as performing a numerical minimization of Equation 4.54. The reason for the slight difference is because Equation 4.50 was derived assuming that the friction factor was not a function of the diameter. This, of course, is not true. The diameter of the pipe will influence the friction factor. However, looking at the results in Example 4.11 compared to the numerical minimization results, one can conclude that although the friction factor depends on the diameter of the pipe, the dependence is very weak. This suggests that the assumptions used to derive the economic diameter, Equation 4.50, are valid.

4.6.4 Estimated Economic Velocities

The economic diameter calculation, Equation 4.50, requires the engineer to estimate several important economic factors such as the multiplier for hardware costs (F), the maintenance percentage (b), and others. Once these factors are known, Equation 4.50 can be solved. Once the economic diameter is known, the corresponding economic velocity can be easily calculated using Equation 4.51.

It is possible to solve Equation 4.50 for a range of values and determine the resulting range of economic velocities. Table 4.11 shows the result of such a calculation for a typical range of economic and system parameters for several different fluids. For all calculations, the common parameters used are continuous yearly operation in the year 2014, commercial steel schedule 40 class 300 pipe, with a liquid temperature of 70°F. The economic parameter ranges considered are the following:

- Cost of energy: $0.04 \leq C_2 \leq \$0.12/kW\cdot h$
- Hardware multiplier: $5 \leq F \leq 7$
- Maintenance fraction: $0.01 \leq b \leq 0.05$
- Yearly inflation rate: $0 \leq e \leq 0.01$
- Capital recovery factor: $0.1 \leq crf \leq 0.275$

The system parameter ranges considered are the following:

- Fluid volumetric flow rate: $10 \leq \dot{V} \leq 2000$ gpm
- Pump efficiency: $0.45 \leq \eta_p \leq 0.80$

Table 4.11 allows the engineer to solve a type 3 (diameter unknown) problem for a pipe network, calculate the resulting fluid velocity, and use Table 4.11 to see if the velocity is within the estimated economic range. It is important to understand that the use of Equation 4.50 is the recommended procedure for finding the economic diameter. However, Table 4.11 can provide a quick check to ensure that a selected pipe falls within a reasonable economic velocity range without having to solve Equation 4.50 directly.

TABLE 4.11

Estimated Minimum and Maximum Economic Velocities (ft/s) Calculated Using the Economic Diameter Equation 4.50 for a Range of Typical Economic and System Parameters

	Min	Max
Common Fluids		
Water	4.6	9.9
Acetone	5.1	10.8
Ammonia	5.5	11.8
Benzene	4.8	10.4
Toluene	4.8	10.4
Hydrocarbons		
Propane	5.9	12.6
Butane	5.6	12

(continued)

TABLE 4.11 (*Continued*)

Estimated Minimum and Maximum Economic Velocities (ft/s)
Calculated Using the Economic Diameter Equation 4.50 for a
Range of Typical Economic and System Parameters

	Min	Max
Pentane	5.5	11.7
Hexane	5.3	11.4
Heptane	5.2	11.3
Octane	5.1	11.1
Nonane	5.1	11
Decane	5	10.8
Dodecane	4.9	10.6
Isobutane	5.7	12.1
Isopentane	5.5	11.7
Cyclohexane	4.9	10.6
Alcohols		
Methanol	5	10.7
Ethanol	4.9	10.5
Refrigerants		
R22	4.5	9.6
R123	4.2	8.9
R134a	4.5	9.5
Brines (20% Solution)		
Sodium chloride	4.3	9.4
Calcium chloride	4.3	9.2
Magnesium chloride	4.2	9.1
Lithium chloride	4.3	9.3
Ethylene glycol	4.5	9.7
Propylene glycol	4.4	9.6
Other Fluids		
Carbon dioxide	5.2	11.1
Sulfur dioxide	4.3	9.1
Sulfur hexafluoride	4.3	9.2
Propylene	5.9	12.5

4.7 Pumps

In the design of a fluid piping system, one of the tasks of the engineer is to select a pump (or pumps) for the system that provides the required flow rate through the system. There are a wide variety of pumps available, which are designed for specific applications. However, they can be classified into one of two categories: *positive displacement* and *dynamic*. The main focus of this chapter is on centrifugal pumps, a type of dynamic pump, because they are most commonly used in industrial applications. Even among centrifugal pumps, there are many different types including end suction, submersible, and split case. These different pump designs are made to handle different types of fluids, system pressures, flows, and other conditions. The final selection of a pump must take into account all of these factors. Once the techniques of selecting a centrifugal pump are understood, they are easily transferrable to other types of pumps.

4.7.1 Types of Pumps

Pumps move fluid through a pipe system by transferring energy from an external source, usually an electric motor, to the fluid. The method of energy transfer to the fluid classifies the pump. Table 4.12 displays the pump classification and the method of energy transfer to the fluid. For more detail on the different types of pumps available, the reader is encouraged to consult the *Pump Handbook* (Karassik et al. 2008).

4.7.2 Dynamic Pump Operation

As indicated in Table 4.12, dynamic pumps transfer energy to the fluid using an impeller. As mentioned earlier, there are many different types of dynamic pumps. An example of an inline centrifugal pump is shown in Figure 4.10.

The purpose of the pump in a fluid system is to increase the fluid's pressure such that friction effects are overcome and the fluid is delivered at the required flow rate. Therefore, the pressure rise across a pump is significant, but the changes in kinetic and potential energy from the inlet to the outlet are often negligible. Figure 4.11 demonstrates how this is accomplished in a dynamic pump. As the fluid enters the pump through the suction port, there is a small drop in pressure because of friction, and the kinetic energy stays constant because the suction line diameter is constant. As the fluid

TABLE 4.12

Pump Classification and Methods of Energy Transfer to the Fluid

Positive Displacement	Dynamic
Reciprocating action	Centrifugal action
Energy transfer by pistons, plungers, diaphragms, or bellows	Energy transfer by rotating impeller in the radial direction
Rotary action	Propeller action
Energy transfer by vanes, screws, or lobes	Energy transfer by rotating impeller in the axial direction
	Mixed action
	Energy transfer by rotating impeller in a combined radial/axial direction

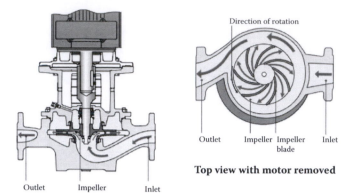

Top view with motor removed

Plan view (Section)

FIGURE 4.10

An inline centrifugal pump. From Grundfos Research and Technology. *The Centrifugal Pump.* Bjerringbro, Denmark: Grundfos, 2014. Reprinted with permission.

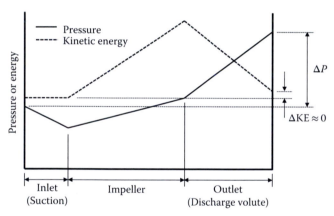

FIGURE 4.11

Operation of a dynamic pump.

moves through the impeller, the fluid experiences a significant increase in velocity and the pressure rises slightly. The largest pressure increase in the pump occurs in the discharge passage of the pump. The discharge passage is called a *discharge volute*. The discharge volute is a liquid diffuser. The cross-sectional flow area of the volute increases in the direction of fluid flow. A thermodynamic analysis of the volute shows that as the fluid flows through the passage, its pressure increases significantly. This pressure increase is offset by decrease in kinetic energy to satisfy the conservation of energy. When the fluid finally leaves the pump, its outlet velocity is nearly the same as its velocity at the suction port, but the pressure has increased significantly.

4.7.2.1 Dynamic Pump Performance

In a fluid system, one of the engineer's tasks is to select a pump to provide the required flow rate. To properly select a pump, the engineer must understand how a pump will perform when connected to the pipe system. Pump manufacturers provide *pump curves* that show how a given pump will perform. These pump curves are determined by the manufacturer based on experimental data. The pump is put on a test stand and the following parameters are measured: rotating shaft torque, impeller rotational speed, the suction and discharge pressures and elevations, the volumetric flow rate of the fluid, and the suction and discharge pipe specifications. With this information, the following performance parameters are calculated: power input to the impeller, power input to the fluid, head increase across the pump, and the pump efficiency. These parameters are presented on *dynamic pump head* versus *capacity* axis. An example of a pump curve is shown in Figure 4.12.

The capacity of a pump is simply the volumetric flow rate delivered. The dynamic pump head can be determined by considering the head at the suction and discharge ports of a pump. On the suction side (subscript s), the total fluid head is given by,

$$H_s = \frac{P_s}{\gamma} + \frac{V_s^2}{2g} + z_s \qquad (4.58)$$

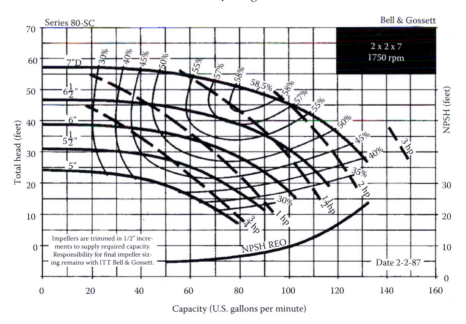

FIGURE 4.12
Centrifugal pump performance curves. (Bell and Gossett. *Series 80-SC Spacer-Coupled Vertical In-Line Centrifugal Pump: 60 Hz Performance Curves.* Morton Grove, IL: Xylem, Inc., 2012. Reprinted with permission.)

Similarly, the fluid head on the discharge side of the pump (subscript d) is given by

$$H_d = \frac{P_d}{\gamma} + \frac{V_d^2}{2g} + z_d \tag{4.59}$$

The dynamic pump head delivered to the fluid is the difference between Equations 4.59 and 4.58.

$$H_p = H_d - H_s = \frac{P_d - P_s}{\gamma} + \frac{V_d^2 - V_s^2}{2g} + (z_d - z_s) \tag{4.60}$$

In most dynamic pumps, the changes in velocity head and elevation head are very small compared to the pressure head delivered by the pump. The value calculated using Equation 4.60 is plotted against the capacity of the pump for several different impeller diameters as shown in Figure 4.12, labeled 5″, 5½″, and so on. Several impeller curves are shown because the pump can be ordered with one of several different impeller diameters.

There are other curves in Figure 4.12 pertinent to the pump's performance. These include the pump efficiency and the power input to the pump. The dynamic pump head is related to the power input to the fluid by

$$H_p = \frac{\dot{W}}{\dot{m}g} \tag{4.61}$$

The power term in Equation 4.61 is the power input to the *fluid*. This power value is sometimes referred to as the *water horsepower* or *hydraulic horsepower*. Because of inefficiencies, the power input to the pump is larger than the hydraulic power input to the fluid. This is quantified by the pump efficiency.

$$\eta_p = \frac{\dot{W}}{\dot{W}_{bhp}} \tag{4.62}$$

The power term in the denominator of Equation 4.62 is the power input to the pump's rotating shaft from an electric motor, an engine, or any other source. This power is often referred to as the *brake horsepower* input to the pump. On the pump performance curves (Figure 4.12), lines of constant pump efficiency and brake horsepower are displayed.

The curve labeled "NPSH REQ" is addressed later. This curve is used to determine if cavitation will occur when the pump is installed in a pipe system.

EXAMPLE 4.12

The pump shown in Figure 4.12 is fitted with a 6½-in. impeller. The pump is operating in a system and delivering 70 gpm. Determine (a) the head delivered by the pump, (b) the pressure rise across the pump assuming that the fluid is water at 70°F, (c) the brake horsepower input to the pump, and (d) the efficiency of the pump.

SOLUTION:

a. From Figure 4.12, at a capacity of 70 gpm, this pump fitted with a 6½-in. impeller is affecting a total head increase of approximately 41.5 ft.
b. In the upper right-hand corner of Figure 4.12, notice the label "2 × 2 × 7." This means that the suction inlet is 2-nom, the discharge port is 2-nom, and the maximum impeller diameter that can be used in this pump is 7 in. Since the suction and discharge ports are the same size, the velocity is the same at both ports. If we also assume that the elevation difference between the suction and discharge ports is small, then the pressure rise across the pump can be calculated using Equation 4.61.

$$H_p = \frac{P_d - P_s}{\gamma} + \frac{V_d^2 - V_s^2}{2g} + (z_d - z_s) = \frac{\Delta P}{\gamma}$$

$$\therefore \quad \Delta P = \gamma H_p = \rho \frac{g}{g_c} H_p = \left(62.298 \, \frac{\text{lbf}}{\text{ft}^3}\right)(41.5 \text{ ft})\left(\frac{\text{ft}^2}{144 \text{ in.}^2}\right) = \underline{18.0 \text{ psi}} \quad \leftarrow$$

Appendix B

c. The brake horsepower input to the pump can be read directly from the pump curve (Figure 4.12) as 1.3 hp.
d. The efficiency of the pump can also be read directly from Figure 4.12 as 57%.

Two interesting extremes can be observed on a pump performance curve. These are indicated on the plot on the left side of Figure 4.13 as the *cutoff head*, point (a), and the *free delivery* condition, point (b).

The cutoff head represents the condition where the pump is not able to provide any flow to the system. A good example of this is a pump being used to move water from a stream to an irrigation ditch. If the elevation of the ditch is higher than the cutoff head, then the pump cannot provide the flow. In fact, there will be no flow at all. For this condition, the efficiency of the pump is zero.

$$\eta_{p,\text{cutoff}} = \frac{\dot{W}}{\dot{W}_{\text{bhp}}} = \frac{\dot{m}gH_p}{\dot{W}_{\text{bhp}}} = \frac{(0)gH_p}{\dot{W}_{\text{bhp}}} = 0 \qquad (4.63)$$

The other extreme point, free delivery, occurs if the pump curves were extrapolated to the zero-head axis. In this case, the pump is providing no

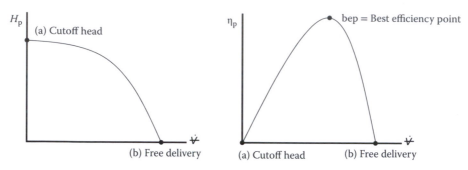

FIGURE 4.13
Pump curve extremes and the resulting pump efficiency.

power input to the fluid, and the fluid is passing through the pump as if it were a pipe. In this case, the efficiency is also zero.

$$\eta_{p,free} = \frac{\dot{W}}{\dot{W}_{bhp}} = \frac{\dot{m}gH_p}{\dot{W}_{bhp}} = \frac{\dot{m}g(0)}{\dot{W}_{bhp}} = 0 \tag{4.64}$$

For volume flow rates between these two extremes, the pump efficiency will increase, achieve a maximum, and then decrease as shown in the plot on the right side of Figure 4.13. The maximum point on the efficiency curve is called the *best efficiency point* (bep). In the selection of a pump for a system, it is good design practice to have the pump operating as near as possible to the bep.

4.7.3 Manufacturer's Pump Curves

There are hundreds of pump manufacturers. However, all manufacturers provide the same information about their pumps similar to the performance curves shown in Figure 4.12. Most of these performance curves are published by the manufacturer in the form of a technical booklet. The booklet clearly indicates what type of a pump the model is (end suction, inline, etc.). For quick reference, these booklets also show performance maps that overlay the operating range of a specific model operating at a given speed, as shown in Figure 4.14. Knowing the system capacity requirement and head that the pump must overcome allows the engineer to quickly identify the proper model within a family of pumps. Once the model is identified, the detailed pump curves can be accessed for design work.

Appendix G contains complete technical booklets for the Bell & Gossett Series 1531 end suction and Series 80-SC inline centrifugal pumps. These booklets are made available for educational purposes. There are many other pump vendors. The reader is encouraged to consult the website www.pump-flo.com for a comprehensive list of vendors and pump curves.

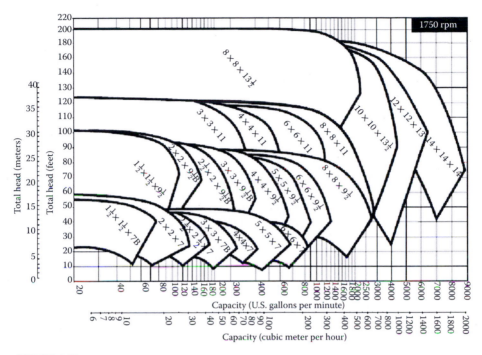

FIGURE 4.14
A performance map for a Bell & Gossett Series 80-SC inline centrifugal pump. (From Xylem Inc. Reprinted with permission.)

4.7.4 System Curve

The pump selected for a specific application must deliver the proper flow rate while overcoming friction effects in the pipe system. Therefore, knowledge of the system's friction effects must be fully understood. This cannot be accomplished until the system is completely designed. The design engineer must know the type of pipe, its diameter, total straight length, number of fittings, and other system details to perform a full energy analysis of the network in order to determine the friction effects.

Systems are often designed to deliver a specified flow rate. For example, in a processing line where six 12-oz bottles are filled every 25 seconds with a beverage, a fixed flow rate must be met. If the flow rate is higher, then the bottles will overfill. If the flow rate is smaller, the bottles will not fill completely. There are cases where a system is retrofitted, which results in an increased or decreased flow rate compared to the initial design. In cases similar to this, replacing the pump may not be necessary, but it is important to know how the pump will respond to changes in the system flow rate. These factors make pump selection a very important step in the complete design of a fluid delivery system. Once a pipe system has been designed, the first step toward selecting a pump is to develop what is known as the *system curve*.

The system curve is a plot of the system head as a function of volumetric flow rate. The system head is analyzed between two points in a pipe system and is made up of any pressure, velocity, and elevation heads that may exist, and the total head loss due to friction.

4.7.4.1 System Curve for a Two-Tank System Open to the Atmosphere

Consider a two-tank system as shown in Figure 4.15. In this system, the two tanks are opened to the atmosphere and the liquid level in each tank is kept constant. Writing the conservation of energy equation between points 1 and 2 (the tank free surfaces) results in

$$\frac{P_1}{\gamma} + \frac{V_1^2}{2g} + z_1 + H_p = \frac{P_2}{\gamma} + \frac{V_2^2}{2g} + z_2 + \left(f\frac{L}{D} + \sum K \right)\frac{V^2}{2g} \tag{4.65}$$

Solving for the required pump head,

$$H_p = \frac{P_2 - P_1}{\gamma} + \frac{V_2^2 - V_1^2}{2g} + (z_2 - z_1) + \left(f\frac{L}{D} + \sum K \right)\frac{V^2}{2g} \tag{4.66}$$

Since the free surfaces of both the tanks are at atmospheric pressure, the change in the pressure head is zero. Similarly, if the diameter of the pipes

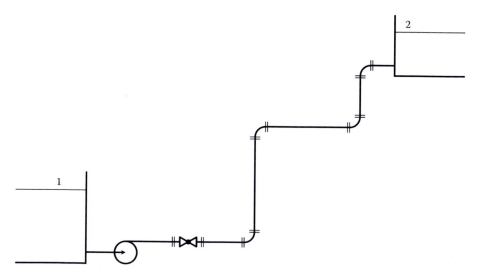

FIGURE 4.15
A two-tank pump and pipe system.

is constant throughout, then the velocity head difference is zero. Therefore, Equation 4.66 reduces to

$$H_p = (z_2 - z_1) + \left(f\frac{L}{D} + \sum K \right)\frac{V^2}{2g} \tag{4.67}$$

As demonstrated previously, the fluid velocity can be written in terms of volumetric flow rate. Making this substitution results in

$$H_p = (z_2 - z_1) + \left(f\frac{L}{D} + \sum K \right)\frac{8\dot{V}^2}{\pi^2 g D^4} \tag{4.68}$$

This equation indicates that the pump head required to overcome the friction effects and the elevation change is a *quadratic* function of the volumetric flow rate, with an offset equal to the elevation difference between the tank surfaces. When this equation is plotted on a head-capacity axis, the result is a parabola, as shown in Figure 4.16. This parabola is known as the system curve. It shows the pump head required to overcome the various head losses in the system for different volumetric flow rates through the system.

4.7.4.2 System Curve for a Closed-Loop System

Another common system in mechanical engineering applications is a closed-loop piping system, as shown in Figure 4.17. Closed-loop systems often have an *expansion tank* for each closed loop. The purpose of the expansion tank is twofold: (1) it allows the system to operate in a steady state even when there are temperature variations that can cause the total fluid volume to increase or decrease, and (2) it creates a point of constant pressure in the system.

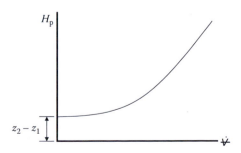

FIGURE 4.16
The system curve for a two-tank system.

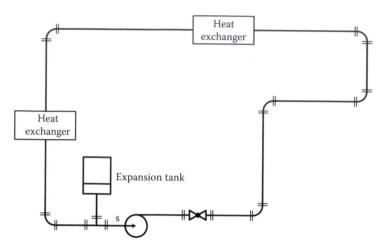

FIGURE 4.17
An example of a closed-loop system.

Assuming that the net energy transfer between the working fluid and the heat exchangers is zero, the conservation of energy equation from the suction side of the pump (labeled "s" in Figure 4.17), all the way around the system and back to the suction side results in

$$\frac{P_s}{\gamma} + \frac{V_s^2}{2g} + z_s + H_p = \frac{P_s}{\gamma} + \frac{V_s^2}{2g} + z_s + \left(f\frac{L}{D} + \sum K \right) \frac{V^2}{2g} \tag{4.69}$$

Solving Equation 4.69 for the pump head shows that the pump head is only dependent on the friction losses in the piping system

$$H_p = \left(f\frac{L}{D} + \sum K \right) \frac{V^2}{2g} \tag{4.70}$$

Introducing the volumetric flow rate gives

$$H_p = \left(f\frac{L}{D} + \sum K \right) \frac{8\dot{V}^2}{\pi^2 gD^4} \tag{4.71}$$

Equation 4.71 indicates that the pump head required to overcome the friction losses in the pipe system is a quadratic function of the volumetric flow rate. In the case of a closed-loop system, there is no offset in the curve. Figure 4.18 shows the system curve for a closed-loop system similar to the one described in Figure 4.17.

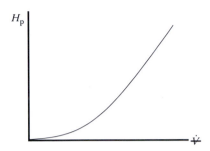

FIGURE 4.18
The system curve for a closed-loop system.

4.7.5 Pump Selection

Notice that the coordinates of the system curve are identical to the coordinates of a pump performance curve. Overlaying a pump performance curve on a system curve, as shown in Figure 4.19, shows that there is a unique intersection of these two curves. This intersection is the *operating point* of the combined pump/pipe system.

To select the right pump for the system, the system curve is overlaid on the pump curves. Then the proper diameter impeller needs to be selected to deliver the required flow rate. As an example, consider a Bell & Gossett Series 1531 2½ BB pump from Appendix G operating at 3550 rpm. The performance curves for this model pump are shown in Figure 4.20. Also included in Figure 4.20 is a system curve for a two-tank system similar to Figure 4.15.

Suppose that the flow requirement for the system is 550 gpm. From Figure 4.20, it can be seen that the 8½-in. impeller will not provide this flow; the intersection of the 8½-in. impeller performance curve with the system curve shows a capacity of only 530 gpm. Therefore, to supply the required 550 gpm, the next larger impeller is selected. The 9-in. impeller pump, when matched with the system, will provide approximately 570 gpm at 284 ft of head. At this operating point, the pump will draw approximately 56 hp and operate with an efficiency of 73.8%. This flow rate is larger than required. However, installation of a control valve downstream of the pump discharge allows the flow to be throttled down if needed.

When a valve is installed in the system, it adds a new friction loss that must be taken into account. The effect of the head loss through the valve shifts the system curve to the left because of the added resistance of the valve, as shown in Figure 4.21. From this figure, it can be seen that adding a control valve allows the pump with the 9-in. impeller to deliver the required 550 gpm at 285 ft of head. At this operating point, the pump will draw approximately 55 hp and operate with an efficiency of 74%.

Matching the pump to the pipe network to deliver the *exact* flow rate required is difficult. It is most common to select a pump with a slightly higher capacity. If the flow requirement is strict, then a valve can be used to

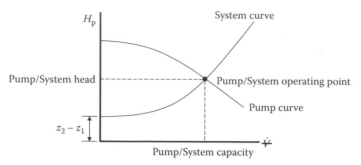

FIGURE 4.19

Operating point of a pump and pipe system with two open tanks.

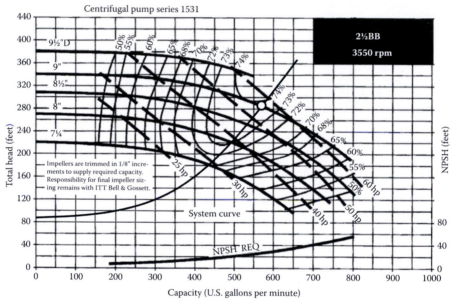

FIGURE 4.20

A system curve overlaid on pump performance curves.

control the flow rate. If the excess flow is acceptable, then the valve can be left wide opened until needed.

It may be entirely possible that there are *no* pumps in a technical booklet that satisfy the required flow rate and provide reasonable operating parameters. In this case, the engineer needs to search other pump models or other vendors for a solution. Many pump manufacturers have technical booklets with pump curves available on their web sites. A comprehensive web site that contains pump curves for *many* manufacturers can be found at www .pump-flo.com. This website features searching capability to find possible pumps for a given operating condition. The performance curves are

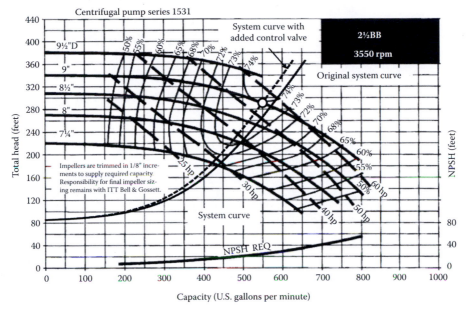

FIGURE 4.21
Achieving the required flow rate with a control valve.

EXAMPLE 4.13

Consider the two-tank open system shown in Figure E4.13A. The fluid being transported from the lower tank to the upper tank is acetone at an average temperature of 60°F. The elevation difference between free surfaces of the tanks is 15 ft. The pipe is 5-nom sch 40 commercial steel. The total length of straight pipe in the system is 40 ft and all fittings are regular. Draw the system curve up to 1000 gpm.

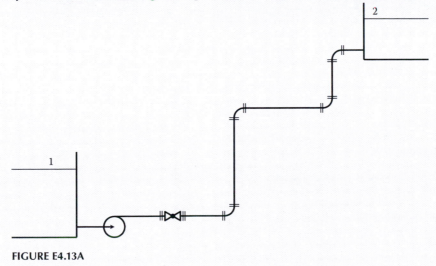

FIGURE E4.13A

SOLUTION: The system curve is determined by conducting a parametric study: varying the system capacity and calculating the required pump head. Therefore, EES will be used to determine the system curve. The given information for this problem is summarized in the EES code as follows:

```
"GIVEN: Transporting acetone in a two-tank pump/pipe system
    as shown"

"The pipe is commercial steel class 300# sch 40"
   epsilon = 0.00015[ft]                "Table 4.3"
   D = 0.42058[ft]                      "5-nom sch 40"
   L = 40[ft]

"System data"
   z_1 = 0[ft]                          "Bottom tank"
   z_2 = 15[ft]                         "Top tank"

"Fluid data"
   f$ = 'acetone'
   T_f = 60[F];      x_f = 0
   rho = density(f$,T=T_f,x=x_f)
   mu = viscosity(f$,T=T_f,x=x_f)*convert(lbm/ft-hr,lbm/ft-s)
$IFNOT Parametric Study
   Vol_dot = 400[gpm]
$ENDIF

"Constants"
   g = 32.174[ft/s^2]
   g_c = 32.174[lbm-ft/lbf-s^2]

"FIND: The system curve to 1000 gpm"
```

Notice that the volumetric flow rate (Vol_dot) is placed inside a Parametric Study directive. This is done so a parametric table of head versus capacity can be calculated and subsequently plotted.

The conservation of energy equation for an open, two-tank system with constant diameter pipe is given by Equation 4.67:

```
"SOLUTION:"
"The Conservation of Energy equation between the tank free
    surfaces is,"
   H_p = (z_2 - z_1) + (f*(L/D) + SUMK)*V^2/(2*g)
```

The friction factor can be found using Swamee–Jain correlation. This requires calculation of the Reynolds number of the flow.

```
"Friction factor calculation"
   f = 0.25*log10((epsilon/D)/3.7 + 5.74/Re^0.9)^(-2)
   Re = rho*V*D/mu
   Vol_dot*convert(gpm,ft^3/s) = A*V
   A = pi*D^2/4
```

The minor losses are determined using the *K* values calculated as shown in Appendix E.

```
"Valves and fittings: Appendix E"
SUMK = K_ent + 4*K_e + K_v + K_exit
   K_e = 30*f_T
   K_v = 340*f_T
   K_exit = 1.0
   K_ent = 0.50
   f_T = 0.25*(log10((epsilon/D)/3.7))^(-2)
```

Once the units have been verified, a Parametric Table can be constructed and the results plotted. Figure E4.13B shows the system curve for this two-tank system up to 1000 gpm.

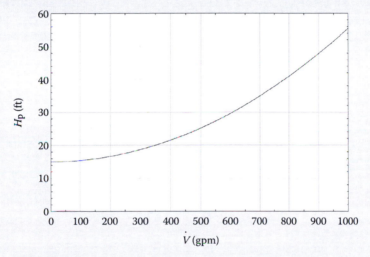

FIGURE E4.13B

Notice that the system curve displays the expected quadratic behavior. The offset on the *y* axis is the difference between the two-tank free surfaces.

EXAMPLE 4.14

Select a pump from Appendix G that will be suitable for the pump and pipe system described in Example 4.13 for a required volumetric flow rate of 400 gpm.

SOLUTION: The pump curves included in Appendix G are for a Bell & Gossett Series 1531, an end suction pump, and a Series 80-SC, an inline pump. In this example, the Series 80-SC pump is considered. The specific model within this series can be found knowing the system curve, as developed in Example 4.13.

From the system curve, the head required at 400 gpm is 21.5 ft. This information allows the design engineer to begin identifying the specific pump model. The cover of the technical booklet for the Series 80-SC pump shows three performance maps. These maps are for three different rotational speeds.

As with any design problem, there are several solutions. Consider the performance map for the 1750 rpm family of pumps shown in Figure E4.14A. This map seems to indicate that a $5 \times 5 \times 7$ pump operating at 1750 rpm may provide a solution. Figure E4.14B shows a portion of the system curve from Example 4.13 superimposed on performance curves for a $5 \times 5 \times 7$ pump operating 1750 rpm. This figure indicates that a 6-in. impeller will deliver 460 gpm at 24 ft of head. The pump will draw approximately 4 hp and operate with an efficiency of 69%. This is a nice choice for several reasons.

- Notice that this is a $5 \times 5 \times 7$ pump. This means the suction and discharge ports are 5-nom, which matches the pipe in the system exactly. Therefore, no reducing fittings are required to connect the pump to the pipeline.
- Look at the location of the pump/system operating point. It lies to the *right* of the best efficiency point. Therefore, using the globe valve to bring the flow back down to 400 gpm will *increase* the efficiency of the pump.

Notice the *fine print* in the bottom left-hand corner of the performance curves in Figure E4.14B. This particular pump manufacturer will trim the impeller in $\frac{1}{8}$-in. increments to provide the required capacity. Therefore, by specifying a trim to a 5¾-in. impeller, the capacity will be approximately 410 gpm at 22 ft of head as shown in Figure E4.14B. This pump will draw about 3.25 hp and operate with an efficiency of 67.5%.

Another possibility is to consider the Series 80-SC pump operating at 1150 rpm. The performance map for the 1150 rpm pumps suggests that the $5 \times 5 \times 9$½ model, shown in Figure E4.14C, may also be a good selection. Overlaying the system curve on the $5 \times 5 \times 9$½ pump curves reveals two other possible selections. From this figure, an 8½-in. impeller will deliver 440 gpm at 22.5 ft of head. This pump will draw about 3.2 hp and operate with an efficiency of 77%. By trimming the impeller to 8¼-in., the capacity will be about 405 gpm at 22 ft of head. This pump will draw about 2.9 hp and operate with an efficiency of 76.5%.

Table E4.14 summarizes the possible pumps that can be selected for this application. Any of these pumps will work for the given application. However, there is a clear trade-off. The higher rpm pumps are smaller, but they tend to draw more power than the lower rpm pumps. The smaller pump may be cheaper because of size, but it may be more expensive to run from an energy-cost point of view.

Other factors need to be taken into account before a final decision is made. For example, it is not clear from the given drawing, but there may be a limited amount of space to install the pump. Oftentimes, space in a processing facility is at a premium. This may sway the decision toward the smaller diameter impeller pumps.

Another issue that may be important is the capacity. By trimming the impellers, the required flow rate of 400 gpm can be very closely achieved. Minimal throttling of the globe valve is required to provide 400 gpm. However, there

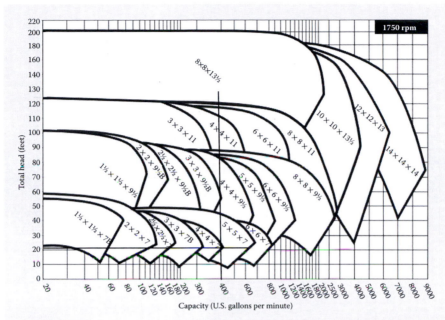

FIGURE E4.14A

may be a possibility of increased capacity in the future. If this is important, perhaps a larger impeller diameter (maybe even larger than those listed here) is desirable. Of course, this means that the throttling will be greater to achieve 400 gpm for current use.

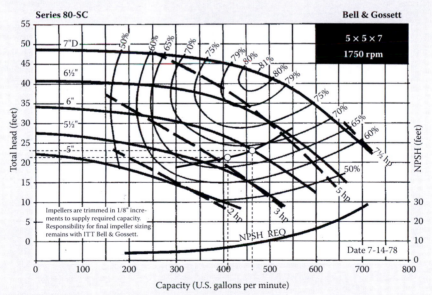

FIGURE E4.14B

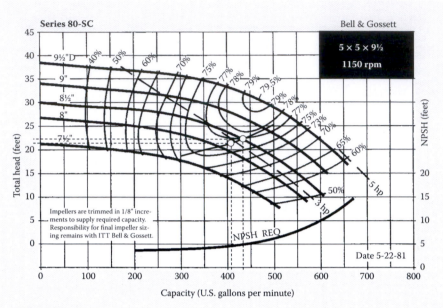

FIGURE E4.14C

TABLE E4.14

Possible Bell & Gossett Series 80-SC Pumps for the Application Shown in Example 4.14

Model	Impeller Diameter (in.)	Capacity (gpm)	Head (ft)	Power Draw (hp)	Efficiency (%)
5 × 5 × 7 (1750 rpm)	5¾	410	22	3.25	67.5
5 × 5 × 7 (1750 rpm)	6	460	24	4	69
5 × 5 × 9½ (1150 rpm)	8¼	405	22	2.9	76.5
5 × 5 × 9½ (1150 rpm)	8½	440	22.5	3.2	77

somewhat interactive in that system curves can be automatically overlaid on the pump curves for quick visual validation that the pump may or may not be an acceptable solution.

4.7.6 Cavitation and the Net Positive Suction Head

Cavitation is a phenomenon that is to be avoided in liquid pump/pipe systems. In pumps, cavitation is prevented by designing the system so that the *net positive suction head* (NPSH) available in the system is greater than the NPSH required by the pump.

Figure 4.22 shows a qualitative pressure trace through a centrifugal pump. If the pressure of the fluid at any point drops below its vapor pressure, then bubbles will form. In a centrifugal pump, the lowest pressure occurs at the eye of the impeller. When the bubbles move to a higher pressure region in the pump, they collapse. This bubble collapse causes a propagation of pressure waves, which results in damage to the pump known as *cavitation erosion*. A pump that is experiencing cavitation makes a loud rattling noise and can experience excessive vibration. To avoid cavitation, calculations involving the NPSH must be performed.

The *net positive suction head available* (NPSHA) is the difference between the pressure head at the suction port of the pump and the vapor pressure head of the fluid. If this value is negative, the pump will experience cavitation. In fact, it is entirely possible that the pump may experience cavitation with a positive NPSHA because of a pressure drop before the fluid gets to the eye of the impeller (Figure 4.22). To determine whether or not cavitation will occur, pump manufacturers provide information on the *net positive suction head required* (NPSHR). This is determined experimentally, and is reported on the pump performance curve as a function of capacity. Refer to any of the pump performance curves in Appendix G to see the NPSHR curves for various pumps. Notice that the NPSHR is read on the right-hand axis for the Bell & Gossett curves.

Cavitation will be prevented if the NPSHA is greater than the NPSHR. It is good design practice to add a safety factor (SF) and actually exceed the NPSHR:

$$\text{NPSHA} - \text{NPSHR} \geq \text{SF} \qquad (4.72)$$

It is common practice to have the NPSHA exceed the NPSHR by SF = 3 ft (≈1 m).

4.7.6.1 Calculating the NPSHA

The NPSHA is a characteristic of the suction side of the pipe system. In a system where fluid is being moved from a tank, there are two possible pumping scenarios, as shown in Figure 4.23: *suction lift* and *flooded suction*. Of these two possibilities, the suction lift is more prone to cavitation.

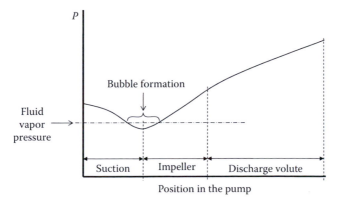

FIGURE 4.22
A pressure trace through a centrifugal pump showing cavitation.

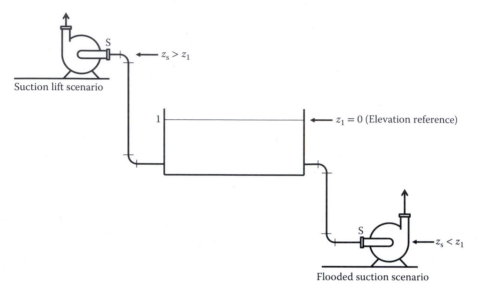

FIGURE 4.23
Suction lift and flooded suction configurations.

In either scenario, the conservation of energy equation written from the free surface of the tank (1) to the suction port of the pump (s) results in

$$\frac{P_1}{\gamma} + z_1 = \frac{P_s}{\gamma} + \frac{V_s^2}{2g} + z_s + \left(f\frac{L}{D} + \sum K \right)\frac{V_s^2}{2g} \tag{4.73}$$

Solving this equation for the pressure head at the suction port of the pump

$$\frac{P_s}{\gamma} = \frac{P_1}{\gamma} + (z_1 - z_s) - \left(f\frac{L}{D} + \sum K + 1 \right)\frac{V_s^2}{2g} \tag{4.74}$$

In Equation 4.74, the elevation difference is referenced to the surface of the fluid in the tank ($z_1 = 0$). This elevation difference is positive for a flooded suction application ($z_s < 0$) and negative for a suction lift application ($z_s > 0$). The NPSHA is the difference between the pressure head at the suction port and the fluid's vapor pressure head. Therefore, the NPSHA can be written as follows:

$$\text{NPSHA} = \frac{P_s}{\gamma} - \frac{P_v}{\gamma} = \frac{P_1}{\gamma} + (z_1 - z_s) - \left(f\frac{L}{D} + \sum K + 1 \right)\frac{V_s^2}{2g} - \frac{P_v}{\gamma} \tag{4.75}$$

Combining Equations 4.75 and 4.72, and solving for the elevation of the suction port of the pump results in

$$z_s = z_1 + \frac{P_1 - P_v}{\gamma} - \left(f\frac{L}{D} + \sum K + 1 \right)\frac{V_s^2}{2g} - (\text{NPSHR} + \text{SF}) \tag{4.76}$$

When forming Equation 4.76, the equality of Equation 4.72 was used. The value of z_s indicates the location of the pump with respect to the free surface of the liquid to prevent cavitation from occurring. In a suction-lift scenario, if the pump is installed at an elevation higher than z_s, cavitation will most likely occur.

EXAMPLE 4.15

Water at 100°F is being pumped vertically upward from an open holding tank by a Bell & Gossett Series 1531 Model 2AC pump with a 6½-in. impeller operating at 3500 rpm, as shown in Figure E4.15A. When the pump is installed in the pipe system, the head delivered by the pump is 150 ft. The suction pipe is 3-nom sch 40 commercial steel, and the fittings and valve are standard. Atmospheric pressure is 14 psia. Determine the maximum height above the water surface in the tank that the pump can be installed to prevent cavitation.

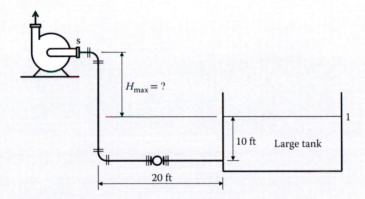

FIGURE E4.15A

SOLUTION: By defining the tank surface to be the elevation reference ($z_1 = 0$ ft), the value of H_{max} is equal to the difference between the elevation of the suction port of the pump, z_s, and the free surface of the water in the tank, z_1. Therefore, Equation 4.76 can be written as follows:

$$H_{max} = \frac{P_1 - P_v}{\gamma} - \left(f\frac{L}{D} + \sum K + 1 \right)\frac{V_s^2}{2g} - (\text{NPSHR} + \text{SF})$$

The water is at a temperature of 100°F. From Appendix B.2, the vapor pressure at this temperature is

$$P_v = 0.9505 \text{ psia}$$

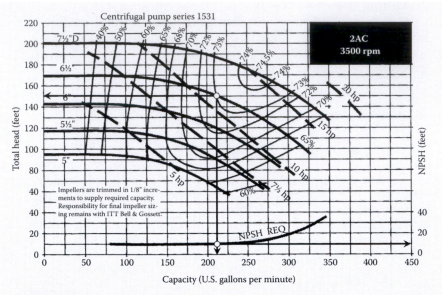

FIGURE E4.15B

The capacity and NPSHR can be found using the pump performance curve shown in Figure E4.15B.

$$\dot{V} = 212 \text{ gpm} \qquad \text{NPSHR} = 10 \text{ ft} \quad \text{(read on the right-hand axis)}$$

The total length of straight pipe on the suction side of the pump is unknown, since it depends on H_{max}. From Figure E4.15A, it can be seen that this relationship is

$$L = (20 + 10)\,\text{ft} + H_{max}$$

Substituting this expression into the conservation of energy equation results in

$$H_{max} = \frac{P_1 - P_v}{\gamma} - \left[f \frac{(30 \text{ ft} + H_{max})}{D} + \sum K + 1 \right] \frac{V_s^2}{2g} - (\text{NPSHR} + \text{SF})$$

The friction factor can be found using any of the correlations in Table 4.2. This requires calculation of the Reynolds number. The velocity of the water in the suction line can be found knowing the volumetric flow rate and the inside diameter of the pipe. Therefore, all parameters in this equation can be determined except H_{max}.

This equation can be rearranged algebraically to solve for H_{max}. However, it is easily solved using an equation solver. Using EES, the given information in the problem statement is shown as follows:

```
"GIVEN:  System shown in the diagram window"
"Pipe Data"
  D = 0.25567[ft]          "3-nom sch-40 - Appendix C"
  epsilon = 0.00015[ft]    "Table 4.3"
"Pump Data - at 150 ft of head for the 6 1/2-inch impeller.
These values are read from the Series 1531 Model 2AC 3500 rpm
pump found in Appendix G"
  Vol_dot = 212[gpm]
  NPSHR = 10[ft]

"Fluid Property Data"
  f$ = 'steam_iapws'
  T = 100[F]
  rho = density(f$,T=T,x=0)
  gamma = rho*g/g_c
  mu = viscosity(f$,T=T,x=0)*convert(lbm/ft-hr,lbm/ft-s)
  P_v = P_sat(steam_iapws,T=T)

"Ambient pressure at the tank surface"
  P_1 = 14.0[psia]

"Cavitation design saftey factor"
  SF = 3[ft]

"Constants"
  g = 32.174[ft/s^2]
  g_c = 32.174[lbm-ft/lbf-s^2]

"FIND:  Pump installation location to prevent cavitation"
```

The pump's suction diameter is 2-nom (specified by the "2" in the 2AC model). Therefore, a reducing fitting is required to connect the suction pipe to the pump. The loss coefficient for the reducer can be determined using Equations 4.18 and 4.20. The loss coefficients for the other fittings can be calculated using the expressions in Appendix E. Using this information and the equations shown above, the rest of the EES code for this solution is as follows:

```
"SOLUTION:"
"The conservation of energy written between the water surface and
the pump suction inlet is"
  H_max = (P_1 - P_v)*convert(psi,lbf/ft^2)/gamma -    &
    (f*L/D + SUMK + 1)*V_s^2/(2*g) - (NPSHR + SF)
  L = 30[ft] + H_max

"Friction factor calculation"
  f = 0.25*log10((epsilon/D)/3.7 + 5.74/Re^0.9)^(-2)
  Re = rho*V_s*D/mu
  Vol_dot*convert(gpm,ft^3/s) = A*V_s
  A = pi*D^2/4
```

```
"Losses due to the fittings and ball valve - Appendix E"
  K_ent = 0.5      "Entrance"
  K_v = 3*f_T      "Ball valve"
  K_e = 30*f_T     "Elbow"
  f_T = 0.25*(log10((epsilon/D)/3.7))^(-2)
"A reducer will be needed to connect the 3-nom pipe to the 2-
   nom pump suction port,"
  SUMK = K_ent + K_v + 2*K_e + K_2
"K for the reducer in based on the smaller diameter"
  K_1 = 0.5*(1 - beta^2)^2
"K for the reducer converted to the larger diameter"
  K_2 = K_1/beta^4
  beta = D_1/D_2 "Ratio of small diameter to large diameter"
  D_1 = 0.17225[ft]    "2-nom sch 40"
  D_2 = D              "3-nom sch 40"
```

The solution to this set of equations is $H_{max} = 9.176$ ft. This calculation shows that the pump can be located no higher than 9.2 ft above the surface of the water in the tank, otherwise cavitation will occur.

It may be tempting to neglect the effects of minor losses when calculating the NPSHA. Although this may not be accurate, it will provide a quick estimate for an initial design. However, since the friction effects reduce z_s, as seen in Equation 4.76, the importance of the minor losses should not be neglected in the final design calculations.

4.7.7 Series and Parallel Pump Configurations

Multiple pump systems are used to provide pressure-boosting stations along a very long pipe line (e.g., the Alaskan Pipeline), or to provide redundancy in a system should one of the pumps fail. Pumps can be connected together in *series* or *parallel*. In multiple pump systems, the design engineer is interested in the *composite* performance of all the pumps operating together.

The most common application for pumps in *series* is for boosting pressure along a long pipeline. In a long pipeline, the friction losses may cause a single pump to be very large and cost-prohibitive. Therefore, boosting the pressure with smaller pumps along the pipeline is a cost-effective way of transporting the fluid through the pipeline. Pumps in series are *head additive*. This means that the composite performance of a series pump combination is equal to the sum of the head produced by each individual pumps. This is illustrated in Figure 4.24 for the case of two identical pumps operating in series.

Parallel pump circuits are used to provide system redundancy in the case of a pump failure. Parallel pump systems are also helpful in using several smaller pumps to provide a large flow rate. Since the head across all pumps in a parallel circuit is the same, parallel pumps are *capacity additive*. This is shown in Figure 4.25 for the case of two identical pumps in parallel.

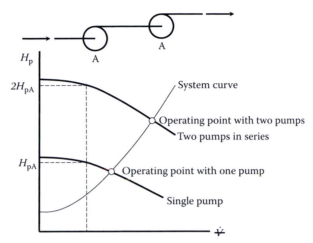

FIGURE 4.24
Performance curves and operating points for two identical pumps operating in series.

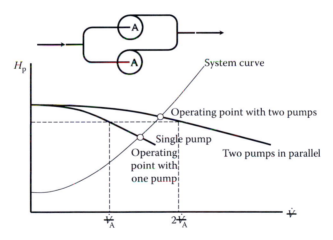

FIGURE 4.25
Performance curves and operating points for two identical pumps operating in parallel.

EXAMPLE 4.16

Two Bell & Gossett Series 80-SC Model 1½ × 1½ × 7B pumps with 6½-in. impellers are operating in parallel at 1150 rpm to deliver 70 gpm of water at 16.9 ft of head with an elevation head increase of 10 ft in the system. Determine (a) the system curve, (b) the composite pump curve, (c) the operating efficiency of each pump, and (d) if one of the pumps fails, determine the capacity, head, and efficiency of the remaining pump.

SOLUTION:

a. There is not much information given about the system. However, there is enough information to draw the system curve. Since the system curve is a parabola centered about the y axis, the system curve can be written as

$$H_p = \Delta z + b\dot{V}^2$$

The constant b can be determined knowing the operating point (70 gpm @ 16.9 ft).

$$b = \frac{H_P - \Delta z}{\dot{V}^2} = \frac{(16.9 - 10)\,\text{ft}}{(70\text{ gpm})^2} = 0.001408\,\frac{\text{ft}}{\text{gpm}^2}$$

This information can be used to construct the system curve by conducting a parametric study. Using EES, the resulting system curve is shown in Figure E4.16A.

b. The composite pump curve can be determined by reading data from the 6½-in. impeller curve, doubling the capacity at the same head, and plotting the results. This can be done for several points along the performance curve. These data can be overlaid on the system curve, as shown in Figure E4.16B.

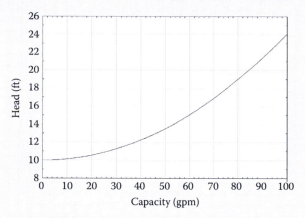

FIGURE E4.16A

c. The pump curves for this model pump from Appendix G are shown in Figure E4.16C. Since parallel pumps are capacity additive, each pump is providing half of the total capacity. Therefore,

$$\dot{V}_{single\ pump} = \frac{70\text{ gpm}}{2} = 35\text{ gpm}$$

The efficiency of each pump can be read directly from the performance curves at 35 gpm for the 6½-in. impeller (point A on the pump curves shown in Figure E4.16C). This point indicates that each pump will operate at an efficiency of 53%. Notice that this point is the same point identified by 35 gpm at 16.9 ft of head. This must be true since the pressure head across each pump in the parallel circuit must be the same.

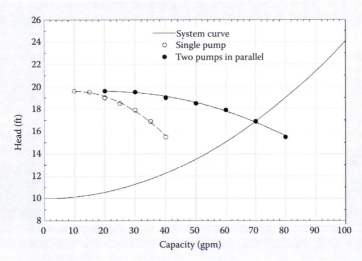

FIGURE E4.16B

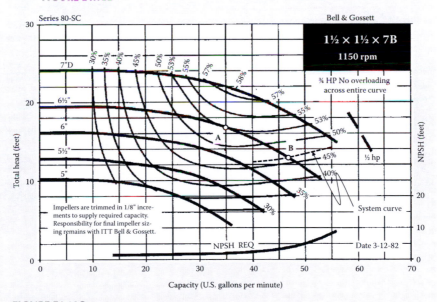

FIGURE E4.16C

d. When one of the pumps fails, the other pump's operating conditions change to match the system curve. Loss of a pump does not change the system curve. However, it does change the operating point. The pump curves in Figure 4.16C show a portion of the system curve (dashed line). The intersection of the system curve with the pump curve for the 6½-in. impeller is the operating point when one of the pumps fails (point B). This operating point indicates that a single pump will deliver 47 gpm at 13.2 ft of head. It can also be seen that the efficiency of the pump drops to 46%.

In Example 4.16, it can be seen that in a parallel pump circuit, the failure of one pump reduces the total capacity and the efficiency of the second pump. In the example, the efficiency dropped from 53% to 46%. This may seem like a small change. However, if the efficiency drops too low, it is possible that the pump's motor would overheat because of increased electrical current flowing through the copper windings. Careful design work must be considered to avoid this unfortunate consequence. It is always a good idea to determine the operating condition of the remaining pump(s), when one or more pumps in a parallel system fail, to avoid possible overheating and burnout of a motor.

4.7.8 Affinity Laws

Pump curves from manufacturers are presented for a specified rotational operating speed of the pump. Many pumps use a single-speed motor because they are reasonably priced. However, as technology has improved over the years, the *variable-speed* motor is becoming very attractive for use with pumps. Varying the rotational speed of the pump changes its performance. Given the pump curves at a single operating speed, the design engineer can use the *affinity laws* to predict its performance at a different operating speed. The affinity laws are also helpful if the impeller diameter is modified along with the rotational speed.

The affinity laws are derived using dimensional analysis and similitude. If all pumps of one type (e.g., centrifugal) behave similarly, then dimensional analysis should reveal how the pump parameters scale. In dimensional analysis, the Buckingham Pi theorem is used to determine pertinent dimensionless groups. When analyzing centrifugal pumps using the Buckingham Pi theorem, the performance parameters of the pump (head, power, and efficiency) are assumed to be a function of the fluid's density (ρ) and viscosity (μ), the rotational speed of the pump (ω), the impeller diameter (D), and the capacity delivered by the pump ($\dot{V}$). Application of the Buckingham Pi theorem to this functional dependence reveals five dimensionless groups shown in Table 4.13. The resulting functional dependence of these dimensionless groups is shown in the following equations:

$$\frac{gH_p}{\omega^2 D^2} = f_1\left(\frac{\dot{V}}{\omega D^3}, \frac{\rho \omega D^2}{\mu}\right)$$

$$\frac{\dot{W}}{\rho \omega^3 D^5} = f_2\left(\frac{\dot{V}}{\omega D^3}, \frac{\rho \omega D^2}{\mu}\right) \qquad (4.77)$$

$$\eta = f_3\left(\frac{\dot{V}}{\omega D^3}, \frac{\rho \omega D^2}{\mu}\right)$$

TABLE 4.13

Dimensionless Groups Related to Centrifugal Pump Performance

Dimensionless Group Name	Definition
Rotational Reynolds number	$\dfrac{\rho \omega D^2}{\mu}$
Volumetric flow coefficient	$\dfrac{\dot{V}}{\omega D^3}$
Head coefficient	$\dfrac{g H_p}{\omega^2 D^2}$
Power coefficient	$\dfrac{\dot{W}}{\rho \omega^3 D^5}$
Efficiency	η

Laboratory experiments indicate that the head coefficient, power coefficient, and efficiency are only weakly dependent on the rotational Reynolds number. Therefore, Equation set 4.77 can be written as

$$
\begin{aligned}
\frac{g H_p}{\omega^2 D^2} &\approx f_1\left(\frac{\dot{V}}{\omega D^3}\right) \\[2mm]
\frac{\dot{W}}{\rho \omega^3 D^5} &\approx f_2\left(\frac{\dot{V}}{\omega D^3}\right) \\[2mm]
\eta &\approx f_3\left(\frac{\dot{V}}{\omega D^3}\right)
\end{aligned}
\tag{4.78}
$$

Equation set 4.78 is known as the *affinity laws*. Given the performance data for a pump operating at a certain condition, the affinity laws can be used to predict the performance of the same pump at different conditions (e.g., changing the rotational speed and/or changing the impeller diameter). To apply the affinity laws, it is assumed that for geometrically similar pumps, the dimensionless groups shown in Equation set 4.78 are equal at different operating conditions.

There is no substitution for manufacturer's, pump curves. However, the affinity laws can provide a reasonable approximation to pump performance. The affinity laws are particularly helpful to the design engineer when variable-speed motors are used to drive a pump.

EXAMPLE 4.17

The performance curves for a Bell & Gossett Series 80-SC 2 × 2 × 7 pump operating at 1750 rpm are shown in Figure E4.17A. Consider a 6-in. impeller pump delivering 80 gpm of water.

    a. Determine the head, power draw, and efficiency of this pump from the pump curves.

    b. Using the affinity laws, determine the capacity, head, power draw, and efficiency if the impeller diameter is changed to 6½ in.

    c. Compare the results of the affinity law calculations to the actual values on the performance curves.

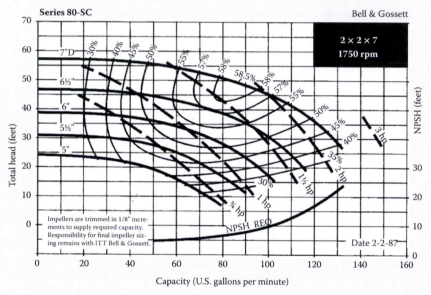

FIGURE E4.17A

SOLUTION:

    a. At a capacity of 80 gpm, the 6-in. impeller pump, the head, power draw, and efficiency can be read from the performance curves.

$$D_o = 6 \text{ in.} \quad \dot{V}_o = 80 \text{ gpm} \quad H_o = 28 \text{ ft} \quad \dot{W}_o = 1.15 \text{ hp} \quad \eta = 51\%$$

The subscript "o" on these values indicates the original operating condition of the pump.

    b. To apply the affinity laws, it is assumed that the dimensionless groups in Equation set 4.78 are the same for geometrically similar pumps. Changing only the impeller diameter should result in a new pump configuration that is geometrically similar to the original. Therefore the new capacity of the pump can be determined by equating the volumetric flow coefficients.

$$\frac{\dot{V}_n}{\omega_n D_n^3} = \frac{\dot{V}_o}{\omega_o D_o^3}$$

In this equation the subscript "n" signifies the new operating condition. Solving this equation for the new capacity,

$$\dot{V}_n = \dot{V}_o \frac{\omega_n D_n^3}{\omega_o D_o^3} = \dot{V}_o \left(\frac{D_n}{D_o}\right)^3 = (80 \text{ gpm})\left(\frac{6.5 \text{ in.}}{6 \text{ in.}}\right)^3 = \underline{101.7 \text{ gpm}} \quad \leftarrow$$

Equating the head coefficients allows for predicting the new head delivered by the pump.

$$\frac{gH_{p,n}}{\omega_n^2 D_n^2} = \frac{gH_{p,o}}{\omega_o^2 D_o^2} \rightarrow H_{p,n} = H_{p,o}\left(\frac{D_n}{D_o}\right)^2 = (28 \text{ ft})\left(\frac{6.5 \text{ in.}}{6 \text{ in.}}\right)^2 = \underline{32.9 \text{ ft}} \quad \leftarrow$$

The power draw under these new operating conditions can be predicted by equating the power coefficients.

$$\frac{\dot{W}_n}{\rho\omega_n^3 D_n^5} = \frac{\dot{W}_o}{\rho\omega_o^3 D_o^5} \rightarrow \dot{W}_n = \dot{W}_o\left(\frac{D_n}{D_o}\right)^5 = (1.15 \text{ hp})\left(\frac{6.5 \text{ in.}}{6 \text{ in.}}\right)^5$$

$$= \underline{1.7 \text{ hp}} \quad \leftarrow$$

Since the efficiency is one of the five dimensionless groups in the affinity laws, the efficiency of the pump should not change.

$$\eta_n = \eta_o = \underline{51\%} \quad \leftarrow$$

c. Figure E4.17.B shows an enlarged section of the performance curves with the original operating condition and the new operating condition predicted by the affinity laws.

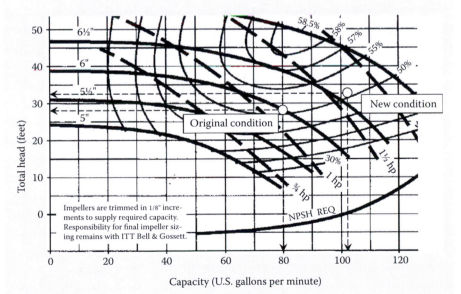

FIGURE E4.17B

As seen in this figure, the affinity laws prediction is not exact (the new condition should be on the 6½-in. impeller curve), but it is reasonably close. This may be attributed to inaccurate reading of the original operating parameters from the pump curves. It can also be attributed to the assumption that the performance parameters are only weakly dependent on the rotational Reynolds number. Equation set 4.78 is *approximate* based on this assumption.

A final comment is in order concerning the affinity laws. These laws scale the pump performance curves. The affinity laws *do not* scale pump/pipe system operating points. This is shown in Figure 4.26. In this figure, only the pump impeller is changed, and the affinity laws are used to predict the new pump performance curve. The operating speed of the pump is the same for the original and new condition. As shown in Figure 4.26, the original operating point, a point on the original pump performance curve, scales to a new performance point that is not necessarily consistent with the new pump/pipe system operating point. The only way to determine the new pump/pipe system operating point is to intersect the system curve with the new pump performance curve predicted from the affinity laws.

4.8 Design Practices for Pump/Pipe Systems

There are many factors that the design engineer must take into account when designing a pump/pipe system. Throughout this chapter, many of these design practices have been mentioned. In this section, a summary of some of the common design practices is presented. This is not intended to be all-inclusive. Nothing substitutes for experience. As the design engineer becomes more experienced in this area, he/she will no doubt build on these practices.

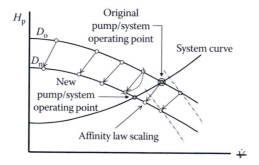

FIGURE 4.26
Demonstration of how the affinity laws scale pump performance but not the pump/pipe system operating point.

4.8.1 Economics

Economics is usually the bottom line in industry. Hydraulic systems need to be designed to have minimum cost. This can be accomplished by carefully laying out the pipe system to minimize the friction losses while maintaining an economic diameter. Careful layout of the system includes wise placement of valves and fittings. Although this may seem obvious, there may be physical limitations. Consider the pipe system shown in Figure 4.27. Good design skills are required to develop an economic pipe system that fits within the confines of the space.

Minimization of friction losses is one way to help reduce cost. Designing systems using the economic diameter is also a good way to minimize costs. As demonstrated in Section 4.6, the economic diameter is one that minimizes the total annual cost per foot of pipe. This includes the amortized capital costs *and* the cost of energy required to move the fluid through the system.

4.8.2 Environmental Impact

There are many ways to drive a pump. However, the predominant method is by using electric motors. According to the International Energy Agency (IEA), in 2011 electric-motor-driven systems (EMDS) accounted for 43%–46%

FIGURE 4.27
A complex pipe network demands good design skills. (Courtesy of Northstar Agri Industries, Fargo, North Dakota. Reprinted with permission.)

of all global electricity use (Waide and Brunner 2011). Electrical power generation using fossil fuels results in the emission of carbon dioxide because of the combustion of the fuel. The IEA estimates that without comprehensive energy policy measures, energy consumption attributed to EMDS is expected to rise to 13,360 TWh/y and carbon dioxide emissions to 8,570 Mt/y on a global scale by 2030. In 2011, end users of EMDS spent $565 billion globally. By 2030, that could rise to nearly $900 billion.

Not all EMDS are pump/motor combinations. However, there are millions of residential and commercial systems that utilize pumps to circulate fluids. The message here is that the real cost of a pump is not just economic; it becomes an environmental issue as well. Selecting the right pump helps minimize the depletion of the natural resources used to generate the electricity required to run the pump. This goes hand in hand with helping to reduce carbon dioxide emissions in the electrical generation process.

4.8.3 Noise and Vibration

In cases where pipes are located in an occupied environment, such as an office, the fluid circulating through the pipes should cause no noise. In addition, excessive pressure drops in pipe systems can lead to vibration, which should also be avoided. Recommended pressure drops to avoid noise and vibration are 25–30 psi/1000 ft for liquids, 10–15 psi/1000 ft for gases, and 4 psi/100 ft for low-pressure steam.

4.8.4 Pump Placement and Flow Control

When placing a pump in a pipe network, it must be located to avoid cavitation. A NPSH analysis should be done for any pump installation to ensure that cavitation will be prevented.

Flow control can be accomplished by using control valves and/or variable-speed motors instead of single-speed motors to drive the pump. If possible, it is recommended that valves be avoided for flow control because they introduce more friction resistance. This may not always be possible, and a flow control valve must be specified. In this case, the valve is to be located at the *discharge* of the pump. Placing the valve in the suction line makes the pump more susceptible to cavitation.

Variable-speed motors with digital control provide an attractive alternative to control valves. In the past, these types of systems may have been cost prohibitive. However, because of rapid improvements in technology, the cost of digitally controlled variable-speed motors is becoming more reasonable. This is a more eco-friendly alternative to flow control compared to a pressure drop across a control valve.

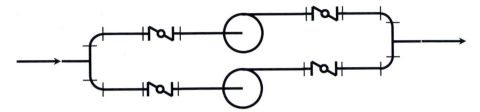

FIGURE 4.28
Using valves to provide an opportunity to isolate a pump for repair or maintenance.

4.8.5 Valves

As discussed in Section 4.8.5, valves can be used for flow control. Another use of valves is to isolate sections of a system for repair and maintenance. An example is shown in Figure 4.28 where four butterfly valves are used. The butterfly valve is used here because it provides positive shutoff, as discussed in Table 4.5. By shutting off the valves on either side of one of the pumps, it can be taken off-line for repair or maintenance while the other pump continues to circulate fluid through the system. The effect of taking one of the pumps off-line can be determined using the procedures outlined in Section 4.7.7.

4.8.6 Expansion Tanks and Entrained Gases

As discussed in Section 4.7.4.2, expansion tanks are often incorporated in closed-loop systems. The expansion tank allows the system to operate in a steady state even when there are temperature variations that can cause the total fluid volume to increase or decrease. The expansion tank also creates a point of constant pressure in the system.

In many cases, particularly startup, there may be nonsoluble gases (e.g., air) entrained in the liquid flowing through the system. These entrained gases can create noise as the liquid circulates through the pipes, and they also cause problems similar to cavitation in pumps. Entrained gases can be removed from a system by providing a high-elevation point in the system with bleed ports. In a closed-loop system, strategic location of the expansion tank can provide an opportunity to bleed off entrained gases.

4.8.7 Other Sources for Design Practices

Listed previously are only some of the more common design practices for pump/pipe system design. There are other sources that the design engineer can consult, including piping handbooks and guides, trade journals (for state-of-the-art information), company design practices, and experienced design engineers.

Problems

4.1 Consider an 8-nom commercial steel pipe carrying propylene at 40°F. The volumetric flow rate of the propylene is 800 gpm. Complete Table P4.1 showing the variation of the propylene parameters inside the pipe as a function of the pipe schedule used.

TABLE P4.1

Schedule	Fluid Parameters Inside the Pipe		
	V (ft/s)	Re	f
20			
40			
80			
120			
160			

4.2 Liquid Refrigerant 123 (R-123) at 50°C is passing through a ½-std copper tube at a flow rate of 1200 kg/h. Complete Table P4.2 showing the variation of the R-123 parameters inside the tube as a function of the type of tubing used.

TABLE P4.2

Type	Fluid Parameters Inside the Tube		
	V (m/s)	Re	f
K			
L			
M			

4.3 A 10-nom sch 40 horizontal commercial steel pipe is being used to transport octane. The pipe is 200 ft long. The octane is at 50°F and is flowing at 600 gpm. Determine the pressure drop (psi) of the octane as it flows through the pipe.

4.4 Cyclohexane is flowing through a 4-std type L copper tube. The tube is 200 m long and the outlet is 10 m higher than the inlet. The pressure drop through the tube is 90 kPa. The cyclohexane is at a temperature of 30°C. Determine the volumetric flow rate (L/s) of the cyclohexane through the tube.

4.5 Water is flowing through a 2½-nom xs cast iron pipe at a rate of 5 L/s. The pipe is 75 m long and its outlet is 6 m higher than the inlet. The water is at a temperature of 25°C. Determine the pressure drop of the water through the pipe.

4.6 A 10% ethylene glycol solution is flowing through a 4-nom sch 80 commercial steel pipe. The pipe is 350 ft long and is horizontal. The pressure drop through the pipe is 3 psi. The ethylene glycol is at a temperature of 20°F.
  a. Determine the volumetric flow rate of the ethylene glycol (gpm).

b. Investigate the effect of the ethylene glycol temperature by plotting the volumetric flow rate (gpm) as a function of temperature for $10°F \leq T \leq 100°F$.

4.7 A fuel line made of type M copper tube is 20 ft long. Diesel fuel (modeled as dodecane) at 80°F is being transported through the line and the line is horizontal. The required flow rate of the fuel is 1300 ft³/h and the allowable pressure drop in the line is 2 psi. Specify the appropriate size (standard diameter) for this fuel line.

4.8 A horizontal pipeline is used to transport ethanol (ethyl alcohol) at 20°C over a distance of 3 km. The pipe is made of commercial steel. The required flow rate of the ethanol is 40 L/s. The pump connected to this pipeline can overcome a pressure drop of 200 kPa in the pipe. Specify the appropriate size (nominal diameter) of a schedule 40 pipe for this pipeline.

4.9 The Alaskan pipeline runs 798 miles from Prudhoe Bay to Valdez. Both cities are at sea level. The design specifications for the pipeline require a capacity of 2.4 million barrels per day (1 barrel = 42 gallons). The crude oil is maintained at 140°F in the pipeline. At this temperature, the oil has a density of 53.7 lbm/ft³ and a viscosity of 0.00257 lbm/ft·s. The pipe is made of commercial steel. Along the pipeline are pumping stations capable of boosting the oil pressure. Each pumping station can provide a pressure increase of 1100 psi.

a. Determine the number of pumping stations required along the pipeline as a function of pipeline diameter (i.e., generate a plot of the number of stations vs. pipeline inside diameter).

b. From the results of (a), specify a pipe size (nominal size and schedule) for the pipeline and defend your selection. Note: The pipe tables in Appendix C may not go high enough in pipe diameter. You may need to do some research to find another source for large-pipe dimensions.

4.10 Fluid flow can cause excessive wear on the inside surface of a pipe, resulting in a much rougher surface over time. This increased surface roughness results in a larger pressure drop, which requires more pumping energy to move the fluid compared to a new pipe. Consider an expensive 14-nom sch 120 pipe that has been in service for 50 years. The pipe is horizontal. It is 1000 ft long and is transporting water at 70°F with a velocity of 8 ft/s. Because of wear, the absolute roughness of the pipe has become $\varepsilon = 0.1$ in.

The replacement of such a large, thick-walled pipe represents a significant cost. So, an alternative solution is being investigated. A ½-in. thick lining can be purchased that would reduce the absolute roughness to $\varepsilon = 0.0004$ in. It is anticipated that installing the lining will provide another 50 years of useful service for the pipe. The pump and motor combination that moves the water through the pipe has an efficiency of 54%, and the cost of energy is 0.08 $/kW·h. The pipeline runs continuously all year long. The mass flow rate of water carried by the original pipe must be the same as that carried through the lined pipe.

a. Determine the annual savings in energy cost for a lined pipe compared to the old, rough pipe.

b. If the cost of the liner is 60 $/ft, determine the rate of return that results from the investment in the liner.

c. On the basis of your analysis in part (b), is lining the pipe a good economic decision?

4.11   A gravity-feed piping system between two large tanks is used to transport a 20% magnesium chloride solution as shown in Figure P4.11. The magnesium chloride is at 80°F and the tanks are open to the atmosphere. The piping system is 2½-nom sch 40 commercial steel. All fittings and valves are 2½ nom. The total straight length of pipe in the system is 60 ft. The elevation difference between the free surfaces of the tanks is 15 ft. Determine the volumetric flow rate (gpm) between the two tanks.

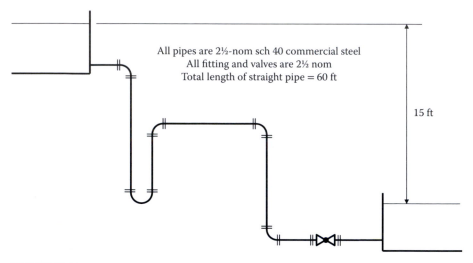

All pipes are 2½-nom sch 40 commercial steel
All fitting and valves are 2½ nom
Total length of straight pipe = 60 ft

15 ft

FIGURE P4.11

4.12   A tilting disc check valve is to be installed in a 4-nom sch 40 steel pipeline. For the valve being considered, the angle $\alpha = 15°$. The fluid being transported through the pipeline is a 20% ethylene glycol solution at a 20°F. The volumetric flow rate of the ethylene glycol is 180 gpm.
   a.   Specify the size and ASME class of the check valve required for this situation.
   b.   Determine the total head loss (ft) associated with the addition of the check valve and any required reducers to the pipe.

4.13   A stop-check angle valve is to be installed in a 3-nom sch 40 steel pipeline. For the valve being considered, $\beta = 1$. The fluid being transported through the pipeline is water at 60°F. The volumetric flow rate of the water is 150 gpm.
   a.   Specify the size and ASME class of the check valve required for this situation.
   b.   Determine the total head loss (ft) associated with the addition of the check valve and any required reducers to the pipe.

4.14   A 1¼-nom class 400 wye fitting with equal leg diameters has 40 gpm of acetone flowing into the fitting. Seventy percent of the flow leaves the fitting through the straight run and the remaining flow leaves through the branch that is at an angle of 30° to the straight run. Determine the total head loss (ft) for this wye fitting.

4.15 Figure P4.15 shows a pump and pipe network being used to transport heptane at 90°F to a large, elevated storage tank that is closed, but vented to ensure that the pressure above the heptane is atmospheric. The volumetric flow rate of the heptane is 500 gpm.
  a. Determine the pressure drop (psi) that the pump must overcome to move the heptane through the discharge pipe.
  b. Investigate the effect of the absolute roughness of the pipe. Construct a plot that shows the variation in pressure drop as a function of the absolute roughness for values ranging from 0 to 0.01 ft.

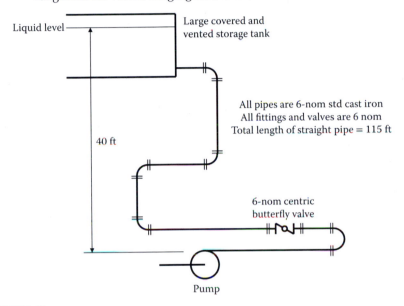

FIGURE P4.15

4.16 A 40% propylene glycol solution at 25°C is being delivered from a large storage tank with the gravity-feed pipe network shown in Figure P4.16. A concentric reducer is used to reduce the pipe size from 3-nom sch 40 to 2-nom sch 40 as shown in the figure. All pipes are commercial steel. Determine the following:

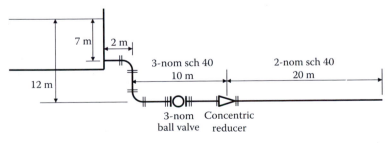

FIGURE P4.16

    a.   The volumetric flow rate of propylene glycol delivered by this system (L/s)
    b.   The velocity of the propylene glycol in each of the pipes (m/s)
The resistance coefficient for the concentric reducer can be estimated using Equations 4.18 and 4.20.

4.17   A large elevated tank (Tank 1) is being used to supply water at 60°F to two tanks at a lower elevation (Tanks 2 and 3) as shown in Figure P4.17. All pipes are 5-nom sch 40 commercial steel and the globe valve is wide opened. The straight lengths of the piping in Sections AB, BC, and BD are indicated in the figure. The elevation of the free surface of the water in each tank is also indicated in the figure. Pumps are used to maintain the water level in each tank.

    a.   Determine the volumetric flow rate (gpm) delivered to each lower tank (Tanks 2 and 3) for the conditions shown.

    b.   The globe valve is used in this application for flow control. Determine the effect of closing the globe valve by constructing two plots; (1) the total volumetric flow from Tank 1, and (2) the volumetric flow delivered to each of the lower tanks (Tanks 2 and 3). Show each of these volume flows as a function of the globe valve constant in the resistance coefficient equation

$$K_{globe} = Cf_T$$

where C varies between 340 (wide opened) and 1000 (partially closed).

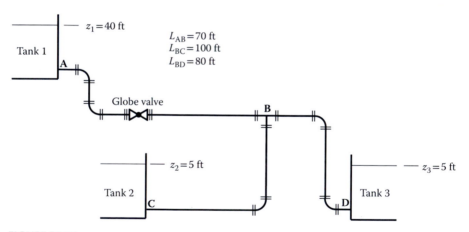

      $z_1 = 40$ ft

      $L_{AB} = 70$ ft
      $L_{BC} = 100$ ft
Tank 1 $L_{BD} = 80$ ft

Globe valve

B

$z_2 = 5$ ft $z_3 = 5$ ft

Tank 2   Tank 3

C   D

FIGURE P4.17

4.18   A pump is being used to deliver water at 65°F to two elevated tanks as shown in Figure P4.18. The pump is delivering a volumetric flow rate of 450 gpm. All pipes are 2½-nom sch 40 commercial steel. The straight lengths of the piping in sections AB, BC, and BD are indicated in the figure. The elevation of the free surface of the water in each tank is also indicated in the figure. Pumps draw water from the tanks, which keep the water level constant in each tank. Determine the pressure required at the pump discharge (psig) and the volumetric flow rate (gpm) delivered to each tank.

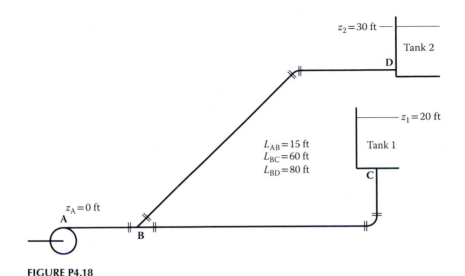

FIGURE P4.18

4.19   A commercial steel pipe is being designed to transport liquid benzene at 185°F, 120 psia to a heat exchanger at a flow rate of 850 gpm. The fittings, valves, supports, and pump are estimated to cost six times the installed pipe cost per foot. Annual maintenance is estimated to be 1.5% of the installed pipe and hardware initial costs per foot. For this service, the pipe classification is determined to be 400 and a sch-80 pipe is required. The pump in the system runs 18 h/day, 260 days per year. The efficiency of the pump is estimated to be 62%. The cost of electrical energy to run the pump is 0.11 $/kW·h. The pipeline is expected to have a 45-year life. An engineering economic analysis reveals that the rate of return because of the investment in the pipeline is 16%. Use an estimated escalation rate of 2% to account for yearly inflation.

   a.   Determine the economic diameter and specify the standard pipe size that results in the minimum system cost if the design work was done for an installation that occurred in 2014.

   b.   Determine the economic velocity of the fluid (ft/s).

   c.   Construct the cost curves for this scenario assuming that the friction factor can be assumed constant at the value computed for the economic diameter from part (a). Your plot should look similar to Figure 4.9.

4.20   A commercial steel schedule 40 pipe (class 300) is being used to transport liquid cyclohexane at 70°F, 80 psia at a volumetric flow rate of 1000 gpm. The fittings, valves, supports, and pump are estimated to cost 6.9 times the installed pipe cost per foot. Annual maintenance is estimated to be 2% of the installed pipe costs per foot. The pump in the system runs 12 h/day, 260 days per year. The efficiency of the pump is estimated to be 68%. The cost of energy is 0.09 $/kW·h. The pipeline is expected to have a 50-year life and a rate of return of 20%. Use an inflation rate of 1.5% to adjust the value of C_1 from Table 4.10 to 2012 dollars.

   a.   Determine the economic diameter and specify the standard pipe size that results in the minimum system cost.

   b.   Determine the economic velocity of the fluid (ft/s).

c.   Construct the cost curves for this scenario assuming that the friction factor can be assumed constant at the value computed for the economic diameter from part (a). Your plot should look similar to Figure 4.9.

4.21   A centrifugal pump is being used to deliver 200 gpm of water at 60°F. The pump head is determined to be 25 ft and its efficiency is 60%. Determine the brake horsepower input to the pump.

4.22   A centrifugal pump delivers 60 L/s of toluene at 30°C. The pump head is found to be 15 m and its efficiency is 48%. Determine the shaft power input to the pump (kW).

4.23   A centrifugal pump is being used to transport 440 gpm of benzene at 80°F. The suction pipe inlet to the pump is 6-nom sch 40 and the pump discharge pipe is 5-nom sch 40. Pressure gauges on the suction and discharge ports of the pump indicate −4 psig and 55 psig, respectively. The elevation difference between the suction and discharge ports of the pump is 0.2 ft. At this condition, the pump has an efficiency of 55%. Determine the following:
   a.   The head produced by this pump (ft)
   b.   The power input required by the electric motor that drives this pump (kW).
   c.   The annual cost required to operate this pump if it runs for 16 h/day for a total of 240 days, and the cost of electrical energy is 0.10 $/kW·h

4.24   A centrifugal pump is being used to transport 20 L/s of a 20% ethylene glycol solution at 20°C. The suction pipe inlet to the pump is 2½-nom sch 40 and the pump discharge pipe is 2-nom sch 40. Pressure gauges on the suction and discharge ports of the pump indicate −10 kPa and 180 kPa, respectively. The elevation difference between the suction and discharge ports of the pump is 0.08 m. At this condition, the pump has an efficiency of 62%. Determine the following:
   a.   The head produced by this pump (m)
   b.   The power input required by the electric motor that drives this pump (kW)
   c.   The annual cost required to operate this pump if it runs for 8 h/day for a total of 260 days, and the cost of electrical energy is 0.12 $/kW·h

4.25   A closed-loop pipe system similar to Figure 4.17 is being designed. At 300 gpm, the system curve indicates that the pump head is 120 ft. A 3 × 3 × 7B Bell & Gossett Series 80-SC pump operating at 3525 rpm is to be specified for this application.
   a.   Specify the impeller diameter for this application.
   b.   What is the operating point of the pump and pipe system with the impeller diameter you specified in part (a)? Specify the capacity (gpm), head (ft), brake horsepower input to the pump (hp), and efficiency of the pump.

4.26   A Bell & Gossett Series 1531 end-suction pump is to be selected for the system shown in Figure P4.26. The system is designed to move water at 60°F from the lower tank to the upper tank. The system must deliver a minimum of 500 gpm of water from the lower tank to the upper tank when the pump is operating. A tilting disc check valve ($\alpha = 15°$) is to be used to prevent the water from flowing backward when the pump is shut off. The total length of straight pipe in this system is 150 ft. The pipe is 4-nom sch 40 commercial steel.
   a.   Plot the system curve up to 800 gpm.
   b.   Select a Series 1531 Bell & Gossett pump for this system.
   c.   What is the operating point of the pump/pipe system based on the pump you selected in (b)? Specify the capacity (gpm), head (ft), brake horsepower input to the pump (hp), and efficiency of the pump.

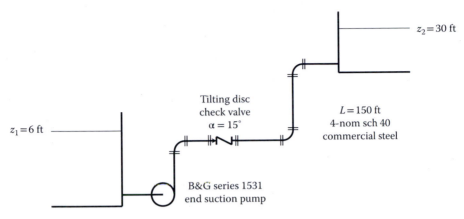

FIGURE P4.26

4.27 Water from a river is to be pumped to an irrigation ditch as shown in Figure P4.27. The water temperature is 55°F. The elevation difference between the river and the surface of the water in the irrigation ditch is 40 ft. A hinged-disc foot valve with a strainer serves two functions: (1) it filters the water entering the suction pipe and (2) it acts like a check valve when the pump is shut down. The pipeline contains 85 ft of 5-nom sch 40 commercial steel pipe. For successful irrigation, 600 gpm of water is required.

   a. Plot the system curve up to 1200 gpm.
   b. Select a Series 80SC Bell & Gossett inline pump for this system.
   c. What is the operating point of the pump/pipe system based on the pump you selected in (b)? Specify the capacity (gpm), head (ft), brake horse-power input to the pump (hp), and efficiency of the pump.

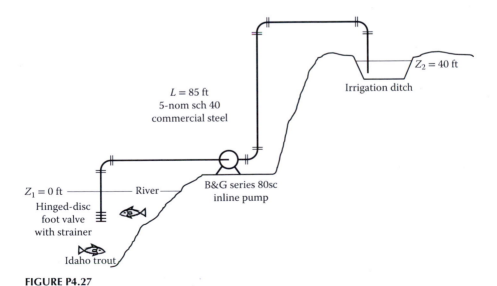

FIGURE P4.27

4.28   Two Bell & Gossett Series 80-SC 3 × 3 × 7B pumps with 5½-in. impellers are connected together in parallel. Both pumps are operating at 3525 rpm. This parallel pump combination is connected to a pipe system that has an elevation difference (outlet to inlet) of 40 ft. At a flow rate of 500 gpm, the head in the system is known to be 100 ft.

   a.   Plot a pump performance curve that shows the performance of each individual pump and the performance of both pumps operating in parallel.
   b.   Overlay the system curve on the plot you developed for part (a) and determine the operating point of this parallel pump/pipe system (head and capacity).
   c.   What is the operating point of each individual pump (head and capacity) when they are both operating in parallel? What is the efficiency and power draw of each pump under this condition?
   d.   If one of the pumps fails, determine the operating point (head and capacity) along with the efficiency and power draw of the remaining operational pump.

4.29   Two Bell & Gossett Series 80-SC 2 × 2 × 7 pumps operating at 1750 rpm are connected in parallel. A third pump, a Bell & Gossett Series 80-SC 2 × 2 × 7 pump operating at 3500 rpm is connected in series with the parallel pumps as shown in Figure P4.29.

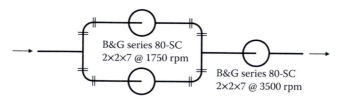

FIGURE P4.29

   This composite pump combination is connected to a pipe network with the following system curve:

$$H_p = 10[\text{ft}] + 0.0115\left[\frac{\text{ft}}{\text{gpm}^2}\right]\dot{V}^2$$

   Determine the following for this system:
   a.   The system curve up to 140 gpm
   b.   The composite pump curve up to 160 gpm
   c.   The operating point (head and capacity) of the combined pump and pipe system

4.30   Water at 70°F is being drawn from a large tank at a rate of 265 gpm as shown in Figure P4.30. The suction pipe is 3-nom sch 40 commercial steel. The total length of the suction line is 50 ft. The pump to be used is an inline pump. The particular pump selected has a required NPSH of 7 ft. Atmospheric pressure is 1 atm.

   a.   Determine the maximum height above the water surface that the pump can be installed to prevent cavitation.
   b.   Investigate the effect of the water temperature by plotting the maximum pump height as a function of the water temperature for 40°F ≤ T ≤ 140°F.

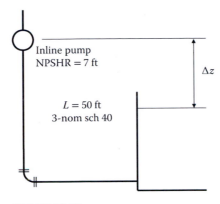

FIGURE P4.30

4.31   An inline pump is being used to transport water from a large tank. The water
is at a temperature of 20°C and is being delivered at a rate of 45 L/s. The suction line is 6-nom sch 40 commercial steel pipe that is 40 m long. The NPSH
required by the pump is 3.2 m. Atmospheric pressure is 1 bar.

a.   Determine the maximum height above the water surface that the pump
can be installed to prevent cavitation.

b.   Investigate the effect of the water temperature by plotting the maximum
pump height as a function of the water temperature for 10°C ≤ T ≤ 80°C.

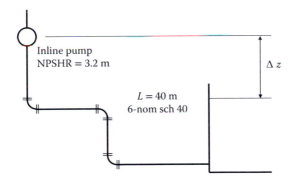

FIGURE P4.31

4.32   A centrifugal pump is tested in the laboratory. The pump has a 6-in. impeller
and is operating at 3500 rpm. With water as the fluid, test results indicate that
the pump delivers 280 gpm at 110 ft of head. Using the affinity laws, predict
what the pump capacity and head would be if the rotational speed was
reduced to 1750 rpm.

4.33   A centrifugal pump is being tested with water at 60°F. The pump is operating
at 1150 rpm and has a 7-in. impeller. Performance test results indicate that the
pump delivers 80 gpm at a head of 32 ft with a power draw of 2.5 hp. Using the
affinity laws, predict what the pump capacity, head, and power draw would
be if the operational speed was increased to 1800 rpm and the fluid

a.   Is water at 130°F

b.   Is toluene at 130°F

4.34   Laboratory experiments indicate that a centrifugal pump operating with a 7-in. impeller at 1150 rpm delivers 60 gpm of water at 60°F with a head of 27 ft. At this condition, the pump draws 1.8 hp. The affinity laws are now used to predict the performance of this pump at 1750 rpm. Conduct a parametric study to determine the capacity, head, and power draw of the pump at this new operating speed as function of the water temperature ranging from 40°F to 190°F. Comment on the results.

4.35   A Bell & Gossett Series 80SC 4 × 4 × 9½ model pump is delivering 400 gpm while operating at 1750 rpm with an impeller diameter of 8½ in. The pump's impeller is now changed to an 8-in. diameter size, and its operational speed is adjusted to 1150 rpm using a variable speed drive.

   a.   Using the affinity laws, determine the capacity, head, input power, and efficiency of the pump operating under these new conditions.

   b.   Using the new capacity calculated from the affinity laws, compare your answers calculated in part (a) to the actual values read from the Series 80-SC 4 × 4 × 9½ model pump curves operating at 1150 rpm with an 8-in. impeller.

   c.   Comment on the validity of the affinity laws for this application.

5

Energy Transport in Thermal Energy Systems

5.1 Introduction

Chapter 4 dealt with fluid transport in thermal energy systems. As fluids move through a pipe system, they are capable of transporting energy because of their pressure and temperature. In many thermal energy systems, fluids are used to transfer energy between different fluid streams. For example, in an air-conditioning system, a cold fluid circulating through tubes can be used to cool down warm air passing over the outside of the tubes. The energy transfer accomplished in this process is *heat transfer*. The heat transfer is accomplished by *conduction* through a metal separating the two fluids (e.g., a pipe or tube) and *convection* between the moving fluids and the metal interface. The device that accomplishes this heat transfer is a *heat exchanger*. Before embarking on a detailed discussion of heat exchangers, a brief review of heat conduction and convection, relative to heat exchangers, is presented.

5.2 Review of Heat Transfer Mechanisms in Heat Exchangers

The purpose of a heat exchanger is to transfer energy in the form of heat between two fluids. In a typical heat exchanger, this occurs by a combination of conduction and convection. Figure 5.1 shows the temperature profile between a hot fluid and a cold fluid in a heat exchanger. In this chapter, an uppercase T indicates the hot fluid temperature, and a lowercase t is used for the cold fluid's temperature. The temperatures T_{hot} and t_{cold} represent the bulk (or average) temperature of the hot and cold fluids, respectively.

Through the convective film, the temperature changes in a nonlinear fashion and the heat transfer rate is governed by *Newton's law of cooling*. For the

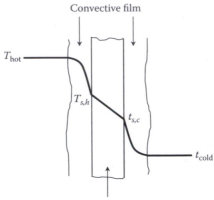

Convective film

T_{hot}

$T_{s,h}$

$t_{s,c}$

t_{cold}

Conduction through metal

FIGURE 5.1
Temperature profile between hot and cold fluids in a heat exchanger.

hot side of the heat exchanger shown in Figure 5.1, Newton's law of cooling is written as

$$\dot{Q} = h_h A_h \left(T_{hot} - T_{s,h} \right) \tag{5.1}$$

In this equation h_h is the *convective heat transfer coefficient*. This is an unfortunate nomenclature conflict when h is used to represent enthalpy. However, this is a common nomenclature and is used throughout this chapter. The convective heat transfer coefficient has units of Btu/h·ft²·°F or W/m²·K. Its value is meant to help quantify the resistance to heat flow through the convective film. Most heat exchangers utilize *forced convection* to transfer the heat between the fluids. Typical convective heat transfer coefficients for forced liquid flow are typically between 18 and 3500 Btu/h·ft²·°F (100–20,000 W/m²·K). Heat exchangers where one or both of the fluids are undergoing a phase change can experience much higher convective heat transfer coefficients: upward of 18,000 Btu/h·ft²·°F (100,000 W/m²·K). Heat exchangers that take advantage of the high heat transfer coefficients offered by a phase change are commonly known as *boilers, evaporators,* or *condensers.*

The *heat flux* is determined by dividing both sides of Equation 5.1 by the area through which the heat is flowing (A_h).

$$q_h'' = \frac{\dot{Q}}{A_h} = h_h \left(T_{hot} - T_{s,h} \right) \tag{5.2}$$

If the heat transfer rate is occurring in a steady state, it is the same through all *layers* shown in Figure 5.1. Therefore, the convective heat transfer rate on the cold-fluid side is given by

$$\dot{Q} = h_c A_c \left(t_{s,c} - t_{cold} \right) \tag{5.3}$$

Notice in Equation 5.3 a lower case "*t*" is used to represent the cold-side temperatures. Similarly, the heat flux through the cold-side convective film is

$$q_c'' = \frac{\dot{Q}}{A_c} = h_c \left(t_{s,c} - t_{cold}\right) \tag{5.4}$$

The heat transfer by conduction through the material separating the hot and cold fluids is governed by *Fourier's law*. For steady-state conduction through a planar surface, Fourier's law is written as

$$\dot{Q} = kA \frac{\left(T_{s,h} - t_{s,c}\right)}{\Delta x} \tag{5.5}$$

In this equation, $T_{s,h}$ is the temperature of the hot surface, $t_{s,c}$ the temperature of the cold surface, Δx the thickness of the material, and k the *thermal conductivity* of the material. Selected values of thermal conductivity can be found in Appendix B.3. If the material is cylindrical rather than planar (e.g., a tube or pipe), then Fourier's law is written in cylindrical coordinates as

$$\dot{Q} = \frac{2\pi L k \left(T_{s,h} - t_{s,c}\right)}{\ln(OD/ID)} \tag{5.6}$$

In this equation, ID and OD are the inside and outside diameters of the tube or pipe, respectively, and L is the length of the tube. Oftentimes, in conduction calculations involving tubes or pipes, the heat transfer rate per unit length is of interest. This can be determined by dividing Equation 5.6 by the tube length, L.

$$q' = \frac{\dot{Q}}{L} = \frac{2\pi k \left(T_{s,h} - t_{s,c}\right)}{\ln(OD/ID)} \tag{5.7}$$

5.2.1 Thermal Resistance

The heat transfer rate can be thought of as being analogous to electrical current flow, as shown in Figure 5.2. The thermal resistances can be determined by rearranging the heat transfer rate equations in the form shown in this figure. For example, rearranging Newton's law of cooling to the form shown in Figure 5.2 for the thermal circuit results in

$$\dot{Q} = hA(T - t) = \frac{T - t}{1/(hA)} \tag{5.8}$$

Therefore, the thermal resistance due to convection is

$$R_{t,conv} = \frac{1}{hA} \tag{5.9}$$

Notice that the thermal resistance has units of h·°F/Btu in the English unit system and K/W in the SI system. Table 5.1 shows a summary of the thermal

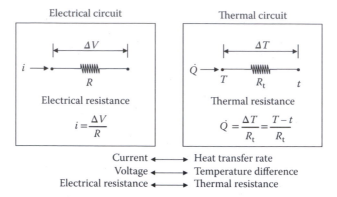

FIGURE 5.2
Heat transfer analogy to electrical current flow.

TABLE 5.1

Thermal Resistance Expressions for Conduction and Convection

Scenario	Thermal Resistance Expressions	
Conduction through a plane wall	$R_t = \dfrac{\Delta x}{kA}$	$R_t'' = AR_t = \dfrac{\Delta x}{k}$
Conduction through a cylindrical wall	$R_t = \dfrac{\ln(OD/ID)}{2\pi kL}$	$R_t' = LR_t = \dfrac{\ln(OD/ID)}{2\pi k}$
Convection on a plane wall	$R_t = \dfrac{1}{hA}$	$R_t'' = AR_t = \dfrac{1}{h}$
Convection on a cylindrical wall	$R_t = \dfrac{1}{\pi DLh}$	$R_t' = LR_t = \dfrac{1}{\pi Dh}$

resistance expressions for convection and conduction pertinent to heat exchangers. For certain geometries, it may be convenient to *weight* the resistance by either an area or a length as shown in Table 5.1. The double prime indicates an area-weighted resistance, whereas a single prime indicates a length-weighted resistance.

Thermal resistances are treated just like electrical resistances. For several thermal resistances in *series*, the total thermal resistance is given by

$$R_{t,series} = \sum_i R_{t,i} \qquad (5.10)$$

For thermal resistances in *parallel*, the total thermal resistance can be determined from

$$\frac{1}{R_{t,parallel}} = \sum_i \frac{1}{R_{t,i}} \qquad (5.11)$$

Using the thermal resistance, the heat transfer rate, heat transfer rate per unit length of tube, and heat flux can be written, respectively, as

$$\dot{Q} = \frac{T-t}{R_t} \qquad q' = \frac{\dot{Q}}{L} = \frac{T-t}{R_t'} \qquad q'' = \frac{\dot{Q}}{A} = \frac{T-t}{R_t''} \tag{5.12}$$

EXAMPLE 5.1

A 10-nom sch 10S AISI 316 stainless steel pipe is carrying acetone at an average temperature of 160°C. The pipe is bare and surrounded by air at 20°C. The convective heat transfer coefficients inside and outside the pipe are 1200 and 75 W/m²·K, respectively. Determine the following:

a. Convective heat transfer rate per unit length of the pipe
b. Convective heat transfer rate per unit length of the pipe if the pipe is sch 80S

SOLUTION: (a) A cross-sectional sketch of the pipe is shown in Figure E5.1. The thermal circuit that describes the heat flow is shown on the left-hand side of the figure. Since the problem asks for the heat transfer per unit length of pipe, it is convenient to use the length-weighted resistance shown in Table 5.1.

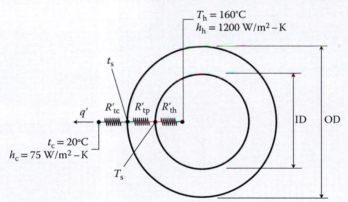

FIGURE E5.1

The heat transfer rate per unit length of pipe is given by

$$q' = \frac{T_h - t_c}{R_t'}$$

The three resistances to heat flow are in series. Therefore, the total length-weighted resistance is

$$R_t' = R_{th}' + R_{tp}' + R_{tc}'$$

Substituting the resistance expressions from Table 5.1,

$$R_t' = \frac{1}{\pi(ID)h_h} + \frac{\ln(OD/ID)}{2\pi k} + \frac{1}{\pi(OD)h_c}$$

For a 10-nom sch 10S pipe, the inside and outside diameters can be found in Appendix C.

$$OD = 0.27305 \text{ m} \quad ID = 0.26467 \text{ m}$$

The thermal conductivity of the pipe needs to be determined to complete the problem. The thermal conductivity can be evaluated at the average pipe temperature. This temperature can be estimated by averaging the hot and cold fluid temperatures.

$$T_{avg} = \frac{T_h + t_c}{2} = \frac{(160 + 20)°C}{2} = 90°C$$

From Appendix B.3, the thermal conductivity of the pipe (AISI 316 stainless steel) at 90°C is

$$k = 14.54 \text{ W/m K}$$

In this example, each individual resistance is calculated, so a comparison can be made between the three values. The resistance of the inside convective film is

$$R'_{th} = \frac{1}{\pi(ID)h_h} = \frac{1}{\pi(0.26467 \text{ m})\left(1200 \dfrac{W}{m K}\right)} = 0.001002 \frac{m K}{W}$$

The resistance due to conduction through the pipe is

$$R'_{tp} = \frac{\ln(OD/ID)}{2\pi k} = \frac{\ln[(0.27305 \text{ m})/(0.26467 \text{ m})]}{2\pi\left(14.54 \dfrac{W}{m K}\right)} = 0.0003413 \frac{m K}{W}$$

The convective film resistance on the outside of the pipe is

$$R'_{tc} = \frac{1}{\pi(OD)h_h} = \frac{1}{\pi(0.27305 \text{ m})\left(75 \dfrac{W}{m K}\right)} = 0.01554 \frac{m K}{W}$$

Summing these three resistances together gives the total length-weighted resistance for the pipe.

$$R'_t = (0.001002 + 0.0003413 + 0.01554)\frac{m K}{W} = 0.01689 \frac{m K}{W}$$

Now, the heat transfer rate per unit length of pipe can be found.

$$q' = \frac{T_h - t_c}{R'_t} = \frac{(160 - 20) K}{0.01689 \dfrac{m K}{W}} = 8,290 \frac{W}{m} = 8.29 \frac{kW}{m} \quad \leftarrow$$

(b) Keeping all conditions the same but moving to a 10-nom sch 80S pipe changes the inside diameter. Therefore, the inside convective film and the conduction resistances will change. The outside convective film resistance stays the same. For the 10-nom sch 80S pipe, the diameters are

$$OD = 0.27305 \text{ m} \qquad ID = 0.24765 \text{ m}$$

Performing similar calculations gives the heat transfer rate per unit length.

$$q' = \frac{T_h - t_c}{R_t'} = \frac{(160 - 20)\text{K}}{0.01768 \dfrac{\text{m K}}{\text{W}}} = 7917 \frac{\text{W}}{\text{m}} = 7.917 \frac{\text{kW}}{\text{m}} \quad \leftarrow$$

Table E5.1 summarizes the results of this calculation.

TABLE E5.1

Summary of Results

Schedule	OD (meter)	ID (meter)	$R_{th}'\left(\dfrac{\text{m K}}{\text{W}}\right)$	$R_{tp}'\left(\dfrac{\text{m K}}{\text{W}}\right)$	$R_{tc}'\left(\dfrac{\text{m K}}{\text{W}}\right)$	$R_t'\left(\dfrac{\text{m K}}{\text{W}}\right)$	$q'\left(\dfrac{\text{kW}}{\text{m}}\right)$
10S	0.27305	0.26467	0.001002	0.0003413	0.01554	0.01689	8.290
Percent of total resistance			5.94	2.02	92.04		
80S	0.27305	0.24765	0.001071	0.001069	0.01554	0.01768	7.917
Percent of total resistance			6.06	6.04	87.90		

This table shows several interesting things about this problem:

- The change of inside diameter has a slight effect on the value of the hot-side convective resistance. The effect is small, but it can be seen that the inside diameter has an influence on the thermal resistance.
- The conduction resistance of the thicker pipe is about three times that of the thinner pipe. This may seem significant until the complete resistance picture is considered.
- The majority of the resistance to heat transfer is the convective film on the cold side (outside) of this tube. Even with the thicker tube, over 88% of the resistance is in the outside convective film.
- Because of the increase in pipe thickness, the heat transfer rate per unit length of pipe reduces by about 4.5%.

5.2.2 Heat Transfer Augmentation with Fins

In some heat exchanger designs, a fluid passes on the outside of tubes that have fins mounted on the outside surface to enhance heat transfer. Finned-tube analysis can be quite complex, especially if the goal is to *design* the fins.

The analysis is made much easier by considering the *surface efficiency* of a fin array.

Figure 5.3 shows two possible thermal circuit models for a section of a fin array transferring heat from hot fins with a base temperature of T_b to the cooler surrounding fluid at a temperature of t_∞. The top circuit is a parallel circuit that accounts for heat that flows through the fin and the unfinned base area. The bottom circuit shows a single resistance that is equivalent to the parallel resistance. This single resistance, R_a, is known as the *fin array resistance*.

Either thermal circuit is valid for analysis of the finned tube. However, it may be considerably easier to analyze the fin array as a single resistance. The fin array resistance is defined by

$$R_a = \frac{1}{\eta_o h A_t} \tag{5.13}$$

In this equation, h is the convective heat transfer coefficient on the surface of the fins. A_t is *total* area of the fin array, including the fins and the unfinned area.

$$A_t = N A_f + A_b \tag{5.14}$$

In Equation 5.14, N is the number of fins in the array, A_f the surface area of a single fin, and A_b the total unfinned area. In Equation 5.13, η_o is called the *surface efficiency* of the fin array. The surface efficiency is defined as

$$\eta_o = 1 - \frac{N A_f}{A_t}(1 - \eta_f) \tag{5.15}$$

In Equation 5.15, η_f is the *fin efficiency*, which is defined in Figure 5.4 for straight rectangular fins mounted on a surface and circular fins with a

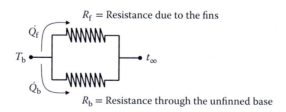

R_f = Resistance due to the fins

R_b = Resistance through the unfinned base

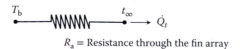

R_a = Resistance through the fin array

FIGURE 5.3
Thermal resistance models for a finned surface.

rectangular cross section mounted on a circular pipe or tube. Fin efficiency expressions for other types of fins can be found in Bergman et al. (2011).

Once the thermal resistance of the fin array is calculated, the heat transfer rate of the finned heat exchanger can be calculated using the electrical analogy.

$$\dot{Q} = \frac{\Delta T}{R_a} \tag{5.16}$$

When using Equation 5.16, the *base temperature* of the fins and the *bulk-temperature* of the fluid flowing over the fins are used to form the temperature difference (Figure 5.3).

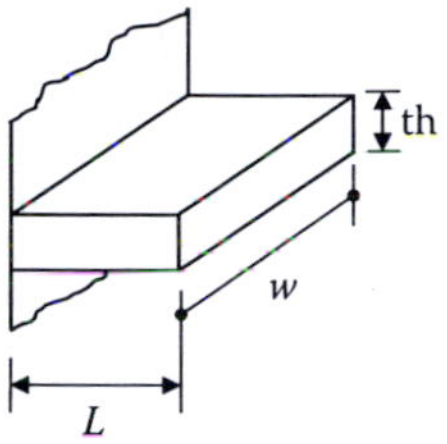

$$\eta_f = \frac{\tanh(mL_c)}{mL_c}$$

$$A_f = 2wL_c$$

$$L_c = L + (th)/2 \qquad m^2 = \frac{2h}{k\,(th)}$$

Rectangular straight fin

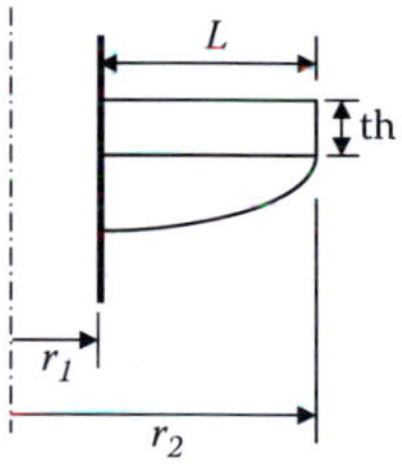

$$\eta_f = \left(\frac{2r_1/m}{r_{2c}^2 - r_1^2}\right) \frac{K_1(mr_1)\,I_1(mr_{2c}) - I_1(mr_1)\,K_1(mr_{2c})}{I_0(mr_1)\,K_1(mr_{2c}) + K_0(mr_1)\,I_1(mr_{2c})}$$

$$A_f = 2\pi\,(r_{2c}^2 - r_1^2)$$

$$r_{2c} = r_2 + (th)/2 \qquad m^2 = \frac{2h}{k\,(th)}$$

$$L = r_2 - r_1$$

Circular fin on a tube

I_0 and K_0 are modified zero-order Bessel functions of the first and second kind.

I_1 and K_1 are modified first-order Bessel functions of the first and second kind.

FIGURE 5.4
Fin efficiency expressions for straight rectangular and circular fins.

EXAMPLE 5.2

A 1-std type L copper tube is fitted with a circular copper fin array, as shown in Figure E5.2A. The fins are 1 in. high and mounted on a $\frac{1}{16}$-in. thick base. A condensing refrigerant inside the tube is at 130°F with a convective heat transfer coefficient of 8,000 Btu/h·ft²·°F. Air moves in cross flow outside the tube at 68°F with a convective heat transfer coefficient of 40 Btu/h·ft²·°F. The contact resistance between the fin array and the tube can be considered negligible. Determine the surface efficiency of this fin array.

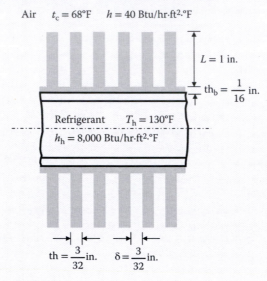

FIGURE E5.2A

SOLUTION: The surface efficiency requires the calculation of the fin efficiency. Referring to Table 5.4, the fin efficiency is a rather complex expression that includes Bessel functions. Therefore, Engineering Equation Solver (EES) is used to solve this problem. The given information is presented as follows:

```
"GIVEN: A finned copper tube"

"1-std Type L Copper"
   OD = 0.09875[ft] "Appendix D"
   ID = 0.08542[ft]

"Fins - Copper"
   L = 1[in]*convert(in,ft)            "1-inch high"
   th_b = 0.0625[ft]                   "1/16-inch base thickness"
   delta = 0.09375[in]*convert(in,ft) "3/32-inch spacing"
   th = delta                          "3/32-inch thickness"

"Condensing refrigerant inside the tube"
   T_h = 130[F]
   h_h = 8000[Btu/hr-ft^2-F]

"Air passing over the finned tubes"
   t_c = 68[F]
   h_c = 40[Btu/hr-ft^2-F]

"FIND: Surface efficiency of the finned-tube"
```

The surface efficiency of this fin array is given by Equation 5.15.

$$\eta_o = 1 - \frac{NA_f}{A_t}(1 - \eta_f)$$

```
"SOLUTION:"
"The surface efficiency of the array is,"
   eta_o = 1 - (N*A_f)*(1 - eta_f)/A_t
```

To determine the surface efficiency, the number of fins (N), the area of a single fin (A_f), the total area of the finned tube (A_t), and the fin efficiency (η_f) need to be determined. The area of a single fin is calculated using the expressions in Figure 5.4.

$$A_f = 2\pi\left(r_{2c}^2 - r_1^2\right) \qquad r_{2c} = r_2 + (th)/2$$

The dimensions r_1 and r_2 are defined in Figure 5.4. Taking into account the $\frac{1}{16}$-in. thick base material, these radii are found by

$$r_1 = \frac{OD}{2} + th_b \qquad L = r_2 - r_1$$

```
"The area of a single fin is,"
   A_f = 2*pi*(r_2c^2 - r_1^2)
   r_2c = r_2 + th/2
   r_1 = OD/2 + th_b
   L = r_2 - r_1
```

The total area of the finned tube is given by

$$A_t = NA_f + A_b$$

```
"The total area is,"
   A_t = N*A_f + A_b
```

To determine the number of fins, a tube length is required. However, in this problem the length of the tube is not given. Therefore, a basis for the tube length must be selected. In this example, the basis is selected to be 1 ft. The number of fins per foot can then be found by

$$N(\delta + th) = 1 \text{ ft}$$

```
"The tube length basis will be 1-foot. Therefore, the areas are
per foot of tube. The number of fins per foot of tube is,"
   N*(delta + th) = 1[ft]
```

In a 1-ft section of tube, the total unfinned area can be found by

$$A_b = N(2\pi r_1)\delta$$

```
"In 1-foot of tube length, the unfinned area is,"
   A_b = N*2*pi*r_1*delta
```

The only thing left is to determine the fin efficiency. From Figure 5.4, the fin efficiency is given by

$$\eta_f = \left(\frac{2r_1/m}{r_{2c}^2 - r_1^2}\right)\frac{K_1(mr_1)I_1(mr_{2c}) - I_1(mr_1)K_1(mr_{2c})}{I_0(mr_1)K_1(mr_{2c}) + K_0(mr_1)I_1(mr_{2c})}$$

In this equation, I_0 and K_0 are zero-order Bessel functions of the first and second kind. I_1 and K_1 are first-order Bessel functions of the first and second kind. These values can be found in EES as follows:

```
"The fin efficiency can be found using Figure 5.4"
   eta_f = ((2*r_1/m)/(r_2c^2 - r_1^2))*(K_11*I_12 - I_11*K_12)&
     /I_01*K_12 + K_01*I_12)
   I_01 = Bessel_I0(m*r_1)
   I_11 = Bessel_I1(m*r_1)
   I_12 = Bessel_I1(m*r_2c)
   K_01 = Bessel_K0(m*r_1)
   K_11 = Bessel_K1(m*r_1)
   K_12 = Bessel_K1(m*r_2c)
   m^2 = (2*h_c)/(k*th)
```

The fin parameter, m, requires the thermal conductivity of the fin material. The fins are made of copper. The fin temperature is estimated using the average temperature of the hot and cold fluids. The following equations complete the problem:

```
"The thermal conductivity of the copper fin is evaluated at the
average temperature between the hot and cold fluids,"
   T_avg = (T_h + t_c)/2
   k = k_('copper',T_avg)
```

When this set of equations is solved, the surface efficiency is found to be $\eta_o = 0.8744$. Once the surface efficiency is found, Equation 5.13 can be used to compute the resistance of the fin array.

5.2.3 Convective Heat Transfer Coefficient

In the majority of heat exchangers, convection is achieved by forced flow. The flow can be inside pipes or tubes, outside pipes or tubes (finned or unfinned), or between two plates. Convective heat transfer coefficients can be determined analytically for simple geometries and laminar flow conditions. However, for turbulent flow, the convective heat transfer coefficients are determined by experiment. The equations that result from experimental studies are called *correlations*.

Equations and correlations describing the convective heat transfer coefficient are often written in terms of the dimensionless quantities, Nusselt number (*Nu*), Reynolds number (*Re*), and Prandtl number (*Pr*). The Nusselt number is defined as

$$Nu = \frac{hL_c}{k} \tag{5.17}$$

In this equation, L_c is a *characteristic length*. In most heat exchanger calculations, this is some type of diameter. Since the Nusselt number is used to quantify the convection effects in the fluid, the thermal conductivity, k, in Equation (5.17) is the fluid's thermal conductivity. The Reynolds number is defined as in fluid-flow calculations by

$$Re = \frac{\rho V L_c}{\mu} \tag{5.18}$$

The Prandtl number is a dimensionless grouping of thermophysical properties of the fluid.

$$Pr = \frac{c_p \mu}{k} \tag{5.19}$$

From this equation, it can be seen that the Prandtl number is a property. Values of the Prandtl number for a variety of substances can be found in Appendix B.1 (liquids) and Appendix B.4 (air at atmospheric pressure).

It is common to express the Nusselt number as a function of the Reynolds number and the Prandtl number.

$$Nu = f(Re, Pr) \tag{5.20}$$

The functional relationship depends on the convection scenario defined by flow conditions (laminar or turbulent), geometry, and other factors. The functional form is important, because it ultimately allows for the calculation of the convective heat transfer coefficient. However, before an equation or correlation is used, it is important to understand the conditions that were used to derive the expression. These conditions place limits on the range of validity of the expression. The engineer must understand what these limits are. Applying the equation beyond its limit of validity may lead to significant errors in the analysis or design.

Correlations derived from experiment can be subject to significant uncertainty. Every effort should be made to use the most current correlation for the situation that is being considered. There are many excellent resources available for convective heat transfer correlations, including *The Heat Transfer Handbook* (Bejan and Kraus 2003) or *The Handbook of Heat Transfer* (Rohsenow et al. 1998). The most up-to-date information concerning convective heat transfer correlations can be found in technical journals such as *The International Journal of Heat and Mass Transfer* and the *Journal of Heat Transfer* (a publication of the American Society of Mechanical Engineers).

The remainder of this section lists equations and correlations for common convective-heat-transfer scenarios found in many different types of heat exchangers. More heat-exchanger-specific correlations are presented in subsequent sections of this chapter.

5.2.3.1 Forced External Cross Flow over a Cylindrical Surface

Churchill and Bernstein (1977) have developed the following correlation for forced external cross flow over a cylindrical surface.

$$\overline{Nu_D} = \frac{\bar{h}D}{k} = 0.3 + \frac{0.62 Re_D^{1/2} Pr^{1/3}}{\left[1+(0.4/Pr)^{2/3}\right]^{1/4}}\left[1+\left(\frac{Re_D}{282,000}\right)^{5/8}\right]^{4/5} \tag{5.21}$$

This correlation is valid for gases and liquids with $Re_D\, Pr \geq 0.2$. The fluid properties are evaluated at the *film temperature*, T_f. The film temperature is a reasonable estimate of the average temperature within the convective film. In the case of external flow, it is calculated using the *free-stream* (T_∞) temperature of the fluid and the *surface temperature* (T_s) of the material in contact with the fluid.

$$T_f = \frac{T_\infty + T_s}{2} \tag{5.22}$$

5.2.3.2 Laminar Flow Inside Circular Pipes or Tubes

As a fluid flows through a tube, *hydrodynamic* and *thermal* boundary layers build inside the tube. The hydrodynamic boundary layer shows how the velocity profile develops, whereas the thermal boundary layer shows how the temperature profile develops. It can be shown that when the thermal boundary layer is fully developed, the heat transfer coefficient remains constant throughout the rest of the tube. However, in the developing region (called the *entry length*), the heat transfer coefficient decreases until the boundary layer is fully developed. This behavior is shown in Figure 5.5 for a situation where the surface of the tube is hotter than the fluid passing through it.

For laminar flow, characterized by $Re_D < 2300$, the thermal entry length (Figure 5.5) can be estimated by

$$x_{fd,thermal} = 0.05 D Re_D\, Pr \tag{5.23}$$

Beyond this point, the thermal boundary layer is fully developed and the heat transfer coefficient is constant.

In the fully developed region, the Nusselt number can be determined analytically. The results are developed in Bergman et al. (2011) and shown below.

$$Nu_D = \frac{hD}{k} = 3.66 \quad \text{(constant tube surface temperature)}$$

$$Nu_D = \frac{hD}{k} = 4.36 \quad \text{(constant heat flux on the tube surface)} \tag{5.24}$$

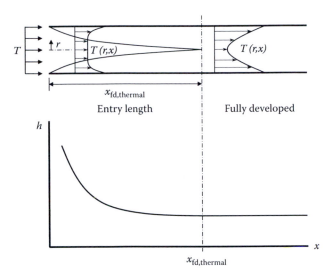

FIGURE 5.5
Thermal boundary layer development and convective heat transfer coefficient behavior for forced flow through a pipe or tube.

The Nusselt number values in Equation set (5.24) are valid for fully developed laminar flow only. The fluid properties are evaluated at the *average mean temperature*, T_m, of the fluid. The average mean temperature of the fluid can be estimated as the average between the mean inlet and outlet temperatures of the fluid entering and leaving the tube.

$$T_m = \frac{T_{m,i} + T_{m,o}}{2} \tag{5.25}$$

The *combined entry length* occurs when the hydrodynamic and thermal boundary layers develop simultaneously. This is the case in most heat exchanger designs. The average heat transfer coefficient from the inlet of the tube to any point along the tube, x, can be determined based on the tube surface condition. Two common tube conditions are as follows: (1) a tube with a constant surface temperature, and (2) a tube experiencing a constant heat flux.

Baehr and Stephan (2006) suggest the following correlation for tubes with a constant surface temperature.

$$Nu_D = \frac{hD}{k} = (A + B)/C$$

$$A = 3.66/\tanh\left[2.264Gz_D^{-1/3} + 1.7Gz_D^{-2/3}\right]$$

$$B = 0.0499Gz_D\tanh\left(Gz_D^{-1}\right) + \frac{0.0668Gz_D}{1 + 0.04Gz_D^{2/3}} \tag{5.26}$$

$$C = \tanh\left(2.432Pr^{1/6}Gz_D^{-1/6}\right)$$

This correlation is valid for the combined entry length with $Pr > 0.1$. The fluid properties are evaluated at the average mean temperature.

Churchill and Ozoe (1973) developed a correlation for the combined entry length where the tube is subject to a constant surface heat flux.

$$Nu_D = 4.364 \left[1 + \left(\frac{Gz_D}{29.6} \right)^2 \right]^{1/6}$$

$$\left\{ 1 + \left[\frac{Gz_D/19.04}{\left[1 + (Pr/0.0207)^{2/3} \right]^{1/2} \left[1 + (Gz_D/29.6)^2 \right]^{1/3}} \right]^{3/2} \right\}^{1/3} \quad (5.27)$$

This correlation is valid for $0.7 \leq Pr \leq 10$. The fluid properties are evaluated at the average mean temperature.

In Equations 5.26 and 5.27, Gz is another dimensionless group known as the Graetz number. It is defined by

$$Gz_D = \frac{Re_D Pr}{x/D} \quad (5.28)$$

5.2.3.3 Turbulent Flow Inside a Circular Tube

The entry length for turbulent flow inside a circular tube can be estimated by

$$x_{fd,turb} = 10D \quad (5.29)$$

In many turbulent flow applications, the entry length is small compared to the total tube length, and the entry effects can often be neglected. However, if the tube is short, then entry effects can be significant. For fully developed, turbulent flow, the recommended correlation is given by Gnielinski (1976).

$$Nu_D = \frac{hD}{k} = \frac{(f/8)(Re_D - 1000)Pr}{1 + 12.7(f/8)^{1/2}(Pr^{2/3} - 1)} \quad (5.30)$$

This correlation is valid for $0.50 \leq Pr \leq 2000$ and $3000 \leq Re_D \leq 5 \times 10^6$. The correlation can be used for any tube surface condition. The fluid properties are evaluated at the average mean temperature.

In Equation 5.30, the friction factor, f, can be determined using any of the empirical expressions in Chapter 4. Even though this correlation was developed for *smooth tubes*, Equation 5.30 can be used to obtain a reasonable estimate of the convective heat transfer coefficient with other values of surface roughness.

5.2.4 Fouling on Heat Exchange Surfaces

Thermal resistances to heat transfer because of conduction and convection are part of any heat exchanger. Heat exchanger surfaces also collect unwanted materials that further inhibit heat transfer. This phenomenon is known as *fouling*. The effects of fouling can be significant and should be included in analysis and design calculations. Figure 5.6 shows a cross-sectional view of a section of a tube and the associated resistances to heat transfer.

Fouling is a complex physical phenomenon that can be modeled analytically. However, for design and analysis calculations, this approach is too complex. In lieu of the analytical treatment, typical fouling resistance values have been determined and tabulated. Table 5.2 lists some representative values for various fluids. The tabulated resistance values are area-weighted, which allows for the total resistance to be calculated once the surface area where the fouling occurs is known (i.e., inside or outside of a tube, such as the one shown in Figure 5.6). These area-weighted fouling resistances are known as *fouling factors*.

$$R_f'' = AR_f \qquad (5.31)$$

Fouling is a time-dependent phenomenon. A brand new heat exchanger has clean surfaces and no fouling is present. After the heat exchanger is put into service, fouling will begin and continue until maintenance is performed to remove the fouling deposits. The fouling factors listed in Table 5.2 represent values after 1 year of normal service with no maintenance.

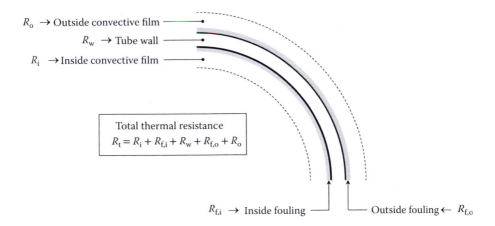

FIGURE 5.6
Cross-sectional view of a tube showing resistances to heat transfer.

TABLE 5.2

Representative Values of Fouling Factors for Heat Exchangers

Industrial Fluids	Fouling Factor R_f''	
	h·ft²·°F/Btu	m²-K/W
Oils		
No. 2 fuel oil	0.002	0.000352
No. 6 fuel oil	0.005	0.000881
Engine oil	0.005	0.000881
Gases and vapors		
Engine exhaust	0.010	0.001761
Steam	0.0005	0.000088
Refrigerants	0.002	0.000352
Compressed air	0.010	0.001761
Ammonia	0.010	0.001761
Carbon dioxide	0.010	0.001761
Coal flue gas	0.010	0.001761
Natural gas flue gas	0.0005	0.000088
Liquids		
Ammonia	0.001	0.000176
Ammonia (oil bearing)	0.003	0.000528
Refrigerants	0.001	0.000176
Organic liquids (e.g., hydrocarbons)	0.002	0.000352
Hydraulic fluid	0.001	0.000176
Carbon dioxide	0.001	0.000176
Methanol or ethanol solutions	0.002	0.000352
Ethylene glycol solutions	0.002	0.000352
Sodium or calcium chloride solutions	0.003	0.000528
Water[a]		
City or well water	0.002	0.000352
River water	0.004	0.000705
Seawater	0.001	0.000176
Brackish water	0.003	0.000528
Distilled water (condensate)	0.0005	0.000088
Treated distilled water	0.001	0.000176

Source: Kakac, S., *Heat Exchangers: Selection, Rating, and Thermal Design*, CRC Press, Boca Raton, FL, 2012.

[a] Extensive information is available for water fouling. The values shown here are worst-case values.

5.2.5 Overall Heat Transfer Coefficient

Consider the tube cross section shown in Figure 5.6. In this heat exchange scenario, the total resistance to heat transfer is made up of five resistances.

Starting on the inside of the tube and working outward, these resistances are as follows:

1. Convective resistance on the inside of the tube
2. Fouling resistance on the inside of the tube
3. Conduction resistance through the tube wall separating the fluids
4. Fouling resistance on the outside of the tube
5. Convection resistance on the outside of the tube

The total thermal resistance can be expressed as

$$R_t = R_i + R_{f,i} + R_w + R_{f,o} + R_o \tag{5.32}$$

If the heat exchange is occurring in a circular tube with internal and external fins, the total thermal resistance can be written as

$$R_t = \frac{1}{\eta_{o,i} h_i A_i} + \frac{R''_{f,i}}{\eta_{o,i} A_i} + \frac{\ln(OD/ID)}{2\pi k L} + \frac{R''_{f,o}}{\eta_{o,o} A_o} + \frac{1}{\eta_{o,o} h_o A_o} \tag{5.33}$$

In heat exchanger calculations, it is more convenient to express the total resistance in terms of the *overall heat transfer coefficient, U*. The definition of the overall heat transfer coefficient is given by

$$UA = \frac{1}{R_t} \tag{5.34}$$

In Equation 5.34, A represents the total heat transfer area in the heat exchanger. Making this substitution into the heat transfer rate expression given in Equation 5.12 results in

$$\dot{Q} = UA(T - t) \tag{5.35}$$

This equation is similar in form to Newton's law of cooling. However, it takes into account *all* of the thermal resistance to heat transfer, not just the convection. Comparing Equation 5.35 to Newton's law of cooling, it can be seen that the units of U are identical to the units of the convective heat transfer coefficient, h. In heat exchanger design and analysis, the UA product of the heat exchanger is an important parameter that must be determined. Therefore, it is important to understand how the overall heat transfer coefficient is calculated.

Consider a circular tube without fins. In Equation 5.34, the area A can represent either the inside or outside surface area of the tube. If the outside area, A_o, is chosen, the value of U corresponds to U_o. Similarly, U_i is based on A_i. Since the inside and outside surface areas are different, the inside and outside U values will be different. However, Equation 5.34 indicates that the *UA product* is the same for both inside and outside surfaces. Therefore,

$$U_o A_o = U_i A_i \qquad (5.36)$$

For a circular tube without fins, the total resistance is given by

$$R_t = \frac{1}{h_i A_i} + \frac{R''_{f,i}}{A_i} + \frac{\ln(OD/ID)}{2\pi kL} + \frac{R''_{f,o}}{A_o} + \frac{1}{h_o A_o} \qquad (5.37)$$

The overall heat transfer coefficient, based on the outside surface area can be found by substituting Equation 5.37 into Equation 5.34 and solving for $1/U_o$.

$$\frac{1}{U_o} = A_o R_t = \frac{1}{h_i}\left(\frac{A_o}{A_i}\right) + R''_{f,i}\left(\frac{A_o}{A_i}\right) + A_o \frac{\ln(OD/ID)}{2\pi kL} + R''_{f,o} + \frac{1}{h_o} \qquad (5.38)$$

The areas are the inside (subscript "i") and outside (subscript "o") surface areas of the tube. Therefore, Equation 5.38 can be written as

$$\frac{1}{U_o} = A_o R_t = \frac{1}{h_i}\left[\frac{\pi(OD)L}{\pi(ID)L}\right] + R''_{f,i}\left[\frac{\pi(OD)L}{\pi(ID)L}\right]$$
$$+ [\pi(OD)L]\frac{\ln(OD/ID)}{2\pi kL} + R''_{f,o} + \frac{1}{h_o} \qquad (5.39)$$

Simplifying this equation gives

$$\frac{1}{U_o} = A_o R_t = \frac{1}{h_i}\left(\frac{OD}{ID}\right) + R''_{f,i}\left(\frac{OD}{ID}\right) + \frac{OD}{2k}\ln\left(\frac{OD}{ID}\right) + R''_{f,o} + \frac{1}{h_o} \qquad (5.40)$$

If the overall heat transfer coefficient was based on the *inside* diameter of the tube, then the resulting expression would be

$$\frac{1}{U_i} = A_i R_t = \frac{1}{h_i} + R''_{f,i} + \frac{ID}{2k}\ln\left(\frac{OD}{ID}\right) + R''_{f,o}\left(\frac{ID}{OD}\right) + \frac{1}{h_o}\left(\frac{ID}{OD}\right) \qquad (5.41)$$

Equations 5.40 and 5.41 show that the magnitude of U_o and U_i are different. However, as shown previously, the *UA* product is the same for both surfaces. In heat exchanger calculations, it is common to base the overall heat transfer coefficient on the *outside diameter* of the tube or pipe.

EXAMPLE 5.3

A 4-std type L copper tube is transporting an ethylene glycol solution. Superheated steam passes over the outside of the tube. The convective heat transfer coefficients are found to be 200 W/m² K on the inside of the tube and 180 W/m² K on the outside of the tube. Under these conditions, the average temperature of the tube is 150°C. The tube is 30 m long. Determine (a) the overall heat transfer coefficients based on the inside and outside surface areas for a fouled condition, (b) the UA product of this tube without fouling, and (c) the UA product of the tube after 1 year of service.

SOLUTION: The data for the copper tube can be found in Appendix D and Appendix B.3.

$$OD = 0.10478 \text{ m} \quad ID = 0.099187 \text{ m} \quad k = 391.4 \text{ W/m K}$$

The fouling factors for the ethylene glycol solution on the inside of the tube and the superheated vapor steam on the outside of the tube are found in Table 5.2.

$$R''_{f,i} = 0.000352 \text{ m}^2 \text{ K/W} \qquad R''_{f,o} = 0.000088 \text{ m}^2 \text{ K/W}$$

a. Equations 5.40 and 5.41 can be used to find the overall heat transfer coefficients based on the outside and inside surface areas, respectively.

$$\frac{1}{U_{o,\text{fouled}}} = \frac{1}{h_i}\left(\frac{OD}{ID}\right) + R''_{f,i}\left(\frac{OD}{ID}\right) + \frac{OD}{2k}\ln\left(\frac{OD}{ID}\right) + R''_{f,o} + \frac{1}{h_o}$$

$$\frac{1}{U_{o,\text{fouled}}} = \frac{1}{200}\frac{\text{m}^2 \text{ K}}{\text{W}}\left(\frac{0.10478 \text{ m}}{0.099187 \text{ m}}\right) + 0.000352\frac{\text{m}^2 \text{ K}}{\text{W}}\left(\frac{0.10478 \text{ m}}{0.099187 \text{ m}}\right)$$

$$+ \frac{0.10478 \text{ m}}{2(391.4)}\frac{\text{m K}}{\text{W}}\ln\left(\frac{0.10478 \text{ m}}{0.099187 \text{ m}}\right) + 0.000088\frac{\text{m}^2 \text{ K}}{\text{W}} + \frac{1}{180}\frac{\text{m}^2 \text{ K}}{\text{W}}$$

$$\frac{1}{U_{o,\text{fouled}}} = 0.01130\frac{\text{m}^2 \text{ K}}{\text{W}}$$

$$\therefore U_{o,\text{fouled}} = 88.46\frac{\text{W}}{\text{m}^2 \text{ K}} \leftarrow$$

$$\frac{1}{U_{i,\text{fouled}}} = \frac{1}{h_i} + R''_{f,i} + \frac{ID}{2k}\ln\left(\frac{OD}{ID}\right) + R''_{f,o}\left(\frac{ID}{OD}\right) + \frac{1}{h_o}\left(\frac{ID}{OD}\right)$$

$$\frac{1}{U_{i,\text{fouled}}} = \frac{1}{200}\frac{\text{m}^2 \text{ K}}{\text{W}} + 0.000352\frac{\text{m}^2 \text{ K}}{\text{W}} + \frac{0.099187 \text{ m}}{2(391.4)}\frac{\text{m K}}{\text{W}}\ln\left(\frac{0.10478 \text{ m}}{0.099187 \text{ m}}\right)$$

$$+ 0.000088\frac{\text{m}^2 \text{ K}}{\text{W}}\left(\frac{0.099187 \text{ m}}{0.10478 \text{ m}}\right) + \frac{1}{180}\frac{\text{m}^2 \text{ K}}{\text{W}}\left(\frac{0.099187 \text{ m}}{0.10478 \text{ m}}\right)$$

$$\frac{1}{U_{i,\text{fouled}}} = 0.01070\frac{\text{m}^2 \text{ K}}{\text{W}}$$

$$\therefore U_{i,\text{fouled}} = 93.45\frac{\text{W}}{\text{m}^2 \text{ K}} \leftarrow$$

This calculation shows that the U value is dependent on which surface is chosen for A. However, in heat exchanger calculations, the UA product is more useful.

b. To determine the UA product for a clean tube, the U values need to be recalculated without the fouling resistance. The results are

$$\frac{1}{U_{o,clean}} = \frac{1}{h_i}\left(\frac{OD}{ID}\right) + \frac{OD}{2k}\ln\left(\frac{OD}{ID}\right) + \frac{1}{h_o}$$

$$\frac{1}{U_{o,clean}} = 0.01084\frac{m^2\ K}{W} \qquad \therefore\ U_{o,clean} = 92.21\frac{W}{m^2\ K} \quad \leftarrow$$

$$\frac{1}{U_{i,clean}} = \frac{1}{h_i} + \frac{ID}{2k}\ln\left(\frac{OD}{ID}\right) + \frac{1}{h_o}\left(\frac{ID}{OD}\right)$$

$$\frac{1}{U_{i,clean}} = 0.01027\frac{m^2\ K}{W} \qquad \therefore\ U_{i,clean} = 97.41\frac{W}{m^2\ K} \quad \leftarrow$$

c. Now, the UA products can be found for both conditions of a clean and fouled tube. For the clean tube,

$$(UA)_{o,clean} = U_{o,clean}A_o = U_{o,clean}\left[\pi(OD)L\right]$$

$$(UA)_{o,clean} = \left(92.21\frac{W}{m^2\ K}\right)\left[\pi(0.10478\ m)(30\ m)\right] = 910.6\frac{W}{K} \quad \leftarrow$$

$$(UA)_{i,clean} = U_{i,clean}A_i = U_{i,clean}\left[\pi(ID)L\right]$$

$$(UA)_{i,clean} = \left(97.41\frac{W}{m^2\ K}\right)\left[\pi(0.099187\ m)(30\ m)\right] = 910.6\frac{W}{K} \quad \leftarrow$$

As expected, the UA product is the same for both surfaces of the tube. After one year of service the UA product of the fouled tube is

$$(UA)_{o,fouled} = \left(88.46\frac{W}{m^2\ K}\right)\left[\pi(0.10478\ m)(30\ m)\right] = 873.6\frac{W}{K} \quad \leftarrow$$

$$(UA)_{i,fouled} = \left(93.45\frac{W}{m^2\ K}\right)\left[\pi(0.099187\ m)(30\ m)\right] = 873.6\frac{W}{K} \quad \leftarrow$$

Since the UA product is directly proportional to the heat transfer rate, it can be seen that fouling reduces the heat transfer rate in this tube by approximately 4% over the course of one year.

5.3 Heat Exchanger Types

There are many types of heat exchangers. The most common used in industry are *double pipe, shell and tube, plate and frame,* and *cross flow.* Detailed analysis and design procedures for each of these types of heat exchangers are presented in Sections 5.6 through 5.9 of this chapter. In this section, a qualitative description of each of these types is presented.

5.3.1 Double-Pipe Heat Exchanger

The *double-pipe* heat exchanger (DPHX) is one of the simplest heat exchanger configurations. They can be purchased from many vendors. It is also possible to build a DPHX. The DPHX consists of two concentric pipes or tubes, as shown in Figure 5.7. This figure indicates that there are two possible fluid-flow configurations for DPHXs: counterflow and parallel flow. Each of these flow configurations results in different thermal performance.

It is possible to connect DPHXs together in what is known as a *hairpin* configuration. The hairpin configuration is useful when a long heat exchanger length is required, but the installation footprint is small. Figure 5.8 shows a schematic of a counterflow hairpin DPHX. Notice that if several hairpins are added, this design allows the engineer to fit a rather large heat exchanger into a small area.

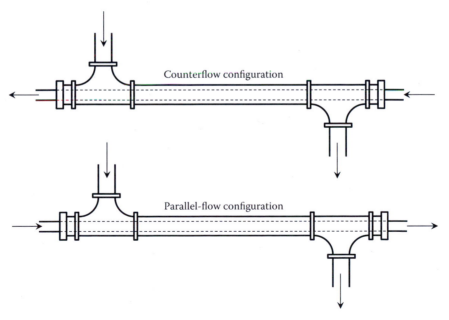

FIGURE 5.7
Counterflow and parallel-flow configurations for double-pipe heat exchangers.

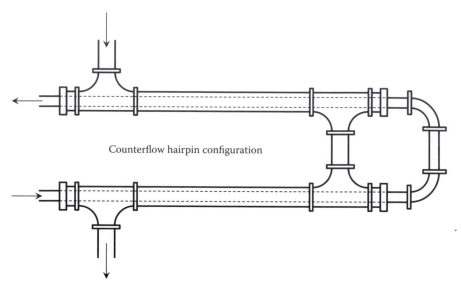

Counterflow hairpin configuration

FIGURE 5.8
A counterflow hairpin double-pipe heat exchanger.

DPHXs are typically used in situations requiring low to moderate heat transfer rates.

5.3.2 Shell and Tube Heat Exchanger

The shell and tube heat exchanger (STHX) provides a substantial amount of heat transfer area in a relatively small volume. It is a much more complex heat exchanger compared to the DPHX. As a result, the STHX is usually more expensive compared to the DPHX for the same heat transfer duty.

The STHX is made up of a large diameter shell that contains multiple tubes inside. The tubes can be arranged for several possible flow configurations. Figure 5.9 shows two possibilities: the *1-tube pass* and the *2-tube pass* STHX.

As shown in Figure 5.9, the shell-side fluid is distributed among the tubes by means of an *inlet plenum*. The shell-side fluid is collected and leaves the STHX through the *outlet plenum*. The plenums are connected to the heat exchanger at the *tube sheet*. The tube sheet is a shell-diameter piece of metal that has holes bored to provide a place to connect the tube bundle. Providing a place for the plenums to connect at the tube sheet allows for disassembly of the heat exchanger for tube maintenance. As the shell-side fluid flows across the tube bundles, its flow is diverted by means of *baffles*. The baffles provide heat transfer enhancement as well as structural support to hold the tube bundle in place.

STHXs are usually used in applications requiring large heat transfer rates. They are often used as condensers or evaporators where one of the fluids in the heat exchanger is undergoing a phase change.

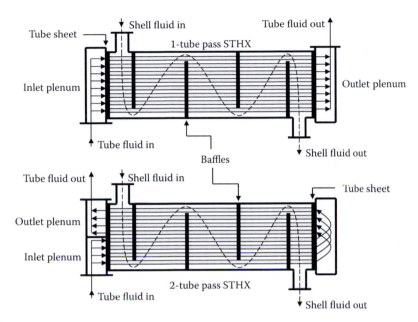

FIGURE 5.9
1-tube pass and 2-tube pass STHXs.

5.3.3 Plate and Frame Heat Exchanger

Plate and frame heat exchangers (PFHXs) were introduced in the 1930s. They are used extensively in the food processing industry and other industrial applications. In the PFHX, the hot and cold fluids circulate through a series of *plates,* as shown in Figure 5.10. The plates are held together with a *frame.* *Gaskets* on the plates keep the fluids sealed inside the heat exchanger and also provide the desired flow path for the hot and cold fluids.

The plates are corrugated with a *chevron* pattern to enhance turbulence that increases the heat transfer rate. The metal plates are usually cold-stamped. Depending on the temperatures and pressures involved, the gaskets can be made of one of several materials, including natural rubber styrene, resin-cured nitrile, silicone rubber, or neoprene. The plates are clamped together in the frame with long bolts, as shown in Figure 5.10. Because of its relatively simple construction, it is easily taken apart for maintenance.

PFHXs are typically used where low to moderate heat transfer rates are required.

5.3.4 Cross-Flow Heat Exchanger

In a *cross-flow* heat exchanger (CFHX), the hot and cold fluids never contact each other directly. However, the fluids can be categorized as *mixed* or *unmixed.* Figure 5.11 shows what is meant by mixed and unmixed fluids.

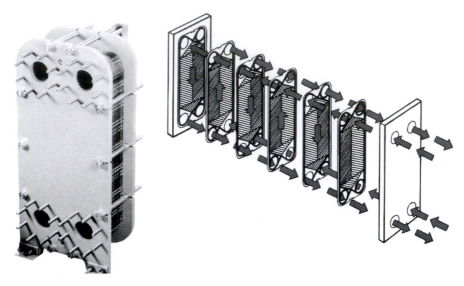

FIGURE 5.10
The plate and frame heat exchanger. (From Bell and Gossett. *GPX Plate and Frame Heat Exchangers*. Buffalo, NY: Xylem, Inc., 2011. Reprinted with permission.)

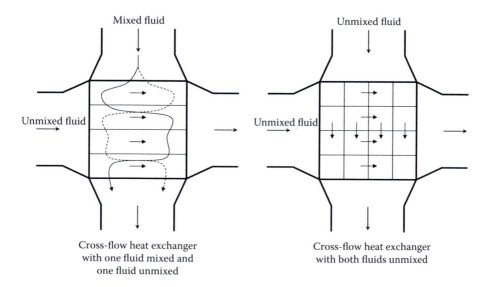

FIGURE 5.11
Cross-flow heat exchangers showing mixed and unmixed fluids.

A fluid is unmixed if it passes through the heat exchanger and its path is controlled by passages created by channels or fins. A fluid is mixed if it enters the heat exchanger and is free to wander about as it works its way through. An example of this is a fluid passing over a bank of tubes in cross flow. Although its main path is *forward* through the heat exchanger, its path can be diverted and therefore mixed. In both cases, the fluids are moving in cross flow to each other.

There are many configurations for CFHXs. Perhaps one of the more common CFHXs is the finned-tube coil found in many heating and air-conditioning systems. These heat exchangers range from very small to quite large. Therefore, the CFHX is used in systems where the heat transfer rate ranges from low to high.

5.4 Design and Analysis of Heat Exchangers

There are two types of problems encountered when considering heat exchangers in a thermal energy system: a *design* problem and an *analysis* problem. In a design problem, the heat exchanger has not yet been selected. It is being designed to meet a specific need. In analysis, the heat exchanger exists (e.g., from a vendor), and the analysis predicts its thermal performance.

As an example of a heat exchanger design problem, consider the case where the following information is known in an application: (1) the mass flow rates of each fluid, (2) the inlet temperature of each fluid, and (3) the outlet temperature of either one of the fluids. A thermodynamic analysis can be used to determine the outlet temperature of the other fluid. However, a heat transfer analysis is required to determine the total heat transfer area, and thus the size, of the heat exchanger. In this case, the heat exchanger is being designed to meet a specific need.

By contrast, consider a case where a heat exchanger has been selected for an application and the following system parameters are known: (1) the mass flow rates of each fluid, and (2) the inlet temperatures of each fluid. In this case, an analysis of the existing heat exchanger is needed to determine the outlet temperatures of each fluid.

5.4.1 Heat Exchanger Design Problem

Figure 5.12 shows a schematic of a heat exchanger. The type of heat exchanger is not important at this point. Included in the figure are three different system boundaries specified with dashed lines.

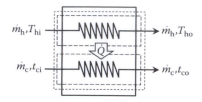

FIGURE 5.12
A heat exchanger with three different system boundaries.

Consider the design problem where the mass flow rates and inlet temperatures of the hot and cold fluids are known. In addition, the outlet temperature of the hot fluid is known. The design problem is to determine the total heat transfer area required and the cold-fluid outlet temperature. Applying the conservation of energy equation to the system boundary surrounding the hot fluid results in

$$\dot{Q} = \dot{m}_h \left(h_{hi} - h_{ho} \right) \tag{5.42}$$

In this equation, h is the hot fluid's enthalpy. If the fluid remains in the single phase (gas or liquid), Equation 5.42 can be rewritten using the heat capacity of the substance as follows:

$$\dot{Q} = \dot{m}_h c_{p,h,avg} \left(T_{hi} - T_{ho} \right) \tag{5.43}$$

In heat exchanger analysis, it is common to introduce the *thermal capacity rate* of the flow, which is defined as

$$\dot{C} = \dot{m} c_p \tag{5.44}$$

Using the thermal capacity rate, Equation 5.43 can be written as

$$\dot{Q} = \dot{C}_h \left(T_{hi} - T_{ho} \right) \tag{5.45}$$

Similarly, for the system boundary surrounding the cold fluid,

$$\dot{Q} = \dot{C}_c \left(t_{co} - t_{ci} \right) \tag{5.46}$$

Equations 5.45 and 5.46 can be combined to solve the outlet temperature of the cold fluid.

$$t_{co} = t_{ci} + \frac{\dot{C}_h}{\dot{C}_c} \left(T_{hi} - T_{ho} \right) \tag{5.47}$$

This equation could have been derived using the larger system boundary in Figure 5.12 that encompasses both fluids. The heat transfer rate between the two fluids can be found from Equations 5.45 or 5.46. The analysis summarized in Equations 5.42 through 5.47 is purely thermodynamic. However, in the heat exchanger design problem, the parameter of interest is the *area* of the heat exchanger required to affect the hot- and cold-fluid temperature changes at the given flow rates. There is no indication of the area required in the thermodynamic analysis. This suggests that an equation is missing to complete the analysis. This equation comes from a heat transfer analysis.

5.4.2 Heat Exchanger Analysis Problem

Consider the same heat exchanger sketch shown in Figure 5.12. For an analysis problem, the heat exchanger is already selected. In other words, the heat exchange area is known. A typical analysis problem would be to predict the outlet fluid temperatures, knowing the mass flow rates and inlet temperatures of each fluid. The thermodynamic analysis is the same as developed earlier. The equation for the cold-fluid outlet temperature, Equation 5.47, cannot be solved because it is a function of the hot-fluid outlet temperature, which is also an unknown. Therefore, Equation 5.47 is one equation with two unknowns. The same conclusion can be drawn as in Section 5.4.1. Something is missing that is required to complete the analysis. The missing equation comes from a heat transfer analysis of the heat exchanger.

5.4.3 Logarithmic Mean Temperature Difference

Thermodynamic analysis does not reveal anything about the *internal* performance of the heat exchanger. This is a limitation of thermodynamics because the heat exchanger is *black-boxed* by the system boundary. This does not diminish the importance of the thermodynamic analysis. As will be shown, the thermodynamic analysis is an important part of the complete heat exchanger modeling strategy.

The heat transfer analysis of a heat exchanger involves the determination of the UA product of the heat exchanger and relating it to a temperature difference. Determination of U captures the effect of the heat transfer modes occurring within the heat exchanger. The size of the heat exchanger is contained in the area A. The question is, "How are these variables related to a temperature difference in a heat exchanger?" Equation 5.35, repeated below, suggests that an equation analogous to Newton's law of cooling may be helpful.

$$\dot{Q} = UA(T - t)$$

This equation could be useful *if* the temperature difference between the hot and cold fluid is constant throughout the length of the heat exchanger. Unfortunately, this only occurs in a specialized heat exchanger under specific flow conditions. In all other cases, the temperature difference between the hot and cold fluids varies as a function of position within the heat exchanger. For example, consider the temperature profiles shown in Figure 5.13. This figure shows the variation in the temperature difference $(T - t)$ as a function of position within the heat exchanger for parallel-flow and counterflow configurations.

To utilize an equation such as Equation 5.35, the temperature difference needs to be modified to account for the variability shown in Figure 5.13. This is done using what is known as the *logarithmic mean temperature difference* (LMTD). Using the LMTD instead of a simple temperature difference, Equation 5.35 can be rewritten as

$$\dot{Q} = UA\,(\text{LMTD}) \tag{5.48}$$

Figure 5.13 shows that the temperature profiles are a function of the *type* of heat exchanger (e.g., parallel flow or counterflow). This implies that the LMTD is *dependent* on the type of heat exchanger.

To develop the LMTD for a parallel-flow heat exchanger, the analysis of a differential area within the heat exchanger is analyzed, as shown in Figure 5.14.

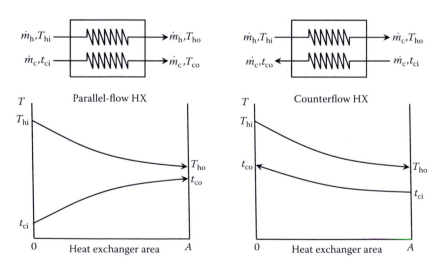

FIGURE 5.13
Temperature profiles inside parallel-flow and counterflow heat exchangers.

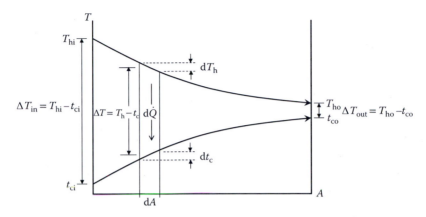

FIGURE 5.14
Differential analysis of a parallel-flow heat exchanger.

The conservation of energy applied to the hot and cold fluids in the differential element dA results in the following expressions:

$$d\dot{Q} = \dot{C}_h \left(-dT_h \right) = -C_h dT_h \qquad (5.49)$$

$$d\dot{Q} = \dot{C}_c dt_c \qquad (5.50)$$

To have the same sign on the heat transfer rate, a negative sign is needed in Equation 5.49 because dT_h is negative. Equations 5.49 and 5.50 are the thermodynamic analysis of the differential area. The differential heat transfer rate can be expressed similar to Equation 5.35 as

$$d\dot{Q} = U\left(T_h - t_c \right) dA = U(\Delta T) dA \qquad (5.51)$$

The temperature difference in Equation 5.51 is the *local* temperature difference across the differential element. The differential of this temperature difference is written as

$$d\Delta T = d\left(T_h - t_c \right) = dT_h - dt_c \qquad (5.52)$$

Substituting Equations 5.49 and 5.50 into 5.52 results in

$$d\Delta T = -d\dot{Q}\left(\frac{1}{\dot{C}_h} + \frac{1}{\dot{C}_c} \right) \qquad (5.53)$$

Substituting Equation 5.51 into 5.53 and rearranging gives

$$\frac{\mathrm{d}\Delta T}{\Delta T} = -U\left(\frac{1}{\dot{C}_h} + \frac{1}{\dot{C}_c}\right)\mathrm{d}A \qquad (5.54)$$

This equation can be integrated from the inlet to the outlet of the heat exchanger.

$$\int_{\mathrm{in}}^{\mathrm{out}} \frac{\mathrm{d}\Delta T}{\Delta T} = -U\left(\frac{1}{\dot{C}_h} + \frac{1}{\dot{C}_c}\right)\int_{\mathrm{in}}^{\mathrm{out}} \mathrm{d}A \qquad (5.55)$$

Performing the integration gives

$$\ln\left(\frac{\Delta T_{\mathrm{out}}}{\Delta T_{\mathrm{in}}}\right) = -UA\left(\frac{1}{\dot{C}_h} + \frac{1}{\dot{C}_c}\right) \qquad (5.56)$$

The thermodynamic energy balances, Equations 5.45 and 5.46 can be substituted into this equation, resulting in

$$\ln\left(\frac{\Delta T_{\mathrm{out}}}{\Delta T_{\mathrm{in}}}\right) = -\frac{UA}{\dot{Q}}[(T_{\mathrm{hi}} - t_{\mathrm{ci}}) - (T_{\mathrm{ho}} - t_{\mathrm{co}})] = -\frac{UA}{\dot{Q}}(\Delta T_{\mathrm{in}} - \Delta T_{\mathrm{out}}) \qquad (5.57)$$

Solving this equation for the heat transfer rate reveals the form of the LMTD.

$$\dot{Q} = UA\left[\frac{\Delta T_{\mathrm{out}} - \Delta T_{\mathrm{in}}}{\ln\left(\Delta T_{\mathrm{out}}/\Delta T_{\mathrm{in}}\right)}\right] \qquad (5.58)$$

Therefore, the LMTD is given by

$$\mathrm{LMTD} = \frac{\Delta T_{\mathrm{out}} - \Delta T_{\mathrm{in}}}{\ln\left(\Delta T_{\mathrm{out}}/\Delta T_{\mathrm{in}}\right)} = \frac{\Delta T_{\mathrm{in}} - \Delta T_{\mathrm{out}}}{\ln\left(\Delta T_{\mathrm{in}}/\Delta T_{\mathrm{out}}\right)} \qquad (5.59)$$

Equation 5.59 shows that the LMTD is a function of the *inlet* and *outlet* temperatures of the hot and cold fluids. This equation was developed for the parallel-flow heat exchanger. However, it is valid for counterflow heat exchangers as well. To apply Equation 5.59 to a counterflow heat exchanger, the inlet and outlet of the heat exchanger need to be defined. This may seem problematic since on one side where a fluid is entering, the other fluid exits as seen in Figure 5.13. This confusion is easily resolved by simply calling one end of the heat exchanger the inlet and the other end the outlet. Once the *ends* of the counterflow heat exchanger are specified, the temperature differences for the LMTD can be formed, as shown in Figure 5.15.

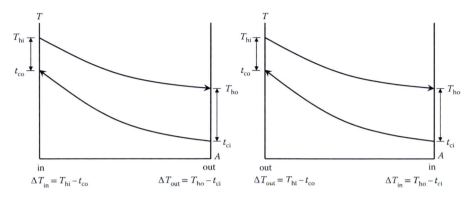

FIGURE 5.15
Possible definitions of ΔT_{out} and ΔT_{in} for a counterflow heat exchanger.

EXAMPLE 5.4

A heat exchanger is being used to transfer heat between water and a 20% ethylene glycol solution. The ethylene glycol enters the heat exchanger at 150 gpm with a temperature of 20°F and exits at 40°F. The water enters the heat exchanger at 130 gpm at a temperature of 80°F. A heat transfer analysis reveals that the overall heat transfer coefficient of the heat exchanger is 400 Btu/h·ft²·°F. Determine the heat transfer area required for (a) a parallel-flow heat exchanger and (b) a counterflow heat exchanger.

SOLUTION: This is a heat exchanger design problem since the area of the heat exchanger is unknown. The area can be determined by solving Equation 5.48 for the unknown area.

$$A = \frac{\dot{Q}}{U(\text{LMTD})}$$

The heat transfer rate can be found from the cold fluid (the ethylene glycol solution).

$$\dot{Q} = \dot{C}_c \left(t_{co} - t_{ci} \right)$$

The thermal capacity rate of the cold fluid is given by

$$\dot{C}_c = \dot{m}_c c_{pc} = \rho_c \dot{V}_c c_{pc}$$

At this point, a decision needs to be made about what temperature should be used to evaluate the properties. Strictly speaking, the density should be evaluated at the inlet condition since the volumetric flow rate is given at that point and volumetric flow is not a conserved quantity. However, the variation in the liquid density is very small over the temperature range considered here. Therefore, volume flow rate can be considered conserved, and the *average* value of the density as well as the average value of the heat capacity will be used. Interpolating in Appendix B.1, the properties of the 20% ethylene glycol mixture are found to be

$$t_{c,avg} = \frac{t_{ci} + t_{co}}{2} = \frac{(20+40)°F}{2} = 30°F \qquad \rho_c = 64.254\frac{lbm}{ft^3} \qquad c_{pc} = 0.9217\frac{Btu}{lbm \cdot R}$$

Therefore, the thermal capacity rate of the cold fluid and the heat transfer rate are

$$\dot{C}_c = \rho_c \dot{V}_c c_{pc} = \left(64.254\frac{lbm}{ft^3}\right)\left(150\frac{gal}{min}\right)\left(0.9217\frac{Btu}{lbm \cdot R}\right)\left(8.0208\frac{ft^3}{h}\frac{min}{gal}\right)$$

Appendix A

$$= 71,251\frac{Btu}{h \cdot R}$$

$$\therefore \quad \dot{Q} = \dot{C}_c\left(t_{co} - t_{ci}\right) = \left(71,251\frac{Btu}{h \cdot R}\right)(40-20)R = 1.425 \times 10^6 \frac{Btu}{h}$$

The LMTD requires both sets of inlet and outlet temperatures. In this case, only three of the temperatures are known. The hot-fluid outlet temperature is not specified. However, it can be calculated by applying the conservation of energy to the hot fluid.

$$\dot{Q} = \dot{C}_h\left(T_{hi} - T_{ho}\right) \quad \rightarrow \quad T_{ho} = T_{hi} - \frac{\dot{Q}}{\dot{C}_h}$$

The hot fluid's thermal capacity rate is given by

$$\dot{C}_h = \dot{m}_h c_{ph} = \rho_h \dot{V}_h c_{ph}$$

The hot-fluid outlet temperature is unknown at this point. Therefore, as a first estimate, the known inlet temperature is used to estimate the properties. From Appendix B.1, for water at 80°F,

$$\rho_h = 62.213\frac{lbm}{ft^3} \qquad c_{pc} = 0.9986\frac{Btu}{lbm \cdot R}$$

Therefore, the thermal capacity rate and outlet temperature of the hot fluid are,

$$\dot{C}_h = \rho_h \dot{V}_h c_{ph} = \left(62.213\,\frac{\text{lbm}}{\text{ft}^3}\right)\left(130\,\frac{\text{gal}}{\text{min}}\right)\left(0.9986\,\frac{\text{Btu}}{\text{lbm}\cdot\text{R}}\right)\left(8.0208\,\frac{\text{ft}^3\,\text{min}}{\text{h}\,\,\text{gal}}\right)$$

Appendix A

$$= 64{,}779\,\frac{\text{Btu}}{\text{h}\cdot\text{R}}$$

$$\therefore\quad T_{ho} = T_{hi} - \frac{\dot{Q}}{\dot{C}_h} = 80°\text{F} - \frac{1.425\times10^6\,\dfrac{\text{Btu}}{\text{h}}}{64{,}779\,\dfrac{\text{Btu}}{\text{h}\cdot°\text{F}}} = 58°\text{F}$$

The value of the LMTD depends on the type of heat exchanger: parallel flow (PF) or counterflow (CF). For each of these configurations

$$\text{LMTD}_{PF} = \frac{\Delta T_{out} - \Delta T_{in}}{\ln\left(\Delta T_{out}/\Delta T_{in}\right)} = \frac{\left(T_{ho} - t_{co}\right) - \left(T_{hi} - t_{ci}\right)}{\ln\left[\left(T_{ho} - t_{co}\right)/\left(T_{hi} - t_{ci}\right)\right]} = 34.89\ \text{R}$$

$$\text{LMTD}_{CF} = \frac{\Delta T_{out} - \Delta T_{in}}{\ln\left(\Delta T_{out}/\Delta T_{in}\right)} = \frac{\left(T_{ho} - t_{ci}\right) - \left(T_{hi} - t_{co}\right)}{\ln\left[\left(T_{ho} - t_{ci}\right)/\left(T_{hi} - t_{co}\right)\right]} = 38.99\ \text{R}$$

Now, the required heat exchange area can be found for each flow configuration,

$$A_{PF} = \frac{\dot{Q}}{U\left(\text{LMTD}_{PF}\right)} = \frac{1.425\times10^6\,\dfrac{\text{Btu}}{\text{h}}}{\left(400\,\dfrac{\text{Btu}}{\text{h}\cdot\text{ft}^2\cdot\text{R}}\right)(34.89\ \text{R})} = 102.1\ \text{ft}^2$$

$$A_{CF} = \frac{\dot{Q}}{U\left(\text{LMTD}_{CF}\right)} = \frac{1.425\times10^6\,\dfrac{\text{Btu}}{\text{h}}}{\left(400\,\dfrac{\text{Btu}}{\text{h}\cdot\text{ft}^2\cdot\text{R}}\right)(38.99\ \text{R})} = 91.36\ \text{ft}^2$$

These values were calculated by evaluating the hot-fluid properties at the inlet temperature (80°F). On the basis of these calculations, the outlet temperature is 58°F, which makes the average temperature of the hot fluid 69°F. To be more accurate, the properties of the hot fluid should be evaluated again at 69°F and the areas recalculated. Clearly, this is an iterative problem, well suited for an equation solver such as EES. Allowing EES to perform the iteration, the results are A_{PF} = 102.0 ft^2, A_{CF} = 91.31 ft^2, and T_{ho} = 58.05°F.

It is interesting to note that the iterative solution using EES produced nearly the same result as the non-iterative calculation. This is because the temperature difference in the hot fluid is relatively small and the properties do not change much since it is a liquid. If the temperature difference is large enough to affect a significant change in properties, then it is recommended that at least one iteration be performed. Alternatively, an equation solver can be used to determine the solution.

5.4.4 LMTD Heat Exchanger Model

In Section 5.4.3, it was shown that thermodynamics *and* heat transfer combine to form a mathematical model of a heat exchanger. Whether the heat exchanger problem is a design or analysis, the complete mathematical description of the heat exchanger is embodied in three equations. For a heat exchanger where both fluids remain in the single phase, the LMTD heat exchanger model can be written as

$$\dot{Q} = \dot{C}_h \left(T_{hi} - T_{ho} \right)$$
$$\dot{Q} = \dot{C}_c \left(t_{co} - t_{ci} \right) \tag{5.60}$$
$$\dot{Q} = UA(\text{LMTD})$$

Although this set of equations seems relatively simple, the complexity can quickly become overwhelming because of the need to determine the overall heat transfer coefficient, U. Subsequent sections of this chapter present the detail in calculation of U for the various types of heat exchangers discussed in Section 5.3. An added level of complexity is the iterative nature of the problem because of unknown temperatures, as shown in Example 5.4.

EXAMPLE 5.5

Liquid hexane flows through a counterflow heat exchanger at 5 gpm, as shown in Figure E5.5A. The hexane enters the heat exchanger at 190°F. Water, flowing at 5 gpm, is used to cool the hexane. The water enters the heat exchanger at 60°F. The UA product of the heat exchanger is found to be 1425 Btu/h·°F. Determine the outlet temperatures of the hot and cold fluids and the heat transfer rate between them.

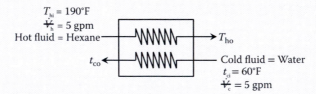

FIGURE E5.5A

SOLUTION: This is a heat exchanger analysis problem. The heat exchanger is already specified (the UA product is known) and analyzed using a heat transfer analysis. The solution to this problem is iterative because neither of the fluid outlet temperatures is known. In addition, the properties are evaluated at the average temperature of the fluid, which introduces another level of iteration. Therefore, this solution will be solved using EES. The given information in the problem is shown as follows:

```
"GIVEN:"
"Hot fluid"
   hot$ = 'n-hexane'
   V_dot_h = 5[gpm]
   T_h_in = 190[F]
"Cold fluid"
   cold$ = 'steam_iapws'
   V_dot_c = 5[gpm]
   t_c_in = 60[F]

"Heat Exchanger Parameters"
   UA = 1425[Btu/hr-R]

"FIND: Fluid outlet temperatures and heat transfer rate between
       the fluids and the heat transfer rate between them"
```

The LMTD heat exchanger model is given by the Equation set (5.60).

```
"SOLUTION:"
"The LMTD heat exchanger model is given by,"
   Q_dot = C_dot_h*(T_h_in - T_h_out)
   Q_dot = C_dot_c*(t_c_out - t_c_in)
   Q_dot = UA*LMTD
```

The thermal capacity rates of the hot and cold fluids are determined by

$$\dot{C}_h = \dot{m}_h c_{ph} \qquad \dot{C}_c = \dot{m}_c c_{pc}$$

The mass flow rates of each fluid can be found since the volumetric flow rates are given.

$$\dot{m}_h = \rho_h \dot{V}_h \qquad \dot{m}_c = \rho_c \dot{V}_c$$

```
"The thermal capacity rates are determined by,"
   C_dot_h = m_dot_h*c_ph
   C_dot_c = m_dot_c*c_pc
   m_dot_h = rho_h*V_dot_h*convert(gpm,ft^3/hr)
   m_dot_c = rho_c*V_dot_c*convert(gpm,ft^3/hr)
```

The LMTD can be calculated using Equation 5.59 once the inlet and outlet of the heat exchanger are identified. However, this equation is somewhat problematic when introduced into an equation solver, because of the natural logarithm function. Solvers use a variety of numerical algorithms to determine a solution. Sometimes, these algorithms can wander into a region where the natural logarithm is undefined. The result is a rather frustrating experience for the user. This problem is quickly remedied by rewriting Equation 5.59 to eliminate the natural logarithm, as follows:

$$\text{LMTD} = \frac{\Delta T_{out} - \Delta T_{in}}{\ln\left(\Delta T_{out}/\Delta T_{in}\right)} \quad \rightarrow \quad \Delta T_{out} = \Delta T_{in} \exp\left(\frac{\Delta T_{out} - \Delta T_{in}}{\text{LMTD}}\right)$$

This form of the equation is much easier for an equation solver to deal with.

Selecting the left-hand side of the heat exchanger to be the inlet (Figure E5.5A), the temperature differences are

$$\Delta T_{in} = T_{hi} - t_{co} \qquad \Delta T_{out} = T_{ho} - t_{ci}$$

```
"The LMTD is determined by,"
   DELTAT_out = DELTAT_in*exp((DELTAT_out - DELTAT_in)/LMTD)
   DELTAT_in = T_h_in - t_c_out
   DELTAT_out = T_h_out - t_c_in
```

The only thing remaining is to determine the properties of the fluids. Since both substances are liquids, the heat capacities are evaluated at the average temperature of each fluid. The volumetric flow rates are given at the inlet of each hot and cold stream. Therefore, the fluid densities are evaluated at their respective inlet temperatures.

```
"PROPERTIES"
"Hot Fluid"
   rho_h = density(hot$,T = T_h_in,x = 0)
   c_ph = cp(hot$,T = T_avg_h,x = 0)
   T_avg_h = (T_h_in + T_h_out)/2
{T_avg_h = T_h_in "Initial Guess to help convergence"}

"Cold Fluid"
   rho_c = density(cold$,T = t_c_in,x = 0)
   c_pc = cp(cold$,T = t_avg_c,x = 0)
   t_avg_c = (t_c_in + t_c_out)/2
{t_avg_c = t_c_in "Initial Guess to help convergence"}
```

At this point, it is worth departing from the topic of heat exchanger analysis to discuss the numerical solution strategy employed by equation-solving software. When a solver embarks on the solution of a system of n equations and n unknowns, an initial guess of the n unknowns is required. As a problem becomes more complex, it is more important for the n unknown guesses to be reasonable. Each software package estimates these initial values differently. The EES software package sets the value of all n unknowns to one. If the user does not adjust these values before attempting a solution, chances of convergence with a complex equation set are not good (e.g., evaluating some property formulations at 1°F may be out of the range of validity for the formulation). One way to avoid this is to provide EES with better initial estimates. This can be done by initially setting the average temperature of the hot and cold fluid to the inlet temperatures that are known in both of these cases. This solution is still iterative (this is the nature of the LMTD model), but it requires no iteration on *properties*. Therefore, the solver should be able to quickly find a solution.

Once a solution is found, the variables can be updated, which resets all guesses to the current solution. Then, the solver can be run again, allowing the average temperature to be calculated. This is the reason for the commented out lines (surrounded by braces) in the property calculation blocks of the EES code.

The resulting outlet temperatures and calculated heat transfer rate between the fluids are shown in Figure E5.5B. This solution indicates that both fluids leave the heat exchanger at nearly the same temperature. In fact, the cold-fluid outlet temperature is slightly *higher* than the hot-fluid outlet temperature. This is a phenomenon that can occur in counterflow heat exchangers, but not in parallel flow.

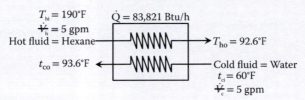

$T_{hi} = 190°F$ $\dot{Q} = 83,821$ Btu/h
$\dot{V}_h = 5$ gpm
Hot fluid = Hexane ⟶ $T_{ho} = 92.6°F$

$t_{co} = 93.6°F$ ⟵ Cold fluid = Water
$t_{ci} = 60°F$
$\dot{V}_c = 5$ gpm

FIGURE E5.5B

Heat exchanger analysis using the LMTD model is iterative because the outlet temperatures of the fluids are not known. A further level of iteration is introduced in determination of the properties. Although this solution method is valid, many engineers prefer a method that can be solved using a calculator and not rely on computer software. Such a method exists and is presented in Section 5.4.5.

5.4.5 Effectiveness–NTU (e-NTU) Heat Exchanger Model

The e-NTU method was developed to eliminate the iterative nature of the LMTD method. The *effectiveness*, ε, of a heat exchanger is defined as the ratio of the actual heat transfer rate in the heat exchanger to the *maximum possible* heat transfer rate that *could* occur.

$$\varepsilon = \frac{\dot{Q}}{\dot{Q}_{max}} \qquad (5.61)$$

The maximum possible heat transfer rate would occur in an infinitely long counterflow heat exchanger. In this case, one of the fluids would experience the maximum possible temperature change, $(T_{hi} - t_{ci})$. To determine which fluid experiences this maximum possible temperature change, consider the following analysis.

Since the heat transfer rate is same for both fluids, it must be true that

$$\dot{C}_h (T_{hi} - T_{ho}) = \dot{C}_c (t_{co} - t_{ci}) \qquad (5.62)$$

Now, consider the case where $\dot{C}_h < \dot{C}_c$. According to Equation 5.62, the hot fluid (the smaller thermal capacity rate) would experience the maximum possible temperature change $(T_{hi} - t_{ci})$ in an infinitely long counterflow heat exchanger. On the other hand, when $\dot{C}_c < \dot{C}_h$, the cold fluid would experience the largest possible temperature difference. This leads to the conclusion that the maximum heat transfer rate is experienced by the fluid with the *minimum* thermal capacity rate.

$$\dot{Q}_{max} = \dot{C}_{min} \left(T_{hi} - t_{ci} \right) \tag{5.63}$$

Therefore, the effectiveness of the heat exchanger can be written in one of two ways:

$$\varepsilon = \frac{\dot{C}_h \left(T_{hi} - T_{ho} \right)}{\dot{C}_{min} \left(T_{hi} - t_{ci} \right)} = \frac{\dot{C}_c \left(t_{co} - t_{ci} \right)}{\dot{C}_{min} \left(T_{hi} - t_{ci} \right)} \tag{5.64}$$

NTU is an acronym for the *number of transfer units*. The NTU is a dimensionless parameter defined by

$$\text{NTU} = \frac{UA}{\dot{C}_{min}} \tag{5.65}$$

For any heat exchanger, it has been shown that (Kays and London 1998)

$$\varepsilon = f \left(\text{NTU}, C_r \right) \quad \text{where} \quad C_r = \frac{\dot{C}_{min}}{\dot{C}_{max}} \tag{5.66}$$

The specific functional relationship is dependent on the type of heat exchanger (e.g., parallel flow and counterflow).

EXAMPLE 5.6

Determine the e-NTU function for a parallel-flow heat exchanger.

SOLUTON: Consider the case where $\dot{C}_c = \dot{C}_{min}$. For this scenario, Equation 5.64 can be written as

$$\varepsilon = \frac{\dot{C}_c \left(t_{co} - t_{ci} \right)}{\dot{C}_{min} \left(T_{hi} - t_{ci} \right)} = \frac{t_{co} - t_{ci}}{T_{hi} - t_{ci}}$$

The ratio of the thermal capacity rates can be written as

$$C_r = \frac{\dot{C}_{min}}{\dot{C}_{max}} = \frac{\dot{Q}}{\left(t_{co} - t_{ci} \right)} \frac{\left(T_{hi} - T_{ho} \right)}{\dot{Q}} = \frac{T_{hi} - T_{ho}}{t_{co} - t_{ci}}$$

Solving this equation for the hot-fluid outlet temperature gives

$$T_{ho} = T_{hi} - C_r \left(t_{co} - t_{ci} \right)$$

In the development of the LMTD for the parallel-flow heat exchanger, Equation 5.56 states that

$$\ln \left(\frac{\Delta T_{out}}{\Delta T_{in}} \right) = -UA \left(\frac{1}{\dot{C}_h} + \frac{1}{\dot{C}_c} \right)$$

For the parallel-flow heat exchanger with the minimum thermal capacity rate for the cold fluid, this equation becomes

$$\ln\left(\frac{T_{ho} - t_{co}}{T_{hi} - t_{ci}}\right) = -UA\left(\frac{1}{\dot{C}_{max}} + \frac{1}{\dot{C}_{min}}\right)$$

This equation can be algebraically rearranged to the following form:

$$\frac{T_{ho} - t_{co}}{T_{hi} - t_{ci}} = \exp\left[-\frac{UA}{\dot{C}_{min}}\left(\frac{\dot{C}_{min}}{\dot{C}_{max}} + 1\right)\right] = \exp\left[-NTU\left(1 + C_r\right)\right]$$

The ratio of temperature differences on the left-hand side of this equation can be algebraically manipulated after substituting the expression for T_{ho} developed earlier.

$$\frac{T_{ho} - t_{co}}{T_{hi} - t_{ci}} = \frac{\left[T_{hi} - C_r\left(t_{co} - t_{ci}\right)\right] - t_{co} + \left(t_{ci} - t_{ci}\right)}{T_{hi} - t_{ci}}$$

$$= \frac{\left(T_{hi} - t_{ci}\right) - C_r\left(t_{co} - t_{ci}\right) - \left(t_{co} - t_{ci}\right)}{T_{hi} - t_{ci}}$$

$$= 1 - C_r\frac{t_{co} - t_{ci}}{T_{hi} - t_{ci}} - \frac{t_{co} - t_{ci}}{T_{hi} - t_{ci}}$$

The temperature ratios on the right-hand side of this equation are the effectiveness of the heat exchanger. Therefore,

$$\frac{T_{ho} - t_{co}}{T_{hi} - t_{ci}} = 1 - \varepsilon C_r - \varepsilon = 1 - \varepsilon\left(1 + C_r\right)$$

Substitution of this equation into the equation that contains the NTU results in

$$1 - \varepsilon\left(1 + C_r\right) = \exp\left[-NTU\left(1 + C_r\right)\right]$$

Solving this equation for the effectiveness gives

$$\varepsilon = \frac{1 - \exp\left[-NTU\left(1 + C_r\right)\right]}{1 + C_r}$$

This equation was developed assuming that the cold fluid has the minimum thermal capacity rate. However, the same result would be found if the hot fluid has the minimum thermal capacity rate. Therefore, this expression reveals the functional relationship of Equation 5.66 for a parallel-flow heat exchanger.

Example 5.6 shows that the e-NTU expressions are *derived* from the LMTD. e-NTU expressions can be derived in a similar way for many other heat exchanger types. Table 5.3 shows the result of the e-NTU relationships derived from the LMTD analysis for several different types of heat exchangers (Kays and London 1998). Some of these expressions can be complex to carry out with a calculator. However, plots of these equations can be drawn, which make the e-NTU calculation easier. Figures 5.16 through 5.19 show the e-NTU plots for several heat exchanger types.

TABLE 5.3

e-NTU Relationships for Several Heat Exchanger Types

Heat Exchanger Type	e-NTU Relationship
Counterflow	$\varepsilon = \dfrac{1-\exp\left[-\text{NTU}(1-C_r)\right]}{1-C_r\exp\left[-\text{NTU}(1-C_r)\right]} \quad C_r < 1$
	$\varepsilon = \dfrac{\text{NTU}}{1+\text{NTU}} \quad C_r = 1 \text{ (regenerative HX)}$
Parallel flow	$\varepsilon = \dfrac{1-\exp\left[-\text{NTU}(1+C_r)\right]}{1+C_r}$
Shell and tube with 1-shell pass and an even number of tube passes	$\varepsilon = 2\left\{1+C_r+\sqrt{(1+C_r^2)}\,\dfrac{1+\exp\left[-\text{NTU}\sqrt{(1+C_r^2)}\right]}{1-\exp\left[-\text{NTU}\sqrt{(1+C_r^2)}\right]}\right\}^{-1}$
Cross flow with both fluids unmixed	$\varepsilon = 1-\exp\left[\left(\dfrac{1}{C_r}\right)(\text{NTU})^{0.23}\left(\exp\left[-C_r(\text{NTU})^{0.73}\right]-1\right)\right]$
Cross flow with $\dot{C}_{\max}$ mixed and $\dot{C}_{\min}$ unmixed	$\varepsilon = \dfrac{1}{C_r}\left\{1-\exp\left[-C_r\left(1-\exp(-\text{NTU})\right)\right]\right\}$
Cross flow with $\dot{C}_{\min}$ mixed and $\dot{C}_{\max}$ unmixed	$\varepsilon = 1-\exp\left\{-\dfrac{1}{C_r}\left[1-\exp(-C_r(\text{NTU}))\right]\right\}$
Boilers, evaporators, and condensers $C_r = 0$	$\varepsilon = 1-\exp(-\text{NTU})$

Source: Kays, W. M. and A. L. London, *Compact Heat Exchangers*, Kreiger Publishing, New York, 1998.

In Table 5.3, the e-NTU relationship for the CFHX with both fluids unmixed is an empirical correlation. This empiricism was developed because the analytical e-NTU relationship is an infinite series (Mason 1954). The functional form of the empirical relationship is taken from Bergman et al. (2011). The NTU exponents were determined by fitting the e-NTU heat exchanger data from Kays and London (1998) for the ranges $0.25 \le C_r \le 1.0$ and $0.25 \le \text{NTU} \le 7$. The resulting NTU exponents (0.23 and 0.73) are slightly different than values shown in Bergman et al. (2011). This is because the NTU exponents provided in Bergman et al. (2011) are valid only for a value of $C_r = 1$.

Using the effectiveness and NTU concepts, the e-NTU heat exchanger model can be summarized in the following set of equations:

$$\dot{Q} = \dot{C}_h\left(T_{hi} - T_{ho}\right)$$
$$\dot{Q} = \dot{C}_c\left(t_{co} - t_{ci}\right)$$
$$\varepsilon = \frac{\dot{Q}}{\dot{Q}_{\max}} \quad \text{where } \varepsilon = f(\text{NTU}, C_r)$$

$$(5.67)$$

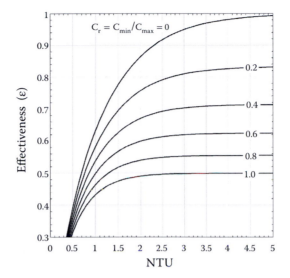

FIGURE 5.16
e-NTU plot for a parallel-flow heat exchanger.

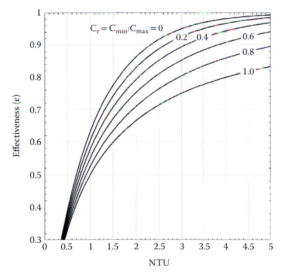

FIGURE 5.17
e-NTU plot for a counterflow heat exchanger.

Comparing the e-NTU model to the LMTD model reveals that the LMTD equation is replaced by two equations: one defining the effectiveness of the heat exchanger and the other defining the e-NTU relationship.

The following example shows that the e-NTU does not require an iterative solution for a heat exchanger analysis problem.

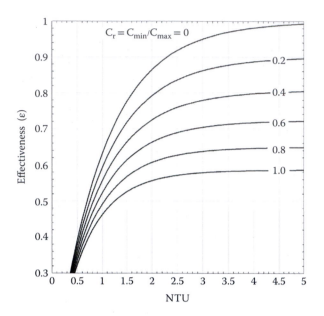

FIGURE 5.18
e-NTU plot for a shell and tube heat exchanger with a single shell pass and 2,4, ... *n* tube passes.

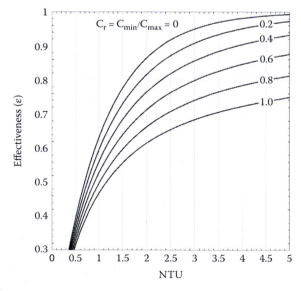

FIGURE 5.19
e-NTU plot for a cross-flow heat exchanger with both fluids unmixed.

EXAMPLE 5.7

Determine the fluid outlet temperatures and the heat transfer rate for the heat exchanger described in Example 5.5, using the e-NTU model. Assume that the properties can be evaluated at the give inlet temperatures. Compare the results to the LMTD model solution.

SOLUTION: The solution to this problem using the LMTD model is shown in Figure E5.7.

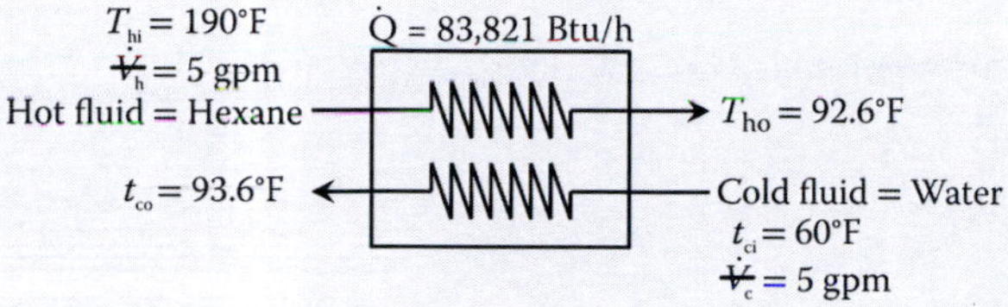

$T_{hi} = 190°F$ $\dot{Q} = 83{,}821$ Btu/h
$\dot{V}_h = 5$ gpm
Hot fluid = Hexane ⟶ $T_{ho} = 92.6°F$

$t_{co} = 93.6°F$ ⟵ Cold fluid = Water
$t_{ci} = 60°F$
$\dot{V}_c = 5$ gpm

FIGURE E5.7.

The effectiveness of this counterflow heat exchanger can be found using the equation in Table 5.3 or from Figure 5.17. The effectiveness equation is used here for more accuracy.

$$\varepsilon = \frac{1 - \exp\left[-NTU\left(1 - C_r\right)\right]}{1 - C_r \exp\left[-NTU\left(1 - C_r\right)\right]}$$

The NTU and C_r values are needed. They are defined as

$$NTU = \frac{UA}{\dot{C}_{min}} \qquad C_r = \frac{\dot{C}_{min}}{\dot{C}_{max}}$$

The first task is to determine which fluids have the maximum and minimum thermal capacity rates. The density and heat capacity values of each fluid can be determined using the fluid *inlet* temperatures. From Appendix B.1,

Hot fluid: hexane $\rho_h = 37.194$ lbm/ft^3 $c_{ph} = 0.6091$ Btu/lbm.R

Cold fluid: water $\rho_c = 62.363$ lbm/ft^3 $c_{pc} = 1.0003$ Btu/lbm.R

The mass flow rate of each fluid can be found once the densities are known:

Hot fluid: hexane $\dot{m}_h = \rho_h \dot{V}_h = \left(37.194\dfrac{\text{lbm}}{\text{ft}^3}\right)\left(5\dfrac{\text{gal}}{\text{min}}\right)\left(8.0208\dfrac{\text{ft}^3 \cdot \text{min}}{\text{h} \cdot \text{gal}}\right) = 1492\dfrac{\text{lbm}}{\text{h}}$

Cold fluid: water $\dot{m}_c = \rho_c \dot{V}_c = \left(62.363\dfrac{\text{lbm}}{\text{ft}^3}\right)\left(5\dfrac{\text{gal}}{\text{min}}\right)\left(8.0208\dfrac{\text{ft}^3 \cdot \text{min}}{\text{h} \cdot \text{gal}}\right) = 2501\dfrac{\text{lbm}}{\text{h}}$

This allows for the calculation of the thermal capacity rate for each fluid:

$$\text{Hot fluid: hexane } \dot{C}_h = \dot{m}_h c_{ph} = \left(1492\,\frac{\text{lbm}}{\text{h}}\right)\left(0.6091\,\frac{\text{Btu}}{\text{lbm}\cdot\text{R}}\right) = 908.6\,\frac{\text{Btu}}{\text{h}\cdot\text{R}}$$

$$\text{Cold fluid: water } \dot{C}_c = \dot{m}_c c_{pc} = \left(2501\,\frac{\text{lbm}}{\text{h}}\right)\left(1.0003\,\frac{\text{Btu}}{\text{lbm}\cdot\text{R}}\right) = 2502\,\frac{\text{Btu}}{\text{h}\cdot\text{R}}$$

Now, the minimum and maximum thermal capacitance rates can be specified, along with the capacitance rate ratio.

$$\dot{C}_{min} = 908.6\,\frac{\text{Btu}}{\text{h}\cdot\text{R}} \qquad \dot{C}_{max} = 2502\,\frac{\text{Btu}}{\text{h}\cdot\text{R}} \qquad \therefore \quad C_r = \frac{908.6\,\text{Btu/h}\cdot\text{R}}{2502\,\text{Btu/h}\cdot\text{R}} = 0.3632$$

Once the minimum thermal capacity rate is found, the NTU can be calculated.

$$\text{NTU} = \frac{UA}{\dot{C}_{min}} = \frac{1425\,\text{Btu/h}\cdot\text{R}}{908.6\,\text{Btu/h}\cdot\text{R}} = 1.568$$

Now, the effectiveness of the heat exchanger can be found.

$$\varepsilon = \frac{1-\exp\left[-\text{NTU}\left(1-C_r\right)\right]}{1-C_r\exp\left[-\text{NTU}\left(1-C_r\right)\right]} = \frac{1-\exp\left[-(1.568)(1-0.3632)\right]}{1-(0.3632)\exp\left[-(1.568)(1-0.3632)\right]} = 0.7292$$

The maximum possible heat transfer rate can be found since both inlet temperatures are known and the value of the minimum thermal capacity rate has been found.

$$\dot{Q}_{max} = \dot{C}_{min}\left(T_{hi} - t_{ci}\right) = \left(908.6\,\frac{\text{Btu}}{\text{h}\cdot\text{R}}\right)(190-60)\text{R} = 118,119\,\frac{\text{Btu}}{\text{h}}$$

Knowing the effectiveness of the heat exchanger and the maximum possible heat transfer rate allows for the calculation of the actual heat transfer rate between the fluids.

$$\varepsilon = \frac{\dot{Q}}{\dot{Q}_{max}} \quad \rightarrow \quad \dot{Q} = \varepsilon\dot{Q}_{max} = (0.7292)\left(118,119\,\frac{\text{Btu}}{\text{h}}\right) = 86,133\,\frac{\text{Btu}}{\text{h}} \quad \leftarrow$$

Once the heat transfer rate is found, the outlet temperatures can be determined using the conservation of energy equations in the heat exchanger model.

$$\dot{Q} = \dot{C}_h \left(T_{hi} - T_{ho} \right) \quad \rightarrow \quad T_{ho} = T_{hi} - \frac{\dot{Q}}{\dot{C}_h} = 190°F - \frac{86{,}133 \text{ Btu/h}}{908.6 \text{ Btu/h} \cdot °F} = \underline{95.2°F} \quad \leftarrow$$

$$\dot{Q} = \dot{C}_c \left(t_{co} - t_{ci} \right) \quad \rightarrow \quad t_{co} = t_{ci} + \frac{\dot{Q}}{\dot{C}_c} = 60°F + \frac{86{,}133 \text{ Btu/h}}{2505 \text{ Btu/h} \cdot °F} = \underline{94.4°F} \quad \leftarrow$$

Notice that this calculation was done without iteration. It can be easily done on a desktop using pencil, paper, and a calculator. Recall that this solution was done by evaluating the properties at the inlet temperature of the fluids. This set of equations can also be programmed into an equation solver, and iteration on the properties can be accomplished. Table E5.7 shows the result of the calculations for this example.

TABLE E5.7

Summary of Results

Parameter	LMTD Model	e-NTU Model[a]	e-NTU Model[b]
Heat transfer rate	83,821 Btu/h	86,133 Btu/h	83,821 Btu/h
Hot fluid outlet temperature	92.6°F	95.2°F	92.6°F
Cold fluid outlet temperature	93.6°F	94.4°F	93.6°F

[a] No iteration on properties—solved *by hand* using the inlet temperatures to evaluate properties.
[b] EES used to iterate on properties.

Notice that the error introduced by using the inlet temperatures to evaluate fluid properties is relatively small for this analysis. The NTU method predicts a heat transfer rate that is 2.8% high. The hot- and cold-fluid outlet temperatures are over predicted by 2.6° and 0.8°, respectively. This shows that a reasonable solution can be obtained using the inlet fluid temperatures to evaluate the fluid properties in an e-NTU analysis problem. If the e-NTU analysis is programmed into a solver which allows for iteration on the fluid properties, then the results are identical to the LMTD model. This is to be expected since the e-NTU method is derived directly from the LMTD analysis.

Table 5.3 shows the e-NTU relations derived from the LMTD analysis. These equations are well suited for heat exchanger analysis problems. However, when performing heat exchanger design calculations, it may be more convenient to work with relationships of the form

$$NTU = f\left(\varepsilon, C_r\right) \tag{5.68}$$

These NTU-explicit (NTU-e) functions are shown in Table 5.4. These equations provide improved accuracy compared to reading e-NTU plots, such as those shown in Figures 5.16 through 5.19.

5.5 Special Application Heat Exchangers

The LMTD and e-NTU heat exchanger models discussed in Sections 5.4.4 and 5.4.5 are valid for many different types of heat exchangers and working fluids. However, there are special cases where the models may be simplified.

5.5.1 Counterflow Regenerative Heat Exchanger

Tables 5.3 and 5.4 show two entries for e-NTU functions for counterflow heat exchangers. One equation is for $C_r < 1$ and the other equation is for $C_r = 1$. If $C_r = 1$, both fluids have the same thermal capacity rate. It is possible for this situation to occur with two different fluids, but it is rather coincidental. However, this situation can occur in a heat exchanger where both hot and cold fluids are the same and they have the same mass flow rate. A common type of heat exchanger where this can occur is a counterflow *regenerative* heat exchanger. An example of a counterflow regenerative heat exchanger used in a gas turbine cycle is shown in Figure 5.20. In this figure, the heat exchanger that is labeled "regenerator" is a counterflow regenerative heat exchanger. In the cycle shown in Figure 5.20, the purpose of the regenerator is to preheat the air entering the combustion chamber by utilizing otherwise-wasted hot turbine exhaust gases. The result is a reduction in fuel requirement for the same power output, resulting in an increased thermal efficiency.

TABLE 5.4

NTU-e Relationships for Several Heat Exchanger Types

Heat Exchanger Type	NTU-e Relationship
Counterflow	$\text{NTU} = \dfrac{1}{C_r - 1} \ln\left(\dfrac{\varepsilon - 1}{\varepsilon C_r - 1}\right) \quad C_r < 1$
	$\text{NTU} = \dfrac{\varepsilon}{1 - \varepsilon} \quad C_r = 1 \text{ (regenerative HX)}$
Parallel flow	$\text{NTU} = -\dfrac{\ln\left[1 - \varepsilon(1 + C_r)\right]}{1 + C_r}$
Shell and tube with 1-shell pass and an even number of tube passes	$\text{NTU} = -\left(1 + C_r^2\right)^{-1/2} \ln\left(\dfrac{\Gamma - 1}{\Gamma + 1}\right) \qquad \Gamma = \dfrac{2/\varepsilon - (1 + C_r)}{\sqrt{(1 + C_r^2)}}$
Cross flow with $\dot{C}_{max}$ mixed and $\dot{C}_{min}$ unmixed	$\text{NTU} = -\ln\left[1 + \left(\dfrac{1}{C_r}\right)\ln(1 - \varepsilon C_r)\right]$
Cross flow with $\dot{C}_{min}$ mixed and $\dot{C}_{max}$ unmixed	$\text{NTU} = -\left(\dfrac{1}{C_r}\right)\ln\left[C_r \ln(1 - \varepsilon) + 1\right]$
Boilers, evaporators, and condensers $C_r = 0$	$\text{NTU} = -\ln(1 - \varepsilon)$

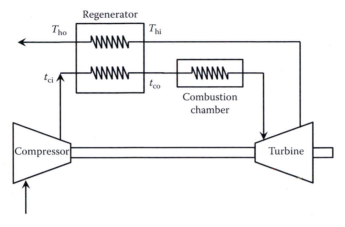

FIGURE 5.20
A gas turbine cycle with a regenerative heat exchanger.

In the gas turbine cycle depicted in Figure 5.20, the fluid leaving the turbine is a mixture of combustion gases while air leaves the compressor. Therefore, the two fluids in the regenerator are different, and they are at different temperatures. However, recall that the *air-standard* gas turbine models are based on the assumption that the working fluid is air throughout. Furthermore, if the *cold air-standard* model is used, the heat capacity of the air is assumed constant throughout the cycle. Therefore, if a cold air-standard assumption is used to analyze the gas turbine shown in Figure 5.20, then the thermal capacity rates of the hot and cold fluids in the regenerator are identical, which means that $C_r = 1$.

Using the e-NTU heat exchanger model to analyze or design a counterflow regenerative heat exchanger is a relatively simple matter, since e-NTU and NTU-e equations have been developed for the counterflow heat exchanger with $C_r = 1$, as shown in Tables 5.3 and 5.4. However, it is interesting to see how the LMTD model is impacted for this special situation.

Consider the regenerative heat exchanger in the gas turbine cycle shown in Figure 5.20. For the counterflow heat exchanger, the LMTD is defined as

$$\text{LMTD} = \frac{\Delta T_{\text{out}} - \Delta T_{\text{in}}}{\ln\left(\Delta T_{\text{out}} / \Delta T_{\text{in}}\right)} = \frac{(T_{\text{hi}} - t_{\text{co}}) - (T_{\text{ho}} - t_{\text{ci}})}{\ln\left[(T_{\text{hi}} - t_{\text{co}}) / (T_{\text{ho}} - t_{\text{ci}})\right]} \tag{5.69}$$

The heat transfer rate between the two-fluid streams is the same (opposite in direction). Therefore,

$$\dot{C}_h \left(T_{\text{hi}} - T_{\text{ho}}\right) = \dot{C}_c \left(t_{\text{co}} - t_{\text{ci}}\right) \tag{5.70}$$

Assuming that the thermal capacitance rates are the same for the hot and cold fluids, Equation 5.70 becomes

$$T_{\text{hi}} - T_{\text{ho}} = t_{\text{co}} - t_{\text{ci}} \tag{5.71}$$

This equation can be algebraically rearranged to consider the temperature differences on each side of the regenerative heat exchanger.

$$T_{hi} - t_{co} = T_{ho} - t_{ci} \tag{5.72}$$

Equation 5.72 indicates that the temperature difference on each side of the heat exchanger is the same. Therefore, the LMTD becomes undefined (0/0). This means that the LMTD cannot be calculated for this type of a heat exchanger. However, to complete the heat exchanger model, a heat transfer equation analogous to Equation 5.58 is needed. To discover the form of the heat transfer equation needed, a differential area inside the regenerative heat exchanger is evaluated, as shown in Figure 5.21. In the differential element dA, the hot and cold fluid temperatures are specified as T and t, respectively.

For the differential area shown in Figure 5.21,

$$d\dot{Q} = \dot{C}dT \quad \text{and} \quad d\dot{Q} = \dot{C}dt \tag{5.73}$$

These are two thermodynamic equations for the heat transfer rate between the fluids: one for each fluid passing through the differential area. The heat transfer equation representing the heat transfer rate between the two fluids is

$$d\dot{Q} = U(T - t)dA \tag{5.74}$$

Equations 5.73 and 5.74 can be combined into two new equations.

$$\frac{dT}{dA} = \frac{U}{\dot{C}}(T - t) \quad \text{and} \quad \frac{dt}{dA} = \frac{U}{\dot{C}}(T - t) \tag{5.75}$$

These equations suggest that the slopes of the temperature profiles are the same for both fluids. In addition, since the slopes through dA are identical, the temperature difference between the hot and cold fluids is the same on both sides of dA, as it is within dA. Therefore, the derivatives in Equation 5.75 are *equal* and *constant*. This implies that the temperature profile of each fluid is linear and the profiles are parallel, as shown in Figure 5.22. Therefore, the temperature difference between the hot and cold fluids is

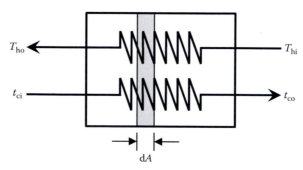

FIGURE 5.21
Differential analysis of a regenerative heat exchanger.

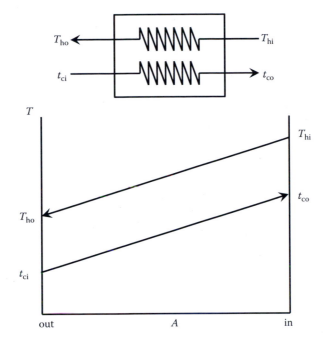

FIGURE 5.22
Temperature profiles in a counterflow regenerative heat exchanger.

constant throughout the heat exchanger. Since the temperature difference is constant throughout this type of heat exchanger, an alternate model for the counterflow regenerative heat exchanger similar to the LMTD method can be developed, as shown in the following equation set:

$$\dot{Q} = \dot{C}\left(T_{hi} - T_{ho}\right)$$
$$\dot{Q} = \dot{C}\left(t_{co} - t_{ci}\right) \tag{5.76}$$
$$\dot{Q} = UA\left(T_{hi} - t_{co}\right) \quad \text{or} \quad \dot{Q} = UA\left(T_{ho} - t_{ci}\right)$$

5.5.2 Heat Exchangers with Phase Change Fluids: Boilers, Evaporators, and Condensers

In many thermal energy systems, one of the working fluids passing through a heat exchanger experiences a phase change. Figure 5.23 is a schematic of two common thermal systems: the vapor power cycle and the refrigeration cycle. In both of these cycles the working fluid undergoes a phase change in the heat exchangers. In the vapor power cycle, liquid water is converted to steam in the boiler and condensed back to liquid in the condenser. In the refrigeration cycle, the saturated refrigerant is converted to vapor in the evaporator and condensed back to liquid in the condenser. These types of heat exchangers are generally categorized as either boilers, evaporators, or condensers. The heat exchangers in Figure 5.23 are shown as counterflow heat exchangers for convenience in the sketch. In applications, the heat exchangers may have several different configurations (e.g., shell and tube).

The e-NTU model for analysis and design of heat exchangers is attractive because of its noniterative nature. Recall that the e-NTU method requires the determination of the thermal capacity rates for each fluid. To determine the thermal capacity rate of a fluid, the mass flow rate and heat capacity need to be known. However, for a pure fluid undergoing an isobaric phase change, the heat capacity is infinite. This can be verified by the following equation:

$$c_p = \left(\frac{\partial h}{\partial T}\right)_P \approx \left(\frac{\Delta h}{\Delta T}\right)_P = \frac{h_{fg}}{T_g - T_f} = \frac{h_{fg}}{0} = \infty \qquad (5.77)$$

The fact that the heat capacity of a phase-change fluid is infinite means that the thermal capacity rate of the fluid must also be infinite. Therefore the maximum heat capacity rate is always associated with the phase-change fluid. This also implies that the ratio of the thermal capacity rates must be zero.

$$C_r = \frac{\dot{C}_{min}}{\dot{C}_{max}} = \frac{\dot{C}_{min}}{\infty} = 0 \qquad (5.78)$$

The last entry in Tables 5.3 and 5.4 shows the e-NTU and NTU-e relationships for this case. Notice that the e-NTU relationship is fairly simple and *independent of the heat exchanger configuration*.

Analysis and design using the LMTD method requires a careful analysis of the temperature profiles within the heat exchanger. Figure 5.24 shows what typical temperature profiles look like for boilers, evaporators, and condensers

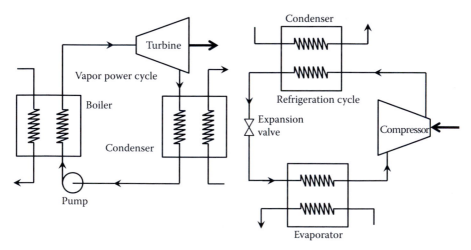

FIGURE 5.23
Vapor power and refrigeration cycles showing heat exchangers where a phase change occurs.

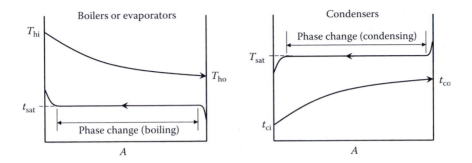

FIGURE 5.24
Typical temperature profiles inside boilers, evaporators, and condensers.

where one fluid undergoes the phase change and the other fluid remains in the single phase. The figures are drawn for a pure fluid undergoing an isobaric phase change. In reality, there are pressure drops through heat exchangers. However in a well-designed heat exchanger, the magnitude of this pressure drop is relatively small. Therefore, Figure 5.24 is a good approximation to the actual behavior that is observed.

Notice that the majority of the heat exchange area in these types of heat exchangers is devoted to the phase change. Only a small percentage is used to subcool or superheat the fluid. Therefore, the LMTD is modified to use the *saturation temperature* of the phase-change fluid for both the inlet and outlet temperatures. Using this idea, the LMTD of an evaporator (or boiler) can be written as

$$\text{LMTD}_E = \frac{\Delta T_{in} - \Delta T_{out}}{\ln(\Delta T_{in}/\Delta T_{out})} = \frac{(T_{hi} - t_{sat}) - (T_{ho} - t_{sat})}{\ln(\Delta T_{in}/\Delta T_{out})} = \frac{T_{hi} - T_{ho}}{\ln(\Delta T_{in}/\Delta T_{out})} \quad (5.79)$$

Similarly, for a condenser, the LMTD is

$$\text{LMTD}_C = \frac{\Delta T_{in} - \Delta T_{out}}{\ln(\Delta T_{in}/\Delta T_{out})} = \frac{(T_{sat} - t_{ci}) - (T_{sat} - t_{co})}{\ln(\Delta T_{in}/\Delta T_{out})} = \frac{t_{co} - t_{ci}}{\ln(\Delta T_{in}/\Delta T_{out})} \quad (5.80)$$

Equations 5.79 and 5.80 were developed for a counterflow heat exchanger. However, the temperature profiles shown in Figure 5.24 are the same for any type of evaporator or condenser. Therefore, Equations 5.79 and 5.80 are independent of heat exchanger configuration.

Once the calculation of the LMTD is determined for the heat exchanger, the complete LMTD model can be developed. For evaporators or boilers where the hot fluid remains in the single phase and the cold fluid is boiling, the LMTD model is given by

$$\dot{Q} = \dot{m}_h c_{ph} (T_{hi} - T_{ho})$$
$$\dot{Q} = \dot{m}_c (h_{co} - h_{ci}) \quad (5.81)$$
$$\dot{Q} = UA(\text{LMTD}_E)$$

Notice that the conservation of energy equation for the cold fluid is written in terms of the boiling fluid's enthalpy change. For the condenser where the hot fluid is undergoing the phase change and the cold fluid remains in the single phase, the LMTD model is

$$\dot{Q} = \dot{m}_h \left(h_{hi} - h_{ho} \right)$$
$$\dot{Q} = \dot{m}_c c_{pc} \left(t_{c,o} - t_{c,i} \right) \qquad (5.82)$$
$$\dot{Q} = UA \left(\mathrm{LMTD}_C \right)$$

In applications where there is very little superheating and subcooling of the phase-change fluid, the enthalpy difference can be approximated as the enthalpy of vaporization of the fluid.

EXAMPLE 5.8

A STHX is being used as a water chiller as shown in Figure E5.8. The chiller is the evaporator of a refrigeration cycle utilizing R-123 as the working fluid. The R-123 flows at a rate of 5000 lbm/h and leaves the condenser of the cycle as a saturated liquid at 200°F. The evaporating pressure is 5 psia, and the pressure drop of the R-123 through the evaporator can be assumed to be negligible. The R-123 leaves the evaporator as a saturated vapor. Water enters the evaporator at 50°F with at a volumetric flow rate of 30 gpm. Determine the following:

  a. Evaporating temperature of the R-123 (°F)
  b. Outlet temperature of the chilled water (°F)
  c. Capacity of the chiller (tons)
  d. UA product of the chiller (Btu/h·°F)

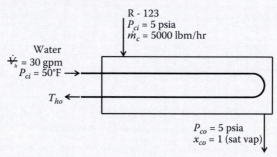

FIGURE E5.8

A schematic of the full refrigeration cycle is seen in Figure 5.23. The exit of the expansion valve in the refrigeration cycle is a two-phase mixture. The R-123 enters at this two-phase condition. It leaves the evaporator as a saturated vapor. Since the pressure is assumed to be constant on the R-123 side of the heat exchanger, the temperature must remain constant. This is the *evaporating temperature* asked for in part (a). Interpolation in Appendix B (Table B.2) gives the value for the evaporating temperature.

$$t_{evap} = \underline{34.1°F} \quad \leftarrow$$

Parts (b) and (c) can be solved purely from the thermodynamic balances on each fluid. The heat transfer rate in the evaporator (also known as its *capacity*) can be found by

$$\dot{Q} = \dot{m}_r \left(h_{co} - h_{ci} \right)$$

The enthalpy of the refrigerant entering the evaporator is the same as the enthalpy leaving the condenser of the refrigeration cycle. This must be true because the expansion valve in the refrigeration cycle is *isenthalpic*. Therefore, the enthalpy into the evaporator, h_{ci}, can be found at the saturated liquid state at 200°F. The enthalpy leaving the evaporator, h_{co}, is a saturated vapor at 5 psia; therefore, its value can also be found. Using Appendix B.2, these enthalpy values are found by interpolation to be

$$h_{co} = 164.8 \text{ Btu/lbm} \qquad h_{ci} = 129.3 \text{ Btu/lbm}$$

Therefore, the capacity of the evaporator is

$$\dot{Q} = \dot{m}_c \left(h_{co} - h_{ci} \right) = \left(5000 \frac{\text{lbm}}{\text{h}} \right) (164.8 - 129.3) \frac{\text{Btu}}{\text{lbm}} \underbrace{\left(\frac{\text{ton} \cdot \text{h}}{12,000 \text{ Btu}} \right)}_{\text{Appendix A}}$$

$$= \underline{14.77 \text{ ton}} \quad \leftarrow$$

Knowing the heat transfer rate allows for the calculation of the outlet temperature of the water.

$$\dot{Q} = \dot{m}_h c_{ph} \left(T_{hi} - T_{ho} \right) \quad \rightarrow \quad T_{ho} = T_{hi} - \frac{\dot{Q}}{\left(\rho_h \dot{V}_h \right) c_{ph}}$$

The density and heat capacity of the water should be evaluated at the average temperature of the water. However, the outlet temperature is not known. As an estimate, the inlet temperature is used to evaluate the properties. From Appendix B.1,

$$\rho_h \approx 62.406 \text{ lbm/ft}^3 \qquad c_{ph} \approx 1.0021 \text{ Btu/lbm} \cdot \text{R}$$

Therefore, the outlet temperature of the chilled water is

$$T_{ho} = 50°F - \frac{(14.77 \text{ ton}) \overbrace{\left(12000 \frac{\text{Btu}}{\text{h} \cdot \text{ton}} \right)}^{\text{Appendix A}}}{\left(62.406 \frac{\text{lbm}}{\text{ft}^3} \right) \left(30 \frac{\text{gal}}{\text{min}} \right) \left(1.0021 \frac{\text{Btu}}{\text{lbm} \cdot °F} \right) \underbrace{\left(8.0208 \frac{\text{ft}^3 \cdot \text{min}}{\text{h} \cdot \text{gal}} \right)}_{\text{Appendix A}}}$$

$$= \underline{38.2°F} \quad \leftarrow$$

The UA product of the evaporator can be found from the third equation in the heat exchanger model.

$$\dot{Q} = UA(\text{LMTD}_E) \quad \rightarrow \quad UA = \frac{\dot{Q}}{(\text{LMTD}_E)}$$

All fluid temperatures are known, the LMTD for the evaporator can be calculated.

$$\text{LMTD}_E = \frac{T_{hi} - T_{ho}}{\ln(\Delta T_{in} / \Delta T_{out})} \frac{(50 - 38.2)\text{R}}{\ln[(50 - 34.1)\text{R}/(38.2 - 34.1)\text{R}]} = 8.74 \text{ R}$$

Therefore, the UA product of the evaporator is

$$UA = \frac{\dot{Q}}{(\text{LMTD}_E)} = \frac{14.77 \text{ ton}}{8.74 \text{ R}} \left|\left(12000 \frac{\text{Btu}}{\text{h} \cdot \text{ton}}\right)\right| = 20,279 \frac{\text{Btu}}{\text{h} \cdot \text{R}} \quad \leftarrow$$

These values were determined assuming that the water temperature was 50°F. Using EES to iterate on the properties results in the following solution: t_{evap} = 34.09°F, $\dot{Q} = 14.77$ ton, T_{ho} = 38.24°F, and UA = 20,256 Btu/h·R. Notice that there is very little difference between the EES solution and the solution non-iterative solution.

This example was solved using the LMTD model for the evaporator. The same results would be found using the e-NTU method.

In Example 5.8, the UA product of the heat exchanger was calculated. If this were a heat exchanger design problem, the overall heat transfer coefficient, U, would need to be determined. Once U is known, then the total heat transfer area required can be found. From this information, the engineer can specify the number of tubes required in the shell.

Convective heat transfer coefficient correlations for boiling and condensing fluids can be quite complex. The boiling and condensing processes are dependent on many things, including gravity effects. Therefore many correlations are dependent on the *orientation* of the tube containing the boiling or condensing liquid. The reader is encouraged to refer to any of the excellent resources available, including *The Heat Transfer Handbook* (Bejan and Kraus 2003) or *The Handbook of Heat Transfer* (Rohsenow et al. 1998). The most up-to-date information concerning boiling and condensing convective heat transfer correlations can also be found in technical journals such as *The International Journal of Heat and Mass Transfer* and the *Journal of Heat Transfer*.

5.6 Double-Pipe Heat Exchanger Design and Analysis

The DPHX was qualitatively discussed in Section 5.3.1. This section details the suggested procedures for design and analysis calculations, including prediction of the heat exchanger's thermal and hydraulic performance. The thermal performance calculations result in determination of a heat exchanger area (a design problem) or outlet fluid temperatures from a given DPHX (an analysis problem). The hydraulic performance of interest is the pressure drop of each fluid as it passes through the heat exchanger.

5.6.1 Double-Pipe Heat Exchanger Diameters

Design and analysis calculations for DPHXs are complicated by the fact that an annular space is created between the two tubes making up the heat exchanger as shown in Figure 5.25. In a DPHX, the heat is transferred between the fluids flowing in the inner tube and the annulus. It is assumed that the outer tube is insulated so there is no heat gain/loss to/from the fluid in the annulus.

For hydraulic calculations involving the pressure drop through a DPHX, the Reynolds number and relative roughness of the flow in the inner tube are based on the inside diameter of tube. The Reynolds number and relative roughness of the fluid in the annulus are based on the *hydraulic* diameter. The hydraulic diameter is defined as

$$D_{\text{hyd}} = \frac{4(\text{cross-sectional flow area})}{\text{wetted perimeter}} = \frac{4A}{P_{\text{wetted}}} \tag{5.83}$$

In the annulus, friction is present on both surfaces. Therefore, the wetted perimeter is the total circumference of the inner and outer tubes making up the annular space. Using the nomenclature in Figure 5.25, the hydraulic diameter of the annulus can be written as

$$D_{\text{hyd}} = \frac{4\left(\dfrac{\pi}{4}\right)\left(\text{ID}_a^2 - \text{OD}_t^2\right)}{\pi\left(\text{ID}_a + \text{OD}_t\right)} = \text{ID}_a - \text{OD}_t \tag{5.84}$$

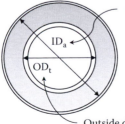

Inside diameter of the outer tube forming the annulus

Outside diameter of the inner tube

FIGURE 5.25
Cross-sectional view of a double-pipe heat exchanger.

For heat transfer calculations, the Nusselt and Reynolds numbers are based on the *equivalent* diameter for the flow in the annulus. Analogous to the hydraulic diameter, the equivalent diameter is defined as

$$D_{equ} = \frac{4(\text{cross-sectional flow area})}{\text{heat transfer perimeter}} = \frac{4A}{P_{heat}} \tag{5.85}$$

In the annular space, the heat transfer is occurring at the outer surface of the inner tube. It is assumed that no heat is transferred through the outer tube. On the basis of this definition, the equivalent diameter of the annulus can be written as

$$D_{equ} = \frac{4\left(\frac{\pi}{4}\right)(ID_a^2 - OD_t^2)}{\pi(OD_t)} = \frac{(ID_a^2 - OD_t^2)}{OD_t} \tag{5.86}$$

On the basis of this discussion, it is seen that there is a single Reynolds number for the flow inside the inner tube. However, for the annulus, there are *two* Reynolds numbers. The Reynolds number based on the hydraulic diameter is used when the pressure drop is calculated. The Reynolds number based on the equivalent diameter is used to determine the overall heat transfer coefficient in the annulus.

5.6.2 Overall Heat Transfer Coefficients for the Double-Pipe Heat Exchanger

In DPHX calculations, it is common to base the overall heat transfer coefficient on the outside surface area of the inner tube. Therefore, Equation 5.40 should be used. In most DPHX designs, the inner tube is a high thermal conductivity material, such as copper. If this is the case, then the thermal resistance of the inner tube is very small compared to the convective and fouling resistances and is often omitted from the calculation.

It is also informative to determine the performance of the heat exchanger when it is brand new (no fouling) and after 1 year of service (fouling). Two overall heat transfer coefficients can be calculated for these conditions as follows:

$$\frac{1}{U_{o,clean}} = \frac{1}{h_i}\left(\frac{OD_t}{ID_t}\right) + \frac{1}{h_o} \tag{5.87}$$

and

$$\frac{1}{U_{o,fouled}} = \frac{1}{h_i}\left(\frac{OD_t}{ID_t}\right) + R_{f,i}''\left(\frac{OD_t}{ID_t}\right) + R_{f,o}'' + \frac{1}{h_o}. \tag{5.88}$$

For turbulent flow, the recommended correlation for the convective heat transfer coefficient is the Gnielinski correlation (Gnielinski 1976) given by Equation 5.30. For laminar flow, the correlation suggested by Baehr and Stephan (2006) given by Equation 5.26 is recommended. This correlation was developed for a tube with a constant surface temperature. As the fluids pass through a DPHX, the inner tube is most likely not at a constant temperature. However, using Equation 5.26 will result in a lower convective heat transfer coefficient compared to Equation 5.27, which was developed for a constant surface heat flux. This results in a heat exchanger that errs on the side of being slightly larger than required. Therefore, Equation 5.26 provides a more conservative estimate of the laminar heat transfer coefficient. Equation 5.26 is quite complex. However, with modern computing software, the complexity can be easily dealt with. Table 5.5 shows the recommended correlations for a DPHX.

As discussed earlier, care must be taken to make sure that the proper diameter is being used when evaluating the Nusselt and Reynolds numbers for the annulus.

5.6.3 Hydraulic Analysis of the Double-Pipe Heat Exchanger

A heat exchanger is part of a larger thermal energy system. The fluids are moved through the heat exchanger by virtue of a pump (for liquid flows) or a compressor (for gas flows). To properly size the pump or compressor, the pressure drop through the heat exchanger needs to be included in the development of the system curve.

TABLE 5.5

Recommended Correlations for the Convective Heat Transfer Coefficient in DPHX Design and Analysis

Flow Regime	Correlation	Equation
Laminar ($Re < 2300$)	$Nu_D = \dfrac{hD}{k} = (A+B)/C$	(5.26)
	$A = 3.66/\tanh\left[2.264 Gz_D^{-1/3} + 1.7 Gz_D^{-2/3}\right]$	
	$B = 0.0499 Gz_D \tanh\left(Gz_D^{-1}\right) + \dfrac{0.0668 Gz_D}{1+0.04 Gz_D^{2/3}}$	
	$C = \tanh\left(2.432 Pr^{1/6} Gz_D^{-1/6}\right)$	
	$Pr \geq 0.1$	
	Properties evaluated at $T_m = (T_i + T_o)/2$	
Turbulent	$Nu_D = \dfrac{hD}{k} = \dfrac{(f/8)(Re_D - 1000) Pr}{1+12.7(f/8)^{1/2}(Pr^{2/3}-1)}$	(5.30)
	$0.5 < Pr < 2000,\ 3000 < Re < 5 \times 10^6$	
	Properties evaluated at $T_m = (T_i + T_o)/2$	

The pressure drop through the inner tube of the DPHX is related to the head loss due to friction through the tube.

$$\frac{\Delta P_t}{\gamma_t} = f_t \frac{L}{ID_t} \frac{V_t^2}{2g} + l_m \qquad (5.89)$$

If the DPHX is of hairpin design, L is the total length of the heat exchanger and l_m the minor loss through the fittings that connect the hairpin(s) together. If the DPHX is a simple tube-in-tube design with no hairpins, then the minor loss in Equation 5.89 is zero. The friction factor in the tube is determined using the methods shown in Chapter 4.

In the annular space, an additional head loss is experienced by the fluid because of the fittings. This is accounted for by adding an additional velocity head to the friction loss. In addition, the hydraulic diameter of the annulus is used to determine the pressure drop.

$$\frac{\Delta P_a}{\gamma_a} = \left(f_a \frac{L}{D_{hyd}} + n \right) \frac{V_a^2}{2g} \qquad (5.90)$$

In this equation, $n = 1$ for a single DPHX. If the heat exchanger is a hairpin design, n represents the total number of DPHXs that make up the complete heat exchanger. The friction factor for turbulent flow in an annulus can be found using the hydraulic diameter and any of the equations listed in Table 4.2. For laminar flow in an annulus, the friction factor is calculated using

$$\frac{1}{f_a} = \frac{64}{Re_{D_{hvd}}} \left[\frac{1+\kappa^2}{(1-\kappa)^2} + \frac{1+\kappa}{(1-\kappa)\ln\kappa} \right] \quad \text{where} \quad \kappa = \frac{OD_t}{ID_a} \qquad (5.91)$$

A well-designed DPHX has pressure drops in either stream that are less than 10 psi ($\approx$70 kPa).

5.6.4 Fluid Placement in a Double-Pipe Heat Exchanger

There are two criteria that dictate which fluid gets placed in the tube and which goes through the annulus. The *hydraulic criterion* is consistent with keeping the pressure drop to a minimum. To meet this criterion, the fluid with the higher mass flow rate should go through the passage with the larger cross-sectional area. However, a sometimes competing criterion is the *fouling* criterion. The idea with the fouling criterion is to extend the life of the heat exchanger by making it easy to mechanically clean. In the DPHX, it is far easier to mechanically clean the inside tube. Therefore, according to this criterion, the fluid with the higher fouling factor should go through the tube. As mentioned earlier, both these criteria may not necessarily be met simultaneously. In instances similar to this, it is up to the design engineer to make the decision, based on which criterion is most important in the heat exchanger application.

5.6.5 Double-Pipe Heat Exchanger Design Considerations

Several design considerations for DPHXs have already been discussed. They are included here along with others. The following list is by no means all-inclusive. However, the major design considerations are summarized.

- The pressure drop through either side of the heat exchanger (tube or annulus) should be no more than 10 psi ($\approx$70 kPa). This will help minimize the cost of moving the fluid through the heat exchanger.

- An optimum DPHX design is one where the fluids in the tube and annulus are within the recommended economic velocity range shown in Table 4.11 or calculated using a reasonable range of economic parameters for the specific application.

- Fluid placement should be based on either the hydraulic criterion (minimizing the pressure drop) or the fouling criterion (easy mechanical cleaning of the heat exchanger).

- The inner tube in a DPHX should be of high thermal conductivity (copper is a good choice). In cases where copper cannot be used, an economically feasible material can be chosen. If the material does not have a large thermal conductivity, the thermal resistance of the material should be included in the calculation of the overall heat transfer coefficients (clean and fouled).

- The material for the outer tube of the DPHX does not need to be made of an expensive material such as copper. In fact, a low thermal conductivity material is more desirable for the outer tube.

- The outer tube of the DPHX should be insulated to avoid stray heat losses to the environment surrounding the heat exchanger.

- When different materials are used for the inner and outer tubes, the absolute roughness values may be different. When calculating the friction factor in the annulus, using the highest value for the absolute roughness will tend to overestimate the pressure drop and lead to a more conservative design. Using the average absolute roughness between the inner and outer tube may give a more reasonable estimate of the pressure drop.

- The flow direction (parallel or counter) is an important consideration. The counterflow DPHX has the advantage of being smaller than the parallel-flow configuration for the same thermal performance. However, in a counterflow heat exchanger, it is possible for the outlet fluid temperatures to *cross*. That is, it is possible for the cold-fluid outlet temperature to be hotter than the hot-fluid outlet temperature. In cases where this is not desirable, a parallel-flow heat exchanger should be used.

5.6.6 Use of Computer Software for Design and Analysis of Heat Exchangers

The complete design or analysis of any heat exchanger involves simultaneous consideration of the heat transfer performance along with the hydraulic performance. These types of calculations can be highly iterative if the LMTD method is chosen for the heat exchanger model. However, the e-NTU method reduces the iteration to only properties. In turbulent flow, the friction factor is required to evaluate the convective heat transfer coefficient. This is another iterative calculation. The message here is that heat exchanger design and analysis calculations can become very complex and iterative on several levels, making them well suited for computer software.

Computer software packages, such as EES, are very capable of solving complex problems. However, as the complete set of equations gets larger and the iterations more complex, the chance of nonconvergence increases. This requires the engineer to select wise starting guesses for the variables in the model. In addition, it may be a good idea to determine a solution using constant values for some of the iterative variables (e.g., a friction factor). Once a solution is found using these constant values, the starting values for all the variables in the model can be updated and the problem re-solved, allowing all variables to iterate to a final solution. Sometimes, solving a heat exchanger design or analysis problems requires several runs of the program.

The following sections show detailed examples of a DPHX design calculation (Section 5.6.7) and a DPHX analysis (Section 5.6.8).

5.6.7 Double-Pipe Heat Exchanger Design Example

Consider the following DPHX design problem.

In an industrial process, 10,000 lbm/h of toluene must be cooled from 120°F to 100°F. A supply of city water at 50°F is available at a flow rate of 12,000 lbm/h to provide the cooling needed. Design a DPHX to accomplish this task. Use type M copper tube for both tubes in the heat exchanger.

5.6.7.1 Fluid Properties

The inlet and outlet temperatures of the hot fluid are known. Therefore, the average properties can be determined using EES. For the toluene (hot fluid) at 110°F,

Fluid	T (°F)	ρ (lbm/ft³)	γ (lbf/ft³)	c_p (Btu/lbm-R)	μ (lbm/ft·s)	k (Btu/h·ft·°F)	Pr
Toluene (hot)	110	52.75	52.75	0.4207	0.0003034	0.07476	6.146

The outlet temperature of the water is unknown, so the average properties cannot be found without iteration. However, looking at the conditions in this problem, the expected temperature rise of the water should be relatively small. Therefore, the properties should not change much from the inlet condition. Adding the water properties evaluated at the inlet temperature from Appendix B.1 to the property table,

Fluid	T (°F)	ρ (lbm/ft³)	γ (lbf/ft³)	c_p (Btu/lbm·R)	μ (lbm/ft·s)	k (Btu/h·ft·°F)	Pr
Toluene (hot)	110	52.75	52.75	0.4207	0.0003034	0.07476	6.146
Water (cold)	50	62.41	62.41	1.002	0.0008776	0.3351	9.447

Notice that the thermal conductivity of the toluene is an order of magnitude smaller than the water. This suggests that it is more difficult to transfer heat to/from toluene. As a result, the heat exchanger might be larger than expected.

5.6.7.2 Fluid Placement

The water flow is larger than the toluene. Therefore, it would be prudent to locate the water flow in the passage with the larger cross-sectional area. However, fouling should also be considered. The fouling factors for each fluid can be found in Table 5.2.

Hot fluid = toluene (organic fluid)	$R_f'' = 0.002$ h · ft² · °F/Btu
Cold fluid = city water	$R_f'' = 0.002$ h · ft² · °F/Btu

Since the fouling factors are the same for each fluid, the hydraulic criterion is used to place the fluids. Initially, the toluene is placed in the tube and the water in the annulus. Once the tube sizes are determined, a check is done to ensure that the hydraulic criterion is satisfied. From this point forward in the design, the hot fluid variables are assigned the subscript "t" to indicate that it is in the tube. Similarly, the subscript "a" is used for the water in the annulus.

Fluid	Placement
Hot fluid = toluene	Tube
Cold fluid = city water	Annulus

5.6.7.3 Determination of Tube Sizes

Keeping the velocities in the tube and annulus in the economic range should result in a heat exchanger that is close to cost-optimized. From Table 4.11, the estimated economic velocity ranges for the hot and cold fluids are

Hot fluid = toluene	4.8–10.4 ft/s
Cold fluid = city water	4.6–9.9 ft/s

The inner tube is selected first. Using an average economic velocity for the toluene, an estimated inside diameter of the inner tube can be found.

$$\dot{m}_t = \rho_t \left(\frac{\pi D_{t,est}^2}{4} \right) V_{t,econ} \rightarrow D_{t,est} = \sqrt{\frac{4\dot{m}_t}{\pi \rho_t V_{t,econ}}} = \sqrt{\frac{4 \left(10,000 \dfrac{lbm}{h} \right) \left(\dfrac{h}{3600\ s} \right)}{\pi \left(52.75 \dfrac{lbm}{ft^3} \right) \left(\dfrac{4.8 + 10.4}{2} \right) \dfrac{ft}{s}}}$$

$$= 0.0939\ ft$$

This value is surrounded by the following type M copper tubes from Appendix D:

Type M Standard Size	Inside Diameter (ft)
1	0.08792
1¼	0.10758

Calculating the velocities that would be experienced inside each of these tubes results in

Type M Standard Size	Inside Diameter (ft)	Fluid Velocity (ft/s)
1	0.08792	8.69
1¼	0.10758	5.79

Both of these velocities are within the economic range. However, the larger-diameter tube will be selected because it will result in smaller pressure drop on the tube side of the heat exchanger. Therefore, the specification for the inner tube is

Tube	Specification	ID (ft)	OD (ft)	A (ft²)	V (ft/s)
Inner	1¼-std type M copper	0.10758	0.11458	0.00909	5.79

Now that the inner tube is specified, the outer tube can be selected. The estimated diameter of the outer tube based on the average economic velocity of the water is

$$\dot{m}_a = \rho_a \left[\frac{\pi}{4} \left(D_{a,est}^2 - OD_t^2 \right) \right] V_{a,econ} \rightarrow D_{a,est} = \sqrt{OD_t^2 + \frac{4\dot{m}_a}{\pi \rho_a V_{a,econ}}}$$

$$D_{a,est} = \sqrt{(0.11458\ ft)^2 + \frac{4 \left(12,000 \dfrac{lbm}{h} \right) \left(\dfrac{h}{3600\ s} \right)}{\pi \left(62.41 \dfrac{lbm}{ft^3} \right) \left(\dfrac{4.6 + 9.9}{2} \right) \dfrac{ft}{s}}} = 0.150\ ft$$

This estimated diameter falls between a 1½ std and 2 std copper tube. Calculating the resulting flow area and velocity in the annulus formed by one of these tubes and the inner tube results in

Tube	Specification	ID (ft)	OD (ft)	A (ft²)	V (ft/s)
Outer	1½-std type M copper	0.12725	0.13542	0.00241	22.2
Outer	2-std type M copper	0.16742	0.17708	0.01170	4.56

This analysis clearly indicates that the outer tube should be 2 std. With the 2-std tube, the hydraulic criterion is met. However, the velocity falls slightly below the estimated minimum economic velocity for water. The final tube sizes for the heat exchanger are given below. This design is specified as a 2 x 1¼ DPHX.

Tube	Specification	ID (ft)	OD (ft)	A (ft²)	V (ft/s)
Inner	1¼-std type M copper	0.10758	0.11458	0.00909	5.793
Outer	2-std type M copper	0.16742	0.17708	0.01170	4.564[a]

[a] Velocity in the annulus formed by the two tubes.

5.6.7.4 Calculation of Annulus Diameters

$$D_{hyd} = ID_a - OD_t = (0.16742 - 0.11458)\,\text{ft} = 0.05284\ \text{ft}$$

$$D_{equ} = \frac{(ID_a^2 - OD_t^2)}{OD_t} = 0.1300\ \text{ft}$$

5.6.7.5 Calculation of the Reynolds Numbers

$$Re_t = \frac{\rho_t V_t (ID_t)}{\mu_t} = \frac{\left(52.75\,\dfrac{\text{lbm}}{\text{ft}^3}\right)\left(5.793\,\dfrac{\text{ft}}{\text{s}}\right)(0.10758\ \text{ft})}{0.0003034\,\dfrac{\text{lbm}}{\text{ft}\cdot\text{s}}} = 108{,}362$$

$$Re_{a,heat} = \frac{\rho_a V_a (D_{equ})}{\mu_a} = \frac{\left(62.41\,\dfrac{\text{lbm}}{\text{ft}^3}\right)\left(4.605\,\dfrac{\text{ft}}{\text{s}}\right)(0.1300\ \text{ft})}{0.0008776\,\dfrac{\text{lbm}}{\text{ft}\cdot\text{s}}} = 42{,}207$$

$$Re_{a,fric} = \frac{\rho_a V_a (D_{hyd})}{\mu_a} = \frac{\left(62.41\,\dfrac{\text{lbm}}{\text{ft}^3}\right)\left(4.605\,\dfrac{\text{ft}}{\text{s}}\right)(0.05284\ \text{ft})}{0.0008776\,\dfrac{\text{lbm}}{\text{ft}\cdot\text{s}}} = 17{,}149$$

5.6.7.6 Calculation of the Friction Factors

$$f_t = \frac{0.25}{\left[\log\left(\dfrac{\varepsilon/ID_t}{3.7} + \dfrac{5.74}{Re_t^{0.9}}\right)\right]^2} = \frac{0.25}{\left[\log\left(\dfrac{0.000005\ \text{ft}/0.10758\ \text{ft}}{3.7} + \dfrac{5.74}{108,362^{0.9}}\right)\right]^2}$$

$$= 0.01786$$

$$f_a = \frac{0.25}{\left[\log\left(\dfrac{\varepsilon/D_{hyd}}{3.7} + \dfrac{5.74}{Re_{a,fric}^{0.9}}\right)\right]^2} = \frac{0.25}{\left[\log\left(\dfrac{0.000005\ \text{ft}/0.05284\ \text{ft}}{3.7} + \dfrac{5.74}{17,149^{0.9}}\right)\right]^2}$$

$$= 0.02706$$

5.6.7.7 Calculation of Nusselt Numbers

The Reynolds numbers indicate that both flows are turbulent. Therefore, the Gnielinski correlation will be used to find the Nusselt numbers.

$$Nu_t = \frac{(f_t/8)(Re_t - 1000)\,Pr_t}{1 + 12.7(f_t/8)^{1/2}\left(Pr_t^{2/3} - 1\right)} = \frac{(0.01786/8)(108,362 - 1000)(6.146)}{1 + 12.7(0.01786/8)^{1/2}\left(6.146^{2/3} - 1\right)}$$

$$= 610.4$$

$$Nu_a = \frac{(f_a/8)(Re_{a,heat} - 1000)\,Pr_a}{1 + 12.7(f_a/8)^{1/2}\left(Pr_a^{2/3} - 1\right)} = \frac{(0.02706/8)(42,207 - 1000)(9.447)}{1 + 12.7(0.02706/8)^{1/2}\left(9.447^{2/3} - 1\right)}$$

$$= 369.6$$

5.6.7.8 Calculation of the Convective Heat Transfer Coefficients

$$Nu_t = \frac{h_t(ID_t)}{k_t} \rightarrow h_t = \frac{Nu_t k_t}{ID_t} = \frac{(610.4)\left(0.07476\,\dfrac{\text{Btu}}{\text{h}\cdot\text{ft}\cdot{}^\circ\text{F}}\right)}{0.10758\ \text{ft}} = 424.2\,\frac{\text{Btu}}{\text{h}\cdot\text{ft}^2\cdot{}^\circ\text{F}}$$

$$Nu_a = \frac{h_a(D_{equ})}{k_a} \rightarrow h_a = \frac{Nu_a k_a}{D_{equ}} = \frac{(369.6)\left(0.3351\,\dfrac{\text{Btu}}{\text{h}\cdot\text{ft}\cdot{}^\circ\text{F}}\right)}{0.05284\ \text{ft}} = 952.6\,\frac{\text{Btu}}{\text{h}\cdot\text{ft}^2\cdot{}^\circ\text{F}}$$

5.6.7.9 Calculation of the Overall Heat Transfer Coefficients

$$\frac{1}{U_{o,clean}} = \frac{1}{h_t}\left(\frac{OD_t}{ID_t}\right) + \frac{1}{h_a} = \left(\frac{1}{424.2}\,\frac{\text{h}\cdot\text{ft}^2\cdot{}^\circ\text{F}}{\text{Btu}}\right)\left(\frac{0.11458\ \text{ft}}{0.10758\ \text{ft}}\right) + \left(\frac{1}{952.6}\,\frac{\text{h}\cdot\text{ft}^2\cdot{}^\circ\text{F}}{\text{Btu}}\right)$$

$$U_{o,clean} = 280.8\,\frac{\text{Btu}}{\text{h}\cdot\text{ft}^2\cdot{}^\circ\text{F}}$$

$$\frac{1}{U_{o,fouled}} = \frac{1}{h_t}\left(\frac{OD_t}{ID_t}\right) + R''_{f,i}\left(\frac{OD_t}{ID_t}\right) + R''_{f,o} + \frac{1}{h_a}$$

$$\frac{1}{U_{o,fouled}} = \left(\frac{1}{424.2}\frac{h \cdot ft^2 \cdot {}^\circ F}{Btu}\right)\left(\frac{0.11458\ ft}{0.10758\ ft}\right) + \left(0.002\frac{h \cdot ft^2 \cdot {}^\circ F}{Btu}\right)\left(\frac{0.11458\ ft}{0.10758\ ft}\right)$$

$$+ \left(0.002\frac{h \cdot ft^2 \cdot {}^\circ F}{Btu}\right) + \left(\frac{1}{952.6}\frac{h \cdot ft^2 \cdot {}^\circ F}{Btu}\right)$$

$$U_{o,fouled} = 130.0\frac{Btu}{h \cdot ft^2 \cdot {}^\circ F}$$

5.6.7.10 Application of the Heat Exchanger Model

At this point, most of the major parameters required in the heat exchanger model have been calculated. A decision now needs to be made concerning the heat exchanger model itself (LMTD or e-NTU). For the design problem, either method turns out to be noniterative. In this solution, the e-NTU method is used. The e-NTU heat exchanger model is

$$\dot{Q} = \dot{C}_h \left(T_{hi} - T_{ho}\right)$$

$$\dot{Q} = \dot{C}_c \left(t_{co} - t_{ci}\right)$$

$$\varepsilon = \frac{\dot{Q}}{\dot{Q}_{max}} \quad \text{where } \dot{Q}_{max} = \dot{C}_{min}\left(T_{hi} - t_{ci}\right)$$

$$\varepsilon = f\left(NTU, C_r\right) \quad \text{where } NTU = \frac{U_o A_o}{\dot{C}_{min}} \quad \text{and } C_r = \frac{\dot{C}_{min}}{\dot{C}_{max}}$$

5.6.7.10.1 Calculation of the Thermal Capacity Rates

$$\dot{C}_h = \dot{m}_h c_{ph} = \left(10,000\frac{lbm}{h}\right)\left(0.4207\frac{Btu}{lbm \cdot R}\right) = 4,207\frac{Btu}{h \cdot R}$$

$$\dot{C}_c = \dot{m}_c c_{pc} = \left(12,000\frac{lbm}{h}\right)\left(1.002\frac{Btu}{lbm \cdot R}\right) = 12,025\frac{Btu}{h \cdot R}$$

$$\therefore \ \dot{C}_{min} = 4,207\frac{Btu}{h \cdot R} \quad \dot{C}_{max} = 12,025\frac{Btu}{h \cdot R} \quad C_r = \frac{4207\ Btu/h \cdot R}{12,025\ Btu/h \cdot R} = 0.3498$$

5.6.7.10.2 Calculation of Heat Exchanger Effectiveness

$$\dot{Q} = \dot{C}_h\left(T_{hi} - T_{ho}\right) = \left(4207\frac{Btu}{h \cdot R}\right)(120 - 100)R = 84,132\frac{Btu}{h}$$

$$\dot{Q}_{max} = \dot{C}_{min}\left(T_{hi} - t_{ci}\right) = \left(4207\frac{Btu}{h \cdot R}\right)(120 - 50)R = 294,461\frac{Btu}{h}$$

$$\therefore \ \varepsilon = \frac{\dot{Q}}{\dot{Q}_{max}} = \frac{84,132\ Btu/h}{294,461\ Btu/h} = 0.2857$$

5.6.7.10.3 Calculation of Heat Exchanger NTU

To calculate the NTU, the type of heat exchanger needs to be specified. There is very little information concerning the actual installation of this heat exchanger. Location of the heat exchanger in the processing line may dictate the type (parallel or counterflow). Since this information is unknown, a counterflow heat exchanger is specified. If the installation requires a parallel configuration, it is a simple enough matter to recalculate the NTU. Using the expression for the NTU-e relationship in Table 5.4,

$$NTU = \frac{1}{C_r - 1} \ln\left(\frac{\varepsilon - 1}{\varepsilon C_r - 1}\right) = \frac{1}{0.3498 - 1} \ln\left[\frac{0.2857 - 1}{(0.2857)(0.3498) - 1}\right] = 0.3555$$

5.6.7.10.4 Calculation of the Heat Exchanger Length

The heat exchanger will experience fouling during the course of its service. Therefore, the length of the heat exchanger should be calculated using the fouled value of the overall heat transfer coefficient.

$$NTU = \frac{U_{o,fouled} A_o}{\dot{C}_{min}} \quad \rightarrow \quad A_o = \frac{\dot{C}_{min}(NTU)}{U_{o,fouled}} = \frac{\left(4207 \dfrac{Btu}{h \cdot R}\right)(0.3555)}{130.0 \dfrac{Btu}{h \cdot ft^2 \cdot R}} = 11.503 \ ft^2$$

$$A_o = \pi(OD_t)L \quad \rightarrow \quad L = \frac{A_o}{\pi(OD_t)} = \frac{11.503 \ ft^2}{\pi(0.11458 \ ft)} = 31.96 \ ft$$

It is also interesting to see what the effect of the fouling is on the heat exchanger length. When the heat exchanger is clean,

$$NTU = \frac{U_{o,clean} A_o}{\dot{C}_{min}} \quad \rightarrow \quad A_o = \frac{\dot{C}_{min}(NTU)}{U_{o,clean}} = \frac{\left(4207 \dfrac{Btu}{h \cdot R}\right)(0.3555)}{280.8 \dfrac{Btu}{h \cdot ft^2 \cdot R}} = 5.326 \ ft^2$$

$$A_o = \pi(OD_t)L \quad \rightarrow \quad L = \frac{A_o}{\pi(OD_t)} = \frac{5.326 \ ft^2}{\pi(0.11458 \ ft)} = 14.79 \ ft$$

This calculation shows the importance of ensuring that the fouling effect is considered in the heat exchanger design. If fouling is ignored, the heat exchanger would perform as expected initially. However, over time, its performance would degrade because of the fouling. Including the fouling effects ensures that the heat exchanger will still meet performance requirements after 1 year of service.

5.6.7.11 Calculation of the Pressure Drops through the Heat Exchanger

The pressure drop is dependent on several heat exchanger parameters, including whether or not a hairpin design is used. In this problem, the total length of 31.96 ft may not require a hairpin design. Without knowing more about the actual installation of the heat exchanger, it will be assumed that no hairpins are used. Therefore,

$$\Delta P_t = \gamma_t \left[f_t \frac{L}{\mathrm{ID}_t} \frac{V_t^2}{2g} + l_m \right]$$

$$= \left(52.75 \frac{\mathrm{lbf}}{\mathrm{ft}^3} \right) \left\{ (0.01786) \left(\frac{31.96 \text{ ft}}{0.10758 \text{ ft}} \right) \left[\frac{(5.793 \text{ ft/s})^2}{2(32.174 \text{ ft/s}^2)} \right] + 0 \text{ ft} \right\} \left(\frac{\mathrm{ft}^2}{144 \text{ in.}^2} \right)$$

$$= 1.01 \text{ psi}$$

$$\Delta P_a = \gamma_a \left[\left(f_a \frac{L}{D_{\mathrm{hyd}}} + n \right) \frac{V_a^2}{2g} \right]$$

$$= \left(61.86 \frac{\mathrm{lbf}}{\mathrm{ft}^3} \right) \left\{ \left[(0.02706) \left(\frac{31.96 \text{ ft}}{0.05284 \text{ ft}} \right) + 1 \right] \left[\frac{(4.605 \text{ ft/s})^2}{2(32.174 \text{ ft/s}^2)} \right] \right\} \left(\frac{\mathrm{ft}^2}{144 \text{ in.}^2} \right)$$

$$= 2.44 \text{ psi}$$

5.6.7.12 Summary of the Final Design

The final specifications for this DPHX are summarized in Table 5.6. The cold-fluid outlet temperature is determined from the cold fluid conservation of energy equation in the heat exchanger model.

$$\dot{Q} = \dot{C}_c \left(t_{co} - t_{ci} \right) \quad \rightarrow \quad t_{co} = t_{ci} + \frac{\dot{Q}}{C_c} = 50°F - \frac{84,132 \text{ Btu/h}}{12,025 \text{ Btu/h} \cdot °F} = 57°F$$

Notice that there is only a 7°F increase in the water's temperature as it passes through the heat exchanger. Therefore, using the water inlet temperature to evaluate the properties should be sufficient for this design calculation. No iteration on the cold fluid properties is necessary.

5.6.8 Double-Pipe Heat Exchanger Analysis Example

Consider the following DPHX analysis problem.

> In an industrial process 10,000 lbm/h of toluene at 120°F is cooled using a supply of 12,000 lbm/h of city water at 50°F. The cooling is done in a 2 × 1¼ double-pipe counterflow heat exchanger. The heat exchanger has a straight length of 32 ft and is made from type M copper tubing. Determine the inlet and outlet fluid temperatures and the pressure drops in the heat exchanger for clean and fouled conditions.

TABLE 5.6

Summary of the Double-Pipe Heat Exchanger Design Results

Heat exchanger	2 × 1¼ type M copper DPHX
Flow configuration	Counterflow
Length	32 ft
Heat exchanger *UA* product	1496 Btu/h·°F
Hot fluid: Toluene	In the tube
Hot fluid inlet temperature	120°F
Hot fluid outlet temperature	100°F
Hot fluid mass flow rate	10,000 lbm/h
Hot fluid (tube) pressure drop	1.0 psi
Cold fluid: city water	In the annulus
Cold fluid inlet temperature	50°F
Cold fluid outlet temperature	57°F
Cold fluid mass flow rate	12,000 lbm/h
Cold fluid (annulus) pressure drop	2.4 psi

Notice that this is the heat exchanger that was designed in the Section 5.6.7. In this case, the heat exchanger is analyzed to determine the outlet fluid temperatures and pressure drops.

In Section 5.6.7, the design problem was solved using the e-NTU model without iteration, which allows the complete design to be done using a calculator. The LMTD model is used here to demonstrate how the other heat exchanger model can be used to analyze the heat exchanger. Using the LMTD model to solve a heat exchanger analysis problem requires iteration, because the outlet temperatures are not known. This makes the LMTD analysis problem a good candidate for computer software. Using EES, the hot and cold fluid information is written as follows:

```
"Hot Fluid = Toluene -> tube"
  h$ = 'toluene'
  T_hi = 120[F]
  m_dot_t = 10000[lbm/hr]
  R_ft_dprime = 0.002[hr-ft^2-F/Btu]
"Properties"
  T_h_avg = (T_hi + T_ho)/2
  {T_h_avg = T_hi}
  rho_t = density(h$,T = T_h_avg,x = 0)
  gamma_t = rho_t*g/g_c
  cp_t = cp(h$,T = T_h_avg,x = 0)
  mu_t = viscosity(h$,T = T_h_avg,x = 0)*convert(lbm/ft-      &
    hr,lbm/ft-s)
  k_t = conductivity(h$,T = T_h_avg,x = 0)
  Pr_t = Prandtl(h$,T = T_h_avg,x = 0)
```

```
"Cold Fluid = City Water -> annulus:"
  c$ = 'steam_iapws'
  t_ci = 50[F]
  m_dot_a = 12000[lbm/hr]
  R_fa_dprime = 0.002[hr-ft^2-F/Btu]
"Properties"
  t_c_avg = (t_ci + t_co)/2
  {t_c_avg = t_ci}
  rho_a = density(c$,T = t_c_avg,x = 0)
  gamma_a = rho_a*g/g_c
  cp_a = cp(c$,T = t_c_avg,x = 0)
  mu_a = viscosity(c$,T = t_c_avg,x = 0)*convert(lbm/ft-    &
    hr,lbm/ft-s)
  k_a = conductivity(c$,T = t_c_avg,x = 0)
  Pr_a = Prandtl(c$,T = t_c_avg,x = 0)
```

Notice that the fluid placement has been assumed to set up the property values. This placement is checked later by calculating the cross-sectional flow areas of the tube and annulus.

In the above EES code, two values of the average temperature of the hot and cold fluids are available. One option is the true average temperature of the fluid. The second option sets the average temperature to the fluid inlet temperature. This is done to avoid convergence problems. The initial solution is found using the fluid inlet temperatures as averages. Once a solution is found, the variables are updated and the program is run again, commenting out the inlet temperatures as averages and allowing EES to calculate the actual average temperature.

The heat exchanger is specified as a 2 × 1¼ using type M copper tubes. This indicates that the outer tube is 2-std type M and the inner tube is 1¼-std type M. The dimensional data of these tubes is found in Appendix D. The absolute roughness of the copper is also needed to determine the friction factors required to determine the Nusselt numbers and the pressure drops.

```
"Heat exchanger tube sizes"
"Inner tube = 1 1/4-std type M copper,"
    ID_t = 0.10758[ft]                    "Appendix D"
    OD_t = 0.11458[ft]                    "Appendix D"
    epsilon_copper = 0.000005[ft]         "Table 4.3"

"Outer tube = 2-std type M copper"
  ID_a = 0.16742[ft]                      "Appendix D"

"Heat exchanger length"
    L = 32[ft]

"Constants"
    g = 32.174[ft/s^2]
    g_c = 32.174[lbm-ft/lbf-s^2]
```

The fouling factors for both fluids are the same. Therefore, the hydraulic criterion is used to determine fluid placement by computing each fluid's velocity and the cross-sectional area of the tube and annulus.

```
"Calculation of velocities and flow areas"
   m_dot_t*convert(lbm/hr,lbm/s) = rho_t*A_t*V_t
   m_dot_a*convert(lbm/hr,lbm/s) = rho_a*A_a*V_a
   A_t = pi*ID_t^2/4
   A_a = (pi/4)*(ID_a^2 - OD_t^2)
```

From this calculation, the cross-sectional areas are found to be $A_t = 0.00909$ ft^2, and $A_a = 0.01170$ ft^2. The water has the larger flow rate. Therefore, it is placed in the annulus, which has a larger cross-sectional area compared to the inner tube. This is consistent with the assumption initially made to set up the properties of the fluids in the tube and annulus.

In complex iterative problems such as the one presented here, there may be several opportunities to solve the existing set of equations, even before a final solution has been determined. For example, with the EES equations developed to this point, there are enough equations and unknowns to solve for the flow areas. This is helpful during the analysis because it is needed for proper fluid placement. In addition, unit analysis can be verified early, making the unit analysis for the final solution much easier and understandable.

Before the heat transfer and hydraulic calculations can proceed, the hydraulic and equivalent diameters of the annulus are needed.

```
"The annulus diameters are,"
   D_hyd = ID_a - OD_t
   D_equ = (ID_a^2 - OD_t^2)/OD_t
```

Now, the Reynolds numbers can be found. There is one Reynolds number for the inner tube and two for the annulus.

```
"Reynolds Numbers"
   Re_t = rho_t*V_t*ID_t/mu_t
   Re_a_heat = rho_a*V_a*D_equ/mu_a
   Re_a_fric = rho_a*V_a*D_hyd/mu_a
```

At this point, the Reynolds numbers need to be known so the proper convective heat transfer correlation can be used to determine the Nusselt numbers. Solving the equation set up to this point reveals $Re_t = 114,339$, $Re_{a,heat} = 42,207$, and $Re_{a,fric} = 17,149$. These numbers indicate that the flow is turbulent in both the tube and the annulus. Therefore, the Gnielinski correlation is used to compute the Nusselt numbers. This correlation requires the friction factors. Using the Swamee–Jain correlation.

```
"Friction factors"
  f_t = 0.25*log10((epsilon_copper/ID_t)/3.7 +            &
    5.74/Re_t^0.9)^(-2)
  f_a = 0.25*log10((epsilon_copper/D_hyd)/3.7 +           &
    5.74/Re_a_fric^0.9)^(-2)
```

Now, the Nusselt numbers and corresponding convective heat transfer coefficients can be found on the inside and outside of the inner tube of the heat exchanger.

```
"Nusselt Numbers"
  Nus _ t = ((f _ t/8)*(Re _ t - 1000)*Pr _ t)/(1 + 12.7     &
    *(f _ t/8)^(1/2)*(Pr _ t^(2/3) - 1))
  Nus _ a = ((f _ a/8)*(Re _ a _ heat - 1000)*Pr _ a)/(1 + 12.7   &
    *(f _ a/8)^(1/2)*(Pr _ a^(2/3) - 1))
  Nus _ t = h _ t*ID _ t/k _ t
  Nus _ a = h _ a*D _ equ/k _ a
```

Notice in these calculations that the friction factor in the annulus is based on the Reynolds number calculated using the hydraulic diameter. However, when calculating the Nusselt number in the annulus, the Reynolds number is based on the equivalent diameter is used.

Once the convective heat transfer coefficients are found, the overall heat transfer coefficient can be calculated. In this case, two values will be found: one for a clean heat exchanger and one that is fouled. Each value is used to determine the performance of the heat exchanger when it is brand new, or just cleaned, and when it is fouled after 1 year of service.

```
"Overall heat transfer coefficients"
  1/U_o_clean = (1/h_t)*(OD_t/ID_t) + 1/h_a
  1/U_o_fouled = (1/h_t)*(OD_t/ID_t) + R_ft_dprime       &
    *(OD_t/ID_t) + R_fa_dprime + 1/h_a
```

Now, the LMTD heat exchanger model can be used to determine the fluid outlet temperatures.

```
"Counter flow heat exchanger model - LMTD"
  Q_dot = C_dot_h*(T_hi - T_ho)
  Q_dot = C_dot_c*(t_co - t_ci)
  Q_dot = UA*LMTD

"Calculate UA based on U_o_clean or U_o_fouled"
  UA = U_o*A_o
  U_o = U_o_fouled

"LMTD Definition for a counter flow HX"
  DELTAT_out = DELTAT_in*exp((DELTAT_out - DELTAT_in)/LMTD)
  DELTAT_in = T_hi - t_co
  DELTAT_out = T_ho - t_ci
```

```
"Thermal capacity ratios"
  C_dot_h = m_dot_t*cp_t
  C_dot_c = m_dot_a*cp_a

"Surface area of the heat exchanger"
  A_o = pi*OD_t*L
```

At this point, the equation set can be solved to determine the fluid outlet temperatures.

The pressure drop on each side of the heat exchanger can be found since the friction factors have been determined.

```
"Pressure Drops"
"Tube side"
  DELTAP_t*convert(psi,lbf/ft^2)/gamma_t =                    &
    f_t*(L/ID_t)*(V_t^2/(2*g)) + l_m
  l_m = 0[ft]    "straight HX with no hairpins"

"Annulus side"
  DELTAP_a*convert(psi,lbf/ft^2)/gamma_a =                    &
    (f_a*(L/D_hyd)+ n)*(V_a^2/(2*g))
  n = 1   "straight HX with no hairpins"
```

This completes the analysis of the heat exchanger. The results for a clean and fouled heat exchanger are summarized in Table 5.7. Notice that the solution for the fouled heat exchanger is consistent with the design results shown in Table 5.6.

This heat exchanger suffers significant degradation in its thermal performance as it becomes more and more fouled. It should also be pointed out that the pressure drop calculation for the fouled heat exchanger does *not* take into account the change in diameter because of the fouling. The effect of fouling will increase the pressure drops, compared to the values calculated here.

TABLE 5.7

Summary of the Double-Pipe Heat Exchanger Analysis for Clean and Fouled Conditions

DPHX Parameter	Clean	Fouled
Hot fluid (toluene) inlet temperature	120°F	120°F
Hot fluid (toluene) outlet temperature	84.79°F	99.91°F
Cold fluid (water) inlet temperature	50°F	50°F
Cold fluid (water) outlet temperature	62.23°F	57.03°F
Overall heat transfer coefficient (U_o)	281.5 Btu/h·ft^2·°F	130.5 Btu/h·ft^2·°F
UA product	3,243 Btu/h·°F	1,504 Btu/h·°F
Heat transfer rate between the fluids	146,942 Btu/h	84,493 Btu/h
Pressure drop in the tube (toluene)	1.02 psi	1.02 psi
Pressure drop in the annulus (water)	2.39 psi	2.41 psi

Any difference in the pressure drops between the clean and fouled scenarios is due to the property variation resulting from different outlet temperatures.

5.7 Shell and Tube Heat Exchanger Design and Analysis

STHXs are typically used where large heat transfer rates are needed. They are often used as condensers or evaporators. The shell is usually made up of wrought iron pipe ranging from 8 to 39 nom. The schedule depends on the pressure inside the shell. Other materials can be used if corrosion can be caused by the shell fluid.

The tubes are specified differently compared to copper tubing. They are standardized to the Birmingham Wire Gage (BWG) specification and often called *condenser tubes*. Table 5.8 shows the dimensional data for several STHX tubes using this BWG specification. When specifying a tube, the outside diameter and BWG are specified. For example, a 1-in. 12 BWG tube has an outside diameter of 1 in. and an inside diameter of 0.782 in. Standard tube

TABLE 5.8

BWG Specification for Tubes in a STHX

OD (inch)	BWG	ID (inch)	ID (centimeter)	OD (inch)	BWG	ID (inch)	ID (centimeter)
¾	10	0.482	1.224	1¼	7	0.890	2.261
	11	0.510	1.295		8	0.920	2.337
	12	0.532	1.351		10	0.982	2.494
	13	0.560	1.422		11	1.010	2.565
	14	0.584	1.483		12	1.032	2.621
	15	0.606	1.539		13	1.060	2.692
	16	0.620	1.575		14	1.084	2.753
	17	0.634	1.610		16	1.120	2.845
	18	0.652	1.656		18	1.152	2.926
	20	0.680	1.727		20	1.180	2.997
1	8	0.670	1.702	1½	10	1.232	3.129
	10	0.732	1.859		12	1.282	3.256
	11	0.760	1.930		14	1.334	3.388
	12	0.782	1.986		16	1.370	3.480
	13	0.810	2.057				
	14	0.834	2.118				
	15	0.856	2.174				
	16	0.870	2.210				
	18	0.902	2.291				
	20	0.930	2.362				

Source: Kakac, S., *Heat Exchangers: Selection, Rating, and Thermal Design*, CRC Press, Boca Raton, FL, 2012.

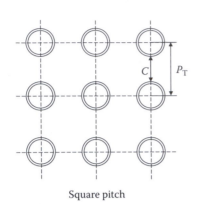

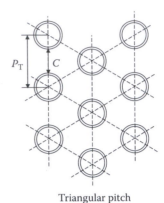

Square pitch   Triangular pitch

FIGURE 5.26
Tube layout for square- and triangular-pitch patterns.

lengths are 8, 12, and 16 ft. However, other lengths can be specified based on the final design of the heat exchanger.

The tubes are laid out in one of two configurations: square pitch or triangular pitch as shown in Figure 5.26. The pitch, P_T, is defined as the tube center-to-center distance. The tube clearance, C, is the distance between tubes.

The maximum number of tubes that can fit inside a given shell depends on the tube size, tube pitch, and number of tube passes in the shell. Table 5.9 shows an abbreviated table of the maximum tube counts for various shell sizes and tube passes (indicated as 1P, 2P, or 4P) to retain the structural integrity of the tube sheet. More comprehensive tables of tube counts and BWG tubing dimensions can be found in Kakac et al. (2012).

Baffles are used to provide support for the tubes inside the shell. The baffles are also designed to influence the flow of the shell fluid. The distance between baffles is known as the *baffle spacing* or *baffle pitch* and given the symbol B. The optimum baffle spacing for segmental baffles is somewhere between 0.4 and 0.6 of the shell diameter.

There are several different types of baffle designs. Unfortunately, many of the baffle designs are proprietary. Manufacturers test their own designs or use an independent contractor to evaluate their designs. To maintain a competitive edge, results are usually kept confidential. The design of the baffle influences the convective heat transfer coefficient on the shell side of the heat exchanger. In this presentation, the performance of the *segmental cut baffle* is given. This is a common type of baffle used in many STHXs. Figure 5.27 shows a sketch of what a segmental cut baffle looks like and how they are typically placed within the shell.

The number of *passes* in a STHX corresponds to the number of passes the tube fluid makes. Figure 5.28 shows the schematic of a 1-shell 1-pass (1-1 or 1P) and a 1-shell 2-pass (1-2 or 2P) STHX. Also available are 1-4, 1-6, and 1-8 STHXs. It is also possible to connect shells together to make a multishell

TABLE 5.9

Tube Counts in STHXs

	¾-in. Tubes						1-in. Tubes					
	1-in. Triangular Pitch			1-in. Square Pitch			1¼-in. Triangular Pitch			1¼-in. Square Pitch		
Shell Diameter (inch)	1P	2P	4P	1P	2P	4P	1P	2P	4P	1P	2P	4P
8	37	30	24	32	26	20	21	16	16	21	16	14
10	61	52	40	52	52	40	32	32	26	32	32	26
12	92	82	76	81	76	68	55	52	48	48	45	40
13¼	109	106	86	97	90	82	68	66	58	61	56	52
15¼	151	138	122	137	124	116	91	86	80	81	76	68
17¼	203	196	178	177	166	158	131	118	106	112	112	96
19¼	262	250	226	224	220	204	163	152	140	138	132	128
21¼	316	302	278	277	270	246	199	188	170	177	166	158
23¼	384	376	352	341	324	308	241	232	212	213	208	192
25	470	452	422	413	394	370	294	282	256	260	252	238
27	559	534	488	481	460	432	349	334	302	300	288	278
29	630	604	556	553	526	480	397	376	338	341	326	300
31	745	728	678	657	640	600	472	454	430	406	398	380
33	856	830	774	749	718	688	538	522	486	465	460	432
35	970	938	882	845	824	780	608	592	562	522	518	488
37	1074	1044	1012	934	914	886	674	664	632	596	574	562
39	1206	1176	1128	1049	1024	982	766	736	700	665	644	624

(Continued)

TABLE 5.9 (*Continued*)

Tube Counts in STHXs

| Shell Diameter (inch) | 1¼-in. Tubes | | | | | | 1½-in. Tubes | | | | | |
| | 1 15/16-in. Triangular Pitch | | | 1 15/16-in. Square Pitch | | | 1 7/8-in. Triangular Pitch | | | 1 7/8-in. Square Pitch | | |
	1P	2P	4P	1P	2P	4P	1P	2P	4P	1P	2P	4P
10	20	18	14	16	12	10	—	—	—	—	—	—
12	32	30	26	30	24	22	18	14	14	16	16	12
13¼	38	36	32	32	30	30	27	22	18	22	22	16
15¼	54	51	45	44	40	37	26	34	32	29	29	24
17¼	69	66	62	56	53	51	48	44	42	29	39	34
19¼	95	91	86	78	73	71	61	58	55	50	48	45
21¼	117	112	105	96	90	86	76	78	70	32	60	57
23¼	140	136	130	127	112	106	95	91	86	78	74	70
25	170	164	155	140	135	127	115	110	105	94	90	86
27	202	196	185	166	160	151	136	131	125	112	108	102
29	235	228	217	193	188	178	160	154	147	131	127	120
31	275	270	255	226	220	209	184	177	172	151	146	141
33	315	305	297	258	252	244	215	206	200	176	170	164
35	357	348	335	293	287	275	246	238	230	202	196	188
37	407	390	380	334	322	311	275	268	260	224	220	217
39	449	436	425	370	362	348	307	299	290	252	246	237

Source: Kakac, S., *Heat Exchangers: Selection, Rating, and Thermal Design*, CRC Press, Boca Raton, FL, 2012.

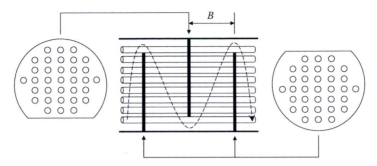

FIGURE 5.27
Segmental cut baffles in a shell and tube heat exchanger.

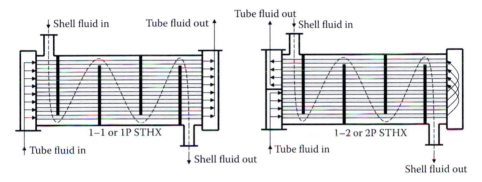

FIGURE 5.28
Single- and multipass shell and tube heat exchangers.

STHX. For example, a 2-4 STHX has 2-shell passes with 2-tube passes in each shell for a total of 4-tube passes.

5.7.1 LMTD for Shell and Tube Heat Exchangers

Because of the complex temperature profiles that occur inside a STHX, the LMTD is difficult to quantify. To simplify this problem, it is common practice to use the fluid temperatures in and out of the heat exchanger to calculate the LMTD for a counterflow heat exchanger and then modify this value using a correction factor. Then, the heat transfer equation in the LMTD model becomes

$$\dot{Q} = UA(F)(\text{LMTD}_{CF})$$ (5.92)

In this equation, LMTD$_{CF}$ is the LMTD for a counterflow heat exchanger. The correction factor, F, depends on the configuration of the heat exchanger. For a STHX with 1-shell pass and an even number of tube passes, F is given by

$$F = \frac{\sqrt{R^2+1}\,\ln\left(\dfrac{1-P}{1-PR}\right)}{(R-1)\ln\left[\dfrac{2-P\left(R+1-\sqrt{R^2+1}\right)}{2-P\left(R+1+\sqrt{R^2+1}\right)}\right]} \tag{5.93}$$

In this equation, P and R are defined as

$$R = \frac{\dot{C}_c}{\dot{C}_h} = \frac{T_{hi}-T_{ho}}{t_{co}-t_{ci}} \tag{5.94}$$

$$P = \frac{t_{co}-t_{ci}}{T_{hi}-t_{ci}} \tag{5.95}$$

The value of F is a modifier to the LMTD for a counterflow heat exchanger. Therefore, the closer F is to 1, the better the heat exchanger is performing. In fact, if F is less than 0.75, the heat exchanger is operating inefficiently, leading to higher operating cost to affect the desired heat transfer rate between the two fluids. Therefore, it is common practice to design STHXs that maintain $F \geq 0.75$.

5.7.2 Tube-Side Analysis of Shell and Tube Heat Exchangers

STHXs are meant for high heat transfer and mass flow rates. Therefore, it is highly unlikely that the flow inside the tubes is laminar. However, it is good practice to check the Reynolds number of the flow. For turbulent flow, the Gnielinski correlation, Equation 5.30, should be used. The Reynolds number is based on the velocity of the flow through one tube. The effect of all tubes together is accounted for when the total outside surface area of the tube bundle is calculated. The velocity through one of the tubes is calculated from the total mass flow rate by

$$V_t = \frac{\dot{m}_t}{\rho_t A_t} \tag{5.96}$$

In this equation, the mass flow rate and the area are *total* values. The total cross-sectional area available in the tubes can be found by

$$A_t = \frac{N_t}{N_p}\left(\frac{\pi ID_t^2}{4}\right) \tag{5.97}$$

Here, N_t is the total number of tubes in the heat exchanger and N_p is the number of tube passes. Using these expressions, the Reynolds number of the flow inside *one* of the tubes is

$$Re_t = \frac{\rho_t V_t ID_t}{\mu_t} \tag{5.98}$$

The pressure drop of the tube-side fluid consists of two parts: the pressure drop as the fluid flows through the tubes, and the pressure drop associated with the abrupt expansion and contraction experienced at the tube sheet and losses associated with any return bends. The pressure drop through the tubes can be found by

$$\frac{\Delta P_{tubes}}{\gamma_t} = N_p \left(f_t \frac{L_t}{ID_t} \frac{V_t^2}{2g} \right) \tag{5.99}$$

The pressure drop due to minor losses in the tube sheet and return bends is given by

$$\frac{\Delta P_m}{\gamma_t} = 4N_p \frac{V_t^2}{2g} \tag{5.100}$$

In this equation, the value "4" was determined experimentally. Adding Equations 5.99 and 5.100 together results in the total pressure drop of the tube-side fluid.

$$\Delta P_t = \Delta P_{tubes} + \Delta P_m = \gamma_t N_p \left(f_t \frac{L_t}{ID_t} + 4 \right) \frac{V_t^2}{2g} \tag{5.101}$$

5.7.3 Shell-Side Analysis of Shell and Tube Heat Exchangers

Analysis of the shell-side fluid is more complicated because of the complex flow path taken as it passes through the heat exchanger. A suitable correlation for the convective heat transfer coefficient between the shell fluid and the outside surface of the tubes for a STHX with segmental baffles is suggested by McAdams (Fraas 1989) and given by

$$Nu_s = 0.36 Re_s^{0.55} Pr_s^{1/3} \left(\frac{\mu}{\mu_w} \right)^{0.14} \tag{5.102}$$

This correlation is valid for Reynolds numbers from 2,000 to 1,000,000. The properties are evaluated at the average temperature of the shell-side fluid, except for μ_w, which is evaluated at the wall temperature of the tubes. The tube wall temperature can be estimated by averaging the average of the hot and cold fluid temperatures.

$$T_w = \frac{1}{2} \left(\frac{T_{hi} + T_{ho}}{2} + \frac{t_{ci} + t_{co}}{2} \right) \tag{5.103}$$

In cases where the tube wall temperature estimated using Equation 5.103 is reasonably close to the shell-fluid's average temperature, the ratio of the dynamic viscosities raised to the 0.14 power in Equation 5.102 is very close to 1. Then Equation 5.102 can be simplified to

$$Nu_s = 0.36 Re_s^{0.55} Pr_s^{1/3} \tag{5.104}$$

In Equations 5.102 and 5.104, the Nusselt and Reynolds numbers are based on the equivalent diameter of the shell.

$$Nu_s = \frac{h_s D_e}{k_s} \quad \text{and} \quad Re_s = \frac{\rho_s V_s D_e}{\mu_s} \tag{5.105}$$

The equivalent diameter is the heat transfer diameter defined in Equation 5.85. For the shell side, the equivalent diameter is dependent on the tube pitch. Figure 5.29 shows square- and triangular-pitch layouts. Using the definition of the equivalent diameter and the variables indicated in Figure 5.29, an expression can be derived for the equivalent diameter of the shell.

For an STHX with square-pitch tubes,

$$D_e = \frac{4(\text{cross-sectional area})}{\text{heat transfer perimeter}} = \frac{4\left[P_T^2 - \dfrac{\pi(OD_t^2)}{4}\right]}{\pi(OD_t)} = \frac{4P_T^2}{\pi(OD_t)} - OD_t \tag{5.106}$$

Similarly, for triangular-pitch tubes,

$$D_e = \frac{4(\text{cross-sectional area})}{\text{heat transfer perimeter}} = \frac{4\left[\dfrac{P_T^2 \sqrt{3}}{4} - \dfrac{\pi}{8}(OD_t^2)\right]}{\dfrac{\pi}{2}(OD_t)} = \frac{2\sqrt{3}(P_T^2)}{\pi(OD_t)} - OD_t \tag{5.107}$$

The shell-side fluid velocity varies as it travels around the labyrinth of tubes and baffles that make up the tube bundle. However, a constant value of

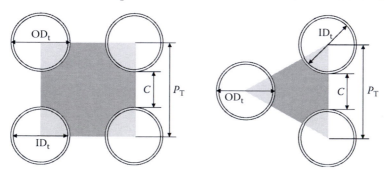

FIGURE 5.29
Square- and triangular-pitch layouts used to determine the equivalent diameter of the shell.

velocity is preferred to calculate the Reynolds number given in Equation 5.105. To do this, a characteristic shell-side flow area is defined as

$$A_s = \frac{D_s CB}{P_T} \qquad (5.108)$$

In this equation, D_s is the actual shell diameter, P_T is the tube pitch, C the tube clearance (see Figure 5.29), and B the baffle spacing. Using this as the characteristic flow area in the shell, the shell-fluid velocity can be found by

$$V_s = \frac{\dot{m}_s}{\rho_s A_s} \qquad (5.109)$$

The pressure drop of the shell-side fluid is difficult to model analytically. Therefore, an empirical expression, based on experimental data, is preferred. The empiricism is cast in the form of a head loss term and is expressed as (Kern 1950),

$$\frac{\Delta P_s}{\gamma_s} = \frac{f_s}{\phi_s}(N_b + 1)\frac{D_s}{D_e}\frac{V_s^2}{2g} \qquad (5.110)$$

In this equation, N_b is the number of baffles in the heat exchanger, $\phi_s = (\mu/\mu_w)^{0.14}$, and f_s is the shell-side friction factor that is determined by the empirical relationship:

$$f_s = \exp[0.576 - 0.19\ln(Re_s)] \qquad (5.111)$$

Equation 5.111 is valid for $400 \le Re_s \le 1{,}000{,}000$. Equation 5.110 is formulated to include minor losses at the inlet and outlet of the shell. As was seen with the shell-side Nusselt number, if the estimated tube wall temperature is reasonably close to the shell-side average temperature, then Equation 5.110 can be estimated as

$$\frac{\Delta P_s}{\gamma_s} = f_s(N_b + 1)\frac{D_s}{D_e}\frac{V_s^2}{2g} \qquad (5.112)$$

As with DPHXs, the pressure drop on either side (tube or shell) should be no greater than 10 psi ($\approx$70 kPa).

5.7.4 Shell and Tube Heat Exchanger Design Considerations

The sections below discuss some of the major design considerations for STHXs. The list is not intended to be all-inclusive. However, most of the major points are addressed here.

5.7.4.1 Tube-Side Considerations

- Higher flow rates through the tubes result in higher heat transfer rates. However, the downside is that the pressure drop across the tubes will increase.

- Multiple tube passes will increase the heat transfer rate between the fluids. However, the pressure drop will increase faster than the heat transfer rate. There is a tradeoff between heat transfer enhancement and increased pumping power that must be considered.

5.7.4.2 Shell-Side Considerations

- The optimum baffle spacing is somewhere between 0.4 and 0.6 of the shell diameter. Keeping the baffle spacing in this range makes the shell fluid flow in more of a cross-flow pattern across the tubes. This increases the heat transfer rate between the two fluids.
- The heat transfer rate can be increased by increasing the number of baffles. However, the drawback is that the pressure drop increases faster than the heat transfer rate. Higher heat transfer rates are desirable. However, the higher pressure drop means more pumping power is required to move the fluid through the shell side.

5.7.4.3 General Considerations

- As with the DPHX, there are two criteria for fluid placement: hydraulic and fouling. It may be more desirable to place the fluid with the higher fouling factor through the tubes. The reason for this is that the tubes are easy to clean mechanically. If the fluids foul about the same, then use the hydraulic criterion and route the higher mass flow rate fluid through the side with the larger flow area.
- In a design problem, the heat exchanger should be kept as small as possible while maintaining the desired performance. This is an economic issue, both from capital cost and operating cost perspectives.
- The economics of the heat exchanger are somewhat reflected in the value of the LMTD correction factor, F. This value should be greater than or equal to 0.75.
- It is desirable to keep the velocity in the tubes in the economic range.
- The heat exchanger should deliver the required heat transfer rate when completely fouled. Recall that the fouling factors represent fouling after 1 year of service.
- The pressure drop on each side of the heat exchanger should not exceed 10 psi ($\approx$70 kPa).

5.7.5 Shell and Tube Heat Exchanger Design and Analysis Example

Consider the following STHX design problem.

Hot distilled water is available at 125°F to heat cyclohexane from 60°F to 110°F. The water flows at 240,000 lbm/h and the cyclohexane flow rate is 200,000 lbm/h. The shell diameter is 27-in. and there are four tube passes. The tubes are ¾-in. 16 BWG placed on a 1-in. triangular pitch. The water is flowing in the shell and the cyclohexane is in the tubes. Determine the required length of the heat exchanger, the number of baffles, and the heat exchanger pressure drops.

This is a design problem because the size of the heat exchanger is to be specified. The size of the heat exchanger is specified by the length of the tubes in the tube bundle.

5.7.5.1 Fluid Properties

The inlet and outlet temperatures of the cold fluid are known. Therefore the average temperature can be used to determine the cold fluid properties. Only the inlet temperature is known for the hot fluid. Since this is a design problem and the length of the tubing is to be specified, the initial calculation is done assuming that the hot fluid average temperature is the same as the inlet temperature. This procedure is well suited for the e-NTU heat exchanger model. The properties of the hot and cold fluids can be found in Appendix B.1.

Fluid	T (°F)	ρ (lbm/ ft³)	γ (lbf/ ft³)	c_p (Btu/ lbm·R)	μ (lbm/ ft·s)	k (Btu/ h·ft·°F)	Pr
Water (hot)	125	61.63	61.63	0.9988	0.0003576	0.3654	3.519
Cyclohexane (cold)	85	48.04	48.04	0.4397	0.0005499	0.07000	12.43

5.7.5.2 Heat Exchanger Parameters

Much is already known about the heat exchanger. This information is summarized in Table 5.10.

5.7.5.3 Calculation of Characteristic Flow Areas and Velocities

Tube side:

$$A_t = \frac{N_t}{N_p}\left(\frac{\pi ID_t^2}{4}\right) = \frac{488}{4}\left[\frac{\pi(0.05167 \text{ ft})^2}{4}\right] = 0.2558 \text{ ft}^2$$

$$V_t = \frac{\dot{m}_t}{\rho_t A_t} = \frac{\left(200,000\frac{\text{lbm}}{\text{h}}\right)\left(\frac{\text{h}}{3600 \text{ s}}\right)}{\left(48.04\frac{\text{lbm}}{\text{ft}^3}\right)(0.2588 \text{ ft}^2)} = 4.52\frac{\text{ft}}{\text{s}}$$

TABLE 5.10

Summary of Known or Calculated Parameters for the STHX

Parameter	Value	Source
Tube OD	$OD_t = 0.75$ in. $= 0.0625$ ft	Given
Tube ID	$ID_t = 0.620$ in. $= 0.05167$ ft	Table 5.8
Tube roughness (copper)	$\varepsilon_t = 0.000005$ ft	Table 4.3
Number of tube passes	$N_p = 4$	Given
Shell diameter	$D_s = 27$ in. $= 2.25$ ft	Given
Tube pitch	$P_T = 1$ in. $= 0.08333$ ft (triangular)	Given
Number of tubes (max)	$N_t = 488$	Table 5.9
Tube clearance	$C = P_T - OD_t = 0.02083$ ft	Calculated
Shell equivalent diameter	$D_e = \dfrac{2\sqrt{3}\left(P_T^2\right)}{\pi\left(OD_t\right)} - OD_t = 0.06002$ ft	Calculated
Baffle spacing (max)	$B = 0.6D_s = 1.35$ ft	Calculated
Fouling factor (water)	$R_{fs}'' = 0.0005$ h $\cdot$ ft$^2 \cdot$ F/Btu	Table 5.2
Fouling factor (cyclohexane)	$R_{ft}'' = 0.002$ h $\cdot$ ft$^2 \cdot$ F/Btu	Table 5.2

Shell side:

$$A_s = \frac{D_s C B}{P_T} = \frac{(2.25\ \text{ft})(0.02083\ \text{ft})(1.35\ \text{ft})}{0.08333\ \text{ft}} = 0.7594\ \text{ft}^2$$

$$V_s = \frac{\dot{m}_s}{\rho_s A_s} = \frac{\left(280{,}000\dfrac{\text{lbm}}{\text{h}}\right)\left(\dfrac{\text{h}}{3600\ \text{s}}\right)}{\left(61.63\dfrac{\text{lbm}}{\text{ft}^3}\right)(0.7594\ \text{ft}^2)} = 1.424\frac{\text{ft}}{\text{s}}$$

The design problem states that the water is in the shell and the cyclohexane is in the tubes. The cyclohexane has a larger fouling factor. Therefore it is placed in the tubes for easier cleaning of the heat exchanger. The water has the higher mass flow rate. Placing it in the larger flow area passage (the shell) should result in reasonable pressure drops for both fluids. In this case, the hydraulic and fouling criteria appear to be complementary.

Notice that the tube velocity is slightly below the estimated minimum economic velocity for cyclohexane (see Table 4.11). This could be resolved by using a thicker-wall tubing (a lower BWG number). However, since the values in Table 4.11 are only estimates, the calculated velocity can be viewed as reasonably close to the economic optimum. Therefore, the design calculations proceed using the given BWG specification.

5.7.5.4 Calculation of the Reynolds Numbers

$$Re_t = \frac{\rho_t V_t (ID_t)}{\mu_t} = \frac{\left(48.04\frac{\text{lbm}}{\text{ft}^3}\right)\left(4.521\frac{\text{ft}}{\text{s}}\right)(0.05167 \text{ ft})}{0.0005499\frac{\text{lbm}}{\text{ft}\cdot\text{s}}} = 20,408$$

$$Re_s = \frac{\rho_s V_s D_e}{\mu_s} = \frac{\left(61.63\frac{\text{lbm}}{\text{ft}^3}\right)\left(1.424\frac{\text{ft}}{\text{s}}\right)(0.06002 \text{ ft})}{0.0003572\frac{\text{lbm}}{\text{ft}\cdot\text{s}}} = 14,736$$

5.7.5.5 Calculation of the Friction Factors

$$f_t = \frac{0.25}{\left[\log\left(\dfrac{\varepsilon/ID_t}{3.7} + \dfrac{5.74}{Re_t^{0.9}}\right)\right]^2} = \frac{0.25}{\left[\log\left(\dfrac{0.000005 \text{ ft}/0.05167 \text{ ft}}{3.7} + \dfrac{5.74}{20,408^{0.9}}\right)\right]^2} = 0.02593$$

$$f_s = \exp\left[0.576 - 0.19\ln(Re_s)\right] = \exp\left[0.576 - 0.19\ln(14,736)\right] = 0.2872$$

5.7.5.6 Calculation of the Nusselt Numbers

$$Nu_t = \frac{(f_t/8)(Pr_t - 1000)Pr_t}{1 + 12.7(f_t/8)^{1/2}(Pr_t^{2/3} - 1)} = \frac{(0.02593/8)(20,408 - 1000)(12.43)}{1 + 12.7(0.02765/8)^{1/2}(12.43^{2/3} - 1)} = 188.1$$

$$Nu_s = 0.36Re_s^{0.55}Pr_s^{1/3} = 0.36(14,736)^{0.55}(3.519)^{1/3} = 107.4$$

Here, the shell-side Nusselt number is estimated using Equation 5.104, assuming that $(\mu/\mu_w)^{0.14}$ is nearly 1.

5.7.5.7 Calculation of the Convective Heat Transfer Coefficients

$$Nu_t = \frac{h_t(ID_t)}{k_t} \rightarrow h_t = \frac{Nu_t k_t}{ID_t} = \frac{(188.1)\left(0.07000\dfrac{\text{Btu}}{\text{h}\cdot\text{ft}\cdot°\text{F}}\right)}{0.05167 \text{ ft}} = 254.9\frac{\text{Btu}}{\text{h}\cdot\text{ft}^2\cdot°\text{F}}$$

$$Nu_s = \frac{h_s D_e}{k_s} \rightarrow h_s = \frac{Nu_s k_s}{D_e} = \frac{(107.4)\left(0.3654\dfrac{\text{Btu}}{\text{h}\cdot\text{ft}\cdot°\text{F}}\right)}{0.06002 \text{ ft}} = 653.9\frac{\text{Btu}}{\text{h}\cdot\text{ft}^2\cdot°\text{F}}$$

5.7.5.8 Calculation of the Overall Heat Transfer Coefficients

$$\frac{1}{U_{o,clean}} = \frac{1}{h_t}\left(\frac{OD_t}{ID_t}\right) + \frac{1}{h_s} = \left(\frac{1}{254.9}\frac{h \cdot ft^2 \cdot °F}{Btu}\right)\left(\frac{0.06250\ ft}{0.05167\ ft}\right) + \left(\frac{1}{653.9}\frac{h \cdot ft^2 \cdot °F}{Btu}\right)$$

$$U_{o,clean} = 159.4\frac{Btu}{h \cdot ft^2 \cdot °F}$$

$$\frac{1}{U_{o,fouled}} = \frac{1}{h_t}\left(\frac{OD_t}{ID_t}\right) + R''_{ft}\left(\frac{OD_t}{ID_t}\right) + R''_{fs} + \frac{1}{h_s}$$

$$\frac{1}{U_{o,fouled}} = \left(\frac{1}{254.9}\frac{h \cdot ft^2 \cdot °F}{Btu}\right)\left(\frac{0.06250\ ft}{0.05167\ ft}\right) + \left(0.002\frac{h \cdot ft^2 \cdot °F}{Btu}\right)\left(\frac{0.06250\ ft}{0.05167\ ft}\right)$$

$$+ \left(0.0005\frac{h \cdot ft^2 \cdot °F}{Btu}\right) + \left(\frac{1}{653.9}\frac{h \cdot ft^2 \cdot °F}{Btu}\right)$$

$$U_{o,fouled} = 108.8\frac{Btu}{h \cdot ft^2 \cdot °F}$$

5.7.5.9 Application of the Heat Exchanger Model

Since the hot fluid temperature has been assumed to be equal to the inlet temperature, the e-NTU heat exchanger model can be used to eliminate the iteration seen in the LMTD model. The e-NTU heat exchanger model is

$$\dot{Q} = \dot{C}_h\left(T_{hi} - T_{ho}\right)$$

$$\dot{Q} = \dot{C}_c\left(t_{co} - t_{ci}\right)$$

$$\varepsilon = \frac{\dot{Q}}{\dot{Q}_{max}} \quad \text{where } \dot{Q}_{max} = \dot{C}_{min}\left(T_{hi} - t_{ci}\right)$$

$$\varepsilon = f\left(NTU, C_r\right) \quad \text{where } NTU = \frac{U_o A_o}{\dot{C}_{min}} \quad \text{and } C_r = \frac{\dot{C}_{min}}{\dot{C}_{max}}$$

5.7.5.9.1 Calculation of the Thermal Capacity Rates

$$\dot{C}_h = \dot{m}_h c_{ph} = \left(240,000\frac{lbm}{h}\right)\left(0.9988\frac{Btu}{lbm \cdot R}\right) = 239,706\frac{Btu}{h \cdot R}$$

$$\dot{C}_c = \dot{m}_c c_{pc} = \left(200,000\frac{lbm}{h}\right)\left(0.4397\frac{Btu}{lbm \cdot R}\right) = 87,939\frac{Btu}{h \cdot R}$$

$$\therefore\ \dot{C}_{min} = 87,939\frac{Btu}{h \cdot R} \quad \dot{C}_{max} = 239,706\frac{Btu}{h \cdot R} \quad C_r = \frac{87,939\ Btu/h \cdot R}{239,706\ Btu/h \cdot R} = 0.3669$$

5.7.5.9.2 Calculation of Heat Exchanger Effectiveness

$$\dot{Q} = \dot{C}_c \left(t_{co} - t_{ci} \right) = \left(87{,}939 \frac{\text{Btu}}{\text{h} \cdot \text{R}} \right) (80 - 60) \text{R} = 4.397 \times 10^6 \frac{\text{Btu}}{\text{h}}$$

$$\dot{Q}_{max} = \dot{C}_{min} \left(T_{hi} - t_{ci} \right) = \left(87{,}939 \frac{\text{Btu}}{\text{h} \cdot \text{R}} \right) (175 - 60) \text{R} = 5.716 \times 10^6 \frac{\text{Btu}}{\text{h}}$$

$$\therefore \quad \varepsilon = \frac{\dot{Q}}{\dot{Q}_{max}} = \frac{4.397 \times 10^6 \ \text{Btu/h}}{5.716 \times 10^6 \ \text{Btu/h}} = 0.7692$$

5.7.5.9.3 Calculation of Heat Exchanger NTU

Using the expression for the NTU-e relationship in Table 5.4 for a single shell-pass STHX,

$$\text{NTU} = -\left(1 + C_r^2 \right)^{-1/2} \ln \left(\frac{\Gamma - 1}{\Gamma + 1} \right) \qquad \Gamma = \frac{2/\varepsilon - (1 + C_r)}{\sqrt{(1 + C_r^2)}}$$

$$\Gamma = \frac{2/\varepsilon - (1 + C_r)}{\sqrt{(1 + C_r^2)}} = \frac{2/0.7692 - (1 + 0.3669)}{\sqrt{(1 + 0.3669^2)}} = 1.1577$$

$$\therefore \quad \text{NTU} = -\left(1 + C_r^2 \right)^{-1/2} \ln \left(\frac{\Gamma - 1}{\Gamma + 1} \right) = -\left(1 + 0.3669^2 \right)^{-1/2} \ln \left(\frac{1.1577 - 1}{1.1577 + 1} \right) = 2.456$$

5.7.5.9.4 Calculation of the Tube Length

The heat exchanger will experience fouling during the course of its service. Therefore, the length of the heat exchanger should be calculated using the fouled value of the overall heat transfer coefficient. The *total* outside tube surface area and *total* tube length needed can be found by

$$\text{NTU} = \frac{U_{o,fouled} A_o}{\dot{C}_{min}} \quad \rightarrow \quad A_o = \frac{\dot{C}_{min} (\text{NTU})}{U_{o,fouled}} = \frac{\left(87{,}939 \dfrac{\text{Btu}}{\text{h} \cdot \text{R}} \right) (2.456)}{108.8 \dfrac{\text{Btu}}{\text{h} \cdot \text{ft}^2 \cdot \text{R}}} = 1985.92 \ \text{ft}^2$$

$$A_o = \pi (OD_t) L \quad \rightarrow \quad L = \frac{A_o}{\pi (OD_t)} = \frac{1985.92 \ \text{ft}^2}{\pi (0.06250 \ \text{ft})} = 10{,}114 \ \text{ft}$$

The length of each tube can now be found by

$$L_t = \frac{L}{N_t} = \frac{10{,}114 \ \text{ft}}{488} = 20.7 \ \text{ft}$$

The next integral length will be specified for the tube length.

$$L_t = \underline{21 \ \text{ft}} \quad \leftarrow$$

The number of baffles and final baffle spacing can now be specified. The number of baffles can first be estimated by

$$N_b = \frac{L_t}{B} - 1 \quad \rightarrow \quad N_b = \frac{21 \text{ ft}}{1.35 \text{ ft}} - 1 = 14.35$$

The baffle spacing listed above is based on the recommended maximum value that is 0.6 of the shell diameter. Since there cannot be 14.35 baffles, the number of baffles is rounded up to 15. Therefore, the final baffle spacing is

$$N_b = \underline{15} \quad \leftarrow \qquad \therefore \quad B = \frac{L_t}{N_b + 1} = \frac{24 \text{ ft}}{15 + 1} = \underline{1.5 \text{ ft}} \quad \leftarrow$$

5.7.5.10 Calculation of the Pressure Drops

Using the properties from above to estimate the pressure drops results in

$$\Delta P_t = \Delta P_{tubes} + \Delta P_m = \gamma_t N_p \left(f_t \frac{L_t}{ID_t} + 4 \right) \frac{V_t^2}{2g}$$

$$\Delta P_t = \left(48.04 \frac{\text{lbf}}{\text{ft}^3} \right) \left(\frac{\text{ft}^2}{144 \text{ in.}^2} \right) (4) \left[(0.02593) \left(\frac{21 \text{ ft}}{0.05167 \text{ ft}} \right) + 4 \right] \left[\frac{(4.521 \text{ ft/s})^2}{2 (32.174 \text{ ft/s}^2)} \right] = \underline{6.1 \text{ psi}} \quad \leftarrow$$

$$\Delta P_s = \gamma_s f_s (N_b + 1) \frac{D_s}{D_e} \frac{V_s^2}{2g}$$

$$\Delta P_s = \left(61.63 \frac{\text{lbf}}{\text{ft}^3} \right) \left(\frac{\text{ft}^2}{144 \text{ in.}^2} \right) (0.2872)(15 + 1) \left(\frac{2.25 \text{ ft}}{0.06002 \text{ ft}} \right) \left[\frac{(1.424 \text{ ft/s})^2}{2 (32.174 \text{ ft/s}^2)} \right] = \underline{2.3 \text{ psi}} \quad \leftarrow$$

In these equations, the shell-side pressure drop is determined using Equation 5.112, assuming that $(\mu/\mu_w)^{0.14}$ is nearly 1.

The calculations shown above were determined assuming that the hot fluid temperature was equal to the inlet temperature. Using EES to iterate on the hot fluid properties, the tube length is found to be 20.8 ft, only 0.1 ft larger. This verifies that the use of the inlet temperature of the hot fluid is valid for this calculation.

The pressure drops are estimates based on the properties initially determined in the design problem. The initial calculations resulted in a tube length of 20.7 ft. This was rounded up to 21 ft. In addition, the original design calculations included a baffle spacing of 0.6D = 1.35 ft. After finding the tube length, the baffle spacing was recalculated and found to be 1.5 ft. Therefore, to accurately predict the performance of this heat exchanger, including the pressure drop, an analysis now needs to be conducted with the length of the heat exchanger fixed at 21 ft and the baffle spacing set to 1.5 ft.

The results of this analysis using EES are shown in Table 5.11 for both a clean heat exchanger and one that is fouled.

Notice that for a 21-ft heat exchanger with a baffle spacing of 1.5 ft, the outlet temperature of the cyclohexane is at the required outlet temperature of 110°F. For a clean heat exchanger, the outlet temperature is 112.4°F. If this causes a problem in the application, the water flow rate can be decreased (using a valve in line with the heat exchanger) to bring the outlet temperature closer to the required 110°F.

In the calculation of the shell-side Nusselt number and pressure drop, it was assumed that $(\mu/\mu_w)^{0.14}$ was nearly 1. This assumption can now be checked using the results shown in Table 5.11. The average of the average hot and cold fluid temperatures is

$$T_w = \frac{1}{2}\left(\frac{T_{hi}+T_{ho}}{2}+\frac{t_{ci}+t_{co}}{2}\right)=\frac{1}{2}\left[\frac{(125+106.7)°F}{2}+\frac{(60+110.0)°F}{2}\right]=100.4°F \approx 100°F$$

From Appendix B.1, the viscosity of the water at this temperature is 0.0004127 lbm/ft·s. Therefore,

$$\left(\frac{\mu}{\mu_w}\right)^{0.14}=\left(\frac{0.0003576 \ \text{lbm/ft}\cdot\text{s}}{0.0004127 \ \text{lbm/ft}\cdot\text{s}}\right)^{0.14}=0.98$$

This shows that approximate Equations 5.104 and 5.112 are valid for this analysis. Notice that the estimated tube wall temperature (100.4°F) is very close to the shell-side fluid average temperature (115.8°F).

TABLE 5.11

Summary of the STHX Analysis for Clean and Fouled Conditions

STHX Parameter	Clean	Fouled
Hot fluid (water) inlet temperature	125°F	125°F
Hot fluid (water) outlet temperature	105.7°F	106.7°F
Cold fluid (cyclohexane) inlet temperature	60°F	60°F
Cold fluid (cyclohexane) outlet temperature	112.4°F	110.0°F
Overall heat transfer coefficient (U_o)	156.7 Btu/h·ft²·°F	107.2 Btu/h·ft²·°F
UA product	315,293 Btu/h·°F	215,737 Btu/h·°F
Heat transfer rate between the fluids	4.62×10^6 Btu/h	4.40×10^6 Btu/h
Tube length	21 ft	21 ft
Pressure drop in the tube (cyclohexane)	6.16 psi	6.16 psi
Pressure drop in the shell (water)	1.95 psi	1.95 psi

5.8 Plate and Frame Heat Exchanger Design and Analysis

The PFHX was briefly introduced in Section 5.3.3. An example of a PFHX is shown in Figure 5.10. The heat exchanger is made up of a series of plates bolted together. The fluids pass between the plates by virtue of gaskets. Some common metals used to manufacture the plates are shown in Table 5.12 along with the material's thermal conductivity at a temperature of 150°F (66°C). The gaskets are used to direct the fluid flow throughout the heat exchanger as well as sealing. They are designed so a gasket failure ends up leaking fluid to the atmosphere, which eliminates possible mixing of the hot and cold fluids.

PFHXs are typically used where low to moderate heat transfer rates are required. The heat exchanger has a distinct advantage of a fairly small footprint compared to other types of heat exchangers. The maximum fluid pressures in the heat exchanger are dictated by the gaskets and plate material. Typical maximum operating pressures can be from 10 to 25 bar (145–360 psig). The maximum fluid temperature is dictated by the gasket material. Depending on the type of gasket material, the maximum fluid temperature can range from 150°C to 260°C (300°F–500°F).

There are many possible flow configurations for PFHXs. In addition, there are many plate chevron patterns available. Therefore, the design of PFHXs turns out to be very specialized. Much of the data for plates and flow configurations is considered proprietary. Manufacturers of PFHXs have developed their own design procedures applicable to their heat exchangers.

In this presentation, the 1/1-U configuration (also known as a *single-pass counterflow arrangement*) is analyzed. Figure 5.30 shows a schematic of the 1/1-U configuration along with the hot and cold fluid flow paths.

The overall heat transfer coefficient for a PFHX can be found knowing the total thermal resistance and the heat transfer area.

$$U = \frac{1}{AR_t} \tag{5.113}$$

TABLE 5.12

Common Materials Used for Plates in PFHX

	Thermal Conductivity at 150°F (66°C)	
Material	Btu/h·ft·°F	W/m K
AISI 304 stainless steel	8.99	15.6
AISI 316 stainless steel	8.16	14.1
Titanium	12.3	21.3
90/10 Copper–nickel	31.5	54.6
70/30 Copper–nickel	17.9	31.0

Note: Values were calculated using EES

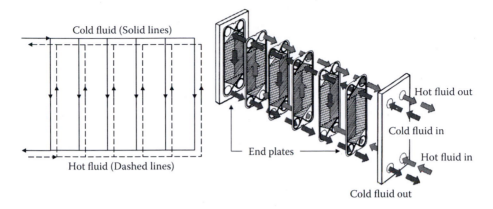

FIGURE 5.30
A 1/1-U configuration plate and frame heat exchanger. (From Bell and Gossett. *GPX Plate and Frame Heat Exchangers.* Buffalo, NY: Xylem, Inc., 2011. Reprinted with permission.)

For the PFHX, the total thermal resistance can be written as

$$R_t = \frac{1}{h_h A_h} + \frac{R''_{fh}}{A_h} + \frac{\Delta x_p}{k_p A_p} + \frac{R''_{fc}}{A_c} + \frac{1}{h_c A_c} \tag{5.114}$$

Notice that the thermal resistance for a plane wall is used to model the conduction resistance through the plate (subscript p). Since the heat transfer is occurring through a plate, all of the areas listed in Equations 5.113 and 5.114 are the same. Multiplying Equation 5.114 through by the plate area and substituting into Equation 5.113 results in an expression for the overall heat transfer coefficient for a PFHX.

$$\frac{1}{U_o} = \frac{1}{h_h} + R''_{fh} + \frac{\Delta x_p}{k_p} + R''_{fc} + \frac{1}{h_c} \tag{5.115}$$

The subscript "o" is used to be consistent with other types of heat exchangers previously analyzed. Since the areas are the same on both sides of the plate, U is same on both sides. The UA product of the PFHX can be found by multiplying U_o by the plate area seen by either fluid in the heat exchanger:

$$UA = U_o \left(N A_p \right) \tag{5.116}$$

The plate area seen by one of the fluids is defined as

$$A_p = bL \tag{5.117}$$

The dimensions b and L for a plate are defined in Figure 5.31. The variable b corresponds to the gasket-to-gasket width of the plate, and L is measured

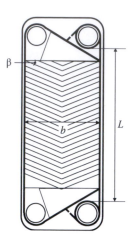

FIGURE 5.31
Plate dimensions.

from port-to-port, as shown in the figure. Even though the plate area is not exactly rectangular, this estimate is sufficient for design and analysis calculations. Also shown in Figure 5.31 is the chevron angle, β.

In Equation 5.116, N is the number of plates in the heat exchanger. When counting the number of plates, the two end plates are not counted since they have fluid flowing on only one side and are not active heat transfer plates. For example, the PFHX shown in Figure 5.30 contains four plates.

Fouling effects for a PFHX are less severe compared to the DPHX and STHX. This is due to several factors, including enhanced turbulence because of the chevron pattern of the plate, very smooth plate surfaces with fewer mineral deposition sites, and other factors. Table 5.13 lists the recommended values of fouling factors for PFHXs. The values in this table were taken from a more comprehensive table in Kakac et al. (2012).

The heat transfer and hydraulic performance of a PFHX are highly dependent on the plate design; in particular, the chevron pattern. There are many different plate designs, leading to a large number of correlations for the convective heat transfer coefficient and the friction factor. A comprehensive review of existing correlations is presented in Kakac et al. (2012). For the purpose of demonstrating the design and analysis calculations, the empirical correlations for the Nusselt number and friction factor developed by Rao et al. (2005) are used.

Rao et al. (2005) developed the following correlations for plates with a chevron pattern where $β = 30°$ (Figure 5.31). The Nusselt number and friction factor are given by

$$Nu = \frac{hD_h}{k} = 0.218Re^{0.65}Pr^{1/3} \quad [600 \le Re \le 5500] \qquad (5.118)$$

TABLE 5.13

Recommended Fouling Factors for PFHX

Fluid	Fouling Factor ($\times 10^5$)	
	h·ft²·°F/Btu	m² K/W
Demineralized or distilled water	1	0.017
Soft water	2	0.034
Hard water	5	0.086
Steam	1	0.017
Lubricating oils	2–5	0.034–0.086
Organic solvents	1–6	0.017–0.103
General process fluids	1–6	0.017–0.103

Source: Kakac, S., *Heat Exchangers: Selection, Rating, and Thermal Design*, CRC Press, Boca Raton, FL, 2012.

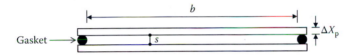

FIGURE 5.32
Cross-sectional view of the flow path between two plates in a plate and frame heat exchanger.

$$f = 21.41 Re^{-0.301} \quad [1000 \leq Re \leq 7000] \tag{5.119}$$

In chevron plate heat exchangers, turbulent conditions generally occur for $Re > 500$. Therefore, Equations 5.118 and 5.119 represent the turbulent flow regime for plates with a chevron pattern of $\beta = 30°$.

In both of these correlations, the Nusselt and the Reynolds numbers are based on the hydraulic diameter of the passage formed by two plates. To determine the hydraulic diameter, consider Figure 5.32. This figure shows a cross-sectional view of the flow channel formed by two plates.

Using the conventional definition of the hydraulic diameter,

$$D_h = \frac{4(\text{cross-sectional area})}{\text{wetted perimeter}} = \frac{4sb}{2s + 2b} \tag{5.120}$$

As shown in Figure 5.32, s is the separation distance between the plates. This value is typically on the order of 2 to 5 mm (0.08–0.2 in.). This is much smaller compared to the plate width, b. Therefore, the hydraulic diameter is often approximated by

$$D_h = \frac{4sb}{2s + 2b} \approx \frac{4sb}{2b} = 2s \tag{5.121}$$

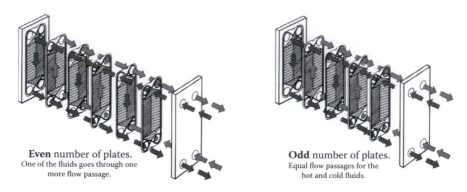

Even number of plates.
One of the fluids goes through one
more flow passage.

Odd number of plates.
Equal flow passages for the
hot and cold fluids.

FIGURE 5.33
Even and odd plate configurations and flow paths. (From Bell and Gossett. *GPX Plate and Frame Heat Exchangers*. Buffalo, NY: Xylem, Inc., 2011. Reprinted with permission.)

To complete the calculation of the Reynolds number, the velocity of the fluid in the passage is required. This value depends on whether there are an even or odd number of plates. As shown in Figure 5.33, if there is an even number of plates, one of the fluids goes through one extra flow passage compared to the other fluid. Conversely, if there are an odd number of plates in the heat exchanger, then both fluids pass through the same number of flow passages.

The velocity of the fluid through a plate passages is given by

$$V = \frac{\dot{m}_p}{\rho A} = \frac{\dot{m}_p}{\rho(sb)} \tag{5.122}$$

In this equation, the subscript "p" indicates the mass flow rate through one plate passage. For a heat exchanger with an odd number of plates, the mass flow rate through one plate passage is given by

$$\dot{m}_p = \frac{\dot{m}}{(N+1)/2} \tag{5.123}$$

In Equation 5.123, $\dot{m}$ is the *total* flow rate of one of the fluids. To envision how Equation 5.123 was derived, consider the PFHX on the right-hand side of Figure 5.33. This heat exchanger has $N = 3$ plates (recall that the end plates are not counted). These three plates form $N + 1 = 4$ flow passages. Each fluid flows through half of these passages, $(N + 1)/2 = 2$. Substituting Equation 5.123 into Equation 5.122 results in the following expression for the velocity of each fluid for a PFHX with an odd number of plates:

$$V_{odd} = \frac{2\dot{m}}{\rho sb(N+1)} \tag{5.124}$$

For a PFHX with an even number of plates (the left-hand side of Figure 5.33), one of the fluids will have a velocity given by

$$V_{even,1} = \frac{2\dot{m}}{\rho s b N} \qquad (5.125)$$

The other fluid travels through one additional passage. Therefore, the velocity of the second fluid is given by

$$V_{even,2} = \frac{2\dot{m}}{\rho s b (N+2)} \qquad (5.126)$$

Using these expressions for the hydraulic diameter and velocity, the Reynolds number is computed by

$$Re = \frac{\rho V D_h}{\mu} \qquad (5.127)$$

The pressure drop of the fluids through a PFHX is dependent on two friction losses: the loss because of friction through the plates and the loss through the inlet and outlet ports of the heat exchanger.

$$\Delta P = \Delta P_{plates} + \Delta P_{ports} \qquad (5.128)$$

The pressure drop through the plates is formulated similar to the friction loss through a pipe, except the hydraulic diameter of the passage is used.

$$\frac{\Delta P_{plates}}{\gamma} = f \frac{L}{D_h} \frac{V^2}{2g} \qquad (5.129)$$

The friction factor in Equation 5.129 is a function of the plate design. The correlation by Rao et al. (2005) in Equation 5.119 is valid for plates with chevron patterns where $\beta = 30°$ and the condition where $1000 \le Re \le 7000$.

The loss through the inlet and outlet ports of the heat exchanger is treated as a minor loss with a loss coefficient of 1.4.

$$\frac{\Delta P_{ports}}{\gamma} = 1.4 \frac{V_{port}^2}{2g} \qquad (5.130)$$

The port velocity is determined from the total flow rate of the fluid and the inside diameter of the port.

$$V_{port} = \frac{\dot{m}}{\rho A_{port}} = \frac{4\dot{m}}{\rho \pi (ID_{port}^2)} \qquad (5.131)$$

Combining the pressure drops because of the plates and the ports results in the total pressure drop the fluid experiences as it passes through the heat exchanger.

$$\frac{\Delta P}{\gamma} = f \frac{L}{D_h} \frac{V^2}{2g} + 1.4 \frac{V_{port}^2}{2g} \qquad (5.132)$$

5.8.1 Plate and Frame Heat Exchanger Model

For true parallel or counterflow PFHXs, the e-NTU model can be used. For other flow configurations, the LMTD model is used. For a PFHX where both fluids are in the single phase, the LMTD model is given by

$$\dot{Q} = \dot{C}_h \left(T_{hi} - T_{ho} \right)$$
$$\dot{Q} = \dot{C}_c \left(t_{co} - t_{ci} \right) \qquad (5.133)$$
$$\dot{Q} = (UA) F \left(\text{LMTD}_{CF} \right)$$

The LMTD in this model is calculated for a counterflow heat exchanger. If the flow pattern in the heat exchanger is true counterflow, then $F = 1$. For parallel flow or multipass flow scenarios, the correction factor, F, must be determined.

5.8.2 Plate and Frame Heat Exchanger Design and Analysis Example

Consider the following design problem.

Hot demineralized water flowing at 80 gpm, 185°F, needs to be cooled to 130°F using a 1/1-U counterflow PFHX with chevron patterns of $\beta = 30°$. Cold city water (hard water), flowing at 65 gpm, 50°F, is used to do the required cooling. The plates are made of AISI 316 stainless steel. Each plate has a heat transfer area of 4.5 ft². The plates are 0.04 in. thick, have a spacing of 0.18 in., and have a width of 18 in. Determine the minimum number of active plates required and the corresponding pressure drop of each fluid.

Because the complexity of the calculations, especially with respect to the fluid velocities, and the highly iterative nature of the LMTD model, it is prudent to use computer software to solve this problem. The number of plates is unknown. However, it is highly unlikely that the value of N will result in an integer value. This implies that the solution to this problem will be a combination of design and analysis. An appropriate strategy to solve this problem is as follows:

1. Start by assuming that the number of plates will be an even number. This allows for the calculation of the fluid velocities.
2. Solve the design problem by determining the value of N, the number of plates. It is likely that the calculated value of N will be a decimal value.
3. Round the value of N up to the next integer. This will be the total number of active plates required in the heat exchanger.
4. Using the rounded-up value of N, solve the analysis problem by fixing N and calculating the cold fluid outlet temperature and pressure drops for each fluid. Make sure that the fluid velocity calculation is correct based on odd or even plates.
5. Verify that the cold fluid outlet temperature meets the requirement.

Using EES, the information concerning the hot and cold fluids can be determined as follows:

```
"GIVEN: A 1/1 U-arrangement counter flow PFHX with 30°
 chevrons"

"Hot Fluid - demineralized water"
  hot$ = 'steam_iapws'
  T_hi = 185[F]
  T_ho = 130[F]
  T_h_avg = (T_hi + T_ho)/2
  Vol_dot_h = 80[gpm]
  m_dot_h = rho_h*Vol_dot_h*convert(gpm,ft^3/hr)
  R_f_h = 1E-05[ft^2-hr-F/Btu] "Table 5.13"
"Properties"
  rho_h = density(hot$,T = T_h_avg,x = 0)
  gamma_h = rho_h*g/g_c
  cp_h = cp(hot$,T = T_h_avg,x = 0)
  mu_h = viscosity(hot$,T = T_h_avg,x = 0)*convert(lbm/    &
    ft-hr,lbm/ft-s)
  k_h = conductivity(hot$,T = T_h_avg,x = 0)
  Pr_h = Prandtl(hot$,T = T_h_avg,x = 0)

"Cold Fluid - hard water"
  cold$ = 'steam_iapws'
  t_ci = 50[F]
  t_c_avg = t_ci
  {t_c_avg = (t_ci + t_co)/2}
  Vol_dot_c = 65[gpm]
  m_dot_c = rho_c*Vol_dot_c*convert(gpm,ft^3/hr)
  R_f_c = 5E-05[ft^2-hr-F/Btu]   "Table 5.13"
"Properties"
  rho_c = density(cold$,T=t_c_avg,x=0)
  gamma_c = rho_c*g/g_c
  cp_c = cp(cold$,T=t_c_avg,x=0)
  mu_c = viscosity(cold$,T=t_c_avg,x=0)*convert(lbm/ft-    &
    hr,lbm/ft-s)
  k_c = conductivity(cold$,T=t_c_avg,x=0)
  Pr_c = Prandtl(cold$,T=t_c_avg,x=0)
```

Notice that initially, the value of the average temperature of the cold fluid is assumed to be equal to the inlet temperature. This is done to avoid convergence problems during iteration on the fluid properties. Once a solution is found, the variables are updated, and the actual cold fluid average temperature is calculated.

The heat exchanger properties and other constants can now be specified.

```
"Heat exchanger data"
  b = 18[in]*convert(in,ft)      "Plate width"
  A_p = 4.5[ft^2]                "Heat transfer area per plate"
  A_p = b*L                      "Calculated the plate length"
```

```
    s = 0.18[in]*convert(in,ft)              "Plate separation"
    DELTAx_p = 0.04[in]*convert(in,ft)    "Plate thickness"
    k_p = 9.5[Btu/hr-ft-F]         "Plate thermal conductivity"


"Constants"
        g = 32.174[ft/s^2]
        g_c = 32.174[lbm-ft/lbf-s^2]


"FIND: The number of active plates required to accomplish the
       cooling and determine the pressure drop for each fluid
       at those conditions"
```

For the initial calculation, the number of plates will be assumed to be even, and the cold fluid is the fluid making one additional pass.

```
"SOLUTION:"
"Fluid velocities - The fluid velocities depend on if the
number of plates is odd or even. In the case of an even number
of plates, the fluid making one more pass is assumed to be the
cold fluid."
"!Comment out either odd or even and then adjust accordingly
based on the calculated N"
{"Odd number of plates"
    V_h*convert(ft/s,ft/hr) = (2*m_dot_h)/(rho_h*s*b*(N+1))
    V_c*convert(ft/s,ft/hr) = (2*m_dot_c)/(rho_c*s*b*(N+1))}
"Even number of plates - cold fluid makes one more pass"
    V_h*convert(ft/s,ft/hr) = (2*m_dot_h)/(rho_h*s*b*N)
    V_c*convert(ft/s,ft/hr) = (2*m_dot_c)/(rho_c*s*b*(N+2))
```

Since the plate separation is very small compared to the width of the plate, the hydraulic diameter can be approximated as,

```
"Hydraulic diameter"
    D_hyd = 2*s
```

The overall heat transfer coefficient can be found by calculating the heat transfer coefficient from the proper Nusselt number correlation. This will require the Reynolds numbers as well.

```
"Reynolds numbers"
    Re_h = rho_h*V_h*D_hyd/mu_h
    Re_c = rho_c*V_c*D_hyd/mu_c

"Nusselt numbers and heat transfer coefficients - Rao et al.
(2005)"
    Nus_h = 0.218*Re_h^0.65*Pr_h^(1/3)
    Nus_c = 0.218*Re_c^0.65*Pr_c^(1/3)
```

```
 Nus_h = h_h*D_hyd/k_h
 Nus_c = h_c*D_hyd/k_c

"Overall heat transfer coefficient"
 1/U_o = 1/h_h + R_f_h + DELTAx_p/k_p + R_f_c + 1/h_c
```

Notice that the heat exchanger is being designed including the fouling effects. The next step is to calculate the performance of the heat exchanger using the LMTD model. Since the flow pattern in the heat exchanger is counterflow, no LMTD correction factor is needed.

```
"Heat exchanger model"
 Q_dot = m_dot_h*cp_h*(T_hi - T_ho)
 Q_dot = m_dot_c*cp_c*(t_co - t_ci)
 Q_dot = UA*LMTD_CF

DELTAT_o = DELTAT_i*exp((DELTAT_o - DELTAT_i)/LMTD_CF)
 DELTAT_o = T_hi - t_co
 DELTAT_i = T_ho - t_ci

 UA = U_o*N*A_p
```

In this problem, there is no indication of the port sizes. Therefore, only the pressure drop through the plates is calculated.

```
"Pressure drop - Only the pressure drop through the plates is
calculated"
 DELTAP_h*convert(psi,psf) = gamma_h*(f_h*L/D_hyd)       &
  *(V_h^2/(2*g))
 DELTAP_c*convert(psi,psf) = gamma_c*(f_c*L/D_hyd)       &
  *(V_c^2/(2*g))

"Friction factors are determined from Rao et al. (2005),"
 f_h = 21.41/(Re_h^0.301)
 f_c = 21.41/(Re_c^0.301)
```

The solution to this problem is shown in Table 5.14. Notice that the Reynolds number of the cold fluid is within the recommended range for the Rao et al. (2005) correlations. However, the hot fluid Reynolds number is slightly out of range on the high side recommended for the Nusselt number correlation. The design is not yet complete since the number of plates has not been finalized. Therefore, even though the hot-fluid Reynolds number is out of range, calculations will proceed. Once the number of plates has been determined, the final Reynolds numbers will be rechecked to ensure that they are within the limits of validity of the Rao et al. (2005) correlations.

As shown in Table 5.14, the number of active plates determined is $N = 18.33$. This will be rounded up to $N = 19$, which is an odd number of plates. To verify that this number of plates will provide the required cooling, the

TABLE 5.14

Solution to the Plate and Frame Heat Exchanger Design Problem

Parameter	Value
Hot water inlet temperature	185°F
Hot water outlet temperature	130°F
Cold water inlet temperature	50°F
Cold water outlet temperature	116.6°F
Number of plates	18.33
Overall heat transfer coefficient (U_o)	353.1 Btu/h·ft²·°F
UA product	29,120 Btu/h·°F
Heat transfer rate between the fluids	2.156×10^6 Btu/h
Pressure drop of the hot water	0.71 psi
Pressure drop of the cold water	0.62 psi
Reynolds number of the hot water	5,511
Reynolds number of the cold water	2,246

program is run again as an analysis. This is easily done by commenting out the hot fluid outlet temperature and inserting the number of active plates.

```
"Hot Fluid - demineralized water"
   hot$ = 'steam_iapws'
   T_hi = 185[F]
{T_ho = 130[F]}
   N = 19
```

Since the number of plates is odd, the velocity calculation needs to be changed as well.

```
"Odd number of plates"
   V_h*convert(ft/s,ft/hr) = (2*m_dot_h)/(rho_h*s*b*(N+1))
   V_c*convert(ft/s,ft/hr) = (2*m_dot_c)/(rho_c*s*b*(N+1))
{"Even number of plates - cold fluid makes one more pass"
   V_h*convert(ft/s,ft/hr) = (2*m_dot_h)/(rho_h*s*b*N)
   V_c*convert(ft/s,ft/hr) = (2*m_dot_c)/(rho_c*s*b*(N+2))}
```

The solution to this analysis problem is shown in Table 5.15. Notice that in this final solution, both the hot- and cold-fluid Reynolds numbers are within the bounds of the Rao et al. (2005) correlations. Table 5.15 indicates that a PFHX with 19 active plates will meet the required cooling need. The hot water will exit the heat exchanger at the desired temperature. The pressure drops of each fluid are relatively low, each about 0.6 psi. Recall the port losses were not calculated in this design and analysis.

This example shows the complexity of the overall design calculation for PFHXs. It is possible to use sophisticated logic in the EES programming to steer the calculations to the proper calculations (e.g., the proper velocity expression to match the number of plates). However, the focus here is on understanding the calculations involved in the overall process.

TABLE 5.15

Solution to the Plate and Frame Heat Exchanger Analysis Problem

Parameter	Value
Hot water inlet temperature	185°F
Hot water outlet temperature	129.5°F
Cold water inlet temperature	50°F
Cold water outlet temperature	117.3°F
Number of plates	19
Overall heat transfer coefficient (U_o)	346.6 Btu/h·ft²·°F
UA product	29,636 Btu/h·°F
Heat transfer rate between the fluids	2.177×10^6 Btu/h
Pressure drop of the hot water	0.67 psi
Pressure drop of the cold water	0.59 psi
Reynolds number of the hot water	5,315
Reynolds number of the cold water	2,178

5.9 Cross-Flow Heat Exchanger Design and Analysis

There are many different types of CFHX. This precludes a detailed presentation in this book. The reader is encouraged to consult Kays and London (1998) for a thorough treatment of many types of CFHXs. However, one particular CFHX used extensively in many industrial applications is discussed here briefly: the plate finned-tube heat exchanger. This type of heat exchanger is used in many heating, ventilating, and air-conditioning applications. In these applications, air is heated or cooled by circulating it on the outside of finned tubes. Hot or cold water is usually in the tubes. It is also possible to use a finned-tube heat exchanger as an evaporator or condenser in a refrigeration system. In either case, the refrigerant flows in the tubes and air is circulated outside the finned tubes. Figure 5.34 shows a sketch of a plate finned-tube CFHX. On the basis of the definitions in Section 5.3.4, this type of heat exchanger is a CFHX with both fluids unmixed.

The design and analysis procedures for the CFHX are similar to the methods developed in the previous sections for other types of heat exchangers. The LMTD and e-NTU models can be used to model the performance of the CFHX. When using the LMTD method, the LMTD is calculated for a counterflow heat exchanger, and then modified using a correction factor, similar to the STHX. The correction factor is dependent on whether the fluids are mixed or unmixed. In lieu of the LMTD correction factor, the e-NTU method can be easily implemented, using the equations shown in Tables 5.3 and 5.4.

The convective heat transfer coefficient on the outside of the finned tubes is usually determined using the *Colburn factor, j*, which is defined as

$$j = \frac{Nu_{D_c}}{Re_{D_c} Pr^{1/3}} \tag{5.134}$$

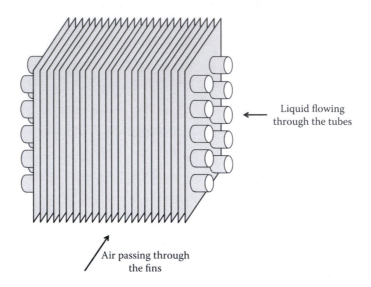

FIGURE 5.34
Example of a plate finned-tube cross-flow heat exchanger.

In this equation, both the Nusselt number and the Reynolds number are based on the collar diameter, D_c, of the fins. The collar diameter is the diameter of the holes punched in the fin plate that accepts the tubes. The correlations developed for j depend on the type of heat exchanger and whether there is water condensing out of the airstream, which is possible during a dehumidification process.

For dry, clean finned tubes, Wang et al. (1996) have demonstrated that Colburn factor can be found using

$$j = 0.394 Re_{D_c}^{-0.392} \left(\frac{th}{D_c} \right)^{-0.0449} N^{-0.0897} \left(\frac{F_p}{D_c} \right)^{-0.212} \tag{5.135}$$

In this equation, "th" is the fin thickness, F_p the fin pitch (spacing), and N the number of rows of tubes in the heat exchanger. This correlation is valid for $800 \leq Re_{D_c} \leq 7500$. The properties are evaluated at the average temperature of the air passing over the finned tubes in the heat exchanger. Wang et al. (1996) reported that 97% of the experimental data collected to develop Equation 5.135 are represented to within ±10%.

Wang et al. (1996) also developed an expression for the friction factor on the finned-tube side of the heat exchanger. This expression is similar to Equation 5.135 and given by

$$f = 1.039 Re_{D_c}^{-0.418} \left(\frac{th}{D_c} \right)^{-0.104} N^{-0.0935} \left(\frac{F_p}{D_c} \right)^{-0.197} \tag{5.136}$$

Eighty-eight percent of the experimental data collected to develop this equation are represented to within ±10%.

There are many other expressions for the Colburn factor and the friction factor for other types of CFHXs. In addition, contact resistance and fouling effects on the finned-tube side can become quite complex.

In lieu of the complex analysis required for CFHXs, it may be possible to extract heat transfer performance information from manufacturer's data. For example, many finned-tube coil manufacturers will provide performance curves that indicate the heat transfer rate within one of their coils for specific fluid flow rates. The calculation of the *UA* product using this type of data is shown in the following example.

EXAMPLE 5.9

Manufacturer's performance curves for a particular finned-tube CFHX utilizing water and air as the working fluids indicate that at a water flow rate of 80 gpm, the heat transferred between the air and the water is 49,860 Btu/h. These results were developed experimentally for the following conditions: (1) the air flows through the heat exchanger at 1000 cfm, and (2) the temperature difference between the inlet temperatures of the hot water and the cold air is 50 R. If the air temperature entering the heat exchanger is 68°F, estimate the *UA* product for this finned-tube CFHX.

SOLUTION: The outlet temperatures of the fluids are not known at this point. Therefore, the properties are estimated using the inlet temperatures. For the air, the inlet temperature is given as 68°F. The water inlet temperature can be found since the inlet temperature difference in the heat exchanger is specified.

$$T_{wi} - t_{ai} = \Delta T \quad \rightarrow \quad T_{wi} = t_{ai} + \Delta T = (68 + 50)°F = 118°F$$

Using these temperatures, the fluid properties can be found in Appendix B.

Fluid	T (°F)	ρ (lbm/ ft^3)	γ (lbf/ft^3)	c_p (Btu/ lbm·R)	μ (lbm/ ft·s)	k (Btu/ h·ft·°F)	Pr
Water (hot)	118	61.74	61.74	0.9986	0.0003816	0.3703	3.704
Air (cold)	68	0.07518	0.07518	0.2404	0.00001223	0.01495	0.7082

Using these properties, the outlet temperatures of the fluids can be found since the heat transfer rate is known:

$$\dot{Q} = \dot{C}_w \left(T_{wi} - T_{wo} \right) \quad \rightarrow \quad T_{wo} = T_{wi} - \frac{\dot{Q}}{\dot{C}_w}$$

$$\dot{C}_w = \dot{m}_w c_{pw} = \left(\rho_w \dot{V}_w \right) c_{pw} = \left(61.74 \frac{\text{lbm}}{\text{ft}^3} \right) (80 \text{ gpm})$$

$$\left(0.9986 \frac{\text{Btu}}{\text{h} \cdot \text{R}} \right) \left(8.0208 \frac{\text{ft}^3}{\text{h} \cdot \text{gpm}} \right)$$

$$\dot{C}_w = 39,561 \frac{\text{Btu}}{\text{h} \cdot \text{R}}$$

$$\therefore \quad T_{wo} = 118°F - \frac{49,860 \text{ Btu/h}}{39,561 \text{ Btu/h} \cdot \text{R}} = 116.7°F$$

Similarly, for the air, the following calculations can be used:

$$\dot{Q} = \dot{C}_a \left(t_{ao} - t_{ai} \right) \quad \rightarrow \quad t_{ao} = t_{ai} + \frac{\dot{Q}}{\dot{C}_a}$$

$$\dot{C}_a = \dot{m}_a c_{pa} = \left(\rho_a \dot{V}_a \right) c_{pa} = \left(0.07518 \frac{\text{lbm}}{\text{ft}^3} \right) \left(1000 \frac{\text{ft}^3}{\text{min}} \right) \left(0.2404 \frac{\text{Btu}}{\text{h} \cdot \text{R}} \right) \left(60 \frac{\text{min}}{\text{h}} \right)$$

$$\dot{C}_w = 1,084.26 \frac{\text{Btu}}{\text{h} \cdot \text{R}}$$

$$\therefore \quad t_{ao} = 68°F + \frac{49,860 \text{ Btu/h}}{1,084.26 \text{ Btu/h} \cdot \text{R}} = 114.0°F$$

Knowing the fluid inlet and outlet temperatures along with the heat transfer rate between the fluids allows for the determination of the UA product of the heat exchanger by solving the heat transfer equation in the heat exchanger model. Using the e-NTU method, the heat transfer equations are represented by,

$$\varepsilon = \dot{Q}/\dot{Q}_{max}$$

$$\dot{Q}_{max} = \dot{C}_{min} \left(T_{wi} - t_{ai} \right)$$

$$NTU = f \left(\varepsilon, C_r \right)$$

$$NTU = \frac{UA}{\dot{C}_{min}}$$

The thermal capacity rates were calculated earlier; therefore, it is a simple matter to pick out the maximum and minimum values and compute the ratio C_r,

$$\dot{C}_{min} = 1,084.26 \text{ Btu/h} \cdot \text{R} \quad \dot{C}_{max} = 39,561 \text{ Btu/h} \cdot \text{R}$$

$$\therefore \quad C_r = \frac{\dot{C}_{min}}{\dot{C}_{max}} = \frac{1,084.26 \text{ Btu/h} \cdot \text{R}}{39,561 \text{ Btu/h} \cdot \text{R}} = 0.02741$$

The maximum heat transfer rate can now be found.

$$\dot{Q}_{max} = \dot{C}_{min}\left(T_{wi} - t_{ai}\right) = \left(1,084.26\frac{\text{Btu}}{\text{h} \cdot \text{R}}\right)(50)\text{R} = 54,213\frac{\text{Btu}}{\text{h}}$$

Therefore, the effectiveness of this heat exchanger is

$$\varepsilon = \frac{\dot{Q}}{\dot{Q}_{max}} = \frac{49,860 \text{ Btu/h}}{54,213 \text{ Btu/h}} = 0.9197$$

In a finned-tube CFHX, both fluids are unmixed. Unfortunately, Table 5.4 does not contain a NTU-e relationship for this configuration. However, Table 5.3 has an e-NTU expression for an unmixed-unmixed CFHX.

$$\varepsilon = 1 - \exp\left[\left(\frac{1}{C_r}\right)(\text{NTU})^{0.23}\left(\exp\left[-C_r\left(\text{NTU}\right)^{0.73}\right] - 1\right)\right]$$

Using EES to solve for the NTU results in a value of NTU = 2.699. Therefore, the heat exchanger's UA product can be estimated as follows:

$$UA = \dot{C}_{min}(\text{NTU}) = \left(1,084.26\frac{\text{Btu}}{\text{h} \cdot \text{R}}\right)(2.699) = 2,927\frac{\text{Btu}}{\text{h} \cdot \text{R}} \quad \leftarrow$$

This is the first estimate of the UA product. Notice that the air temperature change is quite significant in this example (68°F–114°F). Therefore, the use of the inlet air properties may not be very accurate. The above set of equations can be programmed into EES and iteration on the properties can be done. The results are shown in Table E5.9. The results in Table E5.9 indicate that the air properties significantly impact the solution.

The calculated UA value can be used in an analysis of the heat exchanger (e.g., predicting outlet temperatures for known inlet conditions and flow rates) for different conditions than tested by the manufacturer. In reality, the U value of the heat exchanger will change as the flow rates and fluid temperatures

TABLE E5.9

Iterative Solution to the Cross-Flow Heat Exchanger Analysis Problem

Parameter	Value
Hot water inlet temperature	118°F
Hot water outlet temperature	116.7°F
Cold air inlet temperature	68°F
Cold air outlet temperature	116.1°F
Effectiveness of the heat exchanger	0.9612
UA product	3,665 Btu/h·°F

change. However, if the expected change is minimal, the predicted outlet temperature using the estimated UA should be reasonably close.

Example 5.9 shows how manufacturer's performance data can be used to estimate the performance of a CFHX at other conditions. This type of calculation eliminates the need to determine the overall heat transfer coefficient using the complicated Colburn factor. However, if more detail is required, appropriate sources for the Colburn factor should be consulted and an in-depth analysis should be conducted.

Problems

5.1   The plates in a PFHX are made of Inconel 600 ($k = 16$ W/m K) and are 1.5 mm thick. The convective heat transfer coefficients on hot and cold sides of the plate are 750 and 580 W/m^2 K, respectively. Determine the percent error in the heat flux through the plate that would result if the thermal resistance of the plate was neglected in this calculation. Assume that there is no fouling on either side of the plate.

5.2   The plates in a PFHX are made of Monel 200 ($k = 38.1$ Btu/h·ft·°F) and are 0.04 in. thick. The convective heat transfer coefficients on hot and cold sides of the plate are 130 and 105 Btu/h·ft^2·°F, respectively. Determine the percent error in the heat flux through the plate that would result if the thermal resistance of the plate was neglected in this calculation. Assume that there is no fouling on either side of the plate.

5.3   A 2½-nom sch 40S stainless steel pipe ($k = 16.5$ W/m K) is being used in a heat exchanger application. The pipe is 4 m long. The convective heat transfer coefficients on the inside and outside of the pipe are 1575 and 1950 W/m^2 K, respectively. The fouling factor on each side of the pipe is 0.0002 m^2 K/W. Determine the *UA* product for this pipe.

5.4   The convective heat transfer coefficients on the inside and outside of a ½-std type L copper tube are 520 and 180 Btu/h·ft^2·°F, respectively. The tube is 6 ft long. Ethylene glycol is on the outside of the tube and Refrigerant-22 is on the inside. Determine the *UA* product for this pipe. Assume that the average temperature of the copper tube is 100°F.

5.5   Solve Problem 5.4 again assuming that the thermal resistance of the copper tube is negligible. Compare the two answers and comment on the validity of this assumption.

5.6   Because of extreme pressures for a particular application, a 10-ft long pipe within a heat exchanger is made of 2-nom sch 160 pipe ($k_{pipe} = 70$ Btu/h·ft·°F). The convective heat transfer coefficients inside and outside of the pipe have been determined to be $h_i = 1248$ Btu/h·ft^2·°F and $h_o = 652$ Btu/h·ft^2·°F, respectively. The fouling factors on each side of the pipe are 0.002 h·ft^2·°F/Btu. Determine the *UA* value of this pipe for both clean and fouled conditions.

5.7   A copper coil heat exchanger is inserted in a 4-in. (ID) stovepipe to capture
some of the energy normally rejected to the atmosphere from a wood-burning
stove for preheating domestic hot water. The copper coil is made of 6 ft of
½-std type K tubing. Cold water flows at a rate of 5 gpm through the tube. The
average mean temperature of the tube is found to be 88°F. The hot combustion
gases (which can be modeled as air) flow over the outside of the copper coil in
cross flow at a free stream temperature of 450°F.

The flow rate of the combustion gases through the stovepipe is the result of
an induced draft effect. This volumetric flow rate can be estimated using the
following expression:

$$\dot{V}_{gas} = 0.68 A_{chimney} \sqrt{2gh \left(\frac{T_{gas} - T_{amb}}{T_{gas}} \right)}$$

In this equation, $A_{chimney}$ is the cross-sectional area of the stovepipe, h the total
height of the stovepipe, T_{gas} the free-stream combustion gas temperature, and
T_{amb} the outside ambient temperature. Notice that T_{gas} in the denominator of
the term in the square root must be on the *absolute* temperature scale. Consider
the situation where the outdoor temperature is −5°F and the total height of the
stovepipe is 10 ft. Estimate the UA product for the heat exchanger if it is *clean*
(i.e., no fouling).

5.8   A specialized tube designed to enhance heat transfer between a condensing
refrigerant and water is shown in Figure P5.8. The inner tube is ⅜-std type L
copper and the outer tube is 2-std type L copper. Rectangular straight fins made
of copper ($k = 400$ W/m K) with a thickness of 2 mm are attached to the outside
of the inner tube to provide enhanced heat transfer surfaces (fins). The fin array,
fits snugly into the inside diameter of the outer tube. The refrigerant inside the
inner tube, Refrigerant-134a, is condensing at a constant pressure of 11.6 bar.
Cold water flows at a rate of 0.2 gpm through all the channels created by the fins
at an average temperature of 15°C. Because of the phase change of the refrigerant, the convective heat transfer coefficient in the inside tube can be assumed
to be very large. It can also be assumed that the tube is long enough such that
entry effects can be considered negligible.

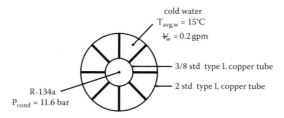

cold water
$T_{avg,w} = 15°C$
$\dot{V}_w = 0.2$ gpm

3/8 std type L copper tube

2 std type L copper tube

R-134a
$P_{cond} = 11.6$ bar

FIGURE P5.8

For the case where the heat exchanger is brand new (i.e., no fouling),
determine the following:

a. UA product of the heat exchanger per unit length (W/m K)
b.   Heat removal rate from the refrigerant per unit length of tube (W/m)

5.9 A heat exchanger is being used to cool 48,000 lbm/h of oil from 150°F to 102°F by using 32,000 lbm/h of water at an inlet temperature of 70°F. The overall heat transfer coefficient is determined to be 136 Btu/h·ft²·°F. The average heat capacities of the oil and water can be taken as 0.5 and 1.0 Btu/lbm·R, respectively. Determine the heat transfer area required for counterflow and parallel-flow configurations.

5.10 A counterflow DPHX is being used to cool hot oil from 320°F to 285°F using cold water. The water, which flows through the inner tube, enters the heat exchanger at 70°F and leaves at 175°F. The inner tube is ¾-std type L copper. The overall heat transfer coefficient based on the outside diameter of the inner tube is 90 Btu/h·ft²·°F. Design conditions call for a total heat transfer between the two fluids of 15,000 Btu/h. Determine the required length of this heat exchanger (ft).

5.11 A flow of cold water enters a parallel-flow heat exchanger at 20°C and leaves at 100°C. The cold water is being heated by a hot water flow that enters the heat exchanger at 160°C and leaves at 120°C. The overall heat transfer coefficient for this heat exchanger is 325 W/m² K. The heat exchanger must transfer heat between the water streams at a rate of 30 kW. Determine the heat exchanger area required (m²).

5.12 A counterflow heat exchanger is being used to cool a flow of hot water using a cold 20% ethylene glycol solution. The hot water enters at 80°F with a volumetric flow rate of 400 gpm. The ethylene glycol solution enters the heat exchanger at –10°F and leaves at 15°F. The ethylene glycol solution is flowing at a volumetric flow rate of 300 gpm. The overall heat transfer coefficient for the heat exchanger is found to be 155 Btu/h·ft²·°F. Determine the following:

   a. Outlet temperature of the water (°F)
   b. Heat transfer rate between the water and ethylene glycol solution (Btu/h)
   c. Required heat transfer area of the heat exchanger (ft²)

5.13 Solve Problem 5.12 for a parallel-flow heat exchanger.

5.14 Solve Problem 5.12 for a STHX with 1-shell pass and 2-tube passes.

5.15 Solve Problem 5.12 for a CFHX where both fluids are unmixed.

5.16 A parallel-flow heat exchanger is utilizing a cold 20% magnesium chloride solution to cool a flow of hexane. The hexane enters the heat exchanger at 30°C and the magnesium chloride solution enters at –20°C. The volumetric flow rates of the hexane and magnesium chloride solution are 25 and 20 L/s, respectively. The heat exchanger has an overall heat transfer coefficient of 1280 W/m² K. The required heat transfer rate between the fluids is 1100 kW. Determine the following:

   a. Outlet temperature of the hexane (°C)
   b. Outlet temperature of the magnesium chloride solution (°C)
   c. Required heat transfer area of the heat exchanger (m²)

5.17 Solve Problem 5.16 for a counterflow heat exchanger.

5.18 Solve Problem 5.16 for a STHX with 1-shell pass and 2-tube passes.

5.19 Solve Problem 5.16 for a CFHX with both fluids unmixed.

5.20 Liquid heptane enters a counterflow heat exchanger at 45°F. The heptane is heated using a flow of hot water entering the heat exchanger at 150°F. The volumetric flow rates of the heptane and water are 280 and 200 gpm, respectively. The *UA* product of the heat exchanger is found to be 25,675 Btu/h-°F.

Determine the outlet temperatures of both fluids (°F) and the heat transfer rate between them (Btu/h).

5.21 Solve Problem 5.20 for a parallel-flow heat exchanger.

5.22 Solve Problem 5.20 for a STHX with 1-shell pass and 2-tube passes.

5.23 Solve Problem 5.20 for a CFHX where the heptane flow is mixed and the water flow is unmixed.

5.24 A counterflow heat exchanger is being used to heat a flow of liquid toluene. The toluene enters the heat exchanger at 5°C with a volumetric flow rate of 18 L/s. Hot water is being used to heat the toluene. The water enters the heat exchanger at 70°C and at a volumetric flow rate of 13 L/s. The UA product of the heat exchanger is 25 kW/K. Determine the outlet temperatures of both fluids (°F) and the heat transfer rate between them (kW).

5.25 Solve Problem 5.24 for a parallel-flow heat exchanger.

5.26 Solve Problem 5.24 for a STHX with 2-shell passes and 4-tube passes.

5.27 Solve Problem 5.24 for a CFHX where the water flow is mixed and the toluene flow is unmixed.

5.28 The series of heat exchangers shown in Figure P5.28 has the purpose of raising a liquid's temperature so a desired chemical reaction can take place in the reactor. The specific heat of the liquid can be assumed constant at an average value of 3.2 kJ/kg·K. The liquid flow rate is 1.5 kg/s. The temperature of the liquid entering HX 1 is 290 K and the temperature leaving the reactor is 390 K. The UA value of HX 1 is 2.88 kW/K. Saturated vapor steam is supplied to HX 2 at 375 K and saturated liquid condensate leaves at 375 K. The UA values of HX 2 and HX 3 are 4.7 and 9.6 kW/K, respectively. Determine the values of the unknown temperatures (K) in the system and the heat transfer rates (kW) in each of the three heat exchangers.

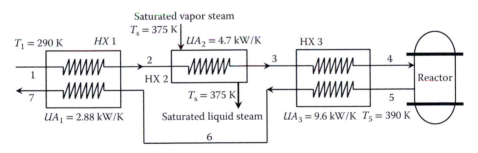

FIGURE P5.28

5.29 A gas turbine cycle utilizes a counterflow regenerative heat exchanger as shown in Figure P5.29. Air enters the compressor at atmospheric conditions, 1 atm and 70°F, with a mass flow rate of 15,000 lbm/h. The exhaust pressure of the compressor is 7 atm. The compressor is operating with an isentropic efficiency of 75%. The regenerator preheats the air entering the combustion chamber, thus reducing the fuel requirement and increasing the thermal efficiency of the cycle. The regenerator in the cycle has a UA product of 20,000 Btu/h·°F. Combustion gases (which can be modeled as air) leave the combustion chamber at 2300°F. The combustion gases then pass through the turbine that has an isentropic efficiency of 72%. Assume that the pressure drops in all heat exchangers and connecting piping is negligible. Even though the temperature varies significantly in this cycle, the average heat capacity evaluated between the highest and lowest temperatures in the cycle can be used for analysis of the regenerator. Use the ideal gas law to model the properties of air and determine the following:

a. Net power delivered by the cycle (hp).
b. Heat transfer rate required at the combustion chamber (Btu/hr).
c. Thermal efficiency of the cycle.
d. Investigate the effect of the size of the regenerator by plotting the thermal efficiency of the cycle as a function of the regenerator UA value for the range $0 \leq UA \leq 50{,}000$ Btu/h·°F. What is the significance of $UA = 0$ Btu/h·°F?

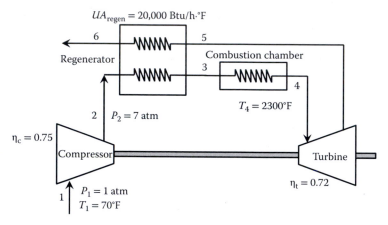

FIGURE P5.29

5.30 Water is being heated in a STHX by condensing steam. The water is flowing in the tubes at a rate of 0.8 kg/s. The water enters the heat exchanger at 15°C and leaves at 58°C. The steam condenses on the shell side of the heat exchanger at a condensing pressure of 110 kPa. Determine the following:
a. Mass flow rate of the condensing steam (kg/s)
b. Heat transfer rate between the water and condensing steam (kW)
c. UA product of the heat exchanger (kW/K)
Consider a situation where the cold water inlet temperature to the heat exchanger is adjusted to 20°C while its flow rate (0.8 kg/s) remains constant. For this scenario, determine the following:
d. Outlet temperature of the cold water (°C)
e. Heat transfer rate between the water and condensing steam (kW)

5.31 The evaporator of a vapor compression refrigeration cycle utilizing R-123 as the refrigerant is being used to chill water. The evaporator is a STHX with the water flowing through the tubes. The water enters the heat exchanger at a temperature of 54°F and leaves at 44°F. The evaporating pressure of the refrigeration cycle is 6 psia and the condensing pressure is 75 psia. The refrigerant is flowing through the cycle with a flow rate of 16,250 lbm/h. The R-123 leaves the evaporator as a saturated vapor, and leaves the condenser as a saturated liquid. Determine the following:

    a.   Volumetric flow rate of the chilled water leaving the evaporator (gpm)

    b.   *UA* product of the evaporator (Btu/h·°F)

    c.   Heat transfer rate between the refrigerant and the water (tons)

5.32   The outer pipe in a DPHX is 4-nom sch 40 commercial steel. The inner tube is 3-std type K copper. The fluid inside the inner tube is flowing at a volumetric rate of 60 gpm and the fluid in the annulus flows at 40 gpm. For this heat exchanger determine

    a.   Hydraulic diameter of the annulus (ft)

    b.   Equivalent diameter of the annulus (ft)

    c.   Velocity of the fluid in the inner tube (ft/s)

    d.   Velocity of the fluid in the annulus (ft/s)

5.33   A DPHX is made of a 6-nom sch 80 outer pipe and a 4-nom sch 40 inner pipe. The fluid in the annular space is liquid water that has a volumetric flow rate of 20 L/s and an average temperature of 40°C. The fluid in the inner tube is a 40% ethylene glycol solution at an average temperature of 20°C and flowing at 15 L/s. Determine the Reynolds numbers of the flow in the pipe and the annulus for

    a.   Friction calculations

    b.   Heat transfer calculations

5.34   A DPHX is made of a 6-nom sch 40 commercial steel outer pipe and a 5-nom sch 40S stainless steel inner pipe. The fluid in the annular space is cyclohexane that has a volumetric flow rate of 120 gpm and an average temperature of 60°F. The fluid in the inner tube is water at an average temperature of 180°F and flowing at 320 gpm. Cyclohexane is an organic liquid and the water can be assumed to be distilled water. The thermal conductivity of the stainless steel inner pipe is 9.5 Btu/h·ft·°F. Determine the following:

    a.   Convective heat transfer coefficients in the inner pipe and the annulus (Btu/h·ft²·°F).

    b.   Overall heat transfer coefficient for the case of a clean heat exchanger (Btu/h·ft²·°F).

    c.   Overall heat transfer coefficient if fouling is considered (Btu/h·ft²·°F).

5.35   A DPHX is made of type M copper tubing. The heat exchanger is a 3 × 1 design (3-std outside tube with a 1-std inside tube). Toluene flows through the inner tube with a volumetric flow rate of 0.5 L/s and an average temperature of 30°C. Water flows in the annulus at a rate of 3 L/s with an average temperature of 60°C. Toluene can be considered an organic liquid and the water can be assumed to be from a city water supply. Determine the following:

    a.   Convective heat transfer coefficients in the inner pipe and the annulus (W/m² K).

    b.   Overall heat transfer coefficient for a clean heat exchanger (W/m² K).

    c.   Overall heat transfer coefficient if fouling is considered (W/m² K).

5.36   A 3 × 1¼ type M copper tube double-pipe counterflow heat exchanger is used to cool a flow of liquid toluene. The heat exchanger consists of three 10-ft-long DPHX hairpinned together. The toluene enters the annulus of the heat exchanger at a flow rate of 25,000 lbm/h and a temperature of 200°F.

The toluene is being cooled using a 20% solution of ethylene glycol. The ethylene glycol enters the inner tube at 20°F with a mass flow rate of 15,000 lbm/h. Determine the following information for this heat exchanger:

a. *UA* product for the heat exchanger for both clean and fouled conditions (Btu/h·°F)
b. Outlet temperatures of each fluid (°F)
c. Pressure drop of each fluid through the heat exchanger (psi)
d. Heat transfer rate between the fluids (Btu/h)

5.37 Solve Problem 5.36 for a double-pipe parallel-flow heat exchanger.

5.38 A 2 × ¾ DPHX is made from sch 40S stainless steel. Acetone enters the heat exchanger at 80°F with a volumetric flow rate of 12 gpm. The other fluid, ethanol, enters the heat exchanger with 10°F with a flow rate of 45 gpm. The required heat exchanger duty (heat transfer rate between the two fluids) is 75,000 Btu/h. The thermal conductivity of the stainless steel tube can be taken as 8.1 Btu/h·ft·°F. The counterflow arrangement will be used to minimize the heat exchanger area required. The footprint for this heat exchanger only allows for a maximum horizontal, straight length of 8 ft.

a. Which fluid should be placed in the annulus? Which fluid should be placed in the tube? Justify your answer with calculations.
b. How many straight, horizontal DPHX must be hairpinned together to meet the heat exchanger duty?
c. Calculate the outlet temperatures of each fluid and the heat exchanger duty when the heat exchanger is first placed in service and it is clean.
d. Calculate the outlet temperatures of each fluid and the heat exchanger duty after the heat exchanger has been in service for 1 year.
e. What are the pressure drops (psi) of each fluid in the heat exchanger?

5.39 A STHX has 2-tube passes inside of a 29 in. shell. The tubes are 1-in. 15 BWG tubes on a 1¼-in. square pitch, and the heat exchanger contains the maximum amount of tubes allowable. Water at an average temperature of 80°F is flowing through the tubes. The total flow rate of the water is 750,000 lbm/h. Determine the following:

a. Clearance (C) between the tubes (in.)
b. Number of tubes in the heat exchanger
c. Reynolds number of the water flow inside of one of the tubes

5.40 The condenser of a steam power plant is a shell and tube design with cooling water flowing through the tubes. The condenser has 4-tube passes inside a 39-in. shell. The tubes are 1½-in. 12 BWG tubes on a 1⅞-in. triangular pitch. The cooling water enters the heat exchanger at a flow rate of 250 kg/s at an average temperature of 30°C. Determine the Reynolds number of the water flow inside one of the tubes of this heat exchanger.

5.41 The tubes in a STHX are 12 ft long. The liquid in the tubes is cyclohexane at an average temperature of 110°F. The flow rate of the cyclohexane entering the heat exchanger is 120,000 lbm/h. The heat exchanger tubes are 1-in. 14 BWG tubes on a 1¼-in. triangular pitch. The shell diameter is 12 in. Determine the pressure drop (psi) through the tube side of the heat exchanger for a 1-pass, 2-pass, and 4-pass tube configuration.

5.42 The evaporator of a refrigeration cycle is a shell and tube design being used to chill water. The water flows in the tubes while the refrigerant boils in the shell. The water enters the heat exchanger at a flow rate of 50 L/s. The average temperature of the water through the heat exchanger is 8°C. The shell of the heat exchanger has a diameter of 21¼ in. The tubes in the heat exchanger have a length of 4 m and are ¾-in. 16 BWG tubes on a 1-in. square pitch. Determine the pressure drop (kPa) on the tube side of the heat exchanger for a 1-pass, 2-pass, and 4-pass tube design.

5.43 A STHX has a single shell and 4-tube passes. The shell diameter is 25 in. The shell contains 10 baffles with a spacing of 0.36 m. Water flows through the shell with a flow rate of 70 kg/s and an average temperature of 85°C. The tubes are 1¼-in. 13 BWG tubes on a 1 $\%_{16}$-in. square pitch. Determine the following:
 a.   Shell-side convective heat transfer coefficient (W/m² K)
 b.   Shell-side pressure drop (kPa)

5.44 A 20% magnesium chloride solution is flowing at a rate of 240,000 lbm/h through the shell side of a STHX. The average temperature of the magnesium chloride is 25°F. The shell has a diameter of 17¼ in. and there is 1-tube pass. The shell contains 12 baffles with a spacing of 10 in. The tubes are ¾ in. 15 BWG tubes on a 1-in. triangular pitch. Determine the following:
 a.   Shell-side convective heat transfer coefficient (Btu/h·ft²·°F)
 b.   Shell-side pressure drop (psi)

5.45 Normal heptane is flowing in the shell of a STHX at a rate of 400,000 lbm/h and an average temperature of 130°F. The shell has a diameter of 27 in. and a length of 16 ft. The tubes in the heat exchanger are ¾-in. 15 BWG tubes on a 1-in. triangular pitch. The purpose of this problem is to investigate how the number of baffles impacts the heat transfer and the pressure drop on the shell side of the heat exchanger. Calculate the shell-side convective heat transfer coefficient and pressure drop for the case where the heat exchanger has 10 baffles. Repeat the calculation for 20 baffles. Then determine the following:
 a.   Ratio of the shell-side convective heat transfer coefficient for the 20-baffle heat exchanger to the 10-baffle heat exchanger.
 b.   Ratio of the shell-side pressure drop for the 20-baffle heat exchanger to the 10-baffle heat exchanger.
 c.   If the optimum baffle spacing is somewhere between $0.4D_s$ and $0.6D_s$, how many baffles would you recommend for this heat exchanger? What are the values of the shell-side convective heat transfer coefficient and pressure drop for the number of baffles you recommended?

5.46 At a certain point in the processing of crude oil, it must be heated. The heating occurs in a 1-4 (1-shell pass, 4-tube pass) STHX. The oil is flowing at a rate of 110,000 lbm/h and it enters the heat exchanger at 80°F. The heating will be accomplished with hot water entering the heat exchanger at 150,000 lbm/h at a temperature of 200°F. The heat exchanger has a shell diameter of 23¼ in. and contains 6 baffles. The tubes are 1-in. OD, 13 BWG tubes on a 1¼-in. square pitch. The tubes are 12 ft long. The oil is routed through the tubes and the water through the shell. The average properties of the crude oil can be taken as

$$\rho_{oil} = 46.2 \text{ lbm/ft}^3 \qquad c_{p,oil} = 0.49 \text{ Btu/lbm} \cdot \text{R}$$
$$\mu_{oil} = 8.22 \text{ lbm/ft} \cdot \text{h} \qquad k_{oil} = 0.0763 \text{ Btu/h} \cdot \text{ft} \cdot °\text{F}$$

 a.   Determine the expected outlet temperatures and pressure drops when the heat exchanger is new and just put into service.
 b.   The engineer overseeing this processing line is interested in knowing the performance of the heat exchanger if the water flow rate is varied.

Construct two plots: one that shows the water and oil outlet temperatures as a function of the water flow rate and one that shows the shell-side pressure drop as a function of the water flow rate. Run the calculations for water flow rates between 50,000 and 300,000 lbm/h.

5.47 Design a STHX that will heat liquid octane from 70°F to 110°F using a 20% propylene glycol solution available at 165°F. The flow rate of both fluids is the same, at 110,000 lbm/h. The fouling factors for both fluids can be considered equal at 0.002 h·ft²·°F/Btu. Specify the following information about the heat exchanger:

   a. Number of shell and tube passes
   b. Diameter of the shell
   c. Placement of the fluids (which fluid is in the shell? in the tubes?)
   d. Type of tubes used in the shell (BWG specification)
   e. Pitch of the tubes
   f. Number of active tubes
   g. Length of the tubes
   h. Outlet temperatures of the octane and propylene glycol solution
   i. Pressure drop of each fluid in the heat exchanger

5.48 A 1-1/U counterflow PFHX has 37 active plates. The plates are made from titanium and stamped with a 30° chevron pattern. Each plate has a length of 32 in. and a width of 17 in. The plate thickness is 0.09 in. and the separation distance is 0.2 in. The hot fluid, hard water, flows at 120,000 lbm/h with an average temperature of 145°F. The cold fluid is hard water flowing at 180,000 lbm/h and an average temperature of 50°F. Determine the following:

   a. Conduction resistance between the two fluids due to the plate (h·ft²·°F/Btu). How does this compare with the fouling factor on each side of the plate?
   b. Pressure drop of the hot and cold water flows due to friction effects of the plates (psi).

5.49 In a distillery, a PFHX with a 1-1/U counterflow plate arrangement is being used to cool down a flow of pure ethanol using cold water. The plates are stamped with a 30° chevron pattern. The plate width is 248 mm and the active heat transfer area of each plate is 0.2 m². The plates are separated by a space of 1.5 mm. The ethanol enters the heat exchanger at a flow rate of 2.9 kg/s at an average temperature of 80°C. The cold water flows at a rate of 3.8 kg/s with an average temperature of 15°C. The diameter of the hot and cold ports on the heat exchanger is 125 mm. The heat exchanger has 54 plates. Determine the total pressure drop of each fluid as it passes through this heat exchanger (kPa).

5.50 Distilled water is being used to cool a 20% ethylene glycol solution. The ethylene glycol solution flows at 4.4 L/s to a 1/1-U counterflow PFHX. The ethylene glycol solution enters the heat exchanger at 32°C where it transfers heat to the distilled water. The water enters the heat exchanger at 4.1 L/s with a temperature of 15°C. The plates are made from titanium and have a 30° chevron pattern. The plates have a height of 1.5 m and a width of 0.5 m. Each plate has a thickness of 0.8 mm and the plates are spaced 4 mm apart. There are 18 active plates in the heat exchanger. The fouling factor for the ethylene glycol solution can be taken as 5×10^{-7} m² K/W. Determine the following:

   a. Outlet temperatures of both fluids (°C)
   b. Pressure drop of each fluid because of plate friction as it passes through the heat exchanger (kPa).

5.51 Hot air is used to heat cold water from 35°C to 95°C in a finned-tube heat exchanger. The water flows in the tubes at a flow rate of 2.5 kg/s. The air, at 1 atm, enters the heat exchanger at 205°C and leaves at 97°C. The overall heat transfer coefficient for the heat exchanger is 185 W/m² K. Determine the air-side area required for this heat exchanger.

5.52 Solve Problem 5.51 for the case where the tubes are not finned, resulting in an overall heat transfer coefficient of 170 W/m² K.

5.53 To increase the thermal efficiency of a gas turbine cycle, the combustion air is often preheated using the hot gases exhausting the turbine. Consider the case where this is being accomplished using a cross-flow (unmixed–unmixed) heat exchanger, as shown in Figure P5.53. Hot combustion gases exit the turbine at a mass flow rate of 51,600 lbm/h with a temperature of 850°F. Air enters the compressor at 1 atm, 70°F, with a mass flow rate of 49,200 lbm/h. The pressure ratio across the compressor is PR = 8 and the isentropic efficiency of the compressor is 85%. The overall heat transfer coefficient for the heat exchanger has been found to be $U = 17.6$ Btu/h·ft²·°F. If the combustion air is to be preheated to 700°F, determine the total heat exchange area required for this CFHX (ft²). Assume that the pressure drops through the connecting lines and the heat exchangers are negligible and that the working fluid can be modeled as air throughout the cycle.

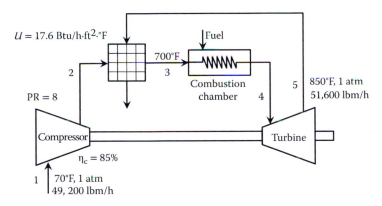

FIGURE P5.53

6

Simulation, Evaluation, and Optimization of Thermal Energy Systems

6.1 Introduction

Before the advent of modern computing technology, thermal energy systems were designed using a combination of lengthy hand calculations, graphs and plots that represent component performance, and experience. The resulting system may not have been an optimum economic design, but it met the required need. Modern engineering practice requires the engineer to consider the environmental impact of his/her designs in addition to thermal performance and cost-effectiveness. This implies that the system should be designed to meet some sort of optimum condition. Modern computing technology allows this to occur through a series of processes known as *simulation*, *evaluation*, and *optimization*.

Simulation is the development and solution of a mathematical model that predicts the operating condition of a thermal energy system. Evaluation involves the use of a simulation to study the effect of changing one or more variables in the simulation. Together, simulation and evaluation provide the design engineer with the means to study several thermal energy systems using computer software and feel confident that the selected thermal system will operate as predicted by the simulation. Optimization considers what configuration the thermal energy system should take to operate at an optimum condition. This optimum condition may be the system configuration with minimum cost, the system with maximum thermal efficiency, the system with the minimum exergy destruction, and so forth. To perform a successful optimization, the mathematical model making up the simulation must be developed first.

6.2 Thermal Energy System Simulation

The simulation problem may be stated in general terms as follows:
Given:

- Thermal energy system configuration (system specifications)
- Thermal energy system performance parameters (i.e., temperatures, flows)
- Component performance equations

Find: The operating *point* of the thermal system.
Solution: The general solution strategy is as follows:

- Write a system of n equations with n unknowns that completely describes the system. This system of equations will include fluid properties, application of appropriate fundamental laws from thermodynamics, fluid mechanics and/or heat transfer, and equations describing component performance.
- Solve the resulting system of equations. The resulting solution is the operating point of the thermal energy system for the given conditions.

The resulting set of equations is an $n \times n$ system. Therefore, computer software can be used to determine the solution. When using computer software to solve a simulation, reasonable initial guesses of the unknown variables may be required for successful convergence of a solution. This can become quite challenging as the simulations become more complex.

6.2.1 Pump and Pipe System Simulation

To begin with, a familiar example is presented. In this example, a simulation is written and solved for a pump and pipe system. Consider the following problem.

A pump and pipe system used to transfer irrigation water in a farming application is shown in Figure 6.1. The pipe system is made of 4-nom sch 40 commercial steel pipes with regular fittings. A globe valve is used in the system for flow control. With the globe valve wide opened, this system must provide a minimum of 300 gpm of water at 60°F. The pump selected to meet this requirement is a Bell & Gossett Series 80-SC $4 \times 4 \times 9\frac{1}{2}$ pump with a 9-in. impeller operating at 1150 rpm. Determine the operating point (volumetric flow rate and head) of the system.

In Chapter 4, these types of problems are solved by determining the system curve and superimposing it on a pump curve to select the pump and determine the operating point graphically. To write a simulation for this system, all component and system parameters need to be represented by equations. When formulating the $n \times n$ equation set that makes up the simulation, it is

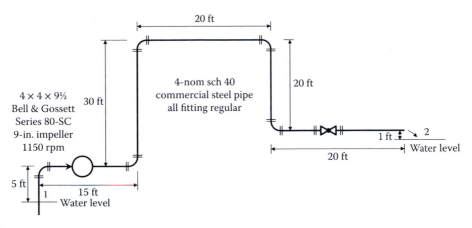

FIGURE 6.1
A pump and pipe network.

important to keep track of equations and unknowns. Once an *n×n* system is determined, it can be solved.

The pipe system, from point 1 to point 2, is represented by the conservation of energy:

$$z_1 + H_p = z_2 + \left(f\frac{L}{D} + 5K_e + K_v + K_{exit} \right)\frac{V^2}{2g} \tag{6.1}$$

This is the first equation of the simulation. Unfortunately, many of the variables are unknown at this point. To help keep track of the known and unknown values, it is helpful to build the simulation equations in tabular form as shown below. The goal is to continue adding equations until the number of equations is equal to the number of unknowns. Some of the unknown values in this first equation are easily determined. For example, the total pipe length and elevations can be found using the dimensions in the system sketch. These *easy-to-find* variables do not need to be included in the equation set. Following this strategy, the first equation in the simulation is shown below.

Equation	New Unknowns	Equations × Unknowns
$z_1 + H_p = z_2 + \left(f\frac{L}{D} + 5K_e + K_v + K_{exit} \right)\frac{V^2}{2g}$	$H_p, f, K_e, K_v, K_{exit}, V$	1×6

Notice that the elevations, pipe length, pipe diameter, and the acceleration due to gravity (*g*) are not counted as unknowns. These are values that are easily determined.

The resistance coefficients for the elbows, valve, and exit can be found in Appendix E and added to the simulation equation set.

Equation	New Unknowns	Equations × Unknowns
$K_e = 30 f_T$	f_T	2×7
$K_v = 340 f_T$		3×7
$K_{exit} = 1.0$		4×7
$f_T = 0.25 \left(\log \dfrac{\varepsilon/D}{3.7} \right)^{-2}$		5×7

Notice that one more *new* unknown is introduced in this set of equations (f_T). The absolute roughness of the pipe, ε, is not counted as an unknown because it can be easily found in Table 4.3.

The friction factor in the conservation of energy equation can be determined by any of the correlations presented in Chapter 4. Using the Swamee–Jain correlation, another set of equations can be added to the simulation as follows:

Equation	New Unknowns	Equations × Unknowns
$f = 0.25 \left[\log \left(\dfrac{\varepsilon/D}{3.7} + \dfrac{5.74}{Re^{0.9}} \right) \right]^{-2}$	Re	6×8
$Re = \dfrac{\rho VD}{\mu}$		7×8
$\dot{V} = AV$	$\dot{V}, A$	8×10
$A = \dfrac{\pi D^2}{4}$		9×10

The density and dynamic viscosity of the water are not treated as unknowns here because they can easily be evaluated because the temperature is known and the water can be treated as an incompressible substance.

At this point, there appears to be at least one more equation that needs to be added to complete the simulation. The behavior of the *system* has been fully identified with the nine equations developed to this point. What is missing is the *performance* of the pump in the system. Therefore, to complete the simulation, an equation needs to be developed that describes how the pump head varies with capacity.

The pump curve is a two-dimensional (2D) type of equation. Equations of this type can be easily fit using a variety of software packages. To determine the equation, a series of *x–y* data needs to be read from the pump performance curve. Figure 6.2 shows the result of fitting several data points from the 9-in. impeller pump curve using a spreadsheet program with curve-fitting capability.

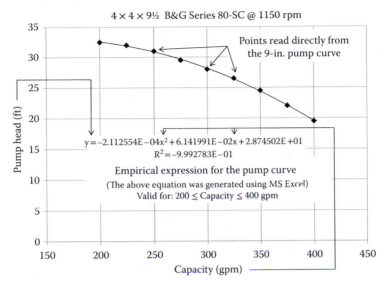

FIGURE 6.2
Determination of an empirical equation to represent the pump performance.

TABLE 6.1

Coefficients of Equation 6.2 Resulting from Curve Fitting the Pump Performance Data Read from Figure 6.1

Coefficient	Value and Unit
a_1	$-2.112554E-04$ ft/gpm^2
a_2	$6.141991E-02$ ft/gpm
a_3	$2.874502E+01$ ft

From this analysis it can be seen that the pump performance can be written as follows:

$$H_p = a_1 \dot{V}^2 + a_2 \dot{V} + a_3 \tag{6.2}$$

The coefficients resulting from the curve-fitting exercise are given in Table 6.1.

The equation developed for the pump curve is empirical. Therefore, it is very important to make certain that the coefficients are assigned the proper units. The units of the coefficients can easily be determined by considering the functional form being fit.

The pump performance curve, Equation 6.2, is the final equation needed to complete the simulation of the system.

Equation	New Unknowns	Equations × Unknowns
$H_p = a_1 \dot{V}^2 + a_2 \dot{V} + a_3$		10×10

The coefficients are not counted as unknowns since they have been determined by the curve-fitting exercise. After adding this equation, the number of equations equals the number of unknowns. This means that the simulation is complete and is ready to be solved. Solving this equation set using Engineering Equation Solver (EES) results in an operating point defined by $H_p = 27.83$ ft and $\dot{V} = 305$ gpm.

All of the variables in the simulation are determined when the system of equations is solved. It is interesting to see the other operating parameters of the system such as the Reynolds number of the flow, the velocity of the fluid in the pipe, and so forth. However, the main parameters of interest from the simulation are the capacity (gpm) and pump head (ft).

To determine the power draw and the pump efficiency, this point needs to be identified on the pump performance curve as shown in Figure 6.3. This figure indicates that the pump will draw approximately 3 hp and operate with a fairly high efficiency of about 73%. This appears to be a good pump selection for this application.

As demonstrated in this familiar example, the process of constructing a system simulation requires the mathematical representation of the performance of the thermal energy system and the components that make up the thermal energy system. For the pipes and fittings, the conservation of energy is required. For the pump, the manufacturer's performance information needs to be converted to an empirical equation through a data-fitting exercise. The simulation results provide the operating point of the pump and pipe system.

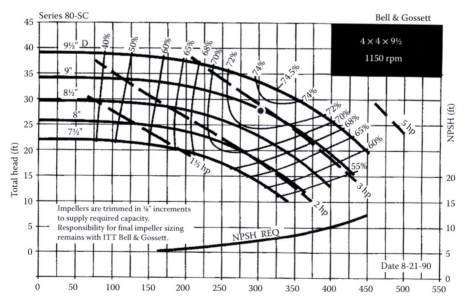

FIGURE 6.3
Determination of pump efficiency and power draw from the pump performance curve.

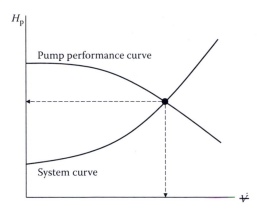

FIGURE 6.4
Graphical representation of the simulation.

A simulation of this type involves two major unknowns: the pump head and the resulting volumetric flow rate through the pipe system. The simulation equations are solving for the intersection of two 2D equations: the system curve and the pump performance curve. This solution is shown graphically in Figure 6.4. Even though this plot indicates that there are two equations and two unknowns, there are really more unknowns involved in the problem that must be determined to establish these curves. For example, the Reynolds number and friction factor must be found to determine the system curve.

Often, engineering components are three-dimensional (3D) in their performance. For example, manufacturers of refrigeration compressors report the power draw by the compressor to be a function of the condensing and evaporating temperatures in the refrigeration cycle.

$$\dot{W} = f\left(T_c, t_e\right) \tag{6.3}$$

As systems and component performance become more complex, the graphical solution is impossible to envision. The solution of the simulation set is in n-dimensional space where n is the number of equations and unknowns. However, writing the simulation equations as shown in this example will lead to a set of equations that can be solved using software.

6.2.2 Modeling Thermal Equipment

In Section 6.2.1, data read from the pump performance curve were fit to a 2D function. However, many thermal components behave three dimensionally. Consider a cooling tower being used to cool condenser water in a refrigeration plant. The water is cooled by direct contact with cool, moist atmospheric air.

A sketch of a cooling tower is shown in Figure 6.5. Manufacturers of cooling towers often present cooling tower performance data showing the *outlet water temperature* (T_{out}) as a function of the *wet bulb temperature* (T_{wb}) of the entering ambient air and the *range* (R). The range is the difference between the inlet and outlet temperatures of the water. Table 6.2 shows an example of typical manufacturer's performance data for a cooling tower such as the one shown in Figure 6.5. Referring to Table 6.2, when the wet bulb temperature is 20°C and the range is 10°C, the temperature of the water leaving the tower is 25.9°C. Therefore, the temperature of the water entering the cooling tower is 25.9 + 10 = 35.9°C.

Table 6.2 represents the real performance of the cooling tower. The data are 3D. Therefore, the mathematical representation of these data is an equation of the form

$$T_{out} = f(R, T_{wb}) \tag{6.4}$$

To convert the numerical performance data from Table 6.2 into a mathematical equation, the functional form of the equation must be determined. Most thermal equipment operates very predictably and usually does not exhibit any strange fluctuating behavior. Therefore, a simple polynomial function is usually sufficient to model the 3D performance. The order of the polynomial may be more difficult to determine. However, for most thermal components with 3D performance characteristics, the quadratic terms should suffice in modeling the performance data. A general quadratic 3D equation can be written as follows:

$$y(x,z) = a_1 + a_2 x + a_3 z + a_4 xz + a_5 x^2 + a_6 z^2 + a_7 x^2 z + a_8 xz^2 + a_9 x^2 z^2 \tag{6.5}$$

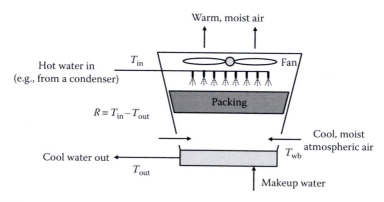

FIGURE 6.5
A cooling tower.

TABLE 6.2

Cooling Tower Performance Data: Outlet Water Temperature, T_{out} (°C) as a Function of the Inlet Air Wet Bulb Temperature, T_{wb} (°C), and the Range, $R = T_{in} - T_{out}$

	T_{wb}(°C)						
R (°C)	20	21	22	23	24	25	26
10	25.90	26.42	26.92	27.50	28.11	28.72	29.40
12	26.24	26.69	27.20	27.76	28.34	28.96	29.62
14	26.62	27.04	27.54	28.05	28.64	29.23	29.91
16	27.00	27.43	27.89	28.40	28.94	29.56	30.20
18	27.45	27.81	28.30	28.77	29.29	29.88	30.54
20	27.92	28.29	28.70	29.18	29.69	30.27	30.89
22	28.40	28.73	29.14	29.60	30.10	30.70	31.30

Using this functional form, the cooling tower performance data in Table 6.2 can be represented by the following equation:

$$T_{out} = a_1 + a_2 R + a_3 T_{wb} + a_4 R T_{wb} + a_5 R^2 + a_6 T_{wb}^2 + a_7 R^2 T_{wb} + a_8 R T_{wb}^2 + a_9 R^2 T_{wb}^2 \quad (6.6)$$

Given this functional form, the task is to find the coefficients a_1–a_9 that best represent the performance data. There are several methods that can be used to determine the coefficients. One of the easier methods to implement is the *exact-fitting* method.

6.2.2.1 Exact-Fitting Method

The idea behind the exact-fitting method is to select a set of data points and determine an equation that passes through these points *exactly*. The set of points may be all, or a subset of the manufacturer's performance data, depending on the functional form chosen to represent the data. When a data set (or subset) is fit exactly, a performance equation is written for every data point. Therefore, if the equation chosen to represent the data has n terms, then n data points are needed to determine the unknown coefficients, a_1, a_2,..., a_n in the function.

Using this reasoning and the functional form proposed in Equation 6.6, nine data points are needed to determine the coefficients a_1–a_9 for the cooling-tower example. Table 6.2 contains 49 data points. With some clever data selection, nine data points can be extracted from Table 6.2 that cover the full performance range. Table 6.3 shows the performance data of Table 6.2 with nine highlighted data points (in **bold** font) that represent the full performance range.

Using the highlighted data from Table 6.3, nine equations can be written using the functional form shown in Equation 6.6. The resulting equations are shown in Equation set 6.7.

TABLE 6.3

Cooling Tower Performance Data: Nine Selected Data Points Representing the Full Range of Performance Data in Table 6.1

R (°C)	T_{wb}(°C)						
	20	21	22	23	24	25	26
10	**25.90**	26.42	26.92	**27.50**	28.11	28.72	**29.40**
12	26.24	26.69	27.20	27.76	28.34	28.96	29.62
14	26.62	27.04	27.54	28.05	28.64	29.23	29.91
16	**27.00**	27.43	27.89	**28.40**	28.94	29.56	**30.20**
18	27.45	27.81	28.30	28.77	29.29	29.88	30.54
20	27.92	28.29	28.70	29.18	29.69	30.27	30.89
22	**28.40**	28.73	29.14	**29.60**	30.10	30.70	**31.30**

$$a_1 + 10a_2 + 20a_3 + 200a_4 + 100a_5 + 400a_6 + 2000a_7 + 4000a_8 + 40000a_9 = 25.9$$
$$a_1 + 10a_2 + 23a_3 + 230a_4 + 100a_5 + 529a_6 + 2300a_7 + 5290a_8 + 52900a_9 = 27.5$$
$$a_1 + 10a_2 + 26a_3 + 260a_4 + 100a_5 + 676a_6 + 2600a_7 + 6760a_8 + 67600a_9 = 29.4$$
$$a_1 + 16a_2 + 20a_3 + 320a_4 + 256a_5 + 400a_6 + 5120a_7 + 6400a_8 + 102400a_9 = 27.0$$
$$a_1 + 16a_2 + 23a_3 + 368a_4 + 256a_5 + 529a_6 + 5888a_7 + 8464a_8 + 135424a_9 = 28.4$$
$$a_1 + 16a_2 + 26a_3 + 416a_4 + 256a_5 + 676a_6 + 6656a_7 + 10816a_8 + 173056a_9 = 30.2$$
$$a_1 + 22a_2 + 20a_3 + 440a_4 + 484a_5 + 400a_6 + 9680a_7 + 8800a_8 + 193600a_9 = 28.4$$
$$a_1 + 22a_2 + 23a_3 + 506a_4 + 484a_5 + 529a_6 + 11132a_7 + 11638a_8 + 256036a_9 = 29.6$$
$$a_1 + 22a_2 + 26a_3 + 572a_4 + 484a_5 + 676a_6 + 12584a_7 + 14872a_8 + 327184a_9 = 31.3$$

$$(6.7)$$

In this set of equations, the units have left out for clarity. However, each number and coefficient must have a consistent set of units. This can be addressed after the coefficients are determined. The solution of this set of equations can easily be accomplished using equation solving software. Table 6.4 shows the resulting numerical coefficients for Equation 6.6 based on this exact-fitting exercise. Notice that the units of each coefficient are included in Table 6.4. Since performance equations are often empirical (resulting from a curve fit to data), it is very important to make certain that coefficients have units that are consistent with the data that were used. In this example, the coefficients must have units that make each term in Equation 6.6 dimensionally homogeneous. In other words, each term of Equation 6.6 must have units of °C.

Once the coefficients are determined, it is always a good idea to make certain that the resulting equation compares favorably to the equipment performance data. The best way to verify the equation is to compute either the deviation or the percent deviation between the data and the values computed

TABLE 6.4

Coefficients of Equation 6.6 Resulting from the Exact
Fitting of the Nine Data Points Highlighted in Table 6.3

Coefficient	Value and Unit
a_1	1.52518526723E+01°C
a_2	7.23148038585E−01
a_3	3.25925838797E−01
a_4	−5.09259142123E−02°C^{-1}
a_5	4.16667100466E−03°C^{-1}
a_6	7.40740961541E−03°C^{-1}
a_7	−4.46773371771E−10°C^{-2}
a_8	9.25925628863E−04°C^{-2}
a_9	1.10013838859E−11°C^{-3}

from the empirical equation. The deviation between the performance data
and the calculated value is

$$\Delta y = y_{data} - y_{calc} \qquad (6.8)$$

The percent deviation is defined as

$$\% \, \Delta y = 100\left(\frac{\Delta y}{y_{data}}\right) = 100\left(\frac{y_{data} - y_{calc}}{y_{data}}\right) \qquad (6.9)$$

When the y values become very small, the percent deviation calculation can
be misleading. In those cases, the deviation, Equation 6.8, is favored over
the percent deviation, Equation 6.9. Using the coefficients in Table 6.4 in
Equation 6.6, the percent deviations in water outlet temperature are calculated and displayed in the plot shown in Figure 6.6. A plot of this type is
called a *deviation plot*.

In the deviation plot shown in Figure 6.6, the nine selected data points
for the exact fit have a zero percent deviation. The other data are all within
±0.1% of the published performance data. On the basis of the percent deviation analysis, the equation developed can be said to be accurate to within
±0.1% over the entire range of the manufacturer's performance data.

In Table 6.4, a_7 and a_9 coefficients are several orders of magnitude smaller
than the rest of the coefficients. Therefore, it appears that they may be insignificant in this equation. To test this thinking, the terms containing the coefficients a_7 and a_9 can be removed from Equation 6.6 to produce the following
equation:

$$T_{out} = a_1 + a_2 R + a_3 T_{wb} + a_4 R T_{wb} + a_5 R^2 + a_6 T_{wb}^2 + a_8 R T_{wb}^2 \qquad (6.10)$$

The coefficients of this equation are the same as in Table 6.4 with a_7 and
a_9 coefficients eliminated. Table 6.5 shows the coefficients of Equation 6.10.

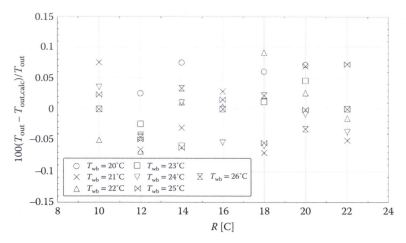

FIGURE 6.6

Deviation plot showing the percent deviation between water outlet temperatures calculated with Equation 6.6 and the manufacturer's performance data.

TABLE 6.5

Coefficients of Equation 6.10 Resulting from the Elimination of a_7 and a_9 Terms from Equation 6.6

Coefficient	Value and Unit
a_1	1.52518526723E+01°C
a_2	7.23148038585E−01
a_3	3.25925838797E−01
a_4	−5.09259142123E−02°C^{-1}
a_5	4.16667100466E−03°C^{-1}
a_6	7.40740961541E−03°C^{-1}
a_8	9.25925628863E−04°C^{-2}

To verify that the terms containing a_7 and a_9 coefficients can be eliminated, the deviation plot can be recalculated using Equation 6.10. Figure 6.7 shows the resulting deviation plot. Notice that there is no perceivable difference between Figures 6.6 and 6.7. Therefore, the equation that represents the manufacturer's performance data for the cooling tower water outlet temperature can be represented by Equation 6.10 with the coefficients from Table 6.5.

The procedure for developing empirical equations from manufacturer's performance data was applied to performance data for a cooling tower. However, the procedure is the same for a wide variety of common components in thermal energy systems.

The exact-fitting method is so popular and easy to implement that some manufacturers supply what is known as "9-point data." For example, Figure 6.8 shows a manufacturer's 9-point data for a refrigeration compressor.

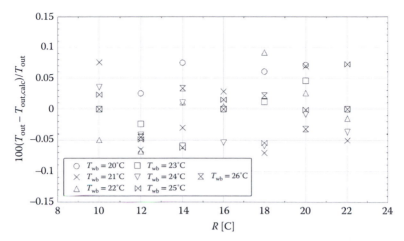

FIGURE 6.7
Deviation plot showing the percent deviation between water outlet temperatures calculated with Equation 6.10 and the manufacturer's performance data.

9-Point Data

Condensing Temp.	Rating	Evaporating Temp.		
		−30°F	**−10°F**	**+10°F**
130°F	Btu/h	500	1,031	1,864
	Watts	141	213	302
	Amps	2.07	2.51	3.16
	E.E.R.	3.54	4.84	6.17
120°F	Btu/h	542	1,080	1,923
	Watts	143	210	292
	Amps	2.08	2.49	3.08
	E.E.R.	3.78	5.14	6.59
110°F	Btu/h	583	1,135	1,958
	Watts	144	207	280
	Amps	2.09	2.47	2.99
	E.E.R.	4.04	5.49	7.00

FIGURE 6.8
An example of 9-point data for a refrigeration compressor.

From the data shown in Figure 6.8, four equations can be developed: one for the capacity the compressor is able to provide (Btu/h), one for the power input to the compressor (Watts), one for the current input to the compressor (Amps), and one for the compressor's energy efficiency ratio (E.E.R.). All of these parameters can be correlated as a function of the evaporating temperature and the condensing temperature.

$$\text{Btu/h, or Watts, or Amps, or E.E.R.} = f\left(T_{evap}, T_{cond}\right) \qquad (6.11)$$

6.2.2.2 Method of Least Squares

Another method that is commonly used to fit performance data is the method of least squares. In this method, the unknown coefficients are determined by performing an optimization. The goal is to *minimize* the sum of the squares of the residuals (SSQR) between the data and the correlated equation. In general, the function to be minimized is

$$\min \quad SSQR(a_1, a_2, \ldots, a_n) = \sum_{i=1}^{m} \left(y_{i,\text{calc}} - y_{i,\text{data}} \right)^2 \tag{6.12}$$

The function y_{calc} contains the unknown coefficients a_1–a_n. The minimization is accomplished by taking the partial derivative of the SSQR function for each unknown coefficient and setting it equal to zero.

$$\frac{\partial(SSQR)}{\partial a_k} = 0 \qquad k = 1, n \tag{6.13}$$

The resulting set of equations is an $n \times n$ system that contains the unknown coefficients. In addition, the n equations contain the performance data. The equation set that results from the partial differentiation described in Equation 6.13 can be written in matrix form as follows:

$$[A][a] = [b] \tag{6.14}$$

The [A] matrix is square ($n \times n$) and contains data. The [b] matrix is a column vector that also contains data. Solving Equation 6.14 for the column vector of unknown coefficients results in the following matrix equation:

$$[a] = [A]^{-1}[b] \tag{6.15}$$

This matrix manipulation allows for the calculation of the unknown coefficients in the column vector [a].

Applying the method of least squares to the cooling tower performance data in Table 6.2, we are interested in determining the coefficients of Equation 6.10 that minimize the SSQR. Mathematically, this can be expressed as follows:

$$\min \quad SSQR(a_1, a_2, a_3, a_4, a_5, a_6, a_8) = \sum_{i=1}^{7} \left(T_{i,\text{calc}} - T_{i,\text{data}} \right) \tag{6.16}$$

Performing the partial differentiation and setting each derivative equal to zero results in the following set of equations:

$$\frac{\partial(SSQR)}{\partial a_1} = 2\left\{ \sum_{i=1}^{n} \left[\left(a_1 + a_2 R_i + a_3 T_{\text{wb},i} + a_4 R_i T_{\text{wb},i} + a_5 R_i^2 + a_6 T_{\text{wb},i}^2 + a_8 R_i T_{\text{wb},i}^2 \right) - T_i \right] \right\} = 0$$

$$\frac{\partial(SSQR)}{\partial a_2} = 2\left\{\sum_{i=1}^{n}\left[\left(a_1 + a_2 R_i + a_3 T_{wb,i} + a_4 R_i T_{wb,i} + a_5 R_i^2 + a_6 T_{wb,i}^2 + a_8 R_i T_{wb,i}^2\right) - T_i\right]R_i\right\} = 0$$

$$\frac{\partial(SSQR)}{\partial a_3} = 2\left\{\sum_{i=1}^{n}\left[\left(a_1 + a_2 R_i + a_3 T_{wb,i} + a_4 R_i T_{wb,i} + a_5 R_i^2 + a_6 T_{wb,i}^2 + a_8 R_i T_{wb,i}^2\right) - T_i\right]T_{wb,i}\right\} = 0$$

$$\frac{\partial(SSQR)}{\partial a_4} = 2\left\{\sum_{i=1}^{n}\left[\left(a_1 + a_2 R_i + a_3 T_{wb,i} + a_4 R_i T_{wb,i} + a_5 R_i^2 + a_6 T_{wb,i}^2 + a_8 R_i T_{wb,i}^2\right) - T_i\right]R_i T_{wb,i}\right\} = 0$$

$$\frac{\partial(SSQR)}{\partial a_5} = 2\left\{\sum_{i=1}^{n}\left[\left(a_1 + a_2 R_i + a_3 T_{wb,i} + a_4 R_i T_{wb,i} + a_5 R_i^2 + a_6 T_{wb,i}^2 + a_8 R_i T_{wb,i}^2\right) - T_i\right]R_i^2\right\} = 0$$

$$\frac{\partial(SSQR)}{\partial a_6} = 2\left\{\sum_{i=1}^{n}\left[\left(a_1 + a_2 R_i + a_3 T_{wb,i} + a_4 R_i T_{wb,i} + a_5 R_i^2 + a_6 T_{wb,i}^2 + a_8 R_i T_{wb,i}^2\right) - T_i\right]T_{wb,i}^2\right\} = 0$$

$$\frac{\partial(SSQR)}{\partial a_8} = 2\left\{\sum_{i=1}^{n}\left[\left(a_1 + a_2 R_i + a_3 T_{wb,i} + a_4 R_i T_{wb,i} + a_5 R_i^2 + a_6 T_{wb,i}^2 + a_8 R_i T_{wb,i}^2\right) - T_i\right]R_i T_{wb,i}^2\right\} = 0$$

These equations are known as the *normal equations*. In this example, there are seven normal equations with seven unknowns (the a_i coefficients). Rewriting these equations in matrix form [**A**][**a**] = [**b**] results in the following:

$$\begin{bmatrix}
\sum 1 & \sum R_i & \sum T_{wb,i} & \sum R_i T_{wb,i} & \sum R_i^2 & \sum T_{wb,i}^2 & \sum R_i T_{wb,i}^2 \\
\sum R_i & \sum R_i^2 & \sum R_i T_{wb,i} & \sum R_i^2 T_{wb,i} & \sum R_i^3 & \sum R_i T_{wb,i}^2 & \sum R_i^2 T_{wb,i}^2 \\
\sum T_{wb,i} & \sum R_i T_{wb,i} & \sum T_{wb,i}^2 & \sum R_i T_{wb,i}^2 & \sum R_i^2 T_{wb,i} & \sum T_{wb,i}^3 & \sum R_i T_{wb,i}^3 \\
\sum R_i T_{wb,i} & \sum R_i^2 T_{wb,i} & \sum R_i T_{wb,i}^2 & \sum R_i^2 T_{wb,i}^2 & \sum R_i^3 T_{wb,i} & \sum R_i T_{wb,i}^3 & \sum R_i^2 T_{wb,i}^3 \\
\sum R_i^2 & \sum R_i^3 & \sum R_i^2 T_{wb,i} & \sum R_i^3 T_{wb,i} & \sum R_i^4 & \sum R_i^2 T_{wb,i}^2 & \sum R_i^3 T_{wb,i}^2 \\
\sum T_{wb,i}^2 & \sum R_i T_{wb,i}^2 & \sum T_{wb,i}^3 & \sum R_i T_{wb,i}^3 & \sum R_i^2 T_{wb,i}^2 & \sum T_{wb,i}^4 & \sum R_i T_{wb,i}^4 \\
\sum R_i T_{wb,i}^2 & \sum R_i^2 T_{wb,i}^2 & \sum R_i T_{wb,i}^3 & \sum R_i^2 T_{wb,i}^3 & \sum R_i^3 T_{wb,i}^2 & \sum R_i T_{wb,i}^4 & \sum R_i^2 T_{wb,i}^4
\end{bmatrix}
\begin{bmatrix} a_1 \\ a_2 \\ a_3 \\ a_4 \\ a_5 \\ a_6 \\ a_8 \end{bmatrix}
=
\begin{bmatrix}
\sum T_i \\
\sum T_i R_i \\
\sum T_i T_{wb,i} \\
\sum T_i R_i T_{wb,i} \\
\sum T_i R_i^2 \\
\sum T_i T_{wb,i}^2 \\
\sum T_i R_i T_{wb,i}^2
\end{bmatrix}$$

The summations in the [**A**] matrix and [**b**] vector can be calculated using *all* of the performance data shown in Table 6.2. The result is a matrix equation where the [**A**] matrix and [**b**] vector are made up completely of numbers, and the [**a**] vector contains the unknown coefficients. When using the least-squares technique, all the performance data can be included in the analysis. This is an advantage over the exact-fitting method. However, the trade-off is that the calculation becomes more complex because of the additional burden of calculating all of the summation terms in the [**A**] and [**b**] matrices. Although forming these matrices requires some calculation, it should be noted in this example that the [**A**] matrix is symmetric. Therefore, only the diagonal and either the upper or lower triangle need to be calculated. Using the matrix algebra shown in Equation 6.15, the coefficients of Equation 6.10 can be found. Table 6.6 shows the resulting coefficients.

Interestingly, the least-squares analysis reveals that the term associated with the a_8 coefficient appears to be insignificant. Therefore, the empirical equation resulting from method of least squares is

$$T_{out} = a_1 + a_2 R + a_3 T_{wb} + a_4 R T_{wb} + a_5 R^2 + a_6 T_{wb}^2 \tag{6.17}$$

The coefficients a_1 through a_6 in Equation 6.17 are given in Table 6.6. Figure 6.9 shows the deviation plot comparing the correlated equation, Equation 6.17, to the manufacturer's performance data for the cooling tower. As Figure 6.9

TABLE 6.6

Coefficients of Equation 6.10 Resulting from a Least-Squares Analysis of the Cooling Tower Performance Data Presented in Table 6.2

Coefficient	Value and Unit
a_1	2.36882920039E+01°C
a_2	2.34938807586E−01
a_3	−4.11792261455E−01
a_4	−8.20718008592E−03°C^{-1}
a_5	4.16879251696E−03°C^{-1}
a_6	2.33826844768E−02°C^{-1}
a_8	−1.96352735633E−09°C^{-2}

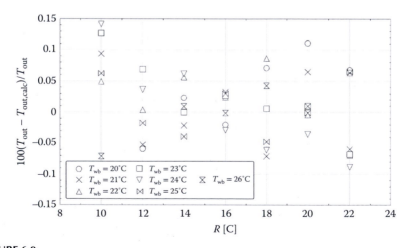

FIGURE 6.9
Deviation plot showing the percent deviation between water outlet temperatures calculated with Equation 6.17 and the manufacturer's performance data.

indicates, all the performance data are represented within ±0.15%. The results are very similar to the exact-fitting method.

Spreadsheet programs are convenient to use when implementing the method of least squares because they have the capability to do matrix algebra. Other software packages have matrix manipulation capability as well.

An alternative way to implement the method of least squares is to use an equation solver that has the ability to perform a numerical minimization of Equation 6.12. EES has this capability and is fairly easy to exploit to conduct a numerical least-squares analysis. In subsequent examples and applications discussed in this chapter, the implementation and results of Min/Max studies will be discussed. An explanation on how to use the Min/Max feature in EES is given in Appendix F.

To use EES for least-squares analysis, the first thing that needs to be done is to build a Lookup Table that contains the performance data. For the cooling tower example that is being considered in this section, *all* the performance data from Table 6.2 can be entered into a Lookup Table. The resulting Lookup Table is shown in Figure 6.10. Notice that the table is named "Data" (refer to the tab at the upper left portion of the table). Row 1 of the table contains the wet bulb temperature, T_{wb}, and Column 1 of the table contains the range, R. The outlet temperatures are contained in Rows and Columns 2 through 8. In other words, the EES Lookup Table looks exactly like the performance data shown in Table 6.2. Once the data are entered into the Lookup Table, they are available for access from the EES Equation Window using the LOOKUP command.

In the Equation Window, the goal is to calculate the SSQR defined in Equation 6.12. For purposes of comparison, this numerical least-squares method is used to determine the coefficients of the quadratic 3D polynomial defined by Equation 6.6 for the cooling tower. This way, a direct comparison to the result of the exact-fitting method and the conventional least-squares method can be made.

	Range	WBT1	WBT2	WBT3	WBT4	WBT5	WBT6	WBT7
Row 1		20.00	21.00	22.00	23.00	24.00	25.00	26.00
Row 2	10	25.90	26.42	26.92	27.50	28.11	28.74	29.40
Row 3	12	26.24	26.69	27.20	27.76	28.34	28.96	29.62
Row 4	14	26.62	27.04	27.54	28.05	28.64	29.23	29.91
Row 5	16	27.00	27.43	27.89	28.40	28.94	29.56	30.20
Row 6	18	27.45	27.81	28.30	28.77	29.29	29.88	30.54
Row 7	20	27.92	28.29	28.70	29.18	29.69	30.27	30.89
Row 8	22	28.40	28.73	29.14	29.60	30.10	30.70	31.30

FIGURE 6.10
EES Lookup Table containing the cooling tower performance data from Table 6.2.

The following EES code is used to calculate the SSQR:

```
DUPLICATE j = 2,8
  WBT[j] = LOOKUP('Data',1,j)
  RNG[j] = LOOKUP('Data',j,1)

    DUPLICATE i = 2,8
      T[i,j] = LOOKUP('Data',i,j)
      T_calc[i,j] = a[1] + a[2]*RNG[i] + a[3]*WBT[j]           &
          + a[4]*RNG[i]*WBT[j] + a[5]*RNG[i]^2                 &
          + a[6]*WBT[j]^2 + a[7]*RNG[i]^2*WBT[j]               &
          + a[8]*RNG[i]*WBT[j]^2 + a[9]*RNG[i]^2*WBT[j]^2
      DELTA[i,j] = T[i,j] - T_calc[i,j]
      DEV[i,j] = 100*DELTA[i,j]/T[i,j]
      SQDELTA[i,j] = DELTA[i,j]^2
    END

  SSQ[j] = SUM(SQDELTA[k,j],k=2,8)

END

SSQR = SUM(SSQ[k],k = 2,8)
```

Notice how the performance data from the Lookup Table are being accessed using the LOOKUP function. This relatively compact set of equations is actually 267 equations with 276 unknowns. This indicates that there are $276 - 267 = 9$ degrees of freedom. These 9 degrees of freedom, or independent variables, are the unknown coefficients, a[1] through a[9]. Figure 6.11 shows the result of using the Min/Max capability in EES (see Appendix F), to determine the coefficients a[1] through a[9] and the SSQR.

The coefficients resulting from this analysis are summarized in Table 6.7. These coefficients are different compared to the values resulting from the exact-fitting method and conventional least-squares examples. However, the deviation plot for the numerically derived least-squares equation, shown in Figure 6.12, indicates that all the data are still represented within ±0.1%.

FIGURE 6.11
Results of an EES Min/Max study to determine the coefficients of Equation 6.6 and the SSQR.

TABLE 6.7

Coefficients of Equation 6.6 Resulting from the Numerical
Least-Squares Analysis of the Data Points in Table 6.2

Coefficient	Value and Unit
a_1	1.331892545E+01°C
a_2	1.067193129E+00
a_3	4.812700824E−01
a_4	−7.904524932E−02°C^{-1}
a_5	−7.396429073E−03°C^{-1}
a_6	4.340913100E−03°C^{-1}
a_7	9.504400026E−04°C^{-2}
a_8	1.491490802E−03°C^{-2}
a_9	−1.922206303E−05°C^{-3}

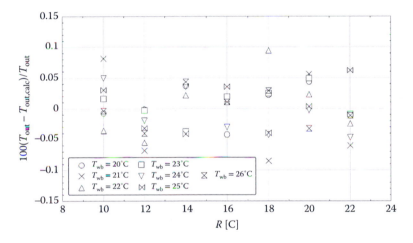

FIGURE 6.12
Deviation plot showing the percent deviation between water outlet temperatures calculated
with Equation 6.6 using the coefficients in Table 6.7 and the manufacturer's performance data.

This should be expected because the minimized value of the SSQR is very
small (0.006246). Therefore, the coefficients in Table 6.7 are as reliable as those
in Table 6.4 or 6.6, even though they are not the same. Any of these sets of
coefficients sufficiently represents the manufacturer's performance data for
the cooling tower.

The method of least squares has several advantages over exact fitting.

- All of the performance data of the component being modeled can be
  used. This can be an advantage when it is difficult to find a representative subset of data for the exact-fitting method.

- The functional form is not limited to a 3D quadratic polynomial. Virtually any functional form can be used when a numerical minimization of the SSQR is being conducted. However, as functions become more and more complicated, the minimization algorithm may fail to find an optimum solution.

6.2.3 Simulation Example: Simulation of an Air Conditioning System

Section 6.2.1 showed how simulation techniques can be used to determine the operating point of a pump and pipe network. In this section, a more complex thermal energy system is simulated; an air conditioning system. In this system, dry air at 28°C with a mass flow rate of 4 kg/s flows through a counterflow heat exchanger as shown in Figure 6.13. Cold water enters the other side of the heat exchanger at 6°C. The UA product for the heat exchanger is 7 kW/K. The suction pressure entering the pump is 80 kPa (gage) and the pressure of the water leaving the heat exchanger is 90 kPa (gage). The performance data for the pump have been correlated to the following function:

$$P_2 - P_1 = a_1 + a_2 \dot{m}_w^2 \tag{6.18}$$

In Equation 6.18, the constants are a_1 = 120 kPa and a_2 = −15.4 kPa·s²/kg². These coefficients result in a pressure drop expressed in kPa. Clearly, the mass flow rate of the water must be in kg/s.

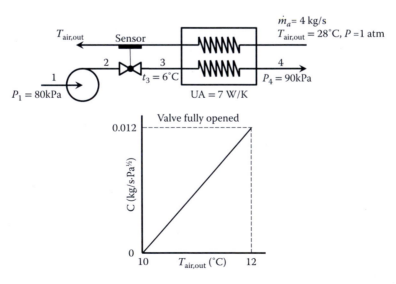

FIGURE 6.13
An air conditioning system.

The heat exchanger manufacturer has provided data that show how the water pressure drop varies with the mass flow rate. These data have been correlated to the following empirical function:

$$P_3 - P_4 = a_3 \dot{m}_w^2 \tag{6.19}$$

In this equation, a_3 = 9.26 kPa·s^2/kg^2 and the water flow rate is in kg/s. The resulting pressure drop calculated with this equation is in kPa.

The air temperature leaving the heat exchanger is used to control the control valve (a globe valve). A sensor placed on the air outlet line senses the air temperature and adjusts the valve to maintain the air outlet temperature somewhere between 10°C and 12°C. The pressure drop through the valve is related to the water flow rate by the following empirical relationship:

$$\dot{m}_w = C\sqrt{P_2 - P_3} \tag{6.20}$$

In Equation 6.20, C has units of kg/s·Pa$^{1/2}$. Therefore, the pressure drop in this equation must be expressed in Pa. C is not a constant. It is a function of the air outlet temperature. The valve manufacturer's performance data for the control valve is shown in the plot accompanying the system sketch in Figure 6.13. The linear relationship shown on this plot can be represented by the following equation:

$$C = a_4 T_{air,out} + a_5 \tag{6.21}$$

In Equation 6.21, a_4 = 0.006 kg/s·Pa$^{1/2}$·°C and a_5 = −0.06 kg/s·Pa$^{1/2}$.

The task is to simulate this thermal energy system and determine the water mass flow rate, the outlet temperatures of the fluids leaving the heat exchanger, the heat transfer rate between the two fluids, the pressures on each side of the control valve, and the value of C for the control valve.

The component performance equations, extracted from manufacturer's performance data, are summarized in Equations 6.18 through 6.21. To fully describe the air conditioning system's operating point, the counterflow heat exchanger needs to be modeled. Using the logarithmic mean temperature difference (LMTD) method, the heat exchanger model is given by

$$\dot{Q}_{HX} = \dot{m}_a c_{pa} \left(T_{air,in} - T_{air,out} \right)$$
$$\dot{Q}_{HX} = \dot{m}_w c_{pw} (t_4 - t_3) \tag{6.22}$$
$$\dot{Q}_{HX} = UA\,(LMTD)$$

The LMTD, written in the equation-solver-friendly version is

$$\Delta T_{out} = \Delta T_{in} \exp\left[\frac{\Delta T_{out} - \Delta T_{in}}{LMTD} \right] \tag{6.23}$$

For a counterflow heat exchanger, the temperature differences are defined by

$$\Delta T_{in} = T_{air,in} - t_4$$
$$\Delta T_{out} = T_{air,out} - t_3 \qquad (6.24)$$

Using the strategy developed in Section 6.2.1, the component performance equations and the heat exchanger model can be formulated into a simulation as shown in Table 6.8. The simulation indicates that there are 14 unknowns. Among these unknowns are the desired operating parameters of the air conditioning system. Using EES, the results of the simulation are shown in Figure 6.14. In this figure, the calculated values from the simulation are shown in boxes.

The simulation results tell the engineer much about the operating point of the thermal energy system. For example, with the equipment specified, a flow rate of 1.85 kg/s of water at 6°C is needed to cool the air to 11.9°C. In addition, the heat transfer rate between the water and air is 64.78 kW. The solution indicates that the control valve is operating in a nearly wide-opened condition as evidenced by the air outlet temperature and the value of the valve coefficient, C (see Figure 6.13).

TABLE 6.8

Simulation Equations for the Air Conditioning System Shown in Figure 6.13

Component	Equation	New Unknowns	Equations × Unknowns
Pump	$P_2 - P_1 = a_1 + a_2 \dot{m}_w^2$	$P_2, \dot{m}_w$	1×2
Cooling coil	$P_3 - P_4 = a_3 \dot{m}_w^2$	P_3	2×3
	$\dot{Q}_{HX} = \dot{m}_a c_{pa}(T_{air,in} - T_{air,out})$	$\dot{Q}_{HX}, T_{air,out}, c_{pa}$	3×6
	$\dot{Q}_{HX} = \dot{m}_w c_{pw}(t_4 - t_3)$	t_4, c_{pw}	4×8
	$\dot{Q}_{HX} = UA(LMTD)$	LMTD	5×9
	$\Delta T_{out} = \Delta T_{in} \exp\left[\dfrac{\Delta T_{out} - \Delta T_{in}}{LMTD}\right]$	$\Delta T_{in}, \Delta T_{out}$	6×11
	$\Delta T_{in} = T_{air,in} - t_4$		7×11
	$\Delta T_{out} = T_{air,out} - t_3$		8×11
Valve	$\dot{m}_w = C\sqrt{P_2 - P_3}$	C	9×12
	$C = a_4 T_{air,out} + a_5$		10×12
Properties	$c_{pa} = c_p(T_{a,avg}, P_{atm})$	$T_{a,avg}$	11×13
	$T_{a,avg} = (T_{air,in} + T_{air,out})/2$		12×13
	$c_{pw} = c_p(t_{w,avg}, x = 0)$	$t_{w,avg}$	13×14
	$t_{w,avg} = (t_3 + t_4)/2$		14×14

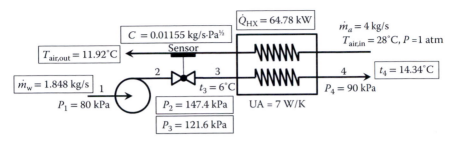

FIGURE 6.14
Operating conditions of the air conditioning system predicted by solving the system simulation equations shown in Table 6.8.

6.2.4 Advantages and Pitfalls of Thermal Energy System Simulation

The advent of modern computing technology has significantly changed the way thermal energy system design and analysis is done. Modern software has allowed the engineer to completely model the thermal energy system to any level of complexity he/she desires. The results of the simulation reveal the system operating conditions to a level of certainty consistent with the assumptions made in building the simulation equation set. The more complex the model becomes, the more realistic the results should be. However, the drawback with complex systems of equations is that they are more difficult to solve.

It is important for the engineer to understand the numerical algorithm used by the software as it seeks a solution to the simulation equations. In particular, it is important to know how the software *seeds* its solution. Any numerical solution of an $n \times n$ system of equations requires an initial guess for the unknown variables. This initial guess seeds the method. The engineer must be familiar with how these initial guesses are established with the software being used. As an example, EES, which is used extensively in this book, defaults the initial guess of all unknown variables to a value of "1." Although this may not cause a convergence problem with simple systems of equations, it can cause very significant convergence problems when thermophysical properties are involved. To alleviate this issue, it is worth the time to do some quick calculations and determine reasonable values for some of the unknown variables and use them to seed the numerical algorithm. It can be very frustrating knowing that the simulation set is correct but the numerical algorithm will not converge. Wise selection of the initial guesses can help alleviate this frustration.

Even with wise initial guesses for the seed of the numerical algorithm, the simulation set may still not converge. When this happens, the following actions are suggested:

- Take as much of the complexity out of the equation set as possible. For example, consider temporarily specifying an unknown variable and see if the equation set can be solved. This will reveal many things. If the answers seem reasonable, they can be used as initial guesses as the complexity is built back into the simulation.

- One of the more common errors in building a complex equation set is overlooking the unit analysis. The units *must* work out correctly. Make sure that *all* the equations in the simulation set are dimensionally homogeneous. Make sure that proper unit conversions are applied. An equation set that has inconsistent units is incorrect and unable to contribute to a meaningful solution for the simulation.

- When dealing with heat exchangers, there will be several temperature differences established. Make sure that these temperatures' differences are consistent with the heat exchanger model as discussed in Chapter 5. For example, a heat exchanger with single-phase fluids can be modeled using the LMTD method by the following equation set:

$$\dot{Q} = \dot{m}_h c_{ph} \left(T_{hi} - T_{ho} \right)$$
$$\dot{Q} = \dot{m}_c c_{pc} \left(t_{co} - t_{ci} \right) \qquad (6.25)$$
$$\dot{Q} = UA\,(\text{LMTD})$$

In this equation set, the subscript "h" represents the hot fluid and "c" the cold fluid. The temperature differences of the hot and cold fluids must be arranged such that the heat transfer rate has the same algebraic sign for both fluids. In addition, the temperature differences needed to define the LMTD need to be written correctly.

- Make sure that the performance equations developed for various components are being applied within the region of validity of the empirical equation that you are using. If an empirical equation is being used outside the bounds of its validity, there is no guarantee that it represents the component's performance. Although the equation may extrapolate smoothly, this does not mean it follows the component behavior. It is good practice to avoid any extrapolation of empirical equations. It is also good practice to avoid extrapolation of thermophysical property formulations used in the simulation.

These are just a few suggestions that may be helpful in arriving at the successful convergence of a complex simulation. Although not very reassuring, it is also a good idea to adopt the credo of the Apollo 13 ground team: *Failure is not an option.*

6.3 Thermal Energy System Evaluation

Simulation is a first step to a much more design-oriented activity called *evaluation*. Once a simulation is developed and successfully solved, the engineer can use the simulation to conduct *parametric studies*. The result of a parametric study tells the engineer how the system will behave if one or more of its operating conditions are changed.

Consider the air conditioning system that was simulated in Section 6.2.3. Without changing the equipment, there are several meaningful evaluations that can be conducted once the original simulation is complete. For example, what would happen to the outlet air temperature and the water flow rate if

- The inlet air temperature changes?
- The air flow rate changes?
- The water temperature entering the heat exchanger changes?

These three parametric studies can be easily conducted using an equation solver. Once the parametric table is solved, the results can be plotted to gain a visual understanding of how the system will respond to the parametric variable(s). However, before embarking on these three parametric studies, special attention needs to be given to the water flow rate. In Section 6.2.3, the solution to the simulation revealed that the control valve was nearly wide opened. Since the control valve is operating over such a narrow range of outlet air temperatures, it is entirely possible that when the parameters discussed above are varied, the valve may end up in a wide-opened state. There are several ways to accomplish this in a computer program. In EES, one possibility is to write a function that contains the logic required to turn the valve on or off. Following is an EES function that will accomplish this:

```
FUNCTION C_valve(T)
  a_4 = 0.006[kg/s-C-Pa^(1/2)]
  a_5 = -0.06[kg/s-Pa^(1/2)]
  C_valve = a_4*T + a_5
  IF (T < 10) THEN C_valve = 0[kg/s-Pa^(1/2)] "valve is closed"
  IF (T > 12) THEN C_valve = 0.012[kg/s-Pa^(1/2)] "valve fully
     opened"
END
```

This function can be called from the simulation equations to determine the value of C for the valve. Incorporating this function, the results of the three parametric studies in the bullet list above are shown in Figures 6.15 through 6.17.

Figures 6.15 through 6.17 indicate that over the range of the parameter being varied, the control valve ends up in a wide-opened position. This is indicated by the mass flow rate of the water suddenly becoming constant. This behavior influences the outlet temperature of the air. The control valve in this system was obviously selected to maintain outlet air temperatures between 10°C and 12°C. The results of the parametric studies tell the engineer about the operating limits to maintain the desired range of outlet temperatures. For example, Figure 6.16 reveals that if the air flow rate is higher than about 4.1 kg/s, the control valve ends up wide opened and the outlet air temperature increases. If the application requires strict control of the air outlet temperature, these parametric studies are of great importance to the

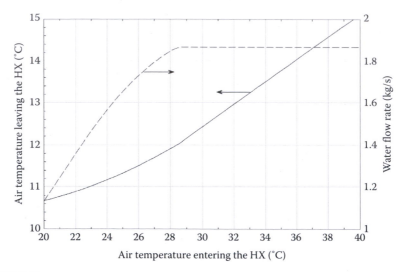

FIGURE 6.15
Outlet air temperature and water flow rate as a function of the inlet air temperature for the air conditioning system shown in Figure 6.13.

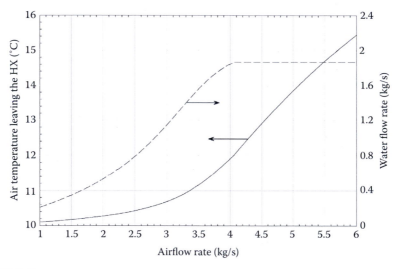

FIGURE 6.16
Outlet air temperature and water flow rate as a function of the airflow rate for the air conditioning system shown in Figure 6.13.

engineer. If the air outlet temperature can exceed 12°C, the parametric studies are still useful in predicting how the system will respond.

This section shows that the evaluation of thermal energy systems requires a successful simulation. In some instances, the simulation equations need to be modified to allow for limiting behavior of equipment (e.g., the control

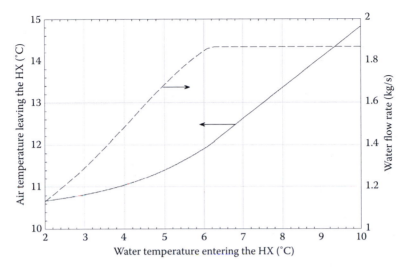

FIGURE 6.17

Outlet air temperature and water flow rate as a function of the water temperature entering the heat exchanger for the air conditioning system shown in Figure 6.13.

valve in the air conditioning example). The evaluation allows the engineer to perform a series of *what if* scenarios to predict how the system will respond.

6.4 Thermal Energy System Optimization

Optimization is the process of finding conditions that give a minimum or maximum value of a function. Optimization has always been an expected role of engineers. However, on small projects, the cost of engineering time may not justify an optimization analysis. As the system grows in complexity, the optimization of the system becomes more difficult. In these cases, it may be possible to consider optimization of subsystems and then combine these optimized subsystems. This approach, however, may not lead to the true system optimum design.

In thermal system design, the engineer is interested in minimizing or maximizing a system parameter. For example, minimizing total system cost, minimizing system irreversibility, minimizing pressure drop, maximizing profit, or maximizing thermal efficiency. Component modeling and system simulation are preliminary steps to optimizing thermal energy systems.

6.4.1 Mathematical Statement of Optimization

Optimization is the process of finding conditions that result in the minimum or maximum value of a function. In the purest mathematical sense, optimization refers to the *minimization* of a function. A function can always be maximized by minimizing the negative of the function. The function minimized is called the *objective function*. Sometimes the objective function is referred to as the *cost function* since it is often written to represent total system cost. The variables involved in the objective function are called the *design variables*.

The formal mathematical statement of optimization is written as follows:

$$\text{minimize} \qquad y(x_1, x_2, \ldots, x_n)$$

$$\text{subject to } \Phi_i(x_1, x_2, \ldots, x_n) = 0 \quad i = 1, 2, \ldots, p \qquad (6.26)$$

$$\Psi_j(x_1, x_2, \ldots, x_n) \leq 0 \quad j = 1, 2, \ldots, q$$

The Φ_i functions are known as *equality constraints* and the Ψ_j functions are called *inequality constraints*. The variables x_1 through x_n are known as the *design variables*. In the optimization problem, the values being sought are the design variables x_1 through x_n that minimize the objective function. Notice that the objective function and the constraints are functions of only design variables.

EXAMPLE 6.1

Formulate the optimization statement required to design a minimum cost cylindrical storage tank, closed at both ends, to contain a specified volume. The cost is found to depend directly on the area of the material used to construct the tank.

SOLUTION: Since the cost is directly proportional to the area of the material used, the objective function is written to represent the total surface area of the enclosed cylindrical tank. Figure E6.1 shows a sketch of the tank and the primary dimensions, r, the tank radius, and H, the tank height.

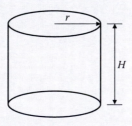

FIGURE E6.1

The total surface area of the tank is given by

$$A = 2\pi rH + 2\pi r^2$$

The design variables are r and H. There is one equality constraint in this problem: the tank must have a specified volume. The tank volume is given by,

$$\forall = \pi r^2 H$$

Now, the formal mathematical statement of this optimization problem can be written:

$$\text{minimize} \quad A(r, H) = 2\pi r H + 2\pi r^2$$
$$\text{subject to} \quad \Phi(r, H) = \pi r^2 H - \mathcal{V} = 0$$

Notice that the constraint is written such that it is exactly equal to zero. The constraint equation also contains the design variables r and H. The tank volume is not a variable. It is a specified constant.

6.4.2 Closed-Form Solution of the Optimization Problem

Consider the cylindrical tank problem in Example 6.1. The formal optimization statement was found to be

$$\text{minimize} \quad A(r, H) = 2\pi r H + 2\pi r^2$$
$$\text{subject to} \quad \Phi(r, H) = \pi r^2 H - \mathcal{V} = 0 \tag{6.27}$$

This problem can be converted to an *unconstrained optimization* by using the single constraint to eliminate one of the design variables from the problem. For example, in Equation set 6.27, the constraint can be solved for the tank height, H, as follows:

$$H = \frac{\mathcal{V}}{\pi r^2} \tag{6.28}$$

Equation 6.28 can be substituted into the objective function to give

$$A = 2\pi \left(\frac{\mathcal{V}}{\pi r} + r^2 \right) \tag{6.29}$$

Notice that Equation 6.29 is only a function of one variable, r. The minimum value of the tank surface area, A, can be found by taking the derivative of A with respect to r and setting it equal to zero. This procedure gives

$$r^* = \left(\frac{\mathcal{V}}{2\pi} \right)^{1/3} \tag{6.30}$$

In Equation 6.30, an asterisk is used to indicate an optimum value. This is consistent nomenclature found in optimization literature. Once r^* is found, the value of H^* can be found from Equation 6.28.

$$H^* = \frac{\mathcal{V}}{\pi (r^*)^2} = \frac{\mathcal{V}}{\pi} \left(\frac{2\pi}{\mathcal{V}} \right)^{2/3} = \left(\frac{4\mathcal{V}}{\pi} \right)^{1/3} \tag{6.31}$$

Notice that the *second derivative* of A with respect to r is

$$\frac{d^2 A}{dr^2} = \frac{2\Psi}{\pi r^3} + r > 0 \tag{6.32}$$

Equation 6.32 is an indication that the values of r^* and H^* produce a *minimum* value of A since the second derivative is positive.

Solution of unconstrained optimization problems using the method shown here is easily done for simple objective functions and equality constraints. However, as the optimization problem becomes more complex and multidimensional, the method may be more difficult to implement. Although the mathematics is sound, the implementation may be too complex. Section 6.4.3 presents a mathematical theorem that can be applied to more complex optimization problems.

6.4.3 Method of Lagrange Multipliers

As optimization problems become multidimensional and more complex, they cannot be solved in a simple closed-form solution as shown in Section 6.4.2. However, a method exists that can be used to solve more complex optimization problems. The method is known as the *Method of Lagrange Multipliers*. Implementation of this method relies on a mathematical theorem.

Consider the problem of minimizing an objective function $f(\mathbf{x})$ subject to equality constraint functions $\Phi_i(\mathbf{x})$ where $i = 1, 2, ..., p$. Let $\mathbf{x}^*$ be a solution that represents a local minimum for the problem. For these conditions, the *Lagrange Multiplier Theorem* states that there exist Lagrange multipliers, λ_j where $j = 1, 2, ..., p$ such that

$$\nabla L(\mathbf{x}^*, \lambda^*) = 0$$

where

$$L(\mathbf{x}, \lambda) = f(\mathbf{x}) + \sum_{j=1}^{p} \lambda_j \Phi_j(\mathbf{x}).$$

The function L is called the *Lagrange function* or the *Lagrangian*.

To demonstrate how this theorem can be applied to an optimization problem, consider the cylindrical tank design problem defined by Example 6.1. The optimization problem considered was to minimize the surface area (and thus the cost) of a cylindrical storage tank. The formal optimization problem is given by Equation set 6.27. To solve the optimization problem using the method of Lagrange multipliers, the Lagrange function needs to be constructed. Applying the Lagrange Multiplier Theorem, the Lagrange function for this problem is given by,

$$L = 2\pi r H + 2\pi r^2 + \lambda \left(\pi r^2 H - \Psi \right) \tag{6.33}$$

There is only one equality constraint in the tank optimization problem. Therefore there is only one Lagrange multiplier. According to the Lagrange Multiplier Theorem, the design variables and Lagrange multiplier are determined by solving the set of equations that arise from setting the gradient of the Lagrange function to zero. For the tank optimization problem, the gradient of the Lagrange function is,

$$\nabla L(r,H,\lambda)=0 \quad \rightarrow \quad \begin{aligned} \frac{\partial L}{\partial r} &= 2\pi H + 4\pi r + 2\pi\lambda rH = 0 \\ \frac{\partial L}{\partial H} &= 2\pi r + \lambda\pi r^2 = 0 \\ \frac{\partial L}{\partial \lambda} &= \pi r^2 H - \Psi = 0 \end{aligned} \tag{6.34}$$

These three equations can be simplified to

$$\begin{aligned} H + 2r + \lambda rH &= 0 \\ 2r + \lambda r^2 &= 0 \\ \pi r^2 H - \Psi &= 0 \end{aligned} \tag{6.35}$$

From the second equation in Equation set 6.35, it can be seen that

$$\lambda^* = -\frac{2}{r^*}. \tag{6.36}$$

Substituting Equation 6.36 into the first equation in Equation set 6.35 gives

$$H^* = 2r^* \tag{6.37}$$

From the third equation in Equation set 6.35,

$$H^* = \frac{\Psi}{\pi(r^*)^2} \tag{6.38}$$

Substituting Equation 6.38 into Equation 6.37 and solving for r gives

$$r^* = \left(\frac{\Psi}{2\pi}\right)^{1/3} \tag{6.39}$$

Finally, substituting Equation 6.39 into Equation 6.37 gives

$$H^* = 2r^* = 2\left(\frac{\Psi}{2\pi}\right)^{1/3} = \left(\frac{4\Psi}{\pi}\right)^{1/3} \tag{6.40}$$

Notice that Equations 6.39 and 6.40 are the same results as developed in Equations 6.30 and 6.31. In addition to the optimum design variables, r^* and H^*, the Lagrange multiplier can also be found. From Equation 6.36

$$\lambda^* = -\frac{2}{r^*} = -2\left(\frac{2\pi}{V}\right)^{1/3} = -\left(\frac{16\pi}{V}\right)^{1/3} \qquad (6.41)$$

The Lagrange multiplier is useful in what is known as *postoptimality stud-ies*, which is discussed later in this section. The application of the Lagrange Multiplier Theorem allows for the solution of complex optimization prob-lems that contain several design variables and equality constraints.

EXAMPLE 6.2

Circular ducts are being used to deliver air to different rooms in an air condi-tioning system as shown in Figure E6.2 (top view). A total volumetric flow rate of 6000 cfm enters the main duct as shown in the figure. As air is drawn off the main duct and delivered to different rooms, the diameter of the main duct is reduced to keep the velocity of the air at an appropriate speed for proper distribution into the room. A total of 850 ft² of sheet metal is available to build the main duct. For the duct system shown, it can be assumed that the fric-tion factor is constant at $f = 0.02$. The specific weight of the air can also be assumed constant at 0.075 lbf/ft³. The lengths of the main duct sections (50, 60, 65, and 40 ft) are fixed by the construction of the space where the duct is to be installed. Using the Method of Lagrange Multipliers, determine the diameters of the main air duct, D_1, D_2, D_3, and D_4 such that the pressure drop between points A and B is minimized.

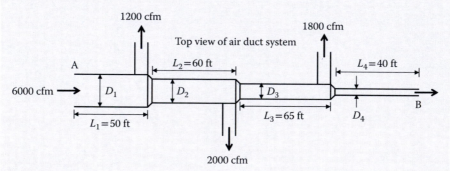

FIGURE E6.2

SOLUTION: The first step is to construct the objective function. The goal is to minimize the pressure drop from point A to B in the system. Therefore, the objective function will be of the form,

$$\text{min} \quad \rightarrow \quad \Delta P_{AB} = \Delta P_1 + \Delta P_2 + \Delta P_3 + \Delta P_4$$

The pressure drops in the individual sections of the main duct can be deter-mined from the conservation of energy applied between the beginning and the end of the section. Neglecting minor losses because of the decreasing diameter from section to section, the individual pressure drops can be written as follows:

$$\frac{\Delta P_{AB}}{\gamma} = \frac{\Delta P_1}{\gamma} + \frac{\Delta P_2}{\gamma} + \frac{\Delta P_3}{\gamma} + \frac{\Delta P_4}{\gamma}$$

$$\frac{\Delta P_{AB}}{\gamma} = f\frac{L_1}{D_1}\frac{V_1^2}{2g} + f\frac{L_2}{D_2}\frac{V_2^2}{2g} + f\frac{L_3}{D_3}\frac{V_3^2}{2g} + f\frac{L_4}{D_4}\frac{V_4^2}{2g}$$

The velocity of the air in each section of the main duct is related to the diameter of the duct and the volumetric flow rate by

$$V = \frac{\dot{V}}{A} = \frac{4\dot{V}}{\pi D^2} \qquad \therefore \quad V^2 = \frac{16\dot{V}^2}{\pi^2 D^4}$$

Using this expression, the individual pressure losses can be written as

$$\Delta P = \gamma f \frac{L}{D}\frac{V^2}{2g} = \left(\frac{8\gamma f L \dot{V}^2}{\pi^2 g}\right)\frac{1}{D^5} = \frac{K}{D^5}$$

In this equation, K is a constant that can be calculated for each section of the main duct. Therefore, the objective function can be written as

$$\text{min} \quad \rightarrow \quad \Delta P_{AB} = \frac{K_1}{D_1^5} + \frac{K_2}{D_2^5} + \frac{K_3}{D_3^5} + \frac{K_4}{D_4^5}$$

From this equation, it can be seen that there are four *design variables*, D_1, D_2, D_3, and D_4. There is one equality constraint in this problem: the total number of square feet of sheet metal available to construct the main duct. This constraint can be written as

$$\pi D_1 L_1 + \pi D_2 L_2 + \pi D_3 L_3 + \pi D_4 L_4 = 850 \text{ ft}^2$$

To use the Lagrange Multiplier Theorem, the optimization problem must be written in its formal mathematical form specified by the Equation 6.27:

$$\text{min} \quad \rightarrow \quad \Delta P_{AB} = \frac{K_1}{D_1^5} + \frac{K_2}{D_2^5} + \frac{K_3}{D_3^5} + \frac{K_4}{D_4^5}$$

subject to $\quad \pi D_1 L_1 + \pi D_2 L_2 + \pi D_3 L_3 + \pi D_4 L_4 - 850 \text{ ft}^2 = 0$

The Lagrange function can now be formed as follows:

$$L(D_1, D_2, D_3, D_4, \lambda) = \frac{K_1}{D_1^5} + \frac{K_2}{D_2^5} + \frac{K_3}{D_3^5} + \frac{K_4}{D_4^5}$$
$$+ \lambda\left(\pi D_1 L_1 + \pi D_2 L_2 + \pi D_3 L_3 + \pi D_4 L_4 - 850 \text{ ft}^2\right)$$

Setting the gradient of the Lagrange function equal to zero results in

$$\frac{\partial L}{\partial D_1} = -\frac{5K_1}{D_1^6} + \lambda \pi L_1 = 0$$

$$\frac{\partial L}{\partial D_2} = -\frac{5K_2}{D_2^6} + \lambda \pi L_2 = 0$$

$$\frac{\partial L}{\partial D_3} = -\frac{5K_3}{D_3^6} + \lambda \pi L_3 = 0$$

$$\frac{\partial L}{\partial D_4} = -\frac{5K_4}{D_4^6} + \lambda \pi L_4 = 0$$

$$\frac{\partial L}{\partial \lambda} = \pi D_1 L_1 + \pi D_2 L_2 + \pi D_3 L_3 + \pi D_4 L_4 - 850 \text{ ft}^2 = 0$$

This is a system of five equations with five unknowns. The solution to these equations results in the diameters D_1, D_2, D_3, and D_4 along with the Lagrange multiplier, λ. Once these diameters are known, the minimum pressure drop can be calculated from the objective function.

The equations can be programmed into EES as follows:

```
"GIVEN: A main duct system for air distribution as shown
    in Figure E6.2"

"Volume flow rates"
  V_dot[1] = 6000[cfm]
  V_dot[2] = V_dot[1] - 1200[cfm]
  V_dot[3] = V_dot[2] - 2000[cfm]
  V_dot[4] = V_dot[3] - 1800[cfm]

"Lengths"
  L[1] = 50[ft]
  L[2] = 60[ft]
  L[3] = 65[ft]
  L[4] = 40[ft]

"Other constants"
  gamma = 0.075[lbf/ft^3]
  f = 0.02
  g = 32.174[ft/s^2]

"FIND: The values of D[1], D[2] and D[3] that minimize
the pressure drop from A to B"

"SOLUTION:"
"Lagrange gradient equations"
  -5*K[1]/D[1]^6 + lambda*pi*L[1] = 0
  -5*K[2]/D[2]^6 + lambda*pi*L[2] = 0
  -5*K[3]/D[3]^6 + lambda*pi*L[3] = 0
```

```
   -5*K[4]/D[4]^6 + lambda*pi*L[4]  =  0
   pi*D[1]*L[1] + pi*D[2]*L[2] + pi*D[3]*L[3] +           &
      pi*D[4]*L[4] - 850[ft^2]  =  0


"K-constants for each branch of the main duct"
   DUPLICATE i = 1,4
      K[i] = 8*gamma*f*L[i]*(V_dot[i]*            &
            convert(cfm,ft^3/s))^2/(pi^2*g)
   END


"Convert the duct diameters to inches"
   DUPLICATE i = 1,4
   D_inch[i] = D[i]*convert(ft,in)
   END


"The minimum pressure drop from A to B is,"
   DELTAP_AB = SUM(K[i]/D[i]^5,i = 1,4)
```

The solution to this set of equations results in the following values for the design variables: $D_1 = 1.519$ ft, $D_2 = 1.410$ ft, $D_3 = 1.178$ ft, and $D_4 = 0.836$ ft. The value of the Lagrange multiplier determined from this solution is $\lambda = 0.049$ lbf/ft⁴. Substituting the calculated design variables into the objective function reveals that the minimum pressure drop from A to B is $\Delta P_{AB} = 8.326$ lbf/ft².

As shown in Example 6.2, the Method of Lagrange Multipliers is fairly easy to implement for multivariate optimization problems. The method ultimately results in a system of n equations and n unknowns, representing the gradient of the Lagrange function. Therefore, it is convenient to use an equation solver for problems similar to this.

Up to this point, not much has been mentioned about the Lagrange multiplier, λ, and its significance. The Lagrange Multiplier Theorem indicates that each Lagrange multiplier is associated with a unique constraint. This would lead one to believe that multiplier has something to do with the constraint.

In Example 6.2, the Lagrange multiplier was found to be $\lambda = 0.049$ lbf/ft⁴. To maintain dimensional homogeneity, the Lagrange multiplier must have *units*; in this case, lbf/ft⁴. This seems like a very odd set of units. However, when investigating the Lagrange function for this problem, the units start to make some sense. The Lagrange function for Example 6.2 is repeated as follows:

$$L\left(D_1, D_2, D_3, D_4, \lambda\right) = \frac{K_1}{D_1^5} + \frac{K_2}{D_2^5} + \frac{K_3}{D_3^5} + \frac{K_4}{D_4^5}$$
$$+ \lambda\left(\pi D_1 L_1 + \pi D_2 L_2 + \pi D_3 L_3 + \pi D_4 L_4 - 850 \text{ ft}^2\right)$$

$$(6.42)$$

Notice that for dimensional homogeneity, each term has units of *pressure*. The Lagrange multiplier is multiplied by the constraint equation, which deals with *area*. Therefore, the Lagrange multiplier for this problem must have units of pressure per unit area. If the pressure is expressed in lbf/ft², then the Lagrange multiplier must have units of lbf/ft²/ft² or lbf/ft⁴. Of course, this unit analysis can all be sorted out without paying attention to the *significance* of the units. However, once the significance of the units is understood, then it is possible to discover the significance of the Lagrange multiplier itself.

Now, consider the following question, "What would the minimum pressure drop from A to B be if one additional square foot of sheet metal is available to construct the ducts?" More sheet metal would allow for larger duct diameters that should lead to a smaller minimized pressure drop. To test this thinking, the optimization problem can be solved again, this time using 851 ft² of sheet metal instead of 850 ft² as was originally specified. For this new optimization problem, the Lagrange function is shown in Equation 6.43:

$$L(D_1, D_2, D_3, D_4, \lambda) = \frac{K_1}{D_1^5} + \frac{K_2}{D_2^5} + \frac{K_3}{D_3^5} + \frac{K_4}{D_4^5}$$
$$+ \lambda \left(\pi D_1 L_1 + \pi D_2 L_2 + \pi D_3 L_3 + \pi D_4 L_4 - 851 \text{ ft}^2 \right)$$
(6.43)

Solving this problem gives a minimum pressure drop of $\Delta P_{AB} = 8.277$ lbf/ft². The difference between this solution and the solution in Example 6.2 is $(8.326 - 8.277)$ lbf/ft² $= 0.049$ lbf/ft². Interestingly, this is the value of the original Lagrange multiplier. This leads to the conclusion that the Lagrange multiplier reveals how the objective function changes per unit change in the constraint. In the duct design problem of Example 6.2, $\lambda = 0.049$ lbf/ft⁴ means that for every additional square foot of sheet metal available, we can expect the minimized pressure drop to decrease by 0.049 lbf/ft².

What has been conducted here is known as a *postoptimality analysis*. These types of analyses use the Lagrange multipliers to determine how the minimized objective function changes without going through the optimization process again. This reveals the *significance* of the Lagrange multiplier and its units.

6.4.4 Formulation and Solution of Optimization Problems Using Software

Optimization of a thermal system can often be much more complex than the examples discussed to this point. For most thermal energy system optimization problems, the development of the objective function is usually straightforward. The difficulty comes in the formulation of the constraints. In most cases, the constraints are related to the system simulation. In other words, an objective function can be minimized for a thermal system, but the optimum configuration of the system must obey the physical behavior of the system (i.e., the simulation). Because of the complexity of such problems, the solutions are determined numerically using software. Most equation-solving software includes the capability to find minimum or maximum conditions. To illustrate the formulation and numerical solution of an optimization problem, the duct design problem of Example 6.2 is considered again in Example 6.3.

EXAMPLE 6.3

Solve the duct design problem presented in Example 6.2 using a numerical optimization algorithm.

SOLUTION: The solution to this optimization problem will be sought using computer software (EES). The reader is referred to Appendix F for an explanation of the software-specific procedures to conduct an optimization study.

The given information for this problem is summarized in the following EES equations:

```
"GIVEN: An air distribution system as shown in Figure
E6.2,"

"Main duct section lengths"
   L[1]  =  50[ft]
   L[2]  =  60[ft]
   L[3]  =  65[ft]
   L[4]  =  40[ft]

"Volumetric flow rates through each section of the main
duct"
   Vol_dot[1]  =  6000[cfm]
   Vol_dot[2]  =  Vol_dot[1]  -  1200[cfm]
   Vol_dot[3]  =  Vol_dot[2]  -  2000[cfm]
   Vol_dot[4]  =  Vol_dot[3]  -  1800[cfm]

"Total
amount of sheet metal available to make the ducts"
   A_total  =  850[ft^2]

"Given constants"
   gamma  =  0.075[lbf/ft^3]
   f  =  0.02
   g  =  32.174[ft/s^2]
"FIND: The diameters that minimize the pressure drop from
A to B"
```

The objective function to be minimized is the pressure drop from points A to B:

$$\Delta P_{AB} = \sum_{k=1}^{4} \Delta P_k$$

In this equation, the subscript k refers to the main duct section. Assuming that the minor losses because of the transitions are negligible, the pressure drops in each section can be written using the conservation of energy between the inlet and outlet of the sections.

$$\Delta P_{AB} = \sum_{k=1}^{4} \gamma f \frac{L_k}{D_k} \frac{V_k^2}{2g}$$

The velocity in each section of the duct is related to the volumetric flow rate and the duct diameter. For any duct section, k,

$$\dot{V}_k = A_k V_k = \left(\frac{\pi D_k^2}{4} \right) V_k.$$

These equations are written in the following EES form:

```
"SOLUTION:"
"The pressure drop from A to B is determined by,"
  DELTAP_AB = SUM(DELTAP[k],k = 1,4)
```

```
"The pressure drops through each section of the main duct
are determined from the conservation of energy equation,"
  DUPLICATE i = 1,4
    DELTAP[i] = gamma*f*(L[i]/D[i])*V[i]^2/(2*g)
    Vol_dot[i]*convert(cfm,ft^3/s) = A[i]*V[i]
    A[i] = pi*D[i]^2/4
  END
```

One additional equation is required to enforce the constraint that a fixed area of sheet metal is available to make the duct.

$$A_{total} = \sum_{k=1}^{4} \pi D_k L_k$$

In EES form, this is written as

```
"The constraint is the total amount of sheet metal
available"
  A_total = SUM(pi*D[k]*L[k],k = 1,4)
```

These equations complete the optimization problem. In this case, there are 3 degrees of freedom. This may seem counterintuitive because there are four unknown diameters. However, the diameters are related through the constraint. Therefore, only three of these diameters are independent.

Performing the optimization using EES reveals the following information for the design variables: $D_1 = 1.519$ ft, $D_2 = 1.410$ ft, $D_3 = 1.178$ ft, and $D_4 = 0.836$ ft. The resulting pressure drop using these design variables is $\Delta P_{AB} = 8.326$ lbf/ft². As expected, this solution is the same as found in Example 6.2, which used the Lagrange multiplier approach.

In Example 6.3, the constraint containing the design variables was easily written in a single equation. However, in a more complex system, the constraint may take the form of a system of n equations and n unknowns: the *simulation* of the system. Optimizations that incorporate simulations as constraints may be extremely difficult to solve numerically unless the simulation is well written and very good starting values are used to seed the optimization process.

6.4.5 Final Comments Regarding Thermal Energy System Optimization

Section 6.4 is not meant to be an all-encompassing treatise on optimization. There are complete courses and books that deal with this subject. It is a mathematically rigorous topic. Mathematics provides the framework for the development of the numerical algorithms that are available in various software packages. The goal in any of these numerical methods is to find the global minimum of the objective function quickly in a computationally efficient way.

Users of numerical optimization algorithms may experience significant frustration due to nonconvergence, especially as the system becomes more complex. This often happens because the algorithm was seeded with poor initial guesses for the design variables. In complex optimization problems with many design variables, it is very helpful to predict final values for design variables and use them as the seed to start the optimization algorithm. As an example, consider a fluid flowing through a heat exchanger. If the inlet temperature of the fluid is known, and the heat is being transferred from the fluid, then the outlet temperature of the fluid must be smaller than the inlet temperature. So, when specifying the initial value for the outlet temperature, specify it to be something less than the inlet temperature. There are many instances where very sensible engineering guesses can be made for design variables. Using sensible and physically correct initial values for any of the unknown design variables will go a long way in successful convergence of the optimization algorithm.

Once a solution has been found to an optimization problem, it is helpful to update the initial guesses for the design variables, and run the optimization again. It may even be helpful to change the optimization algorithm if the software allows the user to choose from several optimization algorithms. If the same solution can be found using several different methods, then the user can be confident that the correct solution has indeed been found. The message here is that it may take *several* runs of an optimization algorithm to ensure that the final solution is the correct optimum solution.

Problems

6.1 Develop an empirical equation that represents the pump head (ft) as a function of capacity (gpm) for a Bell & Gossett Series 80-SC 2 × 2 × 7 pump with a 6½ in. impeller operating at 1750 rpm. The equation should be valid from 40 to 110 gpm. Develop a deviation plot to validate the accuracy of the empirical equation.

6.2 Develop an empirical equation for the power draw (W) of the refrigeration compressor as a function of the saturated evaporating and condensing temperatures (°F) described by the 9-point performance data shown in Figure 6.8.

6.3 Develop and solve a simulation to determine the operating point (head [ft] and capacity [gpm]) of the pump and pipe system shown in Figure P6.3. All fittings are regular and the globe valve is wide opened.

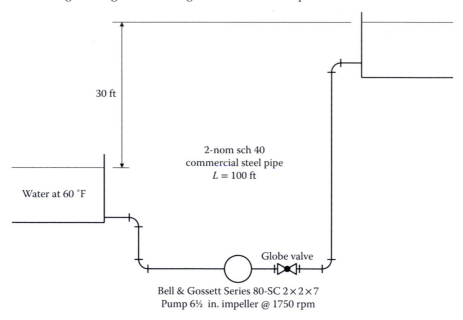

FIGURE P6.3

6.4 Using the simulation developed in Problem 6.3, develop a plot that shows the operating point (head [ft] and capacity [gpm]) as a function of the rotational speed of the 6½ in. impeller over the range 1500 ≤ rpm ≤ 2500 in 50-rpm increments. Plot the pump head as a function of capacity using a symbol to indicate the impeller speed.

6.5 A centrifugal pump and a gear pump are operating in series in a closed loop to deliver a 20% ethylene glycol solution at an average temperature of 10°C through a long pipe system. In the capacity range up to 2 m³/s, the pump performance equations expressed in terms of pressure increase as a function of capacity are given by

$$\text{centrifugal pump: } \Delta P = 50 + 0.2\dot{V} - 10\dot{V}^2$$

$$\text{gear pump: } \qquad \Delta P = 30 - 10\dot{V}$$

A detailed friction analysis reveals that the pressure drop through the closed-loop system is given by

$$\Delta P = 20\dot{V}^2$$

In the pump performance equations and system curve equation, the pressure changes are in kPa and the capacity is in m³/s.

a. On the same plot, show the system curve, the performance curves for each pump, and the performance curve of the combined series pumps in terms of head (m) as a function of capacity (m³/s). Draw the curves from 0 to 2 m³/s.

b. Write a simulation of the system and determine its operating point.

c. If the economic velocity of the ethylene glycol for this system is 10 ft/s, specify the standard schedule steel pipe required for this application.

6.6 A liquid nitrogen cooler is shown in Figure P6.6. The goal of this system is to cool the liquid nitrogen flow at state 1. This is accomplished by diverting a portion of the liquid through an expansion valve. When the liquid passes through the valve, it flashes into saturated liquid and saturated vapor in the flash tank. The saturated vapor is drawn off the top, and the saturated liquid is used to cool the liquid flow using an in-tank heat exchanger. For the conditions shown in Figure P6.6, determine the mass flow rate (lbm/h) and temperature (°F) of the liquid nitrogen at state 4, and the mass flow rate (lbm/h) of the saturated vapor at state 3.

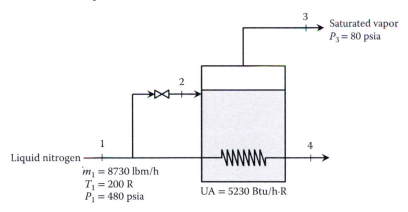

FIGURE P6.6

6.7 One way to liquefy a gas is to flash it through a valve. However, this requires that the inlet conditions to the valve be at a state where an expansion will cause the gas to flash into saturated liquid and saturated vapor. Consider a situation where helium gas is at 10 bar and the desired pressure of the liquid is 1 bar. To produce liquid helium, the inlet state to the throttling process must lie between the two states, identified as *a* and *b* on the P–h diagram shown in Figure P6.7A. For an inlet pressure of 10 bar, this temperature range is 2.53 K $\leq T \leq 7.60$ K for helium (calculated using EES). One possible way to achieve this low temperature at the entrance to the valve is shown in Figure P6.7B. This system is known as a *liquefaction system*. In this system, the helium gas is supplied at 10 bar, 18 K, at a flow rate of 4.6 g/s. This gas is cooled further

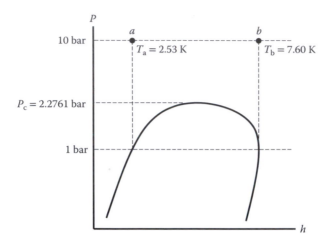

FIGURE P6.7A

using two heat exchangers. Cooling of the gas is accomplished by mixing the
cold saturated vapor leaving the separator and the cold gas at the exhaust of a
small turbine operating between the high and low pressures of the system.

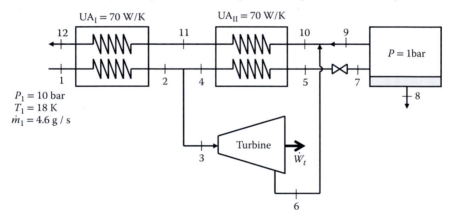

FIGURE P6.7B

The UA product of each heat exchanger is 70 W/K. The separator operates at a
pressure of 1 bar. Manufacturer's performance data for the turbine relate the
turbine temperatures and mass flow rate with two empirical equations:

$$T_6 = a_1 + a_2 T_3$$
$$\dot{m} = b_1 + b_2 T_3$$

In these turbine performance equations $a_1 = 4.6$ K, $a_2 = 0.1$, $b_1 = 3.75$ g/s, and
$b_2 = -0.125$ g/s·K.
Simulate this system and determine the following:
a.   The unknown temperatures in the cycle (K)
b.   The heat transfer rate in each heat exchanger (W)
c.   The liquid helium mass flow rate leaving the separator (g/s)
d.   The power delivered by the turbine (W)

6.8 Use the simulation developed in Problem 6.7 to determine how the inlet helium flow rate at state 1 influences the liquid helium production rate and the turbine output power. Plot the following parameters as a function of the inlet helium mass flow rate at state 1 for the range $3.4 \leq \dot{m}_1 \leq 5.0$ g/s:
   a. The liquid helium mass flow rate leaving the separator (g/s)
   b. The quality of the helium leaving the valve at state 7
   c. The turbine output power (W)

6.9 Process steam boilers often incorporate a continuous blowdown scheme to eliminate impurities that can accumulate over time. One possible way to accomplish this is shown in Figure P6.9A. In this system, the boiler is operating at 300 psia. Saturated vapor steam leaves the boiler as the high-pressure process steam flow at a flow rate of 1,000,000 lbm/h. Saturated liquid water, containing the impurities, exits the boiler at 300 gpm. The dirty water is then flashed into a separator through a valve. The separator pressure is 65 psia. Saturated vapor leaves the separator as the low-pressure process steam flow. The high-pressure and low-pressure process steam flows make their way through the plant and eventually condense, combine, and return to the boiler as a liquid boiler feedwater at 102°F. The saturated liquid at the separator exit is discarded. To maintain steady operation, makeup water must replace the discarded dirty water. This makeup water enters the boiler at 78°F. Simulate this system and determine the following operating parameters:
   a. The mass flow rate (lbm/h) of the low-pressure process steam
   b. The volumetric flow rate (gpm) of the discarded dirty water
   c. The temperature of the discarded dirty water
   d. The boiler heat transfer rate required to maintain continuous, steady-state operation (Btu/h)

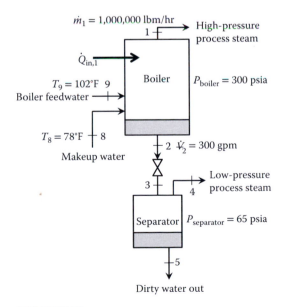

$\dot{m}_1 = 1,000,000$ lbm/hr   High-pressure process steam

$\dot{Q}_{in,1}$

$T_9 = 102°F$ 9   Boiler $P_{boiler} = 300$ psia
Boiler feedwater

$T_8 = 78°F$ 8
Makeup water   2 $\dot{V}_2 = 300$ gpm

3   4   Low-pressure process steam

Separator $P_{separator} = 65$ psia

5

Dirty water out

FIGURE P6.9A

The result of your simulation should reveal that the temperature of the discarded dirty water is quite hot. It is proposed that this hot, dirty water be used to preheat the makeup water going into the boiler in an attempt to reduce fuel cost. This can be accomplished by adding a heat exchanger as shown in Figure P6.9B. It is reasonable to treat the heat exchanger as a counterflow regenerative heat exchanger. The mass flow rates of the water on each side are the same (the makeup water must replace the discarded dirty water). However, the heat capacities of the water on each side of the heat exchanger are slightly different. In reality, this difference is very small. Therefore, the thermal capacitance rate of each flow can be assumed to be equal without introducing too much error into the calculation. A heat capacity value of 1.01 Btu/lbm·R is a reasonable estimate of c_p for the water passing through the heat exchanger. The UA product of the heat exchanger is 9580 Btu/h·R. Simulate this modified system and determine the following operating conditions:

e. The temperature of the discarded dirty water
f. The temperature rise of the makeup water as it flows through the heat exchanger
g. The boiler heat transfer rate required to maintain continuous, steady-state operation (Btu/h)
    This particular system operates continuously throughout the year. The heat transfer rate is provided by the combustion of natural gas that costs $0.70/therm. (A therm is 100,000 Btu.) Determine the following:
h. The annual energy cost savings realized by incorporating the regenerative heat exchanger as shown in Figure P6.9B.

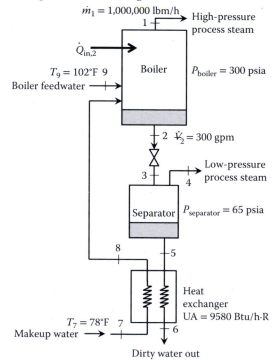

FIGURE P6.9B

i. If management requires a minimum rate of return of 25% and the heat exchanger is expected to have a 50-year life, how much money can be spent on the heat exchanger? Does the heat exchanger appear to be a viable investment?

6.10 Investigate the effect of the separator pressure on the operating parameters of the boiler blowdown system described in Figure P6.9B. For this evaluation, vary the separator pressure from 20 to 100 psia and plot the following curves:

a. The annual energy cost savings as a function of the separator pressure.

b. The mass flow rate of the low-pressure process steam (lbm/h) and the volumetric flow rate of the dirty water (gpm) leaving the separator. Plot these on the same graph (using the plot overlay feature in EES). Use the left-hand axis to plot the low-pressure process steam mass flow rate and the right-hand axis to plot the dirty water volumetric flow rate.

c. Comment on the results of the parametric studies from parts (a) and (b).

6.11 A retrofit HVAC project includes installing rectangular air ducts in braced floor joists. The dimensions of the joist and bracing are shown in Figure P6.11.

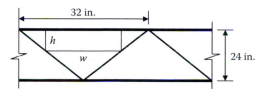

FIGURE P6.11

Using the Method of Lagrange Multipliers, determine the optimal values of h and w that maximize the cross-sectional area of the duct.

6.12 A flow rate of 32,000 cfm of gas at a temperature 120°F and a pressure 25 psia is to be compressed to 2500 psia. Under these conditions, the gas behaves according to the ideal gas law. The choice of compressor type is influenced by the fact that centrifugal compressors can handle high-volume flow rates but develop only low-pressure ratios per stage. A reciprocating compressor, on the other hand, is suited to low-volume flow rates but can develop high-pressure ratios. To combine the advantages of each, the compression will be done using a low-stage centrifugal compressor and a high-stage reciprocating compressor. Between the compressor stages, an intercooler is used to return the gas temperature to 120°F. A sketch of this dual-stage compressor system is shown in Figure P6.12.

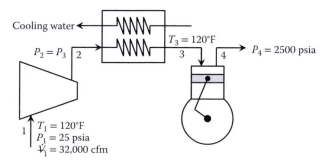

FIGURE P6.12

The first cost of each compressor, expressed in terms of volume flow rates and pressure ratios are given by the following equations:

$$IC_c = a_1 \dot{V_1} + a_2 \frac{P_2}{P_1} \text{ (centrifugal compressor)}$$

$$IC_r = b_1 \dot{V_3} + b_2 \frac{P_4}{P_3} \text{ (reciprocating compressor)}$$

In the abovementioned equations, $a_1 = \$0.03/cfm$, $a_2 = \$1600$, $b_1 = \$0.09/cfm$, and $b_2 = \$800$. Using the Lagrange Multiplier Theorem, determine the following:
a. The minimum first cost of the compressors
b. The pressure ratios across each compressor to achieve this optimum condition
c. Determine the parameters in parts a. and b. using a numerical optimization algorithm. How does the numerical solution compare to the solution using the Lagrange Multiplier Theorem?

6.13 A cascade refrigeration cycle is shown in Figure P6.13. The purpose of a refrigeration cycle of this type is to achieve very low-temperature refrigeration while saving energy costs. Energy costs are reduced because the work required for the two-stage cascade system is lower than the compression for a single-stage operating between the same source and sink conditions. In the interstage heat exchanger, the refrigerant in the low-pressure stage leaves at T_c. The refrigerant enters the interstage heat exchanger from the high-pressure stage at T_e. The overall heat transfer coefficient in the interstage heat exchanger is $U = 800$ W/m²·K and the total heat transferred between the two stages is 160 kW.

A cost analysis reveals that the initial cost of the interstage heat exchanger is related to the heat transfer area by

$$IC_{HX} = a_1 A,$$

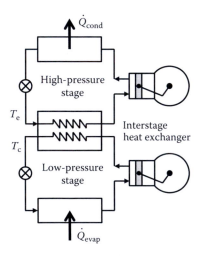

FIGURE P6.13

where $a_1 = \$100/m^2$. The cost of energy required by each compressor is a function of the refrigerant temperatures in the interstage heat exchanger. An economic analysis conducted over the expected life of the system shows that the present value of the lifetime energy costs for each stage is given by

$$PV_L = a_2 + a_3 T_c \quad \text{(low-pressure stage)}$$

$$PV_H = a_4 + a_5 T_e^2 \quad \text{(high-pressure stage)}$$

In these cost equations, $a_2 = \$127,000$, $a_3 = \$150/K$, $a_4 = \$189,000$, and $a_5 = \$-0.3/K^2$. Notice that the temperatures must be on the kelvin (K) scale.
a. Using the Method of Lagrange Multipliers, determine the minimum cost of this cascade refrigeration system.
b. What are the values of the design variables that result in the minimum cost calculated in part (a)?
c. Using the value(s) of the Lagrange multiplier(s), determine the minimum cost of the system if the heat transfer rate between the fluids in the interstage heat exchanger was increased to 170 kW.

6.14 Two heat exchangers in a circulating water loop transfer heat from a fluid condensing at 175°F to a boiling fluid at 68°F, as shown in Figure P6.14. The overall heat transfer coefficients of the evaporator and condenser are $U_E = 5.3$ Btu/h·ft²·°F and $U_C = 7$ Btu/h·ft²·°F, respectively. The first cost of the heat exchangers is $\$21.50/ft^2$, and the present worth of the lifetime pumping costs is $11.75[\$·h/lbm]\dot{m}_w$. This system must accomplish a heat transfer rate of 220,000 Btu/h between the condensing and boiling fluids. Using a numerical optimization algorithm, determine the heat exchanger areas, A_1 and A_2, that minimize the total present worth of costs.

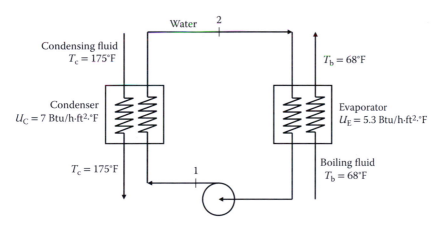

FIGURE P6.14

Appendix A: Conversion Factors

The conversion factors in this appendix are calculated using Engineering Equation Solver (EES). The conversions factors are in *columns*. For example: In the Length table on page 404, 1 m = 3.9370E+01 in = 39.370 in. The conversion factors have been rounded off to five significant figures.

Length

	mm	cm	m	km	in.	ft	yd	mile	
mm	**1.0000E+00**	1.0000E+01	1.0000E+03	1.0000E+06	2.5400E+01	3.0480E+02	9.1440E+02	1.6093E+06	mm
cm	1.0000E−01	**1.0000E+00**	1.0000E+02	1.0000E+05	2.5400E+00	3.0480E+01	9.1440E+01	1.6093E+05	cm
m	1.0000E−03	1.0000E−02	**1.0000E+00**	1.0000E+03	2.5400E−02	3.0480E−01	9.1440E−01	1.6093E+03	m
km	1.0000E−06	1.0000E−05	1.0000E−03	**1.0000E+00**	2.5400E−05	3.0480E−04	9.1440E−04	1.6093E+00	km
in.	3.9370E−02	3.9370E−01	3.9370E+01	3.9370E+04	**1.0000E+00**	1.2000E+01	3.6000E+01	6.3360E+04	in.
ft	3.2808E−03	3.2808E−02	3.2808E+00	3.2808E+03	8.3333E−02	**1.0000E+00**	3.0000E+00	5.2800E+03	ft
yd	1.0936E−03	1.0936E−02	1.0936E+00	1.0936E+03	2.7778E−02	3.3333E−01	**1.0000E+00**	1.7600E+03	yd
mile	6.2137E−07	6.2137E−06	6.2137E−04	6.2137E−01	1.5783E−05	1.8939E−04	5.6818E−04	**1.0000E+00**	mile
	mm	cm	m	km	in.	ft	yd	mile	

Area

	mm²	cm²	m²	km²	in.²	ft²	yd²	mile²	
mm²	**1.0000E+00**	1.0000E+02	1.0000E+06	1.0000E+12	6.4516E+02	9.2903E+04	8.3613E+05	2.5900E+12	mm²
cm²	1.0000E−02	**1.0000E+00**	1.0000E+04	1.0000E+10	6.4516E+00	9.2903E+02	8.3613E+03	2.5900E+10	cm²
m²	1.0000E−06	1.0000E−04	**1.0000E+00**	1.0000E+06	6.4516E−04	9.2903E−02	8.3613E−01	2.5900E+06	m²
km²	1.0000E−12	1.0000E−10	1.0000E−06	**1.0000E+00**	6.4516E−10	9.2903E−08	8.3613E−07	2.5900E+00	km²
in.²	1.5500E−03	1.5500E−01	1.5500E+03	1.5500E+09	**1.0000E+00**	1.4400E+02	1.2960E+03	4.0145E+09	in.²
ft²	1.0764E−05	1.0764E−03	1.0764E+01	1.0764E+07	6.9444E−03	**1.0000E+00**	9.0000E+00	2.7878E+07	ft²
yd²	1.1960E−06	1.1960E−04	1.1960E+00	1.1960E+06	7.7161E−04	1.1111E−01	**1.0000E+00**	3.0976E+06	yd²
mile²	3.8610E−13	3.8610E−11	3.8610E−07	3.8610E−01	2.4910E−10	3.5870E−08	3.2283E−07	**1.0000E+00**	mile²
	mm²	cm²	m²	km²	in.²	ft²	yd²	mile²	

Volume

	mm³	cm³	m³	L	in.³	ft³	yd³	gal	
mm³	**1.0000E+00**	1.0000E+03	1.0000E+09	1.0000E+06	1.6387E+04	2.8317E+07	7.6456E+08	3.7854E+06	mm³
cm³	1.0000E−03	**1.0000E+00**	1.0000E+06	1.0000E+03	1.6387E+01	2.8317E+04	7.6456E+05	3.7854E+03	cm³
m³	1.0000E−09	1.0000E−06	**1.0000E+00**	1.0000E−03	1.6387E−05	2.8317E−02	7.6456E−01	3.7854E−03	m³
L	1.0000E−06	1.0000E−03	1.0000E+03	**1.0000E+00**	1.6387E−02	2.8317E+01	7.6456E+02	3.7854E+00	L
in.³	6.1024E−05	6.1024E−02	6.1024E+04	6.1024E+01	**1.0000E+00**	1.7280E+03	4.6656E+04	2.3100E+02	in.³
ft³	3.5315E−08	3.5315E−05	3.5315E+01	3.5315E−02	5.7870E−04	**1.0000E+00**	2.7000E+01	1.3368E−01	ft³
yd³	1.3080E−09	1.3080E−06	1.3080E+00	1.3080E−03	2.1434E−05	3.7037E−02	**1.0000E+00**	4.9511E−03	yd³
gal	2.6417E−07	2.6417E−04	2.6417E+02	2.6417E−01	4.3290E−03	7.4805E+00	2.0197E+02	**1.0000E+00**	gal
	mm³	cm³	m³	L	in.³	ft³	yd³	gal	

Mass

	mg	g	kg	grains	oz	lbm	slug	ton	
mg	**1.0000E+00**	1.0000E+03	1.0000E+06	6.4799E+01	2.8350E+04	4.5359E+05	1.4594E+07	9.0719E+08	mg
g	1.0000E−03	**1.0000E+00**	1.0000E+03	6.4799E−02	2.8350E+01	4.5359E+02	1.4594E+04	9.0719E+05	g
kg	1.0000E−06	1.0000E−03	**1.0000E+00**	6.4799E−05	2.8350E−02	4.5359E−01	1.4594E+01	9.0719E+02	kg
grains	1.5432E−02	1.5432E+01	1.5432E+04	**1.0000E+00**	4.3750E+02	7.0000E+03	2.2522E+05	1.4000E+07	grains
oz	3.5274E−05	3.5274E−02	3.5274E+01	2.2857E−03	**1.0000E+00**	1.6000E+01	5.1478E+02	3.2000E+04	oz
lbm	2.2046E−06	2.2046E−03	2.2046E+00	1.4286E−04	6.2500E−02	**1.0000E+00**	3.2174E+01	2.0000E+03	lbm
slug	6.8522E−08	6.8522E−05	6.8522E−02	4.4402E−06	1.9426E−03	3.1081E−02	**1.0000E+00**	6.2162E+01	slug
ton	1.1023E−09	1.1023E−06	1.1023E−03	7.1429E−08	3.1250E−05	5.0000E−04	1.6087E−02	**1.0000E+00**	ton
	mg	g	kg	grains	oz	lbm	slug	ton	

Force

	N	kN	lbf	dyne	m·kg/s²	m·kg/h²	ft·lbm/s²	ft·lbm/h²
N	**1.0000E+00**	1.0000E+03	4.4482E+00	1.0000E-05	1.0000E+00	7.7161E-08	1.3826E-01	1.0668E-08
kN	1.0000E-03	**1.0000E+00**	4.4482E-03	1.0000E-08	1.0000E-03	7.7161E-11	1.3826E-04	1.0668E-11
lbf	2.2481E-01	2.2481E+02	**1.0000E+00**	2.2481E-06	2.2481E-01	1.7346E-08	3.1081E-02	2.3982E-09
dyne	1.0000E+05	1.0000E+08	4.4482E+05	**1.0000E+00**	1.0000E+05	7.7161E-03	1.3826E+04	1.0668E-03
m·kg/s²	1.0000E+00	1.0000E+03	4.4482E+00	1.0000E-05	**1.0000E+00**	7.7161E-08	1.3826E-01	1.0668E-08
m·kg/h²	1.2960E+07	1.2960E+10	5.7649E+07	1.2960E+02	1.2960E+07	**1.0000E+00**	1.7918E+06	1.3826E-01
ft·lbm/s²	7.2330E+00	7.2330E+03	3.2174E+01	7.2330E-05	7.2330E+00	5.5810E-07	**1.0000E+00**	7.7161E-08
ft·lbm/h²	9.3740E+07	9.3740E+10	4.1698E+08	9.3740E+02	9.3740E+07	7.2330E+00	1.2960E+07	**1.0000E+00**
	N	kN	lbf	dyne	m·kg/s²	m·kg/h²	ft·lbm/s²	ft·lbm/h²

Pressure

	Pa	kPa	MPa	bar	lbf/in.²	lbf/ft²	atm	in Hg
Pa	**1.0000E+00**	1.0000E+03	1.0000E+06	1.0000E+05	6.8948E+03	4.7880E+01	1.0133E+05	3.3864E+03
kPa	1.0000E-03	**1.0000E+00**	1.0000E+03	1.0000E+02	6.8948E+00	4.7880E-02	1.0133E+02	3.3864E+00
MPa	1.0000E-06	1.0000E-03	**1.0000E+00**	1.0000E-01	6.8948E-03	4.7880E-05	1.0133E-01	3.3864E-03
bar	1.0000E-05	1.0000E-02	1.0000E+01	**1.0000E+00**	6.8948E-02	4.7880E-04	1.0133E+00	3.3864E-02
lbf/in.²	1.4504E-04	1.4504E-01	1.4504E+02	1.4504E+01	**1.0000E+00**	6.9444E-03	1.4696E+01	4.9115E-01
lbf/ft²	2.0885E-02	2.0885E+01	2.0885E+04	2.0885E+03	1.4400E+02	**1.0000E+00**	2.1162E+03	7.0726E+01
atm	9.8692E-06	9.8692E-03	9.8692E+00	9.8692E-01	6.8046E-02	4.7254E-04	**1.0000E+00**	3.3421E-02
in Hg	2.9530E-04	2.9530E-01	2.9530E+02	2.9530E+01	2.0360E+00	1.4139E-02	2.9921E+01	**1.0000E+00**
	Pa	kPa	MPa	bar	lbf/in.²	lbf/ft²	atm	in Hg

Dynamic Viscosity

	centipoise	poise	kg/m·s	kg/m·h	N·s/m²	lbm/ft·s	lbm/ft·h	lbf·s/ft²	
centipoise	1.0000E+00	1.0000E+02	1.0000E+03	2.7778E-01	1.0000E+03	1.4882E+03	4.1338E-01	4.7880E+04	centipoise
poise	1.0000E-02	1.0000E+00	1.0000E+01	2.7778E-03	1.0000E+01	1.4882E+01	4.1338E-03	4.7880E+02	poise
kg/m·s	1.0000E-03	1.0000E-01	1.0000E+00	2.7778E-04	1.0000E+00	1.4882E+00	4.1338E-04	4.7880E+01	kg/m·s
kg/m·h	3.6000E+00	3.6000E+02	3.6000E+03	1.0000E+00	3.6000E+03	5.3574E+03	1.4882E+00	1.7237E+05	kg/m·h
N·s/m²	1.0000E-03	1.0000E-01	1.0000E+00	2.7778E-04	1.0000E+00	1.4882E+00	4.1338E-04	4.7880E+01	N·s/m²
lbm/ft·s	6.7197E-04	6.7197E-02	6.7197E-01	1.8666E-04	6.7197E-01	1.0000E+00	2.7778E-04	3.2174E+01	lbm/ft·s
lbm/ft·h	2.4191E+00	2.4191E+02	2.4191E+03	6.7197E-01	2.4191E+03	3.6000E+03	1.0000E+00	1.1583E+05	lbm/ft·h
lbf·s/ft²	2.0885E-05	2.0885E-03	2.0885E-02	5.8015E-06	2.0885E-02	3.1081E-02	8.6336E-06	1.0000E+00	lbf·s/ft²
	centipoise	poise	kg/m·s	kg/m·h	N·s/m²	lbm/ft·s	lbm/ft·h	lbf·s/ft²	

Kinematic Viscosity

	centistoke	stoke	cm²/s	m²/s	m²/h	in.²/s	ft²/s	ft²/h	
centistoke	1.0000E+00	1.0000E+02	1.0000E+02	1.0000E+06	2.7778E+02	6.4516E+02	9.2903E+04	2.5806E+01	centistoke
stoke	1.0000E-02	1.0000E+00	1.0000E+00	1.0000E+04	2.7778E+00	6.4516E+00	9.2903E+02	2.5806E-01	stoke
cm²/s	1.0000E-02	1.0000E+00	1.0000E+00	1.0000E+04	2.7778E+00	6.4516E+00	9.2903E+02	2.5806E-01	cm²/s
m²/s	1.0000E-06	1.0000E-04	1.0000E-04	1.0000E+00	2.7778E-04	6.4516E-04	9.2903E-02	2.5806E-05	m²/s
m²/h	3.6000E-03	3.6000E-01	3.6000E-01	3.6000E+03	1.0000E+00	2.3226E+00	3.3445E+02	9.2903E-02	m²/h
in.²/s	1.5500E-03	1.5500E-01	1.5500E-01	1.5500E+03	4.3056E-01	1.0000E+00	1.4400E+02	4.0000E-02	in.²/s
ft²/s	1.074E-05	1.0764E-03	1.074E-03	1.0764E+01	2.9900E-03	6.9444E-03	1.0000E+00	2.7778E-04	ft²/s
ft²/h	3.8750E-02	3.8750E+00	3.8750E+00	3.8750E+04	1.0764E+01	2.5000E+01	3.6000E+03	1.0000E+00	ft²/h
	centistoke	stoke	cm²/s	m²/s	m²/h	in.²/s	ft²/s	ft²/h	

Thermal Conductivity

	W/cm K	W/m K	kW/cm K	kW/m K	Btu/s·in.°F	Btu/s·ft·°F	Btu/h·in.°F	Btu/h·ft·°F	
W/cm K	1.0000E+00	1.0000E−02	1.0000E+03	1.0000E+01	7.4768E+02	6.2307E+01	2.0769E−01	1.7307E−02	W/cm K
W/m K	1.0000E+02	1.0000E+00	1.0000E+05	1.0000E+03	7.4768E+04	6.2307E+03	2.0769E+01	1.7307E+00	W/m K
kW/cm K	1.0000E−03	1.0000E−05	1.0000E+00	1.0000E−02	7.4768E−01	6.2307E−02	2.0769E−04	1.7307E−05	kW/cm K
kW/m K	1.0000E−01	1.0000E−03	1.0000E+02	1.0000E+00	7.4768E+01	6.2307E+00	2.0769E−02	1.7307E−03	kW/m K
Btu/s·in.°F	1.3375E−03	1.3375E−05	1.3375E+00	1.3375E−02	1.0000E+00	8.3333E−02	2.7778E−04	2.3148E−05	Btu/s·in.°F
Btu/s·ft·°F	1.6050E−02	1.6050E−04	1.6050E+01	1.6050E−01	1.2000E+01	1.0000E+00	3.3333E−03	2.7778E−04	Btu/s·ft·°F
Btu/h·in.°F	4.8149E+00	4.8149E−02	4.8149E+03	4.8149E+01	3.6000E+03	3.0000E+02	1.0000E+00	8.3333E−02	Btu/h·in.°F
Btu/h·ft·°F	5.7779E+01	5.7779E−01	5.7779E+04	5.7779E+02	4.3200E+04	3.6000E+03	1.2000E+01	1.0000E+00	Btu/h·ft·°F
	W/cm K	W/m K	kW/cm K	kW/m K	Btu/s·in.°F	Btu/s·ft·°F	Btu/h·in.°F	Btu/h·ft·°F	

Energy

	J	kJ	ft·lbf	Btu	cal	kWh	therms	quads	
J	1.0000E+00	1.0000E−03	7.3756E−01	9.4782E−04	2.3885E−01	2.7778E−07	9.4782E−09	9.4782E−19	J
kJ	1.0000E+03	1.0000E+00	7.3756E+02	9.4782E−01	2.3885E+02	2.7778E−04	9.4782E−06	9.4782E−16	kJ
ft·lbf	1.3558E+00	1.3558E−03	1.0000E+00	1.2851E−03	3.2383E−01	3.7662E−07	1.2851E−08	1.2851E−18	ft·lbf
Btu	1.0551E+03	1.0551E+00	7.7817E+02	1.0000E+00	2.5200E+02	2.9307E−04	1.0000E−05	1.0000E−15	Btu
cal	4.1868E+00	4.1868E−03	3.0880E+00	3.9683E−03	1.0000E+00	1.1630E−06	3.9683E−08	3.9683E−18	cal
kWh	3.6000E+06	3.6000E+03	2.6552E+06	3.4121E+03	8.5985E+05	1.0000E+00	3.4121E−02	3.4121E−12	kWh
therms	1.0551E+08	1.0551E+05	7.7817E+07	1.0000E+05	2.5200E+07	2.9307E+01	1.0000E+00	1.0000E−10	therms
quads	1.0551E+18	1.0551E+15	7.7817E+17	1.0000E+15	2.5200E+17	2.9307E+11	1.0000E+10	1.0000E+00	quads
	J	kJ	ft·lbf	Btu	cal	kWh	therms	quads	

Specific Energy

	J/g	kJ/kg	MJ/kg	ft·lbf/lbm	Btu/lbm	cm²/s²	m²/s²	ft²/s²	
J/g	**1.0000E+00**	1.0000E+00	1.0000E-03	2.9891E-03	2.3260E+00	1.0000E-07	1.0000E-03	9.2903E-05	J/g
kJ/kg	1.0000E+00	**1.0000E+00**	1.0000E-03	2.9891E-03	2.3260E+00	1.0000E-07	1.0000E-03	9.2903E-05	kJ/kg
MJ/kg	1.0000E-03	1.0000E-03	**1.0000E+00**	2.9891E-06	2.3260E-03	1.0000E-10	1.0000E-06	9.2903E-08	MJ/kg
ft·lbf/lbm	3.3455E+02	3.3455E+02	3.3455E-05	**1.0000E+00**	7.7817E+02	3.3455E-05	3.3455E-01	3.1081E-02	ft·lbf/lbm
Btu/lbm	4.2992E-01	4.2992E-01	4.2992E+02	1.2851E-03	**1.0000E+00**	4.2992E-08	4.2992E-04	3.9941E-05	Btu/lbm
cm²/s²	1.0000E+07	1.0000E+07	1.0000E+10	2.9891E+04	2.3260E+07	**1.0000E+00**	1.0000E+04	9.2903E+02	cm²/s²
m²/s²	1.0000E+03	1.0000E+03	1.0000E+06	2.9891E+00	2.3260E+03	1.0000E-04	**1.0000E+00**	9.2903E-02	m²/s²
ft²/s²	1.0764E+04	1.0764E+04	1.0764E+07	3.2174E+01	2.5037E+04	1.0764E-03	1.0764E+01	**1.0000E+00**	ft²/s²
	J/g	kJ/kg	MJ/kg	ft·lbf/lbm	Btu/lbm	cm²/s²	m²/s²	ft²/s²	

Energy Transfer Rate

	W	kW	MW	ft·lbf/s	ft·lbf/h	Btu/h	hp	tons	
W	**1.0000E+00**	1.0000E+03	1.0000E+06	1.3558E+00	3.7662E-04	2.9307E-01	7.4570E+02	3.5169E+03	W
kW	1.0000E-03	**1.0000E+00**	1.0000E+03	1.3558E-03	3.7662E-07	2.9307E-04	7.4570E-01	3.5169E+00	kW
MW	1.0000E-06	1.0000E-03	**1.0000E+00**	1.3558E-06	3.7662E-10	2.9307E-07	7.4570E-04	3.5169E-03	MW
ft·lbf/s	7.3756E-01	7.3756E+02	7.3756E+05	**1.0000E+00**	2.7778E-04	2.1616E-01	5.5000E+02	2.5939E+03	ft·lbf/s
ft·lbf/h	2.6552E+03	2.6552E+06	2.6552E+09	3.6000E+03	**1.0000E+00**	7.7817E+02	1.9800E+06	9.3380E+06	ft·lbf/h
Btu/h	3.4121E+00	3.4121E+03	3.4121E+06	4.6262E+00	1.2851E-03	**1.0000E+00**	2.5444E+03	1.2000E+04	Btu/h
hp	1.3410E-03	1.3410E+00	1.3410E+03	1.8182E-03	5.0505E-07	3.9302E-04	**1.0000E+00**	4.7162E+00	hp
tons	2.8435E-04	2.8435E-01	2.8435E+02	3.8552E-04	1.0709E-07	8.3333E-05	2.1204E-01	**1.0000E+00**	tons
	W	kW	MW	ft·lbf/s	ft·lbf/h	Btu/h	hp	tons	

Entropy and Specific Heat Capacity

	J/g K	kJ/kg K	MJ/kg K	ft·lbf/lbm·R	Btu/lbm·R	cal/g K	kWh/kg K	kWh/lbm·R	
J/g K	**1.0000E+00**	1.0000E+00	1.0000E-03	5.3803E-03	4.1868E+00	4.1868E+00	3.6000E+03	1.4286E+04	J/g K
kJ/kg K	1.0000E+00	**1.0000E+00**	1.0000E-03	5.3803E-03	4.1868E+00	4.1868E+00	3.6000E+03	1.4286E+04	kJ/kg K
MJ/kg K	1.0000E-03	1.0000E-03	**1.0000E+00**	5.3803E-06	4.1868E-03	4.1868E-03	3.6000E+00	1.4286E+01	MJ/kg K
ft·lbf/lbm·R	1.8586E+02	1.8586E+02	1.8586E-05	**1.0000E+00**	7.7817E+02	7.7817E+02	6.6911E+05	2.6552E+06	ft·lbf/lbm·R
Btu/lbm·R	2.3885E-01	2.3885E-01	2.3885E-02	1.2851E-03	**1.0000E+00**	1.0000E+00	8.5985E+02	3.4121E+03	Btu/lbm·R
cal/g K	2.3885E-01	2.3885E-01	2.3885E-02	1.2851E-03	1.0000E+00	**1.0000E+00**	8.5985E+02	3.4121E+03	cal/g K
kWh/kg K	2.7778E-04	2.7778E-04	2.7778E-01	1.4945E-06	1.1630E-03	1.1630E-03	**1.0000E+00**	3.9683E+00	kWh/kg K
kWh/lbm·R	6.9999E-05	6.9999E-05	6.9999E-02	3.7662E-07	2.9307E-04	2.9307E-04	2.5200E-01	**1.0000E+00**	kWh/lbm·R
	J/g K	**kJ/kg K**	**MJ/kg K**	**ft·lbf/lbm·R**	**Btu/lbm·R**	**cal/g K**	**kWh/kg K**	**kWh/lbm·R**	

Specific Volume

	cm³/g	L/kg	m³/kg	in.³/lbm	ft³/lbm	gal/lbm	in.³/slug	ft³/slug	
cm³/g	**1.0000E+00**	1.0000E+00	1.0000E-03	3.6127E-02	6.2428E+01	8.3454E+00	1.1229E-03	1.9403E+00	cm³/g
L/kg	1.0000E+00	**1.0000E+00**	1.0000E-03	3.6127E-02	6.2428E+01	8.3454E+00	1.1229E-03	1.9403E+00	L/kg
m³/kg	1.0000E-03	1.0000E-03	**1.0000E+00**	3.6127E-05	6.2428E-02	8.3454E-03	1.1229E-06	1.9403E-03	m³/kg
in.³/lbm	2.7680E+01	2.7680E+01	2.7680E+04	**1.0000E+00**	1.7280E+03	2.3100E+02	3.1081E-02	5.3708E+01	in.³/lbm
ft³/lbm	1.6019E-02	1.6019E-02	1.6019E+01	5.7870E-04	**1.0000E+00**	1.3368E-01	1.7987E-05	3.1081E-02	ft³/lbm
gal/lbm	1.1983E-01	1.1983E-01	1.1983E+02	4.3290E-03	7.4805E+00	**1.0000E+00**	1.3455E-04	2.3250E-01	gal/lbm
in.³/slug	8.9057E+02	8.9057E+02	8.9057E+05	3.2174E+01	5.5597E+04	7.4322E+03	**1.0000E+00**	1.7280E+03	in.³/slug
ft³/slug	5.1538E-01	5.1538E-01	5.1538E+02	1.8619E-02	3.2174E+01	4.3010E+00	5.7870E-04	**1.0000E+00**	ft³/slug
	cm³/g	**L/kg**	**m³/kg**	**in.³/lbm**	**ft³/lbm**	**gal/lbm**	**in.³/slug**	**ft³/slug**	

Mass Flow Rate

	g/s	kg/s	kg/h	lbm/s	lbm/min	lbm/h	slug/s	slug/h
g/s	1.0000E+00	1.0000E+03	2.7778E-01	4.5359E+02	7.5599E+00	1.2600E-01	1.4594E+04	4.0539E+00	g/s
kg/s	1.0000E-03	1.0000E+00	2.7778E-04	4.5359E-01	7.5599E-03	1.2600E-04	1.4594E+01	4.0539E-03	kg/s
kg/h	3.6000E+00	3.6000E+03	1.0000E+00	1.6329E+03	2.7216E+01	4.5359E-01	5.2538E+04	1.4594E+01	kg/h
lbm/s	2.2046E-03	2.2046E+00	6.1240E-04	1.0000E+00	1.6667E-02	2.7778E-04	3.2174E+01	8.9372E-03	lbm/s
lbm/min	1.3228E-01	1.3228E+02	3.6744E-02	6.0000E+01	1.0000E+00	1.6667E-02	1.9304E+03	5.3623E-01	lbm/min
lbm/h	7.9366E+00	7.9366E+03	2.2046E+00	3.6000E+03	6.0000E+01	1.0000E+00	1.1583E+05	3.2174E+01	lbm/h
slug/s	6.8522E-05	6.8522E-02	1.9034E-05	3.1081E-02	5.1802E-04	8.6336E-06	1.0000E+00	2.7778E-04	slug/s
slug/h	2.4668E-01	2.4668E+02	6.8522E-02	1.1189E+02	1.8649E+00	3.1081E-02	3.6000E+03	1.0000E+00	slug/h
	g/s	kg/s	kg/h	lbm/s	lbm/min	lbm/h	slug/s	slug/h

Volume Flow Rate

	cm³/s	m³/s	m³/h	in.³/s	ft³/s	ft³/min	ft³/h	gal/min
cm³/s	1.0000E+00	1.0000E+06	2.7778E+02	1.6387E+01	2.8317E+04	4.7195E+02	7.8658E+00	6.3090E+01	cm³/s
m³/s	1.0000E-06	1.0000E+00	2.7778E-04	1.6387E-05	2.8317E-02	4.7195E-04	7.8658E-06	6.3090E-05	m³/s
m³/h	3.6000E-03	3.6000E+03	1.0000E+00	5.8993E-02	1.0194E+02	1.6990E+00	2.8317E-02	2.2713E-01	m³/h
in.³/s	6.1024E-02	6.1024E+04	1.6951E+01	1.0000E+00	1.7280E+03	2.8800E+01	4.8000E-01	3.8500E+00	in.³/s
ft³/s	3.5315E-05	3.5315E+01	9.8096E-03	5.7870E-04	1.0000E+00	1.6667E-02	2.7778E-04	2.2280E-03	ft³/s
ft³/min	2.1189E-03	2.1189E+03	5.8858E-01	3.4722E-02	6.0000E+01	1.0000E+00	1.6667E-02	1.3368E-01	ft³/min
ft³/h	1.2713E-01	1.2713E+05	3.5315E+01	2.0833E+00	3.6000E+03	6.0000E+01	1.0000E+00	8.0208E+00	ft³/h
gal/min	1.5850E-02	1.5850E+04	4.4029E+00	2.5974E-01	4.4883E+02	7.4805E+00	1.2468E-01	1.0000E+00	gal/min
	cm³/s	m³/s	m³/h	in.³/s	ft³/s	ft³/min	ft³/h	gal/min

Heat Transfer Coefficient

	W/cm^2 K	W/m^2 K	kW/cm^2 K	kW/m^2 K	Btu/s·in.^{2}·°F	Btu/s·ft^2·°F	Btu/h·in.^{2}·°F	Btu/h·ft^2·°F	
W/cm^2 K	1.0000E+00	1.0000E−04	1.0000E−03	1.0000E−01	2.9436E+02	2.0442E+00	8.1767E−02	5.6783E−04	W/cm^2 K
W/m^2 K	1.0000E+04	1.0000E+00	1.0000E+07	1.0000E+03	2.9436E+06	2.0442E+04	8.1767E+02	5.6783E+00	W/m^2 K
kW/cm^2 K	1.0000E−03	1.0000E−07	1.0000E+00	1.0000E−04	2.9436E−01	2.0442E−03	8.1767E−05	5.6783E−07	kW/cm^2 K
kW/m^2 K	1.0000E+01	1.0000E−03	1.0000E+04	1.0000E+00	2.9436E+03	2.0442E+01	8.1767E−01	5.6783E−03	kW/m^2 K
Btu/s·in.^{2}·°F	3.3972E−03	3.3972E−07	3.3972E+00	3.3972E−04	1.0000E+00	6.9444E−03	2.7778E−04	1.9290E−06	Btu/s·in.^{2}·°F
Btu/s·ft^2·°F	4.8920E−01	4.8920E−05	4.8920E+02	4.8920E−02	1.4400E+02	1.0000E+00	4.0000E−02	2.7778E−04	Btu/s·ft^2·°F
Btu/h·in.^{2}·°F	1.2230E+01	1.2230E−03	1.2230E+04	1.2230E+00	3.6000E+03	2.5000E+01	1.0000E+00	6.9444E−03	Btu/h·in.^{2}·°F
Btu/h·ft^2·°F	1.7611E+03	1.7611E−01	1.7611E+06	1.7611E+02	5.1840E+05	3.6000E+03	1.4400E+02	1.0000E+00	Btu/h·ft^2·°F
	W/cm^2 K	W/m^2 K	kW/cm^2 K	kW/m^2 K	Btu/s·in.^{2}·°F	Btu/s·ft^2·°F	Btu/h·in.^{2}·°F	Btu/h·ft^2·°F	

Thermal Resistance

	cm^2 K/W	m^2 K/W	cm^2 K/kW	m^2 K/kW	s·in.^{2}·°F/Btu	s·ft^2·°F/Btu	h·in.^{2}·°F/Btu	h·ft^2·°F/Btu	
cm^2 K/W	1.0000E+00	1.0000E−04	1.0000E+03	1.0000E−01	2.9436E+02	2.0442E+00	8.1767E−02	5.6783E−04	cm^2 K/W
m^2 K/W	1.0000E+04	1.0000E+00	1.0000E+07	1.0000E+03	2.9436E+06	2.0442E+04	8.1767E+02	5.6783E+00	m^2 K/W
cm^2 K/kW	1.0000E−03	1.0000E−07	1.0000E+00	1.0000E−04	2.9436E−01	2.0442E−03	8.1767E−05	5.6783E−07	cm^2 K/kW
m^2 K/kW	1.0000E−01	1.0000E−03	1.0000E+04	1.0000E+00	2.9436E+03	2.0442E+01	8.1767E−01	5.6783E−03	m^2 K/kW
s·in.^{2}·°F/Btu	2.9436E+02	2.9436E−01	2.9436E+06	2.9436E+03	1.0000E+00	6.9444E−03	2.7778E−04	1.9290E−06	s·in.^{2}·°F/Btu
s·ft^2·°F/Btu	2.0442E+00	2.0442E−03	2.0442E+04	2.0442E+01	1.4400E+02	1.0000E+00	4.0000E−02	2.7778E−04	s·ft^2·°F/Btu
h·in.^{2}·°F/Btu	8.1767E−02	8.1767E−05	8.1767E+02	8.1767E−01	3.6000E+03	2.5000E+01	1.0000E+00	6.9444E−03	h·in.^{2}·°F/Btu
h·ft^2·°F/Btu	5.6783E−04	5.6783E−07	5.6783E+00	5.6783E−03	1.9290E−06	2.7778E−04	6.9444E−03	1.0000E+00	h·ft^2·°F/Btu
	cm^2 K/W	m^2 K/W	cm^2 K/kW	m^2 K/kW	s·in.^{2}·°F/Btu	s·ft^2·°F/Btu	h·in.^{2}·°F/Btu	h·ft^2·°F/Btu	

Appendix B: Thermophysical Properties

Throughout this text, various thermophysical properties are introduced. This appendix contains thermophysical properties for several substances: liquids, gases, and solids. The properties contained in this appendix were calculated using Engineering Equation Solver (EES). All tables are presented in both International System (SI) and Inch–Pound (IP) units.

Appendix B.1: Thermophysical Properties of Saturated Liquids

Appendix B.1 lists liquid properties for several substances as a function of temperature. The liquid properties are estimated on the saturated liquid line using temperature (T) and a quality of zero ($x = 0$). Properties listed in the tables include density (ρ), heat capacity (c_p), dynamic viscosity (μ), thermal conductivity (k), and Prandtl number (Pr).

Appendix B.2: Saturation Properties

Appendix B.2 lists saturation properties for several substances as a function of temperature. The properties listed are the saturation pressure (P_{sat}), saturated liquid density (ρ_f), saturated vapor-specific volume (v_g), saturated vapor enthalpy (h_g), and saturated liquid enthalpy (h_f).

Appendix B.3: Properties of Metals and Alloys

Appendix B.3 lists the density (ρ), heat capacity (c), and thermal conductivity (k) as a function of temperature for several common metals and alloys used in various heat transfer applications.

Appendix B.4: Properties of Air at 1 atmosphere

Appendix B.4 contains the thermophysical properties of air as a function of temperature for a pressure of 1 atmosphere. The properties listed are density (ρ), heat capacity (c_p), dynamic viscosity (μ), thermal conductivity (k), and Prandtl number (Pr).

APPENDIX B.1

Thermophysical Properties of Saturated Liquids (SI Units)

T (°C)	ρ (kg/m³)	c_p (kJ/kg·K)	μ (kg/m·h)	k (W/m·K)	Pr
			Water		
10	999.65	4.1956	4.7016	0.5800	9.447
15	999.05	4.1888	4.0954	0.5893	8.086
20	998.16	4.1844	3.6059	0.5984	7.004
25	997.00	4.1816	3.2044	0.6071	6.131
30	995.60	4.1801	2.8705	0.6155	5.416
35	993.99	4.1795	2.5895	0.6233	4.824
40	992.17	4.1797	2.3507	0.6306	4.328
45	990.17	4.1804	2.1458	0.6373	3.910
50	987.99	4.1816	1.9686	0.6435	3.553
55	985.65	4.1832	1.8143	0.6492	3.247
60	983.15	4.1852	1.6790	0.6543	2.983
65	980.51	4.1875	1.5597	0.6590	2.753
70	977.73	4.1903	1.4539	0.6631	2.552
75	974.81	4.1934	1.3598	0.6668	2.376
80	971.76	4.1969	1.2756	0.6700	2.220
85	968.59	4.2009	1.2000	0.6728	2.081
90	965.29	4.2053	1.1318	0.6752	1.958
95	961.88	4.2102	1.0702	0.6773	1.848
100	958.35	4.2157	1.0143	0.6791	1.749
105	954.70	4.2217	0.9634	0.6805	1.660
			Benzene		
15	884.07	1.7099	2.5555	0.1486	8.166
20	878.76	1.7223	2.3557	0.1471	7.661
25	873.44	1.7353	2.1799	0.1456	7.218
30	868.11	1.7489	2.0244	0.1440	6.828
35	862.76	1.7629	1.8861	0.1425	6.482
40	857.39	1.7773	1.7628	0.1409	6.175
45	852.00	1.7921	1.6522	0.1394	5.901
50	846.59	1.8073	1.5526	0.1378	5.656
55	841.15	1.8228	1.4626	0.1363	5.435
60	835.68	1.8387	1.3810	0.1347	5.237
65	830.19	1.8548	1.3067	0.1331	5.058
70	824.66	1.8712	1.2388	0.1315	4.896
75	819.09	1.8879	1.1765	0.1299	4.748
80	813.49	1.9048	1.1192	0.1283	4.614
85	807.85	1.9219	1.0662	0.1267	4.492
90	802.17	1.9393	1.0172	0.1251	4.380
95	796.44	1.9569	0.9717	0.1235	4.277
100	790.66	1.9747	0.9293	0.1219	4.183
105	784.83	1.9928	0.8896	0.1202	4.096
110	778.94	2.0111	0.8524	0.1186	4.016

APPENDIX B.1 (*Continued*)

Thermophysical Properties of Saturated Liquids (SI Units)

T (°C)	ρ (kg/m³)	c_p (kJ/kg·K)	μ (kg/m·h)	k (W/m·K)	Pr
			Decane		
−10	753.72	2.0713	5.6413	0.1426	22.77
−5	749.80	2.0875	5.1158	0.1414	20.98
0	745.90	2.1041	4.6611	0.1402	19.43
5	742.00	2.1211	4.2655	0.1391	18.07
10	738.10	2.1384	3.9194	0.1379	16.89
15	734.21	2.1562	3.6151	0.1367	15.84
20	730.32	2.1742	3.3462	0.1355	14.91
25	726.44	2.1926	3.1075	0.1343	14.09
30	722.55	2.2113	2.8946	0.1332	13.35
35	718.66	2.2302	2.7040	0.1320	12.69
40	714.77	2.2494	2.5326	0.1308	12.10
45	710.88	2.2688	2.3779	0.1296	11.56
50	706.98	2.2884	2.2377	0.1284	11.08
55	703.07	2.3082	2.1102	0.1272	10.63
60	699.15	2.3282	1.9938	0.1260	10.23
65	695.23	2.3483	1.8873	0.1248	9.862
70	691.29	2.3686	1.7894	0.1236	9.523
75	687.34	2.3891	1.6992	0.1224	9.211
80	683.37	2.4096	1.6158	0.1212	8.922
85	679.39	2.4303	1.5385	0.1200	8.655
			Acetone		
−40	854.82	2.0281	2.4949	0.1802	7.802
−35	849.52	2.0338	2.2893	0.1788	7.235
−30	844.22	2.0400	2.1119	0.1774	6.747
−25	838.91	2.0468	1.9578	0.1760	6.325
−20	833.58	2.0540	1.8231	0.1746	5.959
−15	828.24	2.0618	1.7045	0.1731	5.638
−10	822.88	2.0701	1.5996	0.1717	5.357
−5	817.50	2.0790	1.5062	0.1703	5.108
0	812.10	2.0884	1.4226	0.1688	4.888
5	806.67	2.0983	1.3473	0.1674	4.691
10	801.21	2.1087	1.2792	0.1659	4.516
15	795.72	2.1197	1.2173	0.1644	4.358
20	790.20	2.1311	1.1606	0.1630	4.216
25	784.63	2.1431	1.1086	0.1615	4.087
30	779.02	2.1556	1.0606	0.1600	3.970
35	773.37	2.1687	1.0161	0.1584	3.863
40	767.66	2.1822	0.9747	0.1569	3.765
45	761.91	2.1963	0.9359	0.1554	3.675
50	756.09	2.2109	0.8994	0.1538	3.591
55	750.22	2.2260	0.8650	0.1522	3.514

(*Continued*)

APPENDIX B.1   (*Continued*)

Thermophysical Properties of Saturated Liquids (SI Units)

T (°C)	ρ (kg/m³)	c_p (kJ/kg·K)	μ (kg/m·h)	k (W/m·K)	Pr
Cyclohexane					
10	788.02	1.6844	4.0982	0.1246	15.39
15	783.29	1.7282	3.7492	0.1237	14.55
20	778.56	1.7693	3.4400	0.1228	13.76
25	773.82	1.8081	3.1651	0.1220	13.04
30	769.06	1.8450	2.9200	0.1211	12.36
35	764.30	1.8800	2.7007	0.1202	11.74
40	759.51	1.9136	2.5039	0.1193	11.16
45	754.71	1.9458	2.3268	0.1183	10.63
50	749.88	1.9769	2.1670	0.1174	10.13
55	745.02	2.0070	2.0224	0.1165	9.678
60	740.14	2.0362	1.8913	0.1156	9.256
65	735.22	2.0646	1.7720	0.1146	8.866
70	730.27	2.0925	1.6634	0.1137	8.505
75	725.29	2.1198	1.5642	0.1127	8.170
80	720.27	2.1467	1.4733	0.1118	7.860
85	715.20	2.1732	1.3901	0.1108	7.573
90	710.10	2.1994	1.3135	0.1098	7.307
95	704.94	2.2255	1.2431	0.1088	7.060
100	699.74	2.2513	1.1781	0.1079	6.831
105	694.48	2.2771	1.1181	0.1069	6.618
Ethanol (Ethyl alcohol)					
−10	814.29	2.0783	7.9886	0.1790	25.77
−5	810.19	2.1348	7.0926	0.1776	23.68
0	806.10	2.1939	6.3279	0.1763	21.87
5	802.01	2.2547	5.6714	0.1749	20.30
10	797.90	2.3167	5.1043	0.1736	18.92
15	793.76	2.3794	4.6119	0.1722	17.70
20	789.60	2.4423	4.1820	0.1708	16.61
25	785.40	2.5051	3.8048	0.1694	15.62
30	781.15	2.5675	3.4723	0.1680	14.74
35	776.84	2.6293	3.1778	0.1666	13.93
40	772.48	2.6904	2.9160	0.1652	13.19
45	768.05	2.7505	2.6822	0.1638	12.51
50	763.55	2.8097	2.4726	0.1623	11.89
55	758.97	2.8679	2.2840	0.1608	11.31
60	754.30	2.9249	2.1138	0.1594	10.78
65	749.54	2.9809	1.9595	0.1579	10.28
70	744.68	3.0358	1.8193	0.1564	9.812
75	739.72	3.0897	1.6916	0.1548	9.377
80	734.65	3.1426	1.5748	0.1533	8.968
85	729.46	3.1946	1.4677	0.1517	8.584

APPENDIX B.1　　*(Continued)*

Thermophysical Properties of Saturated Liquids (SI Units)

T (°C)	ρ (kg/m³)	c_p (kJ/kg·K)	μ (kg/m·h)	k (W/m·K)	Pr
			Ethylene		
−110	576.87	2.3799	0.6414	0.1946	2.180
−105	569.79	2.3898	0.5871	0.1892	2.060
−100	562.61	2.4032	0.5405	0.1838	1.963
−95	555.29	2.4200	0.5009	0.1785	1.887
−90	547.84	2.4398	0.4675	0.1731	1.830
−85	540.24	2.4627	0.4396	0.1678	1.793
−80	532.47	2.4890	0.4166	0.1624	1.773
−75	524.50	2.5190	0.3976	0.1571	1.771
−70	516.33	2.5533	0.3820	0.1517	1.786
−65	507.92	2.5925	0.3690	0.1464	1.815
−60	499.24	2.6377	0.3580	0.1411	1.859
−55	490.26	2.6901	0.3481	0.1357	1.917
−50	480.93	2.7513	0.3387	0.1304	1.985
−45	471.20	2.8235	0.3291	0.1251	2.064
−40	461.02	2.9096	0.3186	0.1197	2.150
−35	450.43	3.0109	0.3063	0.1144	2.239
−30	439.06	3.1388	0.2916	0.1091	2.330
−25	426.72	3.3054	0.2738	0.1038	2.422
−20	413.55	3.5195	0.2522	0.0985	2.503
−15	399.07	3.8164	0.2260	0.0932	2.571
			Heptane		
−15	712.88	2.1064	2.4084	1.0000	1.409
−10	708.74	2.1212	2.2570	0.1366	9.733
−5	704.60	2.1366	2.1177	0.1353	9.286
0	700.45	2.1526	1.9894	0.1341	8.873
5	696.28	2.1692	1.8709	0.1328	8.491
10	692.11	2.1863	1.7615	0.1315	8.135
15	687.92	2.2039	1.6602	0.1302	7.804
20	683.72	2.2221	1.5663	0.1290	7.496
25	679.50	2.2407	1.4791	0.1277	7.208
30	675.27	2.2597	1.3982	0.1265	6.939
35	671.01	2.2792	1.3229	0.1252	6.688
40	666.72	2.2992	1.2528	0.1240	6.452
45	662.42	2.3195	1.1874	0.1228	6.231
50	658.08	2.3402	1.1263	0.1216	6.023
55	653.72	2.3613	1.0693	0.1204	5.827
60	649.32	2.3827	1.0160	0.1192	5.643
65	644.89	2.4045	0.9660	0.1180	5.470
70	640.42	2.4266	0.9192	0.1168	5.306
75	635.91	2.4490	0.8753	0.1156	5.151
80	631.36	2.4717	0.8341	0.1144	5.005

(Continued)

APPENDIX B.1 (*Continued*)

Thermophysical Properties of Saturated Liquids (SI Units)

T (°C)	ρ (kg/m³)	c_p (kJ/kg·K)	μ (kg/m·h)	k (W/m·K)	Pr
			Propane		
−15	547.92	2.4064	0.5398	0.1162	3.105
−10	541.52	2.4398	0.5120	0.1134	3.061
−5	534.99	2.4749	0.4858	0.1106	3.020
0	528.31	2.5119	0.4610	0.1079	2.982
5	521.47	2.5511	0.4376	0.1052	2.948
10	514.45	2.5927	0.4153	0.1026	2.916
15	507.24	2.6371	0.3941	0.1000	2.888
20	499.80	2.6848	0.3739	0.0975	2.861
25	492.11	2.7363	0.3545	0.0950	2.837
30	484.15	2.7926	0.3360	0.0925	2.816
35	475.87	2.8548	0.3181	0.0901	2.798
40	467.23	2.9246	0.3008	0.0878	2.784
45	458.17	3.0043	0.2841	0.0855	2.774
50	448.63	3.0971	0.2678	0.0832	2.770
55	438.52	3.2080	0.2519	0.0809	2.774
60	427.71	3.3446	0.2362	0.0786	2.791
65	416.04	3.5187	0.2207	0.0764	2.825
70	403.29	3.7512	0.2053	0.0741	2.888
75	389.09	4.0808	0.1896	0.0717	2.997
80	372.85	4.5926	0.1735	0.0694	3.192
			Hexane		
−15	693.24	2.0986	1.7854	0.1390	7.486
−10	688.79	2.1162	1.6751	0.1373	7.173
−5	684.32	2.1343	1.5713	0.1356	6.872
0	679.84	2.1529	1.4738	0.1339	6.584
5	675.34	2.1720	1.3823	0.1322	6.309
10	670.82	2.1916	1.2967	0.1305	6.047
15	666.28	2.2116	1.2167	0.1289	5.799
20	661.71	2.2321	1.1421	0.1273	5.563
25	657.11	2.2530	1.0727	0.1257	5.341
30	652.48	2.2744	1.0083	0.1241	5.133
35	647.81	2.2961	0.9487	0.1226	4.937
40	643.11	2.3183	0.8937	0.1210	4.756
45	638.37	2.3408	0.8430	0.1195	4.587
50	633.59	2.3638	0.7964	0.1180	4.432
55	628.77	2.3871	0.7538	0.1165	4.290
60	623.90	2.4109	0.7149	0.1150	4.162
65	618.97	2.4350	0.6794	0.1136	4.046
70	613.99	2.4596	0.6473	0.1121	3.944
75	608.96	2.4846	0.6182	0.1107	3.854
80	603.86	2.5100	0.5920	0.1093	3.776

APPENDIX B.1 *(Continued)*

Thermophysical Properties of Saturated Liquids (SI Units)

T (°C)	ρ (kg/m³)	c_p (kJ/kg·K)	μ (kg/m·h)	k (W/m·K)	Pr
			Octane		
−15	729.47	2.0998	3.6731	0.1390	15.41
−10	725.48	2.1137	3.4191	0.1376	14.59
−5	721.50	2.1282	3.1869	0.1361	13.84
0	717.45	2.1434	2.9744	0.1347	13.14
5	713.52	2.1590	2.7796	0.1333	12.50
10	709.53	2.1751	2.6006	0.1319	11.91
15	705.53	2.1918	2.4360	0.1306	11.36
20	701.52	2.2090	2.2844	0.1292	10.85
25	697.50	2.2267	2.1445	0.1279	10.37
30	693.47	2.2447	2.0153	0.1265	9.931
35	689.43	2.2632	1.8959	0.1252	9.518
40	685.37	2.2821	1.7852	0.1239	9.133
45	681.29	2.3013	1.6827	0.1226	8.771
50	677.20	2.3208	1.5875	0.1214	8.433
55	673.09	2.3407	1.4990	0.1201	8.116
60	668.95	2.3609	1.4166	0.1188	7.817
65	664.79	2.3814	1.3400	0.1176	7.536
70	660.61	2.4022	1.2685	0.1164	7.272
75	656.39	2.4232	1.2017	0.1152	7.022
80	652.15	2.4444	1.1394	0.1140	6.786
			Propylene		
−60	624.13	2.1350	0.7341	0.1553	2.803
−55	618.06	2.1525	0.6857	0.1529	2.682
−50	611.92	2.1711	0.6419	0.1504	2.574
−45	605.70	2.1909	0.6024	0.1479	2.478
−40	599.39	2.2121	0.5667	0.1455	2.394
−35	592.99	2.2346	0.5343	0.1430	2.320
−30	586.48	2.2586	0.5050	0.1405	2.255
−25	579.86	2.2842	0.4784	0.1380	2.200
−20	573.11	2.3115	0.4544	0.1355	2.153
−15	566.23	2.3406	0.4325	0.1330	2.114
−10	559.20	2.3716	0.4127	0.1305	2.083
−5	552.01	2.4049	0.3947	0.1280	2.060
0	544.64	2.4406	0.3784	0.1255	2.045
5	537.08	2.4791	0.3635	0.1229	2.037
10	529.30	2.5207	0.3501	0.1203	2.037
15	521.29	2.5659	0.3379	0.1177	2.045
20	513.02	2.6155	0.3269	0.1151	2.063
25	504.46	2.6701	0.3169	0.1125	2.089
30	495.58	2.7310	0.3079	0.1099	2.126
35	486.33	2.7996	0.2997	0.1072	2.174

(Continued)

APPENDIX B.1   (*Continued*)

Thermophysical Properties of Saturated Liquids (SI Units)

T (°C)	ρ (kg/m³)	c_p (kJ/kg·K)	μ (kg/m·h)	k (W/m·K)	Pr
		Refrigerant 22			
−40	1396.3	1.1122	1.1468	0.1124	3.151
−35	1381.7	1.1194	1.0772	0.1102	3.040
−30	1366.9	1.1273	1.0145	0.1079	2.944
−25	1351.8	1.1358	0.9577	0.1057	2.859
−20	1336.5	1.1451	0.9059	0.1035	2.785
−15	1330.7	1.1413	0.8876	0.1032	2.727
−10	1305.0	1.1663	0.8144	0.0990	2.664
−5	1288.8	1.1783	0.7736	0.0968	2.615
0	1272.2	1.1915	0.7356	0.0946	2.572
5	1255.2	1.2059	0.6998	0.0924	2.536
10	1237.7	1.2218	0.6661	0.0902	2.506
15	1219.8	1.2393	0.6341	0.0880	2.481
20	1201.3	1.2589	0.6036	0.0858	2.461
25	1182.3	1.2809	0.5744	0.0835	2.448
30	1171.2	1.2834	0.5603	0.0826	2.418
35	1150.6	1.3100	0.5320	0.0802	2.414
40	1129.0	1.3408	0.5045	0.0777	2.417
45	1098.1	1.4057	0.4667	0.0740	2.463
50	1074.5	1.4520	0.4412	0.0714	2.491
55	1049.5	1.5086	0.4161	0.0688	2.534
		Refrigerant 123			
−40	1619.1	0.9494	3.5306	0.0949	9.809
−35	1607.7	0.9551	3.2714	0.0935	9.286
−30	1597.0	0.9608	3.0521	0.0921	8.840
−25	1585.4	0.9671	2.8400	0.0907	8.410
−20	1573.1	0.9740	2.6371	0.0892	7.997
−15	1561.4	0.9806	2.4636	0.0878	7.640
−10	1550.2	0.9869	2.3131	0.0865	7.328
−5	1537.6	0.9941	2.1601	0.0851	7.010
0	1525.6	1.0009	2.0270	0.0837	6.730
5	1514.0	1.0074	1.9095	0.0825	6.479
10	1501.6	1.0143	1.7958	0.0811	6.236
15	1489.2	1.0212	1.6907	0.0798	6.009
20	1476.6	1.0283	1.5934	0.0785	5.797
25	1463.9	1.0353	1.5032	0.0772	5.599
30	1451.0	1.0425	1.4194	0.0759	5.414
35	1437.7	1.0501	1.3394	0.0746	5.237
40	1424.8	1.0573	1.2686	0.0734	5.077
45	1411.4	1.0650	1.2006	0.0721	4.925
50	1397.5	1.0731	1.1359	0.0708	4.779
55	1384.0	1.0810	1.0776	0.0696	4.647

APPENDIX B.1 (*Continued*)

Thermophysical Properties of Saturated Liquids (SI Units)

T (°C)	ρ (kg/m³)	c_p (kJ/kg·K)	μ (kg/m·h)	k (W/m·K)	Pr
			Refrigerant 407C		
−40	1370.6	1.2753	1.2461	0.1138	3.877
−35	1354.7	1.2942	1.1669	0.1114	3.767
−30	1338.5	1.3131	1.0927	0.1089	3.660
−25	1321.9	1.3318	1.0232	0.1065	3.555
−20	1305.1	1.3507	0.9581	0.1041	3.454
−15	1287.9	1.3697	0.8975	0.1017	3.357
−10	1270.3	1.3889	0.8409	0.0994	3.264
−5	1252.4	1.4085	0.7883	0.0971	3.176
0	1234.1	1.4286	0.7394	0.0948	3.094
5	1215.4	1.4494	0.6940	0.0926	3.017
10	1196.3	1.4712	0.6518	0.0904	2.946
15	1176.8	1.4941	0.6128	0.0883	2.881
20	1156.7	1.5186	0.5765	0.0861	2.824
25	1136.1	1.5452	0.5429	0.0840	2.773
30	1115.0	1.5746	0.5116	0.0819	2.731
35	1093.2	1.6078	0.4825	0.0799	2.698
40	1070.6	1.6463	0.4553	0.0778	2.676
45	1047.3	1.6926	0.4297	0.0758	2.666
50	1022.9	1.7506	0.4055	0.0737	2.675
55	997.3	1.8273	0.3824	0.0716	2.709
			Refrigerant 23		
−100	1501.5	1.2066	1.4397	0.1502	3.212
−95	1482.4	1.2141	1.3086	0.1447	3.050
−90	1464.8	1.2207	1.2029	0.1400	2.915
−85	1446.9	1.2281	1.1101	0.1355	2.794
−80	1428.9	1.2363	1.0279	0.1314	2.687
−75	1410.5	1.2453	0.9546	0.1274	2.591
−70	1391.7	1.2554	0.8886	0.1237	2.504
−65	1372.6	1.2666	0.8287	0.1202	2.426
−60	1352.9	1.2793	0.7742	0.1168	2.356
−55	1332.6	1.2937	0.7240	0.1135	2.293
−50	1311.7	1.3102	0.6776	0.1102	2.237
−45	1290.0	1.3293	0.6344	0.1071	2.187
−40	1267.4	1.3515	0.5939	0.1040	2.144
−35	1243.8	1.3777	0.5557	0.1009	2.108
−30	1218.9	1.4091	0.5195	0.0978	2.079
−25	1192.6	1.4471	0.4850	0.0947	2.058
−20	1164.7	1.4938	0.4518	0.0916	2.048
−15	1132.0	1.5633	0.4169	0.0880	2.057
−10	1101.0	1.6346	0.3872	0.0848	2.072
−5	1067.3	1.7276	0.3581	0.0815	2.108

(*Continued*)

APPENDIX B.1   (*Continued*)

Thermophysical Properties of Saturated Liquids (SI Units)

T (°C)	ρ (kg/m³)	c_p (kJ/kg·K)	μ (kg/m·h)	*k* (W/m·K)	Pr
			Refrigerant 134A		
−40	1417.8	1.2545	1.6760	0.1101	5.305
−35	1403.3	1.2634	1.5508	0.1084	5.022
−30	1388.6	1.2727	1.4389	0.1066	4.773
−25	1373.7	1.2824	1.3384	0.1047	4.552
−20	1358.5	1.2927	1.2475	0.1028	4.356
−15	1343.1	1.3036	1.1649	0.1009	4.182
−10	1327.4	1.3152	1.0896	0.0989	4.026
−5	1311.4	1.3275	1.0205	0.0968	3.887
0	1295.1	1.3406	0.9568	0.0947	3.764
5	1278.3	1.3547	0.8980	0.0925	3.653
10	1261.2	1.3700	0.8433	0.0903	3.555
15	1243.6	1.3865	0.7924	0.0880	3.468
20	1225.5	1.4045	0.7447	0.0856	3.392
25	1206.9	1.4243	0.7000	0.0833	3.327
30	1187.6	1.4463	0.6579	0.0808	3.271
35	1167.6	1.4708	0.6181	0.0783	3.225
40	1146.8	1.4984	0.5803	0.0757	3.190
45	1125.0	1.5299	0.5443	0.0731	3.165
50	1102.3	1.5663	0.5098	0.0704	3.152
55	1078.3	1.6090	0.4767	0.0676	3.152
			Refrigerant 507A		
−40	1294.910	1.2357	1.1553	0.0904	4.387
−35	1279.020	1.2469	1.0766	0.0884	4.218
−30	1262.760	1.2577	1.0029	0.0864	4.053
−25	1246.110	1.2682	0.9339	0.0845	3.893
−20	1229.020	1.2788	0.8694	0.0826	3.740
−15	1211.470	1.2896	0.8094	0.0807	3.594
−10	1193.410	1.3009	0.7534	0.0788	3.456
−5	1174.780	1.3133	0.7014	0.0769	3.327
0	1155.520	1.3270	0.6531	0.0750	3.208
5	1135.560	1.3428	0.6083	0.0732	3.100
10	1114.810	1.3613	0.5668	0.0713	3.005
15	1093.170	1.3833	0.5284	0.0695	2.922
20	1070.500	1.4102	0.4927	0.0676	2.854
25	1046.630	1.4433	0.4596	0.0657	2.803
30	1021.330	1.4848	0.4288	0.0638	2.770
35	994.330	1.5377	0.3999	0.0619	2.760
40	965.190	1.6066	0.3727	0.0599	2.776
45	933.340	1.6991	0.3466	0.0578	2.829
50	897.840	1.8288	0.3211	0.0556	2.933
55	857.12	2.0233	0.2953	0.0532	3.119

APPENDIX B.1   (*Continued*)

Thermophysical Properties of Saturated Liquids (SI Units)

T (°C)	ρ (kg/m³)	c_p (kJ/kg·K)	μ (kg/m·h)	k (W/m·K)	Pr
			40% Ethylene glycol		
−40	1070.5	3.2500	182.94	0.3793	435.5
−35	1069.8	3.2739	131.12	0.3830	311.3
−30	1068.9	3.2976	95.796	0.3868	226.8
−25	1067.9	3.3211	71.291	0.3907	168.3
−20	1066.7	3.3443	54.002	0.3945	127.2
−15	1065.3	3.3672	41.607	0.3983	97.70
−10	1063.8	3.3898	32.582	0.4022	76.28
−5	1062.2	3.4122	25.914	0.4060	60.49
0	1060.4	3.4342	20.918	0.4099	48.68
5	1058.5	3.4559	17.125	0.4138	39.73
10	1056.4	3.4773	14.207	0.4176	32.86
15	1054.2	3.4984	11.937	0.4215	27.52
20	1051.9	3.5190	10.149	0.4253	23.33
25	1049.4	3.5393	8.7256	0.4291	19.99
30	1046.8	3.5592	7.5806	0.4329	17.31
35	1044.2	3.5787	6.6501	0.4367	15.14
40	1041.4	3.5978	5.8864	0.4405	13.35
45	1038.4	3.6165	5.2535	0.4443	11.88
50	1035.4	3.6347	4.7241	0.4480	10.65
55	1032.3	3.6524	4.2769	0.4517	9.607
			40% Propylene glycol		
−40	1053.7	3.5082	984.92	0.3653	2627
−35	1053.0	3.5251	603.02	0.3679	1605
−30	1052.2	3.5419	380.64	0.3706	1011
−25	1051.0	3.5587	247.41	0.3733	655.1
−20	1049.7	3.5754	165.41	0.3761	436.8
−15	1048.2	3.5920	113.61	0.3790	299.1
−10	1046.4	3.6086	80.069	0.3819	210.2
−5	1044.5	3.6251	57.836	0.3848	151.3
0	1042.4	3.6416	42.766	0.3878	111.5
5	1040.1	3.6580	32.334	0.3909	84.05
10	1037.6	3.6743	24.967	0.3940	64.68
15	1035.0	3.6905	19.665	0.3971	50.77
20	1032.3	3.7067	15.782	0.4003	40.60
25	1029.4	3.7229	12.889	0.4035	33.04
30	1026.4	3.7389	10.699	0.4067	27.32
35	1023.3	3.7549	9.0175	0.4099	22.94
40	1020.1	3.7708	7.7068	0.4132	19.54
45	1016.7	3.7867	6.6714	0.4165	16.85
50	1013.3	3.8025	5.8425	0.4198	14.70
55	1009.9	3.8182	5.1702	0.4232	12.96

(*Continued*)

APPENDIX B.1 (*Continued*)

Thermophysical Properties of Saturated Liquids (SI Units)

T (°C)	ρ (kg/m³)	c_p (kJ/kg·K)	μ (kg/m·h)	k (W/m·K)	Pr
20% Magnesium chloride					
−30	1187.6	2.9727	56.621	0.4665	100.2
−27	1187.6	2.9807	49.899	0.4708	87.75
−24	1187.5	2.9888	44.085	0.4752	77.01
−21	1187.2	2.9969	39.044	0.4798	67.75
−18	1186.8	3.0051	34.665	0.4844	59.73
−15	1186.3	3.0133	30.850	0.4892	52.79
−12	1185.8	3.0215	27.521	0.4940	46.76
−9	1185.1	3.0297	24.607	0.4989	41.51
−6	1184.3	3.0380	22.053	0.5039	36.93
−3	1183.4	3.0462	19.808	0.5089	32.94
0	1182.5	3.0545	17.830	0.5140	29.44
3	1181.5	3.0628	16.086	0.5190	26.37
6	1180.4	3.0711	14.542	0.5242	23.67
9	1179.3	3.0795	13.175	0.5293	21.29
12	1178.2	3.0878	11.961	0.5344	19.20
15	1177.0	3.0961	10.880	0.5395	17.34
18	1175.7	3.1044	9.9169	0.5446	15.70
21	1174.5	3.1128	9.0565	0.5496	14.25
24	1173.2	3.1211	8.2866	0.5546	12.95
27	1171.9	3.1294	7.5963	0.5596	11.80
20% Ethylene glycol					
−40	1029.8	3.7984	75.102	0.4310	183.8
−35	1030.4	3.8053	56.513	0.4381	136.4
−30	1030.8	3.8126	43.170	0.4450	102.7
−25	1031.0	3.8201	33.460	0.4519	78.58
−20	1031.0	3.8278	26.301	0.4586	60.98
−15	1030.8	3.8358	20.956	0.4652	48.00
−10	1030.4	3.8440	16.916	0.4716	38.30
−5	1029.9	3.8523	13.827	0.4779	30.96
0	1029.1	3.8608	11.439	0.4841	25.34
5	1028.1	3.8695	9.5738	0.4902	20.99
10	1026.9	3.8783	8.1014	0.4962	17.59
15	1025.6	3.8872	6.9282	0.5020	14.90
20	1024.1	3.8962	5.9847	0.5077	12.76
25	1022.4	3.9053	5.2193	0.5132	11.03
30	1020.6	3.9144	4.5931	0.5186	9.630
35	1018.6	3.9236	4.0768	0.5239	8.481
40	1016.4	3.9328	3.6477	0.5291	7.532
45	1014.1	3.9419	3.2886	0.5341	6.742
50	1011.6	3.9511	2.9857	0.5390	6.080
55	1009.0	3.9603	2.7286	0.5437	5.521

APPENDIX B.1 *(Continued)*

Thermophysical Properties of Saturated Liquids (SI Units)

T (°C)	ρ (kg/m³)	c_p (kJ/kg·K)	μ (kg/m·h)	k (W/m·K)	Pr
			20% Propylene glycol		
−40	1018.4	3.8647	161.73	0.4271	406.5
−35	1019.7	3.8725	112.21	0.4327	279.0
−30	1020.6	3.8807	79.631	0.4382	195.9
−25	1021.2	3.8892	57.746	0.4438	140.6
−20	1021.6	3.8981	42.754	0.4493	103.0
−15	1021.6	3.9072	32.288	0.4548	77.05
−10	1021.3	3.9165	24.851	0.4603	58.74
−5	1020.8	3.9261	19.474	0.4657	45.60
0	1020.1	3.9359	15.525	0.4711	36.03
5	1019.1	3.9459	12.579	0.4765	28.94
10	1017.9	3.9561	10.349	0.4818	23.61
15	1016.4	3.9664	8.6382	0.4870	19.54
20	1014.8	3.9768	7.3083	0.4922	16.40
25	1012.9	3.9873	6.2616	0.4973	13.94
30	1010.9	3.9978	5.4280	0.5024	12.00
35	1008.7	4.0084	4.7565	0.5074	10.44
40	1006.4	4.0190	4.2095	0.5123	9.174
45	1003.9	4.0296	3.7591	0.5171	8.137
50	1001.2	4.0402	3.3841	0.5218	7.278
55	998.47	4.0507	3.0685	0.5265	6.558
			20% Sodium chloride		
−30	1168.1	3.3152	32.020	0.5034	58.57
−27	1167.0	3.3235	27.891	0.5078	50.71
−24	1166.0	3.3314	24.396	0.5121	44.08
−21	1164.9	3.3390	21.428	0.5165	38.48
−18	1163.8	3.3463	18.899	0.5209	33.72
−15	1162.7	3.3533	16.738	0.5253	29.68
−12	1161.5	3.3600	14.885	0.5298	26.22
−9	1160.3	3.3663	13.293	0.5342	23.27
−6	1159.1	3.3723	11.920	0.5387	20.73
−3	1157.9	3.3780	10.734	0.5432	18.54
0	1156.7	3.3833	9.7057	0.5477	16.65
3	1155.4	3.3883	8.8125	0.5522	15.02
6	1154.1	3.3930	8.0347	0.5567	13.60
9	1152.8	3.3974	7.3561	0.5613	12.37
12	1151.5	3.4014	6.7628	0.5658	11.29
15	1150.1	3.4052	6.2431	0.5704	10.35
18	1148.7	3.4086	5.7874	0.5750	9.529
21	1147.3	3.4116	5.3873	0.5796	8.808
24	1145.9	3.4144	5.0356	0.5843	8.174
27	1144.4	3.4168	4.7265	0.5889	7.617

(Continued)

APPENDIX B.1 (*Continued*)

Thermophysical Properties of Saturated Liquids (IP Units)

T (°F)	ρ (lbm/ft³)	c_p (Btu/lbm-R)	μ (lbm/ft-h)	k (Btu/hr-ft-°F)	Pr
			Water		
40	62.423	1.0048	3.7382	0.3290	11.41
50	62.406	1.0021	3.1593	0.3351	9.447
60	62.363	1.0003	2.7120	0.3411	7.953
70	62.298	0.9992	2.3586	0.3469	6.793
80	62.213	0.9986	2.0740	0.3525	5.876
90	62.110	0.9983	1.8411	0.3577	5.139
100	61.991	0.9982	1.6478	0.3625	4.537
110	61.857	0.9984	1.4856	0.3670	4.041
120	61.709	0.9987	1.3479	0.3711	3.627
130	61.549	0.9991	1.2300	0.3748	3.279
140	61.376	0.9996	1.1282	0.3781	2.983
150	61.192	1.0002	1.0397	0.3810	2.729
160	60.998	1.0010	0.9623	0.3836	2.511
170	60.793	1.0018	0.8942	0.3859	2.321
180	60.578	1.0028	0.8339	0.3879	2.156
190	60.354	1.0039	0.7803	0.3896	2.011
200	60.120	1.0052	0.7325	0.3910	1.883
210	59.877	1.0066	0.6896	0.3922	1.770
220	59.626	1.0082	0.6510	0.3931	1.669
230	59.366	1.0099	0.6161	0.3939	1.580
			Benzene		
60	55.154	0.4087	1.7014	0.0858	8.106
70	54.786	0.4120	1.5554	0.0848	7.557
80	54.416	0.4155	1.4286	0.0838	7.082
90	54.046	0.4192	1.3177	0.0828	6.668
100	53.674	0.4230	1.2202	0.0818	6.307
110	53.301	0.4268	1.1341	0.0808	5.988
120	52.926	0.4309	1.0576	0.0798	5.707
130	52.549	0.4350	0.9893	0.0788	5.458
140	52.170	0.4392	0.9280	0.0778	5.237
150	51.789	0.4434	0.8728	0.0768	5.039
160	51.405	0.4478	0.8228	0.0758	4.861
170	51.018	0.4522	0.7774	0.0748	4.702
180	50.628	0.4568	0.7359	0.0737	4.558
190	50.236	0.4613	0.6979	0.0727	4.428
200	49.839	0.4660	0.6629	0.0717	4.310
210	49.440	0.4707	0.6306	0.0706	4.203
220	49.036	0.4755	0.6007	0.0696	4.105
230	48.628	0.4803	0.5728	0.0685	4.016
240	48.215	0.4853	0.5468	0.0675	3.934
250	47.798	0.4903	0.5225	0.0664	3.859

APPENDIX B.1 (*Continued*)

Thermophysical Properties of Saturated Liquids (IP Units)

T (°F)	ρ (lbm/ft³)	c_p (Btu/lbm-R)	μ (lbm/ft-h)	k (Btu/hr-ft-°F)	Pr
			Decane		
10	47.162	0.4930	3.9655	0.0827	23.65
20	46.890	0.4973	3.5496	0.0819	21.55
30	46.619	0.5016	3.1964	0.0812	19.75
40	46.348	0.5061	2.8941	0.0804	18.21
50	46.078	0.5108	2.6337	0.0797	16.88
60	45.808	0.5155	2.4081	0.0789	15.73
70	45.539	0.5203	2.2113	0.0782	14.72
80	45.269	0.5252	2.0387	0.0774	13.83
90	45.000	0.5301	1.8865	0.0766	13.05
100	44.730	0.5352	1.7516	0.0759	12.35
110	44.460	0.5403	1.6314	0.0751	11.73
120	44.189	0.5455	1.5238	0.0744	11.18
130	43.918	0.5508	1.4271	0.0736	10.68
140	43.647	0.5561	1.3398	0.0728	10.23
150	43.374	0.5614	1.2606	0.0721	9.822
160	43.101	0.5668	1.1886	0.0713	9.451
170	42.827	0.5722	1.1227	0.0705	9.111
180	42.551	0.5777	1.0622	0.0697	8.800
190	42.275	0.5832	1.0066	0.0690	8.514
200	41.997	0.5887	0.9552	0.0682	8.249
			Acetone		
−40	53.365	0.4844	1.6765	0.1041	7.802
−30	52.997	0.4859	1.5242	0.1032	7.177
−20	52.629	0.4876	1.3949	0.1023	6.648
−10	52.261	0.4894	1.2841	0.1014	6.197
0	51.891	0.4914	1.1884	0.1005	5.811
10	51.520	0.4935	1.1052	0.0996	5.477
20	51.147	0.4958	1.0322	0.0987	5.187
30	50.773	0.4983	0.9679	0.0977	4.934
40	50.397	0.5009	0.9107	0.0968	4.712
50	50.018	0.5036	0.8596	0.0959	4.516
60	49.637	0.5066	0.8136	0.0949	4.341
70	49.253	0.5096	0.7719	0.0940	4.186
80	48.866	0.5129	0.7339	0.0930	4.047
90	48.476	0.5162	0.6992	0.0920	3.921
100	48.082	0.5198	0.6671	0.0911	3.808
110	47.684	0.5234	0.6374	0.0901	3.704
120	47.282	0.5273	0.6097	0.0891	3.609
130	46.891	0.5311	0.5838	0.0881	3.521
140	46.464	0.5354	0.5594	0.0870	3.441
150	46.047	0.53973	0.5363	0.0860	3.365

(*Continued*)

APPENDIX B.1   (*Continued*)

Thermophysical Properties of Saturated Liquids (IP Units)

T (°F)	ρ (lbm/ft³)	c_p (Btu/lbm-R)	μ (lbm/ft-h)	k (Btu/hr-ft-°F)	Pr
		Cyclohexane			
50	49.194	0.4023	2.7539	0.0720	15.39
60	48.866	0.4139	2.4950	0.0714	14.46
70	48.538	0.4247	2.2687	0.0709	13.60
80	48.209	0.4348	2.0699	0.0703	12.80
90	47.879	0.4444	1.8947	0.0697	12.08
100	47.548	0.4535	1.7396	0.0691	11.41
110	47.215	0.4622	1.6018	0.0686	10.80
120	46.880	0.4705	1.4791	0.0680	10.24
130	46.544	0.4786	1.3693	0.0674	9.726
140	46.205	0.4863	1.2709	0.0668	9.255
150	45.864	0.4939	1.1823	0.0662	8.824
160	45.521	0.5012	1.1024	0.0656	8.428
170	45.174	0.5084	1.0301	0.0650	8.064
180	44.825	0.5155	0.9646	0.0643	7.730
190	44.472	0.5225	0.9050	0.0637	7.422
200	44.116	0.5295	0.8507	0.0631	7.140
210	43.756	0.5363	0.8011	0.0625	6.880
220	43.392	0.5432	0.7556	0.0618	6.640
230	43.023	0.5500	0.7140	0.0612	6.420
240	42.650	0.5569	0.6757	0.0605	6.217
		Ethanol (Ethyl alcohol)			
10	50.948	0.4906	5.6691	0.1037	26.81
20	50.664	0.5053	4.9560	0.1029	24.34
30	50.380	0.5208	4.3596	0.1020	22.25
40	50.096	0.5369	3.8568	0.1012	20.47
50	49.811	0.5533	3.4300	0.1003	18.92
60	49.524	0.5700	3.0650	0.0994	17.57
70	49.235	0.5867	2.7510	0.0985	16.38
80	48.943	0.6033	2.4791	0.0976	15.32
90	48.647	0.6198	2.2424	0.0967	14.37
100	48.346	0.6361	2.0352	0.0958	13.51
110	48.041	0.6522	1.8528	0.0949	12.73
120	47.730	0.6680	1.6915	0.0940	12.02
130	47.413	0.6834	1.5482	0.0930	11.37
140	47.089	0.6986	1.4204	0.0921	10.78
150	46.759	0.7134	1.3058	0.0911	10.22
160	46.421	0.7280	1.2028	0.0902	9.712
170	46.074	0.7422	1.1098	0.0892	9.237
180	45.720	0.7561	1.0255	0.0882	8.794
190	45.356	0.7698	0.9489	0.0872	8.380
200	44.982	0.7833	0.8790	0.0861	7.993

APPENDIX B.1 (*Continued*)

Thermophysical Properties of Saturated Liquids (IP Units)

T (°F)	ρ (lbm/ft³)	c_p (Btu/lbm-R)	μ (lbm/ft-h)	k (Btu/hr-ft-°F)	Pr
			Ethylene		
−170	36.207	0.5677	0.4490	0.1138	2.240
−160	35.719	0.5699	0.4061	0.1104	2.097
−150	35.223	0.5732	0.3698	0.1069	1.983
−140	34.717	0.5775	0.3393	0.1035	1.894
−130	34.201	0.5827	0.3141	0.1000	1.830
−120	33.673	0.5889	0.2935	0.0966	1.790
−110	33.131	0.5960	0.2769	0.0932	1.771
−100	32.575	0.6043	0.2635	0.0897	1.774
−90	32.002	0.6138	0.2526	0.0863	1.797
−80	31.411	0.6249	0.2437	0.0829	1.838
−70	30.795	0.6381	0.2361	0.0794	1.896
−60	30.155	0.6537	0.2290	0.0760	1.969
−50	29.485	0.6723	0.2219	0.0726	2.055
−40	28.780	0.6949	0.2141	0.0692	2.150
−30	28.034	0.7229	0.2048	0.0658	2.251
−20	27.246	0.7575	0.1935	0.0624	2.350
−10	26.373	0.8049	0.1795	0.0590	2.450
0	25.441	0.8671	0.1620	0.0555	2.529
10	24.366	0.9647	0.1406	0.0521	2.600
20	23.138	1.1257	0.1143	0.0487	2.641
			Heptane		
10	44.357	0.5051	1.5608	0.0793	9.942
20	44.073	0.5091	1.4534	0.0785	9.431
30	43.785	0.5133	1.3554	0.0776	8.962
40	43.497	0.5176	1.2657	0.0768	8.531
50	43.207	0.5222	1.1837	0.0760	8.134
60	42.917	0.5269	1.1083	0.0752	7.768
70	42.625	0.5317	1.0391	0.0744	7.430
80	42.332	0.5367	0.9754	0.0736	7.116
90	42.037	0.5418	0.9166	0.0728	6.825
100	41.741	0.5470	0.8624	0.0720	6.554
110	41.443	0.5524	0.8122	0.0712	6.302
120	41.143	0.5578	0.7658	0.0704	6.068
130	40.841	0.5634	0.7227	0.0696	5.848
140	40.536	0.5691	0.6827	0.0689	5.643
150	40.228	0.5749	0.6456	0.0681	5.451
160	39.918	0.5808	0.6110	0.0673	5.270
170	39.604	0.5867	0.5788	0.0666	5.101
180	39.287	0.5928	0.5487	0.0658	4.942
190	38.967	0.5990	0.5206	0.0651	4.793
200	38.643	0.6052	0.4944	0.0643	4.652

(*Continued*)

APPENDIX B.1　　(*Continued*)

Thermophysical Properties of Saturated Liquids (IP Units)

T (°F)	ρ (lbm/ft³)	c_p (Btu/lbm-R)	μ (lbm/ft-h)	k (Btu/hr-ft-°F)	Pr
			Propane		
10	33.985	0.5791	0.3522	0.0662	3.080
20	33.535	0.5883	0.3322	0.0644	3.033
30	33.075	0.5979	0.3134	0.0627	2.990
40	32.603	0.6082	0.2958	0.0610	2.951
50	32.116	0.6193	0.2791	0.0593	2.916
60	31.615	0.6311	0.2633	0.0576	2.884
70	31.096	0.6439	0.2483	0.0560	2.856
80	30.558	0.6579	0.2340	0.0544	2.830
90	29.990	0.6737	0.2202	0.0528	2.808
100	29.411	0.6909	0.2072	0.0513	2.789
110	28.785	0.7114	0.1944	0.0498	2.777
120	28.135	0.7349	0.1822	0.0483	2.771
130	27.448	0.7630	0.1704	0.0469	2.773
140	26.701	0.7988	0.1587	0.0454	2.791
150	25.888	0.8458	0.1472	0.0440	2.830
160	25.018	0.9069	0.1360	0.0426	2.898
170	24.011	1.0011	0.1244	0.0411	3.032
180	22.822	1.1618	0.1122	0.0396	3.294
190	21.257	1.5468	0.0982	0.0384	3.958
200	19.022	3.1013	0.0818	0.0451	5.623
			Hexane		
10	43.123	0.5035	1.1580	0.0798	7.310
20	42.814	0.5083	1.0787	0.0787	6.970
30	42.503	0.5132	1.0045	0.0776	6.647
40	42.192	0.5183	0.9355	0.0765	6.339
50	41.878	0.5234	0.8713	0.0754	6.047
60	41.563	0.5288	0.8118	0.0744	5.772
70	41.245	0.5342	0.7568	0.0733	5.513
80	40.926	0.5398	0.7060	0.0723	5.270
90	40.604	0.5455	0.6594	0.0713	5.044
100	40.279	0.5513	0.6166	0.0703	4.835
110	39.951	0.5573	0.5775	0.0693	4.642
120	39.621	0.5633	0.5419	0.0684	4.465
130	39.286	0.5695	0.5096	0.0674	4.305
140	38.949	0.5758	0.4804	0.0665	4.162
150	38.607	0.5822	0.4541	0.0655	4.034
160	38.261	0.5888	0.4304	0.0646	3.923
170	37.910	0.5954	0.4093	0.0637	3.826
180	37.555	0.6022	0.3905	0.0628	3.745
190	37.194	0.6091	0.3738	0.0619	3.678
200	36.827	0.6162	0.3590	0.0610	3.625

APPENDIX B.1 (*Continued*)

Thermophysical Properties of Saturated Liquids (IP Units)

T (°F)	ρ (lbm/ft³)	c_p (Btu/lbm-R)	μ (lbm/ft-h)	k (Btu/hr-ft-°F)	Pr
			Octane		
10	45.401	0.5034	2.3715	0.0799	14.95
20	45.125	0.5071	2.1920	0.0789	14.08
30	44.848	0.5111	2.0294	0.0780	13.29
40	44.571	0.5152	1.8818	0.0771	12.57
50	44.294	0.5195	1.7475	0.0762	11.91
60	44.017	0.5240	1.6252	0.0754	11.30
70	43.739	0.5285	1.5135	0.0745	10.74
80	43.460	0.5332	1.4114	0.0736	10.22
90	43.180	0.5381	1.3178	0.0728	9.743
100	42.899	0.5430	1.2320	0.0719	9.300
110	42.617	0.5481	1.1531	0.0711	8.889
120	42.333	0.5533	1.0805	0.0703	8.506
130	42.048	0.5585	1.0137	0.0695	8.149
140	41.761	0.5639	0.9519	0.0687	7.817
150	41.473	0.5693	0.8949	0.0679	7.505
160	41.182	0.5748	0.8421	0.0671	7.214
170	40.889	0.5804	0.7933	0.0663	6.941
180	40.594	0.5861	0.7479	0.0656	6.685
190	40.296	0.5918	0.7058	0.0648	6.444
200	39.996	0.5976	0.6667	0.0641	6.217
			Propylene		
−70	38.711	0.5127	0.4712	0.0888	2.720
−60	38.286	0.5175	0.4376	0.0872	2.597
−50	37.856	0.5227	0.4076	0.0856	2.488
−40	37.419	0.5283	0.3808	0.0840	2.394
−30	36.974	0.5343	0.3568	0.0825	2.312
−20	36.522	0.5408	0.3352	0.0809	2.242
−10	36.060	0.5477	0.3159	0.0793	2.183
0	35.588	0.5551	0.2986	0.0777	2.134
10	35.106	0.5631	0.2831	0.0761	2.096
20	34.612	0.5717	0.2691	0.0744	2.067
30	34.104	0.5810	0.2566	0.0728	2.047
40	33.582	0.5910	0.2453	0.0712	2.037
50	33.043	0.6020	0.2353	0.0695	2.037
60	32.487	0.6141	0.2262	0.0679	2.047
70	31.910	0.6275	0.2181	0.0662	2.067
80	31.310	0.6424	0.2109	0.0645	2.100
90	30.684	0.6593	0.2044	0.0628	2.145
100	30.029	0.6788	0.1986	0.0611	2.207
110	29.338	0.7016	0.1934	0.0593	2.287
120	28.606	0.7290	0.1889	0.0576	2.391

(*Continued*)

APPENDIX B.1　(*Continued*)

Thermophysical Properties of Saturated Liquids (IP Units)

T (°F)	ρ (lbm/ft³)	c_p (Btu/lbm-R)	μ (lbm/ft-h)	k (Btu/hr-ft-°F)	Pr
			Refrigerant 22		
−40	87.166	0.2656	0.7706	0.0650	3.151
−30	86.153	0.2676	0.7189	0.0635	3.029
−20	85.122	0.2697	0.6729	0.0621	2.924
−10	84.073	0.2720	0.6316	0.0606	2.833
0	83.004	0.2746	0.5942	0.0592	2.755
10	81.912	0.2774	0.5601	0.0578	2.688
20	80.796	0.2804	0.5288	0.0564	2.630
30	79.652	0.2838	0.4998	0.0550	2.581
40	79.060	0.2837	0.4870	0.0546	2.532
50	77.269	0.2918	0.4476	0.0521	2.505
60	76.591	0.2921	0.4356	0.0516	2.464
70	74.735	0.3018	0.4012	0.0493	2.458
80	73.399	0.3078	0.3796	0.0478	2.445
90	72.551	0.3092	0.3680	0.0471	2.415
100	70.555	0.3229	0.3388	0.0448	2.443
110	69.553	0.3258	0.3269	0.0440	2.423
120	67.930	0.3367	0.3072	0.0423	2.445
130	65.694	0.3587	0.2815	0.0399	2.528
140	63.846	0.3773	0.2628	0.0382	2.597
150	62.317	0.3905	0.2490	0.0369	2.637
			Refrigerant 123		
−40	101.124	0.2267	2.3845	0.0549	9.843
−30	100.289	0.2283	2.1801	0.0539	9.230
−20	99.537	0.2298	2.0180	0.0531	8.741
−10	98.734	0.2315	1.8639	0.0521	8.274
0	97.882	0.2333	1.7189	0.0512	7.834
10	97.104	0.2350	1.6008	0.0504	7.471
20	96.278	0.2368	1.4879	0.0495	7.121
30	95.408	0.2387	1.3813	0.0486	6.790
40	94.598	0.2404	1.2920	0.0477	6.507
50	93.744	0.2423	1.2067	0.0469	6.236
60	92.881	0.2441	1.1286	0.0460	5.985
70	92.006	0.2460	1.0569	0.0452	5.752
80	91.121	0.2479	0.9909	0.0444	5.536
90	90.203	0.2498	0.9286	0.0435	5.331
100	89.294	0.2518	0.8726	0.0427	5.145
110	88.389	0.2537	0.8217	0.0419	4.974
120	87.434	0.2559	0.7725	0.0411	4.810
130	86.496	0.2580	0.7285	0.0403	4.661
140	85.508	0.2603	0.6860	0.0395	4.519
150	84.533	0.2626	0.6477	0.0387	4.390

APPENDIX B.1 (*Continued*)

Thermophysical Properties of Saturated Liquids (IP Units)

T (°F)	ρ (lbm/ft³)	*c*ₚ (Btu/lbm-R)	μ (lbm/ft-h)	*k* (Btu/hr-ft-°F)	Pr
			Refrigerant 407C		
−40	85.565	0.3046	0.8373	0.0658	3.877
−30	84.459	0.3096	0.7785	0.0642	3.755
−20	83.330	0.3146	0.7236	0.0626	3.636
−10	82.177	0.3196	0.6726	0.0611	3.521
0	80.998	0.3246	0.6254	0.0595	3.410
10	79.793	0.3297	0.5816	0.0580	3.305
20	78.561	0.3348	0.5412	0.0565	3.205
30	77.300	0.3401	0.5039	0.0551	3.111
40	76.008	0.3456	0.4696	0.0537	3.025
50	74.684	0.3514	0.4380	0.0522	2.946
60	73.325	0.3575	0.4090	0.0509	2.874
70	71.928	0.3641	0.3822	0.0495	2.812
80	70.490	0.3713	0.3576	0.0481	2.758
90	69.005	0.3795	0.3349	0.0468	2.715
100	67.469	0.3889	0.3139	0.0455	2.684
110	65.871	0.4003	0.2944	0.0442	2.668
120	64.200	0.4147	0.2760	0.0429	2.671
130	62.439	0.4341	0.2587	0.0415	2.704
140	60.558	0.4624	0.2420	0.0402	2.785
150	58.511	0.5080	0.2256	0.0388	2.957
			Refrigerant 23		
−150	93.879	0.2883	0.9807	0.0873	3.239
−140	92.667	0.2898	0.8879	0.0839	3.066
−130	91.443	0.2916	0.8083	0.0809	2.914
−120	90.205	0.2935	0.7395	0.0780	2.782
−110	88.947	0.2957	0.6793	0.0754	2.665
−100	87.665	0.2982	0.6262	0.0729	2.561
−90	86.354	0.3010	0.5788	0.0706	2.469
−80	85.007	0.3042	0.5361	0.0683	2.386
−70	83.620	0.3078	0.4974	0.0662	2.313
−60	81.926	0.3138	0.4563	0.0637	2.247
−50	80.687	0.3169	0.4294	0.0621	2.192
−40	79.123	0.3228	0.3991	0.0601	2.144
−30	77.478	0.3298	0.3707	0.0581	2.104
−20	75.738	0.3384	0.3439	0.0561	2.073
−10	73.884	0.3491	0.3184	0.0541	2.053
0	71.894	0.3626	0.2940	0.0521	2.047
10	69.740	0.3802	0.2704	0.0500	2.057
20	67.381	0.4039	0.2474	0.0478	2.091
30	64.874	0.4345	0.2258	0.0456	2.152
40	61.808	0.4874	0.2024	0.0430	2.296

(*Continued*)

APPENDIX B.1　(*Continued*)

Thermophysical Properties of Saturated Liquids (IP Units)

T (°F)	ρ (lbm/ft³)	c_p (Btu/lbm-R)	μ (lbm/ft-h)	k (Btu/hr-ft-°F)	Pr
			Refrigerant 134A		
−40	88.512	0.2996	1.1262	0.0636	5.305
−30	87.504	0.3020	1.0333	0.0625	4.993
−20	86.482	0.3045	0.9513	0.0613	4.721
−10	85.442	0.3071	0.8783	0.0602	4.484
0	84.384	0.3099	0.8130	0.0589	4.276
10	83.306	0.3129	0.7541	0.0576	4.093
20	82.204	0.3161	0.7008	0.0563	3.932
30	81.077	0.3195	0.6522	0.0550	3.790
40	79.921	0.3232	0.6076	0.0536	3.665
50	78.734	0.3272	0.5667	0.0522	3.555
60	77.512	0.3316	0.5288	0.0507	3.459
70	76.250	0.3365	0.4936	0.0492	3.377
80	74.944	0.3419	0.4608	0.0476	3.307
90	73.588	0.3479	0.4300	0.0460	3.249
100	72.173	0.3548	0.4010	0.0444	3.204
110	70.693	0.3628	0.3737	0.0427	3.171
120	69.134	0.3720	0.3476	0.0410	3.153
130	67.484	0.3831	0.3228	0.0392	3.151
140	65.724	0.3966	0.2989	0.0374	3.169
150	63.829	0.4136	0.2758	0.0355	3.213
			Refrigerant 507A		
−40	80.839	0.2951	0.7764	0.0522	4.387
−30	79.735	0.2981	0.7178	0.0510	4.199
−20	78.603	0.3010	0.6633	0.0497	4.017
−10	77.440	0.3038	0.6128	0.0485	3.841
0	76.242	0.3066	0.5659	0.0472	3.674
10	75.007	0.3095	0.5227	0.0460	3.516
20	73.731	0.3127	0.4827	0.0448	3.369
30	72.407	0.3162	0.4459	0.0436	3.233
40	71.031	0.3203	0.4120	0.0424	3.111
50	69.596	0.3251	0.3809	0.0412	3.004
60	68.091	0.3311	0.3523	0.0400	2.914
70	66.505	0.3384	0.3260	0.0388	2.841
80	64.823	0.3478	0.3018	0.0376	2.790
90	63.025	0.3599	0.2794	0.0364	2.762
100	61.082	0.3758	0.2584	0.0351	2.765
110	58.951	0.3977	0.2387	0.0338	2.807
120	56.566	0.4289	0.2196	0.0324	2.904
130	53.811	0.4770	0.2004	0.0309	3.093
140	50.446	0.5617	0.1800	0.0292	3.468
150	45.795	0.7639	0.1554	0.0269	4.415

APPENDIX B.1 (*Continued*)

Thermophysical Properties of Saturated Liquids (IP Units)

T (°F)	ρ (lbm/ft³)	c_p (Btu/lbm-R)	μ (lbm/ft-h)	k (Btu/hr-ft-°F)	Pr
		40% Ethylene glycol			
−40	66.827	0.7762	122.93	0.2191	435.4
−30	66.778	0.7826	85.009	0.2216	300.2
−20	66.716	0.7889	60.187	0.2240	211.9
−10	66.641	0.7951	43.585	0.2265	153.0
0	66.554	0.8012	32.251	0.2289	112.9
10	66.455	0.8072	24.360	0.2314	84.98
20	66.345	0.8132	18.763	0.2339	65.24
30	66.223	0.8191	14.723	0.2364	51.02
40	66.091	0.8248	11.758	0.2388	40.61
50	65.948	0.8305	9.5470	0.2413	32.86
60	65.795	0.8361	7.8733	0.2438	27.00
70	65.632	0.8416	6.5884	0.2462	22.52
80	65.460	0.8469	5.5885	0.2487	19.03
90	65.278	0.8522	4.8004	0.2511	16.29
100	65.088	0.8573	4.1714	0.2536	14.10
110	64.889	0.8623	3.6634	0.2560	12.34
120	64.682	0.8672	3.2483	0.2584	10.90
130	64.467	0.8719	2.9050	0.2608	9.71
140	64.245	0.8765	2.6178	0.2631	8.720
150	64.016	0.8809	2.3745	0.2654	7.880
		40% Propylene glycol			
−40	65.780	0.8379	661.8	0.2111	2627
−30	65.734	0.8424	384.44	0.2128	1522
−20	65.669	0.8468	231.85	0.2145	915.4
−10	65.587	0.8513	144.93	0.2163	570.5
0	65.489	0.8557	93.754	0.2181	367.9
10	65.375	0.8601	62.661	0.2199	245.1
20	65.245	0.8645	43.200	0.2218	168.4
30	65.102	0.8689	30.671	0.2237	119.1
40	64.945	0.8732	22.389	0.2257	86.64
50	64.776	0.8776	16.777	0.2276	64.67
60	64.595	0.8819	12.884	0.2297	49.47
70	64.403	0.8862	10.123	0.2317	38.72
80	64.202	0.8905	8.1254	0.2337	30.95
90	63.990	0.8947	6.6515	0.2358	25.24
100	63.770	0.8989	5.5440	0.2379	20.95
110	63.543	0.9032	4.6975	0.2400	17.68
120	63.308	0.9074	4.0397	0.2422	15.14
130	63.068	0.9115	3.5200	0.2443	13.13
140	62.822	0.9157	3.1029	0.2464	11.53
150	62.572	0.9198	2.7626	0.2486	10.22

(*Continued*)

APPENDIX B.1 (*Continued*)

Thermophysical Properties of Saturated Liquids (IP Units)

T (°F)	ρ (lbm/ft³)	c_p (Btu/lbm-R)	μ (lbm/ft-h)	k (Btu/hr-ft-°F)	Pr
		20% Magnesium chloride			
−20	74.141	0.7107	36.297	0.2705	95.37
−15	74.138	0.7125	32.314	0.2728	84.39
−10	74.128	0.7143	28.829	0.2752	74.83
−5	74.112	0.7161	25.774	0.2776	66.48
0	74.089	0.7179	23.092	0.2801	59.18
5	74.061	0.7197	20.731	0.2827	52.78
10	74.027	0.7215	18.649	0.2852	47.17
15	73.988	0.7233	16.809	0.2879	42.24
20	73.944	0.7252	15.181	0.2905	37.89
25	73.895	0.7270	13.737	0.2932	34.06
30	73.843	0.7288	12.454	0.2959	30.68
35	73.786	0.7307	11.312	0.2986	27.68
40	73.727	0.7325	10.294	0.3013	25.02
45	73.664	0.7343	9.3844	0.3041	22.66
50	73.599	0.7362	8.5704	0.3068	20.56
55	73.531	0.7380	7.8408	0.3096	18.69
60	73.461	0.7398	7.1856	0.3123	17.02
65	73.390	0.7417	6.5964	0.3150	15.53
70	73.318	0.7435	6.0655	0.3177	14.20
75	73.244	0.7454	5.5865	0.3204	13.00
		20% Ethylene glycol			
−40	64.287	0.9072	50.466	0.2490	183.8
−30	64.330	0.9091	36.829	0.2536	132.0
−20	64.357	0.9110	27.378	0.2580	96.66
−10	64.367	0.9130	20.719	0.2624	72.09
0	64.362	0.9151	15.950	0.2667	54.73
10	64.341	0.9172	12.482	0.2709	42.27
20	64.305	0.9194	9.924	0.2750	33.18
30	64.254	0.9217	8.0095	0.2790	26.46
40	64.189	0.9240	6.5582	0.2829	21.42
50	64.110	0.9263	5.4439	0.2867	17.59
60	64.017	0.9287	4.5782	0.2904	14.64
70	63.910	0.9311	3.8980	0.2941	12.34
80	63.791	0.9335	3.3577	0.2976	10.53
90	63.658	0.9359	2.9241	0.3010	9.091
100	63.513	0.9383	2.5729	0.3044	7.931
110	63.356	0.9408	2.2857	0.3077	6.989
120	63.187	0.9432	2.0487	0.3108	6.217
130	63.007	0.9456	1.8514	0.3139	5.578
140	62.815	0.9480	1.6858	0.3168	5.044
150	62.613	0.9504	1.5456	0.3197	4.595

APPENDIX B.1 (*Continued*)

Thermophysical Properties of Saturated Liquids (IP Units)

T (°F)	ρ (lbm/ft³)	c_p (Btu/lbm-R)	μ (lbm/ft-h)	k (Btu/hr-ft-°F)	Pr
			20% Propylene glycol		
−40	63.579	0.9231	108.68	0.2468	406.5
−30	63.664	0.9251	72.504	0.2504	267.9
−20	63.725	0.9273	49.730	0.2539	181.6
−10	63.762	0.9296	35.025	0.2575	126.4
0	63.776	0.9320	25.299	0.2610	90.32
10	63.769	0.9344	18.717	0.2646	66.11
20	63.741	0.9370	14.166	0.2681	49.52
30	63.693	0.9395	10.955	0.2715	37.91
40	63.627	0.9422	8.6458	0.2750	29.62
50	63.543	0.9449	6.9542	0.2784	23.60
60	63.442	0.9476	5.6941	0.2817	19.15
70	63.326	0.9504	4.7401	0.2851	15.80
80	63.195	0.9532	4.0069	0.2883	13.25
90	63.050	0.9560	3.4351	0.2916	11.26
100	62.892	0.9588	2.9830	0.2948	9.703
110	62.722	0.9616	2.6206	0.2979	8.460
120	62.542	0.9644	2.3262	0.3009	7.455
130	62.352	0.9672	2.0838	0.3039	6.632
140	62.153	0.9700	1.8815	0.3068	5.948
150	61.946	0.9727	1.7101	0.3097	5.372
			20% Sodium chloride		
−20	72.895	0.7925	20.434	0.2918	55.50
−15	72.835	0.7944	18.005	0.2942	48.62
−10	72.774	0.7961	15.922	0.2965	42.75
−5	72.711	0.7978	14.130	0.2988	37.72
0	72.648	0.7994	12.584	0.3012	33.40
5	72.582	0.8009	11.247	0.3036	29.68
10	72.516	0.8024	10.088	0.3059	26.46
15	72.448	0.8038	9.0811	0.3083	23.68
20	72.379	0.8051	8.2035	0.3107	21.26
25	72.309	0.8064	7.4372	0.3131	19.16
30	72.238	0.8076	6.7665	0.3155	17.32
35	72.165	0.8088	6.1782	0.3179	15.72
40	72.091	0.8098	5.6611	0.3203	14.31
45	72.015	0.8108	5.2058	0.3228	13.08
50	71.939	0.8118	4.8042	0.3252	11.99
55	71.861	0.8126	4.4494	0.3276	11.04
60	71.782	0.8135	4.1355	0.3301	10.19
65	71.701	0.8142	3.8574	0.3326	9.444
70	71.619	0.8149	3.6108	0.3350	8.782
75	71.536	0.8155	3.3920	0.3375	8.196

APPENDIX B.2

Saturation Properties (SI Units)

T (°C)	P_{sat} (kPa)	ρ_f (kg/m³)	v_g (m³/kg)	h_f (kJ/kg)	h_g (kJ/kg)
			Water		
10	1.2281	999.65	106.32	42.022	2519.2
20	2.3392	998.16	57.762	83.915	2537.4
30	4.2469	995.60	32.879	125.74	2555.6
40	7.3851	992.17	19.515	167.53	2573.5
50	12.352	987.99	12.026	209.34	2591.3
60	19.947	983.15	7.6670	251.18	2608.8
70	31.202	977.73	5.0396	293.07	2626.1
80	47.416	971.76	3.4053	335.02	2643.0
90	70.183	965.29	2.3593	377.04	2659.6
100	101.42	958.35	1.6720	419.17	2675.6
110	143.38	950.95	1.2095	461.42	2691.1
120	198.67	943.11	0.8913	503.81	2706.0
130	270.28	934.84	0.6681	546.38	2720.1
140	361.53	926.14	0.5085	589.16	2733.5
150	476.16	917.01	0.3925	632.18	2745.9
160	618.23	907.45	0.3068	675.47	2757.5
170	792.18	897.45	0.2426	719.08	2767.9
180	1002.8	887.00	0.1938	763.05	2777.2
190	1255.2	876.08	0.1564	807.43	2785.3
200	1554.9	864.67	0.1272	852.26	2792.0
			Benzene		
30	15.919	868.11	2.2684	−91.391	342.76
40	24.389	857.39	1.6786	−73.756	357.21
50	36.206	846.59	1.1354	−55.826	367.26
60	52.252	835.68	0.7639	−37.586	376.22
70	73.523	824.66	0.5277	−19.023	385.52
80	101.12	813.49	0.3766	−0.1265	395.49
90	136.24	802.17	0.2771	19.114	406.10
100	180.17	790.66	0.2094	38.709	417.26
110	234.29	778.94	0.1618	58.667	428.85
120	300.03	766.99	0.1273	78.997	440.80
130	378.91	754.77	0.1018	99.710	453.01
140	472.49	742.25	0.0824	120.82	465.42
150	582.40	729.38	0.0674	142.33	477.95
160	710.32	716.12	0.0557	164.27	490.54
170	857.98	702.41	0.0463	186.66	503.13
180	1027.2	688.16	0.0387	209.51	515.65
190	1219.8	673.28	0.0326	232.86	528.01
200	1437.8	657.65	0.0274	256.74	540.13
210	1683.2	641.10	0.0232	281.22	551.88
220	1958.3	623.40	0.0196	306.36	563.14

APPENDIX B.2 (*Continued*)

Saturation Properties (SI Units)

T (°C)	P_{sat} (kPa)	ρ_f (kg/m³)	v_g (m³/kg)	h_f (kJ/kg)	h_g (kJ/kg)
			Decane		
0	0.0261	745.90	611.45	−425.74	−50.551
15	0.0877	734.21	192.02	−393.79	−27.157
30	0.2556	722.55	69.198	−361.04	−2.8008
45	0.6597	710.88	28.105	−327.44	22.530
60	1.532	699.15	12.668	−292.97	48.847
75	3.249	687.34	6.2282	−257.59	76.075
90	6.370	675.40	3.3054	−221.29	104.23
105	11.674	663.29	1.8704	−184.04	133.25
120	20.191	650.97	1.1148	−145.85	163.02
135	33.222	638.38	0.6971	−106.70	193.59
150	52.364	625.48	0.4533	−66.576	224.86
165	79.516	612.19	0.3047	−25.474	256.75
180	116.89	598.44	0.2102	16.618	289.15
195	166.98	584.12	0.1491	59.714	322.11
210	232.56	569.11	0.1076	103.83	355.38
225	316.62	553.26	0.0790	149.00	388.89
240	422.25	536.33	0.0587	195.27	422.49
255	552.62	518.02	0.0442	242.69	456.13
270	710.95	497.86	0.0334	291.38	489.20
285	900.60	475.12	0.0255	341.49	522.19
			Acetone		
−20	2.9410	833.58	12.2345	158.59	736.49
−10	5.4004	822.88	6.9138	179.21	747.65
0	9.3961	812.10	4.1098	200.00	758.84
10	15.590	801.21	2.5632	220.99	770.21
20	24.809	790.19	1.6537	242.19	781.39
30	38.057	779.02	1.1118	263.63	792.91
40	56.527	767.66	0.7668	285.34	804.26
50	81.609	756.09	0.5425	307.32	815.57
60	114.91	744.28	0.3918	329.60	826.76
70	158.25	732.19	0.2892	352.20	837.98
80	213.67	719.78	0.2168	375.16	849.03
90	283.45	707.02	0.1658	398.48	860.26
100	370.06	693.84	0.1280	422.21	871.11
110	476.13	680.19	0.0992	446.36	881.18
120	604.43	666.00	0.0788	470.97	891.97
130	757.77	651.19	0.0627	496.08	901.84
140	938.92	635.65	0.0502	521.73	911.14
150	1150.5	619.26	0.0405	547.99	920.02
160	1395.0	601.85	0.0328	574.91	928.13
170	1674.8	583.18	0.0267	602.59	935.43

(*Continued*)

APPENDIX B.2 (*Continued*)

Saturation Properties (SI Units)

T (°C)	P_{sat} (kPa)	ρ_f (kg/m³)	v_g (m³/kg)	h_f (kJ/kg)	h_g (kJ/kg)
			Cyclohexane		
30	16.239	769.06	1.8272	41.500	430.74
40	24.629	759.51	1.2402	60.305	443.47
50	36.239	749.88	0.8660	79.770	456.62
60	51.889	740.14	0.6201	99.850	470.15
70	72.497	730.27	0.4541	120.51	484.05
80	99.075	720.27	0.3389	141.73	498.24
90	132.71	710.10	0.2577	163.48	512.75
100	174.58	699.74	0.1990	185.76	527.54
110	225.90	689.16	0.1559	208.56	542.56
120	287.97	678.35	0.1236	231.89	557.78
130	362.13	667.25	0.0991	255.73	573.17
140	449.77	655.82	0.0803	280.11	588.68
150	552.33	644.02	0.0664	305.02	604.99
160	671.31	631.78	0.0539	330.50	619.89
170	808.25	619.03	0.0446	356.55	635.48
180	964.77	605.66	0.0370	383.21	650.99
190	1142.6	591.57	0.0316	410.52	667.82
200	1343.4	576.61	0.0258	438.51	681.44
210	1569.2	560.60	0.0216	467.23	696.19
220	1822.0	543.33	0.0181	496.77	710.46
			Ethanol (Ethyl alcohol)		
−20	0.3494	822.53	130.31	8.3961	965.60
−10	0.7708	814.29	62.090	28.654	980.08
0	1.599	806.10	31.175	50.007	994.50
10	3.139	797.90	16.437	72.557	1008.9
20	5.861	789.60	9.0705	96.354	1023.4
30	10.455	781.15	5.2215	121.41	1038.0
40	17.892	772.48	3.1258	147.71	1052.5
50	29.483	763.55	1.9401	175.22	1067.2
60	46.931	754.30	1.2447	203.91	1081.7
70	72.383	744.68	0.8232	233.74	1096.2
80	108.46	734.65	0.5596	264.66	1110.5
90	158.27	724.16	0.3899	296.63	1124.5
100	225.42	713.15	0.2778	329.64	1138.0
110	314.00	701.59	0.2019	363.64	1151.1
120	428.57	689.39	0.1492	398.62	1163.4
130	574.11	676.51	0.1118	434.57	1174.9
140	756.02	662.83	0.0849	471.52	1185.2
150	980.05	648.27	0.0650	509.48	1194.2
160	1252.3	632.67	0.0501	548.52	1201.6
170	1579.4	615.84	0.0387	588.73	1206.7

APPENDIX B.2 (*Continued*)

Saturation Properties (SI Units)

T (°C)	P_{sat} (kPa)	ρ_f (kg/m³)	v_g (m³/kg)	h_f (kJ/kg)	h_g (kJ/kg)
			Ethylene		
−30	1936.9	439.06	0.0274	−466.11	−138.43
−28	2046.2	434.14	0.0258	−459.78	−138.76
−26	2160.0	429.23	0.0243	−453.47	−139.34
−24	2278.2	424.36	0.0228	−447.17	−140.03
−22	2401.0	418.95	0.0215	−440.51	−140.89
−20	2528.7	413.55	0.0202	−433.84	−141.91
−18	2661.2	408.15	0.0190	−427.19	−143.18
−16	2798.7	402.09	0.0178	−420.08	−144.65
−14	2941.4	395.99	0.0167	−412.94	−146.40
−12	3089.4	389.58	0.0157	−405.61	−148.43
−10	3242.9	382.83	0.0147	−398.05	−150.78
−8	3402.0	376.08	0.0137	−390.48	−153.52
−6	3567.0	368.04	0.0128	−382.09	−156.71
−4	3738.0	360.31	0.0119	−373.90	−160.48
−2	3915.3	351.37	0.0110	−364.95	−164.93
0	4099.1	341.44	0.0102	−355.39	−170.04
2	4289.7	330.14	0.0094	−345.00	−176.41
4	4487.5	316.85	0.0085	−333.42	−184.48
6	4693.1	300.29	0.0076	−319.88	−195.42
8	4907.2	276.22	0.0065	−301.77	−213.33
			Heptane		
−20	0.3942	717.02	53.237	−97.415	295.69
−10	0.7988	708.74	27.293	−76.350	310.47
0	1.5220	700.45	14.851	−54.982	325.60
10	2.7462	692.11	8.5209	−33.288	341.10
20	4.7213	683.72	5.1219	−11.246	356.96
30	7.7756	675.27	3.2076	11.164	373.18
40	12.324	666.72	2.0839	33.961	389.74
50	18.876	658.08	1.3980	57.163	406.63
60	28.036	649.32	0.9648	80.784	423.82
70	40.508	640.42	0.6833	104.84	441.32
80	57.090	631.36	0.4950	129.34	459.11
90	78.676	622.12	0.3660	154.31	477.16
100	106.25	612.67	0.2754	179.75	495.46
110	140.86	602.99	0.2106	205.67	513.97
120	183.67	593.03	0.1634	232.09	532.67
130	235.89	582.75	0.1283	259.03	551.53
140	298.79	572.12	0.1019	286.50	570.52
150	373.75	561.07	0.0816	314.52	589.59
160	462.18	549.53	0.0660	343.10	608.69
170	565.58	537.40	0.0536	372.29	627.76

(*Continued*)

APPENDIX B.2 *(Continued)*

Saturation Properties (SI Units)

T (°C)	P_{sat} (kPa)	ρ_f (kg/m³)	v_g (m³/kg)	h_f (kJ/kg)	h_g (kJ/kg)
			Propane		
−100	2.8882	643.97	11.271	−23.484	456.88
−90	6.4240	633.47	5.3419	−2.8359	468.52
−80	13.007	622.82	2.7706	18.087	480.33
−70	24.339	612.01	1.5491	39.318	492.26
−60	42.606	601.00	0.9224	60.895	504.28
−50	70.469	589.76	0.5790	82.860	516.32
−40	111.03	578.24	0.3799	105.26	528.33
−30	167.76	566.41	0.2588	128.14	540.25
−20	244.50	554.19	0.1818	151.56	552.01
−10	345.32	541.52	0.1311	175.58	563.55
0	474.56	528.31	0.0967	200.26	574.78
10	636.74	514.45	0.0726	225.71	585.61
20	836.61	499.80	0.0553	252.00	595.94
30	1079.1	484.15	0.0426	279.28	605.60
40	1369.5	467.23	0.0331	307.71	614.40
50	1713.4	448.63	0.0259	337.53	622.00
60	2116.9	427.71	0.0202	369.07	627.90
70	2587.2	403.29	0.0157	402.94	631.19
80	3132.7	372.85	0.0119	440.38	629.95
90	3765.4	328.27	0.0084	485.36	618.07
			Hexane		
−20	1.8757	708.70	12.989	−111.12	292.53
−10	3.4908	697.16	7.2424	−85.046	307.28
0	6.1210	685.94	4.2771	−60.027	322.34
10	10.202	675.00	2.6517	−35.725	337.72
20	16.275	664.28	1.7137	−11.881	353.39
30	24.998	653.73	1.1477	11.708	369.35
40	37.142	643.31	0.7928	35.198	385.57
50	53.598	632.98	0.5625	58.717	402.04
60	75.375	622.69	0.4086	82.366	418.74
70	103.60	612.38	0.3029	106.23	435.65
80	139.52	602.01	0.2286	130.39	452.75
90	184.50	591.52	0.1752	154.89	470.00
100	240.01	580.86	0.1362	179.81	487.38
110	307.65	569.96	0.1071	205.19	504.86
120	389.15	558.73	0.0851	231.09	522.39
130	486.33	547.07	0.0682	257.56	539.93
140	601.15	534.88	0.0550	284.67	557.42
150	735.69	521.99	0.0447	312.48	574.79
160	892.15	508.19	0.0364	341.08	591.94
170	1072.9	493.18	0.0298	370.61	608.76

APPENDIX B.2   (*Continued*)

Saturation Properties (SI Units)

T (°C)	P_{sat} (kPa)	ρ_f (kg/m³)	v_g (m³/kg)	h_f (kJ/kg)	h_g (kJ/kg)
			Octane		
−20	0.0768	733.46	222.37	−96.976	293.66
−10	0.1763	725.48	103.57	−75.977	308.86
0	0.3735	717.45	51.676	−54.660	324.32
10	0.7389	709.53	27.396	−33.103	340.08
20	1.3772	701.52	15.325	−11.183	356.16
30	2.4372	693.47	8.9909	11.085	372.59
40	4.1198	685.37	5.5027	33.719	389.37
50	6.6877	677.20	3.4971	56.735	406.50
60	10.471	668.95	2.2984	80.146	423.99
70	15.875	660.61	1.5566	103.97	441.82
80	23.381	652.15	1.0829	128.20	459.99
90	33.549	643.56	0.7717	152.87	478.48
100	47.019	634.83	0.5618	177.98	497.27
110	64.509	625.92	0.4170	203.53	516.36
120	86.808	616.83	0.3149	229.54	535.71
130	114.78	607.51	0.2415	256.02	555.31
140	149.35	597.94	0.1878	282.97	575.13
150	191.51	588.09	0.1478	310.40	595.15
160	242.30	577.91	0.1176	338.33	615.34
170	302.82	567.36	0.0945	366.77	635.65
			Propylene		
−100	4.0341	670.83	8.4375	−655.47	−164.22
−90	8.8068	659.38	4.0797	−635.08	−153.25
−80	17.537	647.81	2.1551	−614.45	−142.16
−70	32.328	636.07	1.2226	−593.54	−131.09
−60	55.826	624.13	0.7372	−572.33	−120.05
−50	91.193	611.92	0.4677	−550.77	−109.10
−40	142.05	599.39	0.3097	−528.82	−98.299
−30	212.43	586.48	0.2127	−506.41	−87.719
−20	306.68	573.11	0.1505	−483.50	−77.429
−10	429.44	559.20	0.1092	−460.01	−67.508
0	585.61	544.64	0.0810	−435.88	−58.050
10	780.34	529.30	0.0611	−411.02	−49.167
20	1019.0	513.02	0.0467	−385.33	−41.005
30	1307.4	495.58	0.0361	−358.67	−33.759
40	1651.5	476.66	0.0280	−330.88	−27.700
50	2057.9	455.79	0.0218	−301.67	−23.235
60	2533.8	432.16	0.0170	−270.63	−21.010
70	3087.2	404.17	0.0130	−236.89	−22.213
80	3727.4	367.62	0.0096	−198.28	−29.628
90	4467.6	299.09	0.0063	−142.55	−56.624

(*Continued*)

APPENDIX B.2   (*Continued*)

Saturation Properties (SI Units)

T (°C)	P_{sat} (kPa)	ρ_f (kg/m³)	v_g (m³/kg)	h_f (kJ/kg)	h_g (kJ/kg)
			Refrigerant 22		
−100	2.0202	1559.9	8.2526	91.797	358.87
−90	4.8222	1533.6	3.6512	102.37	363.72
−80	10.377	1507.1	1.7844	112.97	368.62
−70	20.467	1480.2	0.9477	123.62	373.55
−60	37.497	1452.8	0.5395	134.34	378.46
−50	64.522	1424.9	0.3257	145.15	383.32
−40	105.23	1396.3	0.2064	156.08	388.08
−30	163.91	1366.9	0.1364	167.13	392.69
−20	245.36	1336.5	0.0933	178.35	397.12
−10	354.88	1305.0	0.0657	189.74	401.31
0	498.14	1272.2	0.0474	201.36	405.21
10	681.17	1237.7	0.0349	213.24	408.76
20	910.33	1201.3	0.0262	225.41	411.90
30	1192.3	1171.2	0.0199	236.76	414.55
40	1534.1	1129.0	0.0152	249.80	416.58
50	1943.3	1074.5	0.0117	264.48	417.81
60	2428.1	1030.7	0.0090	277.78	417.97
70	2997.7	962.29	0.0069	294.13	416.53
80	3663.4	893.33	0.0051	310.51	412.04
90	4441.5	779.78	0.0036	331.96	401.84
			Refrigerant 123		
−50	1.7735	1642.4	6.8269	152.30	352.97
−40	3.5863	1619.1	3.5218	161.84	358.71
−30	6.7676	1597.0	1.9418	171.29	364.56
−20	12.028	1573.1	1.1337	181.05	370.50
−10	20.291	1550.2	0.6954	190.77	376.52
0	32.705	1525.6	0.4453	200.78	382.60
10	50.647	1501.6	0.2959	210.79	388.72
20	75.714	1476.6	0.2030	221.01	394.87
30	109.72	1451.0	0.1433	231.38	401.01
40	154.66	1424.8	0.1037	241.89	407.15
50	212.72	1397.5	0.0766	252.59	413.24
60	286.25	1370.0	0.0577	263.38	419.28
70	377.73	1341.2	0.0441	274.39	425.25
80	489.80	1311.2	0.0342	285.58	431.12
90	625.20	1279.9	0.0268	296.98	436.87
100	786.85	1246.9	0.0212	308.62	442.45
110	977.80	1211.6	0.0169	320.54	447.84
120	1201.3	1174.1	0.0136	332.79	452.97
130	1460.8	1133.4	0.0109	345.38	457.77
140	1760.1	1088.2	0.0088	358.45	462.12

APPENDIX B.2 (*Continued*)

Saturation Properties (SI Units)

T (°C)	P_{sat} (kPa)	ρ_f (kg/m³)	v_g (m³/kg)	h_f (kJ/kg)	h_g (kJ/kg)
			Carbon dioxide		
−50	682.35	1154.6	0.0558	−413.83	−74.100
−46	800.16	1139.7	0.0478	−405.92	−72.918
−42	932.54	1124.3	0.0412	−397.89	−71.900
−38	1080.5	1108.6	0.0356	−389.85	−71.057
−34	1245.3	1092.4	0.0309	−381.70	−70.405
−30	1427.8	1075.7	0.0270	−373.45	−69.961
−26	1629.3	1058.6	0.0236	−365.09	−69.742
−22	1850.9	1040.8	0.0206	−356.62	−69.772
−18	2093.8	1022.3	0.0181	−348.00	−70.077
−14	2359.3	1003.1	0.0160	−339.22	−70.688
−10	2648.6	982.93	0.0141	−330.25	−71.642
−6	2963.1	961.70	0.0124	−321.06	−72.988
−2	3304.1	939.22	0.0109	−311.61	−74.788
2	3673.2	915.22	0.0096	−301.85	−77.123
6	4072.0	889.34	0.0085	−291.69	−80.105
10	4502.2	861.11	0.0074	−281.04	−83.894
14	4965.9	829.69	0.0065	−269.74	−88.730
18	5465.3	793.82	0.0056	−257.53	−95.015
22	6003.2	751.20	0.0047	−243.94	−103.51
26	6583.7	696.40	0.0039	−227.90	−115.98
			Refrigerant 23		
−70	193.54	1391.7	0.1156	99.915	328.19
−65	247.29	1372.5	0.0916	106.16	329.89
−60	311.96	1352.9	0.0733	112.47	331.48
−55	388.94	1332.6	0.0593	118.87	332.95
−50	479.68	1311.7	0.0484	125.37	334.30
−45	585.68	1290.0	0.0398	131.98	335.52
−40	708.48	1267.4	0.0330	138.72	336.57
−35	849.68	1243.8	0.0275	145.61	337.45
−30	1010.9	1218.9	0.0230	152.66	338.14
−25	1193.8	1192.6	0.0193	159.90	338.60
−20	1400.2	1164.7	0.0163	167.36	338.81
−15	1631.7	1132.0	0.0138	175.42	338.72
−10	1890.5	1101.0	0.0117	183.27	338.27
−5	2178.4	1067.2	0.0099	191.46	337.40
0	2497.9	1030.7	0.0084	200.00	335.98
5	2851.5	989.79	0.0070	209.07	333.93
10	3242.5	942.91	0.0059	218.85	330.82
15	3674.9	885.99	0.0048	229.75	326.36
20	4154.9	807.06	0.0039	242.89	319.34
25	4693.9	677.17	0.0026	261.28	300.44

(*Continued*)

APPENDIX B.2   (*Continued*)

Saturation Properties (SI Units)

T (°C)	P_{sat} (kPa)	ρ_f (kg/m³)	v_g (m³/kg)	h_f (kJ/kg)	h_g (kJ/kg)
		Refrigerant 134A			
−20	132.82	1358.5	0.1474	25.472	238.43
−15	164.05	1343.1	0.1207	31.971	241.51
−10	200.74	1327.4	0.0996	38.529	244.55
−5	243.51	1311.4	0.0828	45.149	247.55
0	293.01	1295.1	0.0693	51.834	250.50
5	349.90	1278.3	0.0584	58.588	253.39
10	414.89	1261.2	0.0495	65.417	256.22
15	488.70	1243.6	0.0421	72.325	258.97
20	572.07	1225.5	0.0360	79.317	261.64
25	665.78	1206.8	0.0309	86.401	264.23
30	770.64	1187.6	0.0267	93.583	266.71
35	887.47	1167.6	0.0230	100.87	269.08
40	1017.1	1146.8	0.0200	108.28	271.31
45	1160.5	1125.0	0.0173	115.82	273.40
50	1318.6	1102.3	0.0151	123.50	275.32
55	1492.3	1078.3	0.0131	131.35	277.03
60	1682.8	1052.8	0.0114	139.38	278.47
65	1891.0	1025.6	0.0100	147.64	279.69
70	2118.2	996.28	0.0087	156.15	280.52
75	2365.8	964.02	0.0075	165.01	280.88
		Sulfur dioxide			
−50	11.509	1559.7	2.5010	−54.467	368.86
−40	21.561	1535.6	1.3883	−40.844	374.11
−30	38.023	1511.1	0.8167	−27.234	379.25
−20	63.598	1486.3	0.5053	−13.618	384.26
−10	101.56	1461.1	0.3263	0.0244	389.06
0	155.75	1435.3	0.2187	13.716	393.65
10	230.54	1408.8	0.1514	27.481	397.98
20	330.82	1381.6	0.1078	41.346	402.05
30	461.98	1353.5	0.0785	55.340	405.79
40	629.85	1324.4	0.0583	69.498	409.10
50	840.72	1294.1	0.0441	83.856	412.01
60	1101.2	1262.5	0.0338	98.461	414.43
70	1418.3	1229.2	0.0263	113.36	416.27
80	1799.2	1193.9	0.0206	128.63	417.43
90	2251.2	1156.3	0.0163	144.35	417.70
100	2781.6	1115.8	0.0129	160.62	417.14
110	3398.3	1071.4	0.0103	177.62	415.29
120	4109.5	1021.8	0.0082	195.59	411.83
130	4924.9	964.66	0.0064	214.99	406.12
140	5858.9	893.83	0.0050	236.80	396.75

APPENDIX B.2 (*Continued*)

Saturation Properties (IP Units)

T (°F)	P_{sat} (psia)	ρ_f (lbm/ft³)	v_g (ft³/lbm)	h_f (Btu/lbm)	h_g (Btu/lbm)
			Water		
40	0.1217	62.423	2443.6	8.0324	1078.7
60	0.2564	62.364	1206.1	28.078	1087.4
80	0.5075	62.213	632.41	48.065	1096.1
100	0.9505	61.991	349.83	68.034	1104.7
120	1.6951	61.709	202.94	88.004	1113.2
140	2.8931	61.376	122.81	107.99	1121.6
160	4.7474	60.998	77.185	128.00	1129.8
180	7.5197	60.578	50.172	148.04	1137.9
200	11.538	60.120	33.613	168.13	1145.7
220	17.201	59.626	23.136	188.28	1153.3
240	24.985	59.097	16.316	208.49	1160.5
260	35.447	58.535	11.760	228.79	1167.4
280	49.222	57.940	8.6439	249.20	1173.9
300	67.028	57.312	6.4663	269.73	1180.0
320	89.667	56.650	4.9144	290.40	1185.5
340	118.02	55.955	3.7885	311.24	1190.5
360	153.03	55.225	2.9580	332.28	1194.8
380	195.74	54.458	2.3361	353.53	1198.5
400	247.26	53.653	1.8639	375.04	1201.4
420	308.76	52.806	1.5006	396.84	1203.6
			Benzene		
50	0.8809	55.521	33.643	−54.107	129.40
70	1.5342	54.786	39.750	−45.930	139.54
90	2.5458	54.046	34.476	−37.617	148.99
110	4.0487	53.301	23.707	−29.154	155.14
130	6.2017	52.549	15.218	−20.532	159.61
150	9.1890	51.789	9.9243	−11.743	163.93
170	13.219	51.018	6.7267	−2.7787	168.57
190	18.524	50.236	4.7409	6.3665	173.56
210	25.354	49.440	3.4563	15.698	178.85
230	33.981	48.628	2.5915	25.222	184.37
250	44.689	47.798	1.9881	34.944	190.09
270	57.780	47.031	1.5540	44.607	195.94
290	73.567	46.072	1.2334	55.006	201.88
310	92.378	45.169	0.9912	65.362	207.89
330	114.55	44.234	0.8045	75.947	213.90
350	140.44	43.261	0.6583	86.775	219.90
370	170.40	42.242	0.5419	97.861	225.83
390	204.83	41.167	0.4480	109.23	231.64
410	244.13	40.022	0.3712	120.90	237.27
430	288.74	38.790	0.3078	132.93	242.63

(*Continued*)

APPENDIX B.2 (*Continued*)

Saturation Properties (IP Units)

T (°F)	P_{sat} (psia)	ρ_f (lbm/ft³)	v_g (ft³/lbm)	h_f (Btu/lbm)	h_g (Btu/lbm)
			Decane		
0	0.0007	47.435	47,460	−198.90	−33.137
30	0.0034	46.619	10,736	−184.04	−22.462
60	0.0133	45.808	2954.9	−168.79	−11.295
90	0.0430	45.000	963.04	−153.10	0.3828
120	0.1202	44.189	362.70	−136.97	12.578
150	0.2965	43.374	154.42	−120.37	25.284
180	0.6576	42.551	72.876	−103.28	38.484
210	1.3328	41.717	37.517	−85.697	52.160
240	2.5034	40.868	20.699	−67.613	66.254
270	4.4091	40.000	12.168	−49.021	80.795
300	7.3529	39.108	7.4880	−29.915	95.667
330	11.705	38.187	4.8126	−10.289	110.90
360	17.904	37.229	3.2011	9.8619	126.42
390	26.452	36.227	2.1904	30.547	142.18
420	37.910	35.168	1.5342	51.780	158.11
450	52.879	34.039	1.0950	73.577	174.15
480	71.984	32.817	0.7930	95.971	190.19
510	95.860	31.468	0.5801	119.01	206.11
540	125.16	29.939	0.4261	142.77	221.74
570	160.64	28.131	0.3117	167.42	236.78
			Acetone		
−100	0.0048	55.573	13,830	21.784	291.94
−80	0.0151	54.834	4638.7	31.359	296.95
−60	0.0420	54.099	1752.5	40.973	302.01
−40	0.1045	53.365	740.36	50.635	307.21
−20	0.2362	52.629	341.90	60.354	312.40
0	0.4907	51.891	171.94	70.144	317.71
20	0.9474	51.147	92.617	80.016	323.03
40	1.7161	50.397	53.027	89.984	328.39
60	2.9411	49.637	32.000	100.06	333.77
80	4.8042	48.866	20.283	110.26	339.26
100	7.5279	48.082	13.259	120.59	344.60
120	11.378	47.282	9.0203	131.07	350.09
140	16.666	46.464	6.2767	141.70	355.44
160	23.753	45.624	4.4873	152.51	360.83
180	33.047	44.760	3.2750	163.50	366.16
200	45.008	43.866	2.4189	174.70	371.25
220	60.138	42.940	1.8346	186.11	376.55
240	78.971	41.975	1.3917	197.75	381.31
260	102.06	40.965	1.0828	209.65	386.33
280	129.97	39.902	0.8444	221.83	390.88

APPENDIX B.2   (*Continued*)

Saturation Properties (IP Units)

T (°F)	P_{sat} (psia)	ρ_f (lbm/ft³)	v_g (ft³/lbm)	h_f (Btu/lbm)	h_g (Btu/lbm)
			Cyclohexane		
50	0.9210	49.194	70.268	2.6383	174.82
70	1.5810	48.538	42.437	10.915	180.48
90	2.5909	47.879	26.784	19.613	186.38
110	4.0760	47.215	17.567	28.685	192.52
130	6.1849	46.544	11.917	38.100	198.87
150	9.0885	45.864	8.3266	47.832	205.42
170	12.979	45.174	5.9709	57.865	212.15
190	18.067	44.472	4.3803	68.186	219.05
210	24.581	43.756	3.2784	78.788	226.09
230	32.764	43.023	2.4972	89.667	233.26
250	42.871	42.272	1.9441	100.82	240.66
270	55.171	41.498	1.5139	112.25	247.90
290	69.941	40.699	1.2004	123.97	255.32
310	87.471	39.869	0.9611	135.97	262.77
330	108.06	39.003	0.7759	148.28	270.23
350	132.03	38.093	0.6305	160.90	277.66
370	159.71	37.130	0.5149	173.86	285.02
390	191.44	36.103	0.4220	187.17	292.25
410	227.59	34.997	0.3464	200.87	299.31
430	268.58	33.793	0.2841	215.01	306.11
			Ethanol (Ethyl alcohol)		
−20	0.0237	51.814	4227.8	−3.7964	409.56
0	0.0608	51.234	1762.0	5.5077	416.52
20	0.1435	50.664	786.06	15.324	423.43
40	0.3151	50.096	373.55	25.743	430.32
60	0.6479	49.524	188.22	36.811	437.22
80	1.2564	48.943	100.09	48.546	444.15
100	2.3108	48.346	55.937	60.944	451.12
120	4.0524	47.730	32.713	73.991	458.10
140	6.8068	47.089	19.938	87.665	465.06
160	10.996	46.421	12.616	101.94	471.97
180	17.147	45.720	8.2552	116.80	478.77
200	25.894	44.982	5.5659	132.21	485.40
220	37.983	44.204	3.8526	148.16	491.80
240	54.264	43.381	2.7280	164.64	497.86
260	75.685	42.506	1.9691	181.63	503.51
280	103.29	41.574	1.4439	199.15	508.62
300	138.20	40.574	1.0717	217.20	513.04
320	181.64	39.496	0.8021	235.82	516.57
340	234.87	38.323	0.6031	255.06	518.97
360	299.28	37.032	0.4535	275.02	519.87

(Continued)

APPENDIX B.2 (*Continued*)

Saturation Properties (IP Units)

T (°F)	P_{sat} (psia)	ρ_f (lbm/ft³)	v_g (ft³/lbm)	h_f (Btu/lbm)	h_g (Btu/lbm)
			Ethylene		
−150	17.169	35.223	6.6433	−281.32	−75.659
−140	23.290	34.717	5.0087	−275.54	−73.432
−130	30.957	34.201	3.8444	−269.72	−71.315
−120	40.405	33.673	2.9977	−263.83	−69.320
−110	51.875	33.131	2.3703	−257.87	−67.459
−100	65.620	32.575	1.8974	−251.84	−65.743
−90	81.896	32.002	1.5353	−245.71	−64.187
−80	100.97	31.411	1.2541	−239.48	−62.806
−70	123.10	30.795	1.0329	−233.13	−61.620
−60	148.57	30.155	0.8566	−226.64	−60.650
−50	177.67	29.485	0.7146	−220.00	−59.922
−40	210.68	28.780	0.5990	−213.18	−59.473
−30	247.90	28.034	0.5042	−206.15	−59.315
−20	289.66	27.246	0.4246	−198.91	−59.599
−10	336.29	26.373	0.3581	−191.26	−60.317
0	388.15	25.441	0.3015	−183.33	−61.617
10	445.66	24.366	0.2526	−174.74	−63.709
20	509.28	23.138	0.2096	−165.45	−66.892
30	579.59	21.669	0.1707	−155.11	−71.803
40	657.39	19.573	0.1330	−142.14	−80.221
			Heptane		
−20	0.0288	45.221	1635.4	−49.831	121.61
0	0.0673	44.647	730.97	−39.879	128.52
20	0.1446	44.073	354.63	−29.777	135.63
40	0.2888	43.497	184.71	−19.511	142.93
60	0.5412	42.917	102.33	−9.0657	150.42
80	0.9593	42.332	59.808	1.5701	158.10
100	1.6186	41.741	36.623	12.408	165.96
120	2.6152	41.143	23.375	23.458	174.00
140	4.0663	40.536	15.459	34.731	182.21
160	6.1107	39.918	10.551	46.234	190.58
180	8.9084	39.287	7.4024	57.975	199.10
200	12.640	38.643	5.3217	69.963	207.75
220	17.505	37.981	3.9091	82.204	216.53
240	23.722	37.300	2.9252	94.706	225.42
260	31.525	36.596	2.2249	107.48	234.41
280	41.165	35.863	1.7162	120.54	243.46
300	52.907	35.105	1.3398	133.87	252.56
320	67.033	34.306	1.0566	147.51	261.69
340	83.841	33.462	0.8400	161.47	270.80
360	103.65	32.564	0.6719	175.76	279.84

APPENDIX B.2 (*Continued*)

Saturation Properties (IP Units)

T (°F)	P_{sat} (psia)	ρ_f (lbm/ft³)	v_g (ft³/lbm)	h_f (Btu/lbm)	h_g (Btu/lbm)
			Propane		
−180	0.0754	41.351	903.37	−25.626	187.75
−160	0.2312	40.635	315.05	−15.955	193.14
−140	0.6052	39.911	127.99	−6.1646	198.64
−120	1.3934	39.178	58.820	3.7626	204.24
−100	2.8876	38.433	29.894	13.845	209.92
−80	5.4851	37.673	16.499	24.105	215.65
−60	9.6894	36.896	9.7447	34.565	221.40
−40	16.103	36.099	6.0859	45.253	227.14
−20	25.415	35.276	3.9796	56.197	232.83
0	38.388	34.424	2.7027	67.429	238.43
20	55.843	33.535	1.8935	78.988	243.91
40	78.654	32.603	1.3608	90.917	249.20
60	107.74	31.615	0.9982	103.27	254.27
80	144.05	30.558	0.7441	116.11	259.01
100	188.63	29.411	0.5612	129.53	263.34
120	242.55	28.135	0.4261	143.68	267.08
140	307.03	26.701	0.3237	158.67	269.95
160	383.46	25.018	0.2435	174.82	271.42
180	473.59	22.822	0.1775	193.03	270.27
200	579.99	19.022	0.1146	217.04	261.20
			Hexane		
−20	0.1477	44.903	370.25	−58.277	120.24
0	0.3141	44.081	181.73	−45.234	127.16
20	0.6143	43.287	96.760	−32.936	134.25
40	1.1205	42.516	55.091	−21.133	141.50
60	1.9276	41.765	33.171	−9.6457	148.92
80	3.1547	41.030	20.937	1.6608	156.49
100	4.9463	40.305	13.754	12.889	164.21
120	7.4732	39.587	9.3498	24.118	172.06
140	10.932	38.873	6.5451	35.411	180.03
160	15.546	38.158	4.6988	46.819	188.11
180	21.565	37.438	3.4473	58.384	196.29
200	29.262	36.707	2.5769	70.143	204.55
220	38.939	35.962	1.9575	82.129	212.87
240	50.924	35.195	1.5074	94.374	221.24
260	65.568	34.399	1.1743	106.91	229.62
280	83.253	33.564	0.9235	119.77	237.98
300	104.38	32.679	0.7318	133.00	246.29
320	129.40	31.725	0.5832	146.64	254.49
340	158.75	30.678	0.4662	160.77	262.51
360	192.95	29.497	0.3728	175.50	270.24

(*Continued*)

APPENDIX B.2 (*Continued*)

Saturation Properties (IP Units)

T (°F)	P_sat (psia)	ρ_f (lbm/ft³)	v_g (ft³/lbm)	h_f (Btu/lbm)	h_g (Btu/lbm)
			Octane		
−20	0.0049	46.231	7499.7	−49.624	120.52
0	0.0135	45.677	2986.5	−39.696	127.69
20	0.0331	45.125	1306.4	−29.628	134.99
40	0.0739	44.571	619.96	−19.405	142.43
60	0.1525	44.017	315.83	−9.0141	150.03
80	0.2938	43.460	171.19	1.5575	157.81
100	0.5337	42.899	97.971	12.320	165.78
120	0.9208	42.333	58.810	23.284	173.94
140	1.5187	41.761	36.817	34.457	182.28
160	2.4070	41.182	23.919	45.846	190.81
180	3.6824	40.594	16.056	57.458	199.51
200	5.4590	39.996	11.095	69.300	208.39
220	7.8684	39.385	7.8649	81.376	217.42
240	11.059	38.761	5.7032	93.691	226.60
260	15.195	38.121	4.2193	106.25	235.92
280	20.457	37.463	3.1772	119.06	245.36
300	27.037	36.783	2.4299	132.13	254.91
320	35.142	36.078	1.8838	145.45	264.55
340	44.990	35.344	1.4777	159.05	274.25
360	56.812	34.577	1.1708	172.93	284.00
			Propylene		
−200	0.0303	43.928	2187.1	−306.50	−83.815
−180	0.1096	43.140	650.14	−297.12	−78.798
−160	0.3275	42.353	232.92	−287.59	−73.683
−140	0.8382	41.562	96.864	−277.92	−68.486
−120	1.8923	40.763	45.381	−268.12	−63.246
−100	3.8540	39.954	23.486	−258.19	−57.946
−80	7.2080	39.130	13.151	−248.10	−52.665
−60	12.556	38.286	7.8625	−237.83	−47.424
−40	20.603	37.419	4.9615	−227.35	−42.261
−20	32.146	36.522	3.2735	−216.64	−37.214
0	48.055	35.588	2.2405	−205.65	−32.325
20	69.261	34.612	1.5804	−194.34	−27.644
40	96.752	33.582	1.1422	−182.69	−23.226
60	131.57	32.487	0.8415	−170.62	−19.146
80	174.83	31.310	0.6278	−158.07	−15.564
100	227.69	30.029	0.4746	−144.95	−12.436
120	291.45	28.606	0.3600	−131.13	−10.115
140	367.50	26.979	0.2714	−116.35	−9.0327
160	457.42	25.012	0.2018	−100.13	−9.7371
180	563.14	22.299	0.1432	−81.019	−14.109

APPENDIX B.2 (*Continued*)

Saturation Properties (IP Units)

T (°F)	P_{sat} (psia)	ρ_f (lbm/ft³)	v_g (ft³/lbm)	h_f (Btu/lbm)	h_g (Btu/lbm)
			Refrigerant 22		
−200	0.0108	102.05	2993.3	26.346	148.47
−180	0.0452	100.26	768.86	31.396	150.67
−160	0.1534	98.464	242.97	36.439	152.92
−140	0.4374	96.654	90.774	41.484	155.21
−120	1.0832	94.824	38.857	46.540	157.54
−100	2.3880	92.968	18.589	51.618	159.89
−80	4.7814	91.079	9.7422	56.730	162.24
−60	8.8352	89.148	5.5033	61.887	164.57
−40	15.263	87.166	3.3069	67.101	166.84
−20	24.909	85.122	2.0911	72.386	169.04
0	38.737	83.004	1.3792	77.756	171.14
20	57.811	80.796	0.9418	83.229	173.10
40	83.281	79.060	0.6617	88.248	174.91
60	116.37	76.591	0.4757	94.015	176.51
80	158.38	73.399	0.3482	100.48	177.87
100	210.69	70.555	0.2583	106.62	178.93
120	274.74	67.930	0.1931	112.58	179.59
140	352.16	63.846	0.1445	119.85	179.69
160	444.78	60.070	0.1071	126.85	178.95
180	554.89	54.486	0.0767	135.32	176.63
			Refrigerant 123		
−120	0.0113	107.31	2118.5	51.717	143.72
−100	0.0356	105.70	709.40	56.207	146.22
−80	0.0974	104.23	273.23	60.573	148.81
−60	0.2367	102.64	118.24	65.074	151.48
−40	0.5202	101.12	56.413	69.537	154.22
−20	1.0493	99.537	29.221	74.101	157.01
0	1.9679	97.882	16.226	78.769	159.86
20	3.4671	96.278	9.5586	83.435	162.74
40	5.7880	94.598	5.9220	88.209	165.66
60	9.2218	92.881	3.8305	93.058	168.59
80	14.108	91.121	2.5709	97.982	171.52
100	20.830	89.294	1.7810	103.00	174.46
120	29.813	87.434	1.2677	108.08	177.37
140	41.517	85.508	0.9235	113.25	180.26
160	56.437	83.522	0.6861	118.50	183.11
180	75.096	81.413	0.5183	123.87	185.90
200	98.049	79.210	0.3969	129.35	188.63
220	125.88	76.888	0.3072	134.95	191.26
240	159.21	74.361	0.2398	140.72	193.78
260	198.71	71.640	0.1881	146.66	196.14

(*Continued*)

APPENDIX B.2 (*Continued*)

Saturation Properties (IP Units)

T (°F)	P_{sat} (psia)	ρ_f (lbm/ft³)	v_g (ft³/lbm)	h_f (Btu/lbm)	h_g (Btu/lbm)
			Carbon dioxide		
−50	118.08	71.042	0.7532	−174.13	−31.297
−44	134.13	70.243	0.6651	−171.27	−30.934
−38	151.75	69.428	0.5891	−168.38	−30.623
−32	171.02	68.594	0.5232	−165.47	−30.368
−26	192.05	67.736	0.4659	−162.52	−30.172
−20	214.92	66.865	0.4158	−159.56	−30.041
−14	239.73	65.963	0.3718	−156.56	−29.980
−8	266.58	65.037	0.3331	−153.52	−29.993
−2	295.57	64.082	0.2989	−150.44	−30.088
4	326.81	63.094	0.2685	−147.32	−30.272
10	360.40	62.067	0.2415	−144.13	−30.553
16	396.44	61.001	0.2173	−140.90	−30.943
22	435.06	59.885	0.1956	−137.59	−31.455
28	476.37	58.714	0.1760	−134.20	−32.104
34	520.49	57.476	0.1584	−130.71	−32.912
40	567.57	56.163	0.1423	−127.12	−33.904
46	617.75	54.758	0.1276	−123.40	−35.115
52	671.17	53.234	0.1141	−119.51	−36.594
58	728.02	51.562	0.1017	−115.40	−38.412
64	788.50	49.691	0.0900	−111.02	−40.680
			Refrigerant 23		
−120	12.731	90.205	3.9028	35.338	138.76
−110	17.530	88.947	2.8882	38.248	139.70
−100	23.658	87.665	2.1764	41.182	140.59
−90	31.352	86.354	1.6666	44.145	141.43
−80	40.869	85.007	1.2947	47.144	142.21
−70	52.478	83.620	1.0187	50.185	142.94
−60	66.463	81.926	0.8106	53.515	143.60
−50	83.121	80.687	0.6514	56.424	144.19
−40	102.76	79.123	0.5280	59.639	144.70
−30	125.69	77.478	0.4312	62.932	145.12
−20	152.23	75.738	0.3544	66.315	145.43
−10	182.73	73.884	0.2927	69.801	145.62
0	217.54	71.894	0.2426	73.411	145.66
10	257.01	69.740	0.2016	77.164	145.54
20	301.54	67.381	0.1676	81.094	145.20
30	351.57	64.874	0.1391	85.150	144.60
40	407.62	61.808	0.1149	89.666	143.66
50	470.28	58.864	0.0941	94.090	142.23
60	540.38	54.854	0.0756	99.342	140.04
70	619.16	48.632	0.0578	106.12	136.18

APPENDIX B.2   (*Continued*)

Saturation Properties (IP Units)

T (°F)	P_{sat} (psia)	ρ_f (lbm/ft³)	v_g (ft³/lbm)	h_f (Btu/lbm)	h_g (Btu/lbm)
			Refrigerant 134A		
−40	7.4325	88.512	5.7769	0.0000	97.104
−30	9.8686	87.504	4.4286	3.0081	98.616
−20	12.906	86.482	3.4424	6.0415	100.12
−10	16.642	85.442	2.7097	9.1015	101.62
0	21.185	84.384	2.1575	12.190	103.10
10	26.646	83.306	1.7358	15.308	104.56
20	33.147	82.204	1.4097	18.459	106.00
30	40.813	81.077	1.1548	21.643	107.42
40	49.776	79.921	0.9532	24.864	108.80
50	60.175	78.734	0.7924	28.124	110.15
60	72.152	77.512	0.6628	31.426	111.47
70	85.858	76.250	0.5575	34.773	112.74
80	101.45	74.944	0.4712	38.170	113.96
90	119.08	73.588	0.4000	41.620	115.12
100	138.93	72.173	0.3407	45.130	116.22
110	161.16	70.693	0.2911	48.706	117.25
120	185.96	69.134	0.2493	52.356	118.19
130	213.53	67.484	0.2137	56.091	119.03
140	244.06	65.724	0.1833	59.925	119.74
150	277.79	63.829	0.1571	63.875	120.29
			Sulfur dioxide		
−80	0.7052	99.188	89.903	−30.592	155.74
−60	1.5504	97.533	42.934	−24.068	158.32
−40	3.1272	95.862	22.247	−17.560	160.84
−20	5.8537	94.166	12.372	−11.059	163.29
0	10.274	92.441	7.3128	−4.5527	165.65
20	17.059	90.679	4.5568	1.9698	167.93
40	27.001	88.872	2.9630	8.5224	170.07
60	41.012	87.012	2.0006	15.120	172.08
80	60.114	85.088	1.3943	21.779	173.94
100	85.436	83.090	0.9970	28.519	175.59
120	118.20	81.004	0.7280	35.361	177.00
140	159.72	78.814	0.5418	42.330	178.17
160	211.37	76.497	0.4093	49.458	179.03
180	274.59	74.027	0.3128	56.784	179.53
200	350.83	71.365	0.2412	64.360	179.57
220	441.58	68.457	0.1869	72.259	179.06
240	548.41	65.214	0.1450	80.594	177.81
260	673.09	61.479	0.1119	89.563	175.55
280	818.02	56.914	0.0847	99.583	171.66
300	987.52	50.437	0.0605	111.97	164.23

APPENDIX B.3

Properties of Metals and Alloys (SI Units)

T (°C)	ρ (kg/ m³)	c (kJ/ kg·K)	k (W/ m·K)	ρ (kg/ m³)	c (kJ/ kg·K)	k (W/ m·K)	ρ (kg/ m³)	c (kJ/ kg·K)	k (W/ m·K)	T (°C)
		Aluminum			Brass			Bronze		
30	2701	0.9044	236.1	8530	0.3781	116.7	8799	0.4213	52.00	30
90	2690	0.9320	238.5	8530	0.3886	129.3	8771	0.4453	52.00	90
150	2678	0.9587	239.1	8544	0.3985	138.4	8742	0.4698	52.81	150
210	2665	0.9839	236.7	8581	0.4075	142.0	8712	0.4953	54.91	210
270	2652	1.0090	234.3	8618	0.4165	145.6	8682	0.5208	57.01	270
330	2640	1.0350	231.8	8655	0.4255	149.2	8652	0.5463	59.11	330
		Chromium			Copper			Iron		
30	7160	0.4501	93.61	8932	0.3854	400.7	7869	0.4484	79.86	30
90	7151	0.4711	91.93	8904	0.3926	395.9	7851	0.4742	73.44	90
150	7141	0.4907	89.72	8876	0.3993	391.4	7833	0.4966	67.79	150
210	7129	0.5081	86.66	8847	0.4053	387.2	7813	0.5137	63.35	210
270	7118	0.5255	83.60	8818	0.4113	383.0	7793	0.5308	58.91	270
330	7106	0.5426	80.55	8789	0.4173	378.8	7772	0.5491	54.52	330
		Iron-Armco			AISI 1010 Carbon Steel			AISI 302 Stainless Steel		
30	7867	0.4484	72.48	7831	0.4354	63.74	8052	0.4825	15.27	30
90	7815	0.4742	68.28	7814	0.4618	60.62	8031	0.5008	16.53	90
150	7782	0.4966	64.24	7797	0.4874	57.55	8008	0.5174	17.61	150
210	7782	0.5137	60.46	7777	0.5117	54.58	7983	0.5315	18.42	210
270	7781	0.5308	56.68	7757	0.5360	51.61	7958	0.5456	19.23	270
330	7779	0.5491	52.93	7737	0.5610	48.65	7932	0.5594	20.04	330
		AISI 304 Stainless Steel			AISI 316 Stainless Steel			AISI 347 Stainless Steel		
30	7899	0.4782	14.95	8237	0.4691	13.46	7975	0.4826	14.32	30
90	7877	0.5010	15.97	8214	0.4907	14.54	7954	0.5014	15.24	90
150	7854	0.5199	16.97	8190	0.5093	15.56	7931	0.5183	16.16	150
210	7829	0.5325	17.93	8164	0.5231	16.49	7907	0.5321	17.09	210
270	7805	0.5451	18.89	8139	0.5369	17.42	7882	0.5459	18.02	270
330	7780	0.5574	19.84	8113	0.5504	18.35	7857	0.5594	18.95	330
		Lead			Nickel			Inconel X-750		
30	11,337	0.1291	35.26	8899	0.4453	90.37	8509	0.4401	11.76	30
90	11,277	0.1309	34.48	8876	0.4699	84.07	8490	0.4605	12.84	90
150	11,214	0.1332	33.70	8853	0.4974	78.51	8470	0.4773	13.91	150
210	11,147	0.1362	32.92	8829	0.5295	74.13	8448	0.4884	14.96	210
270	11,081	0.1392	32.14	8805	0.5616	69.75	8426	0.4995	16.01	270
330	11,014	0.1422	31.36	8781	0.5910	65.63	8405	0.5106	17.06	330
		Silver			Titanium			Zirconium		
30	10,498	0.2351	428.9	4500	0.5229	21.85	6570	0.2787	22.67	30
90	10,462	0.2375	426.5	4492	0.5403	20.95	6563	0.2919	22.01	90
150	10,424	0.2403	423.5	4485	0.5556	20.28	6556	0.3025	21.50	150
210	10,386	0.2436	419.6	4477	0.5676	19.98	6548	0.3091	21.23	210
270	10,348	0.2469	415.7	4469	0.5796	19.68	6540	0.3157	20.96	270
330	10,309	0.2502	411.7	4461	0.5917	19.40	6533	0.3223	20.71	330

APPENDIX B.3 (*Continued*)

Properties of Metals and Alloys (IP Units)

T (°F)	ρ (lbm/ ft³)	c (Btu/ lbm·R)	k (Btu/ h·ft·°F)	ρ (lbm/ ft³)	c (Btu/ lbm·R)	k (Btu/ h·ft·°F)	ρ (lbm/ ft³)	c (Btu/ lbm·R)	k (Btu/ h·ft·°F)	T (°F)
	Aluminum			Brass			Bronze			
100	168.5	0.2169	136.6	532.5	0.09062	68.35	549.0	0.1014	30.05	100
200	167.9	0.2230	137.9	532.5	0.09294	75.09	547.4	0.1067	30.05	200
300	167.2	0.2289	138.2	533.4	0.09513	79.92	545.8	0.1121	30.49	300
400	166.4	0.2344	136.9	535.5	0.09712	81.85	544.0	0.1177	31.61	400
500	165.7	0.2400	135.6	537.6	0.09911	83.77	542.3	0.1234	32.74	500
600	165.0	0.2456	134.3	539.8	0.1011	85.70	540.6	0.1290	33.86	600
	Chromium			Copper			Iron			
100	446.9	0.1082	53.96	557.4	0.09227	231.2	491.1	0.1079	45.66	100
200	446.4	0.1128	53.06	555.8	0.09386	228.6	490.1	0.1136	42.23	200
300	445.8	0.1171	51.87	554.2	0.09535	226.2	489.0	0.1185	39.21	300
400	445.1	0.1210	50.23	552.5	0.09668	223.9	487.8	0.1223	36.84	400
500	444.5	0.1248	48.60	550.8	0.09800	221.7	486.7	0.1261	34.46	500
600	443.8	0.1287	46.96	549.1	0.09933	219.4	485.5	0.1299	32.09	600
	Iron-Armco			AISI 1010 Carbon Steel			AISI 302 Stainless Steel			
100	490.7	0.1079	41.56	488.7	0.1048	36.59	502.5	0.1158	8.917	100
200	487.7	0.1136	39.32	487.8	0.1106	34.92	501.3	0.1198	9.590	200
300	485.8	0.1185	37.16	486.7	0.1163	33.29	499.9	0.1235	10.170	300
400	485.8	0.1223	35.14	485.6	0.1217	31.70	498.5	0.1266	10.600	400
500	485.8	0.1261	33.11	484.5	0.1270	30.11	497.0	0.1298	11.030	500
600	485.7	0.1299	31.09	483.3	0.1324	28.52	495.6	0.1329	11.470	600
	AISI 304 Stainless Steel			AISI 316 Stainless Steel			AISI 347 Stainless Steel			
100	492.9	0.1149	8.716	514.0	0.1127	7.856	497.7	0.1158	8.345	100
200	491.7	0.1200	9.262	512.7	0.1175	8.434	496.5	0.1200	8.834	200
300	490.3	0.1241	9.795	511.3	0.1216	8.980	495.2	0.1237	9.326	300
400	488.9	0.1269	10.31	509.8	0.1246	9.477	493.7	0.1268	9.824	400
500	487.5	0.1297	10.82	508.3	0.1277	9.975	492.3	0.1298	10.32	500
600	486.1	0.1325	11.34	506.9	0.1307	10.47	490.9	0.1329	10.82	600
	Lead			Nickel			Inconel X-750			
100	707.2	0.0309	20.31	555.4	0.1071	51.74	531.0	0.1057	6.874	100
200	703.8	0.0313	19.90	554.1	0.1126	48.37	529.9	0.1103	7.452	200
300	700.1	0.0318	19.48	552.7	0.1187	45.41	528.8	0.1139	8.023	300
400	696.3	0.0325	19.06	551.3	0.1258	43.07	527.5	0.1164	8.585	400
500	692.4	0.0331	18.64	549.9	0.1329	40.72	526.3	0.1189	9.146	500
600	688.6	0.0338	18.23	548.5	0.1400	38.38	525.0	0.1213	9.708	600
	Silver			Titanium			Zirconium			
100	655.1	0.0562	247.6	280.8	0.1254	12.56	410.1	0.06697	13.05	100
200	653.0	0.0568	246.3	280.4	0.1293	12.08	409.7	0.06989	12.69	200
300	650.8	0.0574	244.7	280.0	0.1327	11.72	409.3	0.07223	12.42	300
400	648.6	0.0581	242.6	279.5	0.1353	11.56	408.8	0.07369	12.28	400
500	646.4	0.0588	240.6	279.1	0.1380	11.40	408.4	0.07515	12.13	500
600	644.2	0.0596	238.5	278.6	0.1406	11.24	407.9	0.07661	11.99	600

APPENDIX B.4

Properties of Air at 1 atmosphere

	IP UNITS				
T (°F)	ρ (lbm/ft³)	c_p (Btu/lbm·R)	μ (lbm/ft·h)	k (Btu/h·ft·°F)	Pr
−40	0.09463	0.2403	0.03665	0.01226	0.7183
−30	0.09242	0.2403	0.03737	0.01252	0.7172
−20	0.09030	0.2403	0.03807	0.01278	0.7161
−10	0.08828	0.2403	0.03877	0.01303	0.7151
0	0.08635	0.2403	0.03947	0.01328	0.7141
10	0.08451	0.2403	0.04016	0.01353	0.7131
20	0.08274	0.2403	0.04084	0.01378	0.7122
30	0.08104	0.2403	0.04152	0.01403	0.7113
40	0.07941	0.2403	0.04219	0.01427	0.7104
50	0.07785	0.2404	0.04286	0.01452	0.7096
60	0.07634	0.2404	0.04352	0.01476	0.7088
70	0.07490	0.2404	0.04417	0.01500	0.7081
80	0.07351	0.2405	0.04482	0.01524	0.7073
90	0.07216	0.2405	0.04547	0.01548	0.7066
100	0.07087	0.2406	0.04611	0.01571	0.7060
110	0.06962	0.2406	0.04674	0.01595	0.7053
120	0.06842	0.2407	0.04737	0.01618	0.7047
130	0.06726	0.2408	0.04800	0.01641	0.7041
140	0.06613	0.2408	0.04862	0.01664	0.7036
150	0.06505	0.2409	0.04924	0.01687	0.7031
160	0.06399	0.2410	0.04985	0.01710	0.7026
170	0.06298	0.2411	0.05046	0.01733	0.7021
180	0.06199	0.2412	0.05106	0.01756	0.7017
190	0.06103	0.2413	0.05166	0.01778	0.7013
200	0.06011	0.2415	0.05226	0.01800	0.7009
210	0.05921	0.2416	0.05285	0.01823	0.7005
220	0.05834	0.2417	0.05344	0.01845	0.7002
230	0.05749	0.2418	0.05402	0.01867	0.6999
240	0.05667	0.2420	0.05460	0.01889	0.6996
250	0.05587	0.2421	0.05518	0.01911	0.6994
260	0.05509	0.2423	0.05575	0.01932	0.6991
270	0.05433	0.2425	0.05632	0.01954	0.6989
280	0.05360	0.2426	0.05689	0.01975	0.6987
290	0.05288	0.2428	0.05745	0.01997	0.6986
300	0.05219	0.2430	0.05801	0.02018	0.6984
310	0.05151	0.2432	0.05857	0.02039	0.6983
320	0.05085	0.2433	0.05912	0.02061	0.6982
330	0.05020	0.2435	0.05967	0.02082	0.6981
340	0.04957	0.2437	0.06022	0.02103	0.6981

APPENDIX B.4 (*Continued*)

Properties of Air at 1 atmosphere

	IP UNITS				
T (°F)	ρ (lbm/ft³)	c_p (Btu/lbm·R)	μ (lbm/ft·h)	k (Btu/h·ft·°F)	Pr
350	0.04896	0.2440	0.06076	0.02123	0.6981
360	0.04836	0.2442	0.06130	0.02144	0.6980
370	0.04778	0.2444	0.06184	0.02165	0.6980
380	0.04721	0.2446	0.06237	0.02185	0.6981
390	0.04666	0.2448	0.06290	0.02206	0.6981
400	0.04611	0.2451	0.06343	0.02226	0.6982
410	0.04558	0.2453	0.06396	0.02247	0.6983
420	0.04506	0.2455	0.06448	0.02267	0.6984
430	0.04456	0.2458	0.06500	0.02287	0.6985
440	0.04406	0.2460	0.06552	0.02307	0.6986
450	0.04358	0.2463	0.06603	0.02327	0.6987

	SI UNITS				
T (°F)	ρ (kg/m³)	c_p (kJ/kg·K)	μ (kg/m·h)	k (W/m·K)	Pr
−40	1.516	1.006	0.05455	0.02122	0.7184
−30	1.453	1.006	0.05645	0.02202	0.7164
−20	1.395	1.006	0.05832	0.02281	0.7145
−10	1.342	1.006	0.06017	0.02359	0.7128
0	1.293	1.006	0.06199	0.02436	0.7112
10	1.247	1.006	0.06378	0.02512	0.7097
20	1.204	1.007	0.06554	0.02587	0.7083
30	1.164	1.007	0.06728	0.02662	0.7070
40	1.127	1.007	0.06899	0.02735	0.7058
50	1.092	1.008	0.07069	0.02808	0.7047
60	1.059	1.008	0.07236	0.02880	0.7036
70	1.028	1.009	0.07400	0.02952	0.7027
80	0.9993	1.010	0.07563	0.03023	0.7019
90	0.9717	1.011	0.07724	0.03093	0.7012
100	0.9456	1.012	0.07883	0.03162	0.7005
110	0.9209	1.013	0.08040	0.03231	0.6999
120	0.8974	1.014	0.08195	0.03299	0.6995
130	0.8752	1.015	0.08348	0.03367	0.6990
140	0.8540	1.016	0.08500	0.03434	0.6987
150	0.8338	1.018	0.08650	0.03500	0.6984
160	0.8145	1.019	0.08798	0.03566	0.6983
170	0.7961	1.020	0.08945	0.03631	0.6981
180	0.7785	1.022	0.09090	0.03696	0.6981
190	0.7617	1.024	0.09234	0.03761	0.6981
200	0.7456	1.025	0.09377	0.03825	0.6982
210	0.7302	1.027	0.09518	0.03888	0.6983

(*Continued*)

APPENDIX B.4   (*Continued*)

Properties of Air at 1 atmosphere

		SI UNITS			
T (°F)	ρ (kg/m³)	c_p (kJ/kg·K)	μ (kg/m·h)	k (W/m·K)	Pr
220	0.7153	1.029	0.09657	0.03951	0.6985
230	0.7011	1.031	0.09796	0.04014	0.6987
240	0.6874	1.033	0.09933	0.04076	0.6990
250	0.6743	1.035	0.1007	0.04138	0.6994
260	0.6617	1.037	0.1020	0.04200	0.6997
270	0.6495	1.039	0.1034	0.04261	0.7001
280	0.6377	1.041	0.1047	0.04321	0.7006
290	0.6264	1.043	0.1060	0.04382	0.7011
300	0.6155	1.045	0.1073	0.04442	0.7016
310	0.6049	1.048	0.1086	0.04501	0.7022
320	0.5947	1.050	0.1099	0.04561	0.7027
330	0.5849	1.052	0.1112	0.04620	0.7033
340	0.5753	1.055	0.1124	0.04678	0.7040
350	0.5661	1.057	0.1137	0.04737	0.7046
360	0.5571	1.059	0.1149	0.04795	0.7053
370	0.5485	1.062	0.1162	0.04852	0.7060
380	0.5401	1.064	0.1174	0.04910	0.7067
390	0.5319	1.066	0.1186	0.04967	0.7074
400	0.5240	1.069	0.1198	0.05024	0.7081
410	0.5164	1.071	0.1210	0.05081	0.7088
420	0.5089	1.074	0.1222	0.05137	0.7095
430	0.5017	1.076	0.1234	0.05193	0.7103
440	0.4946	1.078	0.1246	0.05249	0.7110
450	0.4878	1.081	0.1258	0.05305	0.7117

Appendix C: Standard Pipe Dimensions

This appendix shows standard dimensions for steel, iron, and stainless steel pipes. In the column labeled "Schedule," steel pipes are designated by a number (e.g., 40 means sch 40). Iron pipes are designated in parenthesis (e.g., "std" means standard). Stainless steel pipe schedules are listed in the column labeled "SS." The data shown here are a subset of a much more comprehensive table published by The Crane Company (2013).

Nominal Diameter	Outside Diameter				Schedule		Inside Diameter			
in.	in.	ft.	cm	m	Steel (Iron)	SS	in.	ft.	cm	m
1/8	0.4050	0.03375	1.0287	0.010287		10S	0.307	0.02558	0.7798	0.007798
					40 (std)	40S	0.269	0.02242	0.6833	0.006833
					80 (xs)	80S	0.215	0.01792	0.5461	0.005461
1/4	0.5400	0.04500	1.3716	0.013716		10S	0.410	0.03417	1.0414	0.010414
					40	40S	0.364	0.03033	0.9246	0.009246
					80	80S	0.302	0.02517	0.7671	0.007671
3/8	0.6750	0.05625	1.7145	0.017145		10S	0.545	0.04542	1.3843	0.013843
					10	40S	0.493	0.04108	1.2522	0.012522
					40 (std)	80S	0.423	0.03525	1.0744	0.010744
					80 (xs)					
1/2	0.8400	0.07000	2.1336	0.021336		5S	0.710	0.05917	1.8034	0.018034
						10S	0.674	0.05617	1.7120	0.017120
					40 (std)	40S	0.622	0.05183	1.5799	0.015799
					80 (xs)	80S	0.546	0.04550	1.3868	0.013868
					160		0.466	0.03883	1.1836	0.011836
					(xxs)		0.252	0.02100	0.6401	0.006401
3/4	1.0500	0.08750	2.6670	0.026670		5S	0.920	0.07667	2.3368	0.023368
						10S	0.884	0.07367	2.2454	0.022454
					40 (std)	40S	0.824	0.06867	2.0930	0.020930
					80 (xs)	80S	0.742	0.06183	1.8847	0.018847
					160		0.612	0.05100	1.5545	0.015545
					(xxs)		0.434	0.03617	1.1024	0.011024
1	1.3150	0.10958	3.3401	0.033401		5S	1.185	0.09875	3.0099	0.030099
						10S	1.097	0.09142	2.7864	0.027864
					40 (std)	40S	1.049	0.08742	2.6645	0.026645
					80 (xs)	80S	0.957	0.07975	2.4308	0.024308
					160		0.815	0.06792	2.0701	0.020701

Nominal Diameter	Outside Diameter				Schedule			Inside Diameter			
	in.	ft.	cm	m	Steel (Iron)	SS	in.	ft.	cm	m
1	1.3150	0.10958	3.3401	0.033401	(xxs)		0.599	0.04992	1.5215	0.015215
1 1/4	1.6600	0.13833	4.2164	0.042164		5S	1.530	0.12750	3.8862	0.038862
						10S	1.442	0.12017	3.6627	0.036627
					40 (std)	40S	1.380	0.11500	3.5052	0.035052
					80 (xs)	80S	1.278	0.10650	3.2461	0.032461
					160		1.160	0.09667	2.9464	0.029464
					(xxs)		0.896	0.07467	2.2758	0.022758
1 1/2	1.9000	0.15833	4.8260	0.048260		5S	1.770	0.14750	4.4958	0.044958
						10S	1.682	0.14017	4.2723	0.042723
					40 (std)	40S	1.610	0.13417	4.0894	0.040894
					80 (xs)	80S	1.500	0.12500	3.8100	0.038100
					160		1.338	0.11150	3.3985	0.033985
					(xxs)		1.100	0.09167	2.7940	0.027940
2	2.3750	0.19792	6.0325	0.060325		5S	2.245	0.18708	5.7023	0.057023
						10S	2.157	0.17975	5.4788	0.054788
					40 (std)	40S	2.067	0.17225	5.2502	0.052502
					80 (xs)	80S	1.939	0.16158	4.9251	0.049251
					160		1.687	0.14058	4.2850	0.042850
					(xxs)		1.503	0.12525	3.8176	0.038176
2 1/2	2.8750	0.23958	7.3025	0.073025		5S	2.709	0.22575	6.8809	0.068809
						10S	2.635	0.21958	6.6929	0.066929
					40 (std)	40S	2.469	0.20575	6.2713	0.062713
					80 (xs)	80S	2.323	0.19358	5.9004	0.059004
					160		2.125	0.17708	5.3975	0.053975
					(xxs)		1.771	0.14758	4.4983	0.044983

(Continued)

Nominal Diameter	Outside Diameter				Schedule			Inside Diameter			
	in.	ft.	cm	m	Steel (Iron)	SS	in.	ft.	cm	m
3	3.500	0.29167	8.8900	0.088900		5S	3.334	0.27783	8.4684	0.084684
						10S	3.260	0.27167	8.2804	0.082804
					40 (std)	40S	3.068	0.25567	7.7927	0.077927
					80 (xs)	80S	2.900	0.24167	7.3660	0.073660
					160		2.624	0.21867	6.6650	0.066650
					(xxs)		2.300	0.19167	5.8420	0.058420
3 1/2	4.000	0.33333	10.160	0.10160		5S	3.834	0.31950	9.7384	0.097384
						10S	3.760	0.31333	9.5504	0.095504
					40 (std)	40S	3.548	0.29567	9.0119	0.090119
					80 (xs)	80S	3.364	0.28033	8.5446	0.085446
4	4.500	0.37500	11.430	0.11430		5S	4.334	0.36117	11.008	0.11008
						10S	4.260	0.35500	10.820	0.10820
					40 (std)	40S	4.026	0.33550	10.226	0.10226
					80 (xs)	80S	3.826	0.31883	9.7180	0.097180
					120		3.624	0.30200	9.2050	0.092050
					160		3.438	0.28650	8.7325	0.087325
					(xxs)		3.152	0.26267	8.0061	0.080061
5	5.563	0.46358	14.130	0.14130		5S	5.345	0.44542	13.576	0.13576
						10S	5.295	0.44125	13.449	0.13449
					40 (std)	40S	5.047	0.42058	12.819	0.12819
					80 (xs)	80S	4.813	0.40108	12.225	0.12225
					120		4.563	0.38025	11.590	0.11590
					160		4.313	0.35942	10.955	0.10955
					(xxs)		4.063	0.33858	10.320	0.10320
6	6.625	0.55208	16.828	0.16828	5	5S	6.407	0.53392	16.274	0.16274

Nominal Diameter in.	Outside Diameter in.	ft.	cm	m	Schedule Steel (Iron)	SS	Inside Diameter in.	ft.	cm	m
6	6.625	0.55208	16.828	0.16828	10	10S	6.357	0.52975	16.147	0.16147
					40 (std)	40S	6.065	0.50542	15.405	0.15405
					80 (xs)	80S	5.761	0.48008	14.633	0.14633
					120		5.501	0.45842	13.973	0.13973
					160		5.187	0.43225	13.175	0.13175
					xxs		4.897	0.40808	12.438	0.12438
8	8.625	0.71875	21.908	0.21908		5S	8.407	0.70058	21.354	0.21354
						10S	8.329	0.69408	21.156	0.21156
					20		8.125	0.67708	20.638	0.20638
					30		8.071	0.67258	20.500	0.20500
					40 (std)	40S	7.981	0.66508	20.272	0.20272
					60		7.813	0.65108	19.845	0.19845
					80 (xs)	80S	7.625	0.63542	19.368	0.19368
					100		7.437	0.61975	18.890	0.18890
					120		7.187	0.59892	18.255	0.18255
					140		7.001	0.58342	17.783	0.17783
					(xxs)		6.875	0.57292	17.463	0.17463
					160		6.813	0.56775	17.305	0.17305
10	10.750	0.89583	27.305	0.27305		5S	10.482	0.87350	26.624	0.26624
						10S	10.420	0.86833	26.467	0.26467
					20		10.250	0.85417	26.035	0.26035
					30		10.136	0.84467	25.745	0.25745
					40 (std)	40S	10.020	0.83500	25.451	0.25451
					60 (std)	80S	9.750	0.81250	24.765	0.24765
					80		9.562	0.79683	24.287	0.24287
					100		9.312	0.77600	23.652	0.23652

(Continued)

Nominal Diameter	Outside Diameter				Schedule		Inside Diameter			
	in.	ft.	cm	m	Steel (Iron)	SS	in.	ft.	cm	m
10	10.750	0.89583	27.305	0.27305	120		9.062	0.75517	23.017	0.23017
					140 (xxs)		8.750	0.72917	22.225	0.22225
					160		8.500	0.70833	21.590	0.21590
12	12.750	1.0625	32.385	0.32385		5S	12.438	1.0365	31.593	0.31593
						10S	12.390	1.0325	31.471	0.31471
					20		12.250	1.0208	31.115	0.31115
					30		12.090	1.0075	30.709	0.30709
					(std)	40S	12.000	1.0000	30.480	0.30480
					40		11.983	0.99858	30.437	0.30437
					(xs)	80S	11.750	0.97917	29.845	0.29845
					60		11.626	0.96883	29.530	0.29530
					80		11.374	0.94783	28.890	0.28890
					100		11.062	0.92183	28.097	0.28097
					120 (xxs)		10.750	0.89583	27.305	0.27305
					140		10.500	0.87500	26.670	0.26670
					160		10.126	0.84383	25.720	0.25720
14	14	1.1667	35.560	0.35560		5S	13.688	1.1407	34.768	0.34768
						10S	13.624	1.1353	34.605	0.34605
					10		13.500	1.1250	34.290	0.34290
					20		13.376	1.1147	33.975	0.33975
					30 (std)		13.250	1.1042	33.655	0.33655
					40		13.124	1.0937	33.335	0.33335
					(xs)		13.000	1.0833	33.020	0.33020
					60		12.812	1.0677	32.542	0.32542

Nominal Diameter	Outside Diameter			Schedule		Inside Diameter			
in.	ft.	cm	m	Steel (Iron)	SS	in.	ft.	cm	m
14	1.1667	35.560	0.35560	80		12.500	1.0417	31.750	0.31750
				100		12.124	1.0103	30.795	0.30795
				120		11.812	0.98433	30.002	0.30002
				140		11.500	0.95833	29.210	0.29210
				160		11.188	0.93233	28.418	0.28418
16	1.3333	40.640	0.40640		5S	15.670	1.3058	39.802	0.39802
					10S	15.624	1.3020	39.685	0.39685
				10		15.500	1.2917	39.370	0.39370
				20		15.376	1.2813	39.055	0.39055
				30 (std)		15.250	1.2708	38.735	0.38735
				40 (xs)		15.000	1.2500	38.100	0.38100
				60		14.688	1.2240	37.308	0.37308
				80		14.312	1.1927	36.352	0.36352
				100		13.938	1.1615	35.403	0.35403
				120		13.562	1.1302	34.447	0.34447
				140		13.124	1.0937	33.335	0.33335
				160		12.182	1.0152	30.942	0.30942
18	1.5000	45.720	0.45720		5S	17.670	1.4725	44.882	0.44882
					10S	17.624	1.4687	44.765	0.44765
				10		17.500	1.4583	44.450	0.44450
				20		17.376	1.4480	44.135	0.44135
				(std)		17.250	1.4375	43.815	0.43815
				30		17.124	1.4270	43.495	0.43495
				(xs)		17.000	1.4167	43.180	0.43180

(Continued)

Nominal Diameter	Outside Diameter			Schedule		Inside Diameter			
in.	ft.	cm	m	Steel (Iron)	SS	in.	ft.	cm	m
18	1.5000	45.720	0.45720	40		16.876	1.4063	42.865	0.42865
				60		16.500	1.3750	41.910	0.41910
				80		16.124	1.3437	40.955	0.40955
				100		15.688	1.3073	39.848	0.39848
				120		15.250	1.2708	38.735	0.38735
				140		14.876	1.2397	37.785	0.37785
				160		14.438	1.2032	36.673	0.36673
20	1.6667	50.800	0.50800		5S	19.624	1.6353	49.845	0.49845
					10S	19.564	1.6303	49.693	0.49693
				10		19.500	1.6250	49.530	0.49530
				20 (std)		19.250	1.6042	48.895	0.48895
				30 (xs)		19.000	1.5833	48.260	0.48260
				40		18.812	1.5677	47.782	0.47782
				60		18.376	1.5313	46.675	0.46675
				80		17.938	1.4948	45.563	0.45563
				100		17.438	1.4532	44.293	0.44293
				120		17.000	1.4167	43.180	0.43180
				140		16.500	1.3750	41.910	0.41910
				160		16.062	1.3385	40.797	0.40797
22	1.8333	55.880	0.55880		5S	21.624	1.8020	54.925	0.54925
					10S	21.564	1.7970	54.773	0.54773
				10		21.500	1.7917	54.610	0.54610
				20 (std)		21.250	1.7708	53.975	0.53975
				30 (xs)		21.000	1.7500	53.340	0.53340
				60		20.250	1.6875	51.435	0.51435

Nominal Diameter	Outside Diameter			Schedule		Inside Diameter			
in.	ft.	cm	m	Steel (Iron)	SS	in.	ft.	cm	m
22	1.8333	55.880	0.55880	80		19.750	1.6458	50.165	0.50165
				100		19.250	1.6042	48.895	0.48895
				120		18.750	1.5625	47.625	0.47625
				140		18.250	1.5208	46.355	0.46355
				160		17.750	1.4792	45.085	0.45085
24	2.0000	60.960	0.60960		5S	23.564	1.9637	59.853	0.59853
				10	10S	23.500	1.9583	59.690	0.59690
				20 (std)		23.250	1.9375	59.055	0.59055
				(xs)		23.000	1.9167	58.420	0.58420
				30		22.876	1.9063	58.105	0.58105
				40		22.624	1.8853	57.465	0.57465
				60		22.062	1.8385	56.037	0.56037
				80		21.562	1.7968	54.767	0.54767
				100		20.938	1.7448	53.183	0.53183
				120		20.376	1.6980	51.755	0.51755
				140		19.876	1.6563	50.485	0.50485
				160		19.312	1.6093	49.052	0.49052
26	2.1667	66.040	0.66040	10		25.376	2.1147	64.455	0.64455
				(std)		25.250	2.1042	64.135	0.64135
				20 (xs)		25.000	2.0833	63.500	0.63500
28	2.3333	71.120	0.71120	10		27.376	2.2813	69.535	0.69535
				(std)		27.250	2.2708	69.215	0.69215
				20 (xs)		27.000	2.2500	68.580	0.68580
				30		26.750	2.2292	67.945	0.67945
30	2.5000	76.200	0.76200		5S	29.500	2.4583	74.930	0.74930
				10	10S	29.376	2.4480	74.615	0.74615

(Continued)

Nominal Diameter	Outside Diameter				Schedule		Inside Diameter			
in.	in.	ft.	cm	m	Steel (Iron)	SS	in.	ft.	cm	m
30	30	2.5000	76.200	0.76200	(std)		29.250	2.4375	74.295	0.74295
					20 (xs)		29.000	2.4167	73.660	0.73660
					30		28.750	2.3958	73.025	0.73025
32	32	2.6667	81.280	0.81280	10		31.376	2.6147	79.695	0.79695
					(std)		31.250	2.6042	79.375	0.79375
					20 (xs)		31.000	2.5833	78.740	0.78740
					30		30.750	2.5625	78.105	0.78105
					40		30.624	2.5520	77.785	0.77785
34	34	2.8333	86.360	0.86360	10		33.376	2.7813	84.775	0.84775
					(std)		33.250	2.7708	84.455	0.84455
					20 (xs)		33.000	2.7500	83.820	0.83820
					30		32.750	2.7292	83.185	0.83185
					40		32.624	2.7187	82.865	0.82865
36	36	3.0000	91.440	0.91440	10		35.376	2.9480	89.855	0.89855
					(std)		35.250	2.9375	89.535	0.89535
					20 (xs)		35.000	2.9167	88.900	0.88900
					30		34.750	2.8958	88.265	0.88265
					40		34.500	2.8750	87.630	0.87630

Source: The Crane Company, *Flow of Fluids through Valves, Fittings and Pipe*, The Crane Company, Stamford, CT, 2013.

Appendix D: Standard Copper Tubing Dimensions

This appendix shows dimensional data for Type K, L, and M standard copper tubing.

Standard Size	Outside Diameter				Type	Inside Diameter			
	in.	ft.	cm	m		in.	ft.	cm	m
1/4	0.375	0.03125	0.9525	0.009525	K	0.305	0.02542	0.7747	0.007747
					L	0.315	0.02625	0.8001	0.008001
3/8	0.500	0.041667	1.27	0.0127	K	0.402	0.03350	1.0211	0.010211
					L	0.430	0.03583	1.0922	0.010922
					M	0.450	0.03750	1.1430	0.011430
1/2	0.625	0.05208	1.5875	0.015875	K	0.527	0.04392	1.3386	0.013386
					L	0.545	0.04542	1.3843	0.013843
					M	0.569	0.04742	1.4453	0.014453
5/8	0.750	0.06250	1.905	0.01905	K	0.652	0.05433	1.6561	0.016561
					L	0.666	0.05550	1.6916	0.016916
3/4	0.875	0.07292	2.2225	0.022225	K	0.745	0.06208	1.8923	0.018923
					L	0.785	0.06542	1.9939	0.019939
					M	0.811	0.06758	2.0599	0.020599
1	1.125	0.09375	2.8575	0.028575	K	0.995	0.08292	2.5273	0.025273
					L	1.025	0.08542	2.6035	0.026035
					M	1.055	0.08792	2.6797	0.026797
1 1/4	1.375	0.1146	3.4925	0.034925	K	1.245	0.10375	3.1623	0.031623
					L	1.265	0.10542	3.2131	0.032131
					M	1.291	0.10758	3.2791	0.032791
1 1/2	1.625	0.1354	4.1275	0.041275	K	1.481	0.12342	3.7617	0.037617
					L	1.505	0.12542	3.8227	0.038227
					M	1.527	0.12725	3.8786	0.038786
2	2.125	0.1771	5.3975	0.053975	K	1.959	0.16325	4.9759	0.049759
					L	1.985	0.16542	5.0419	0.050419
					M	2.009	0.16742	5.1029	0.051029
2 1/2	2.625	0.2188	6.6675	0.066675	K	2.435	0.20292	6.1849	0.061849
					L	2.465	0.20542	6.2611	0.062611
					M	2.495	0.20792	6.3373	0.063373

Standard Size	Outside Diameter				Type	Inside Diameter			
	in.	ft.	cm	m		in.	ft.	cm	m
3	3.125	0.2604	7.9375	0.079375	K	2.907	0.24225	7.3838	0.073838
					L	2.945	0.24542	7.4803	0.074803
					M	2.981	0.24842	7.5717	0.075717
3 1/2	3.625	0.3021	9.208	0.09208	K	3.385	0.28208	8.5979	0.085979
					L	3.425	0.28542	8.6995	0.086995
					M	3.459	0.28825	8.7859	0.087859
4	4.125	0.3438	10.478	0.10478	K	3.857	0.32142	9.7968	0.097968
					L	3.905	0.32542	9.9187	0.099187
					M	3.935	0.32792	9.9949	0.099949
5	5.125	0.4271	13.018	0.13018	K	4.805	0.40042	12.2047	0.122047
					L	4.875	0.40625	12.3825	0.123825
					M	4.907	0.40892	12.4638	0.124638
6	6.125	0.5104	15.558	0.15558	K	5.741	0.47842	14.5821	0.145821
					L	5.845	0.48708	14.8463	0.148463
					M	5.881	0.49008	14.9377	0.149377
8	8.125	0.6771	20.638	0.20638	K	7.583	0.63192	19.2608	0.192608
					L	7.725	0.64375	19.6215	0.196215
					M	7.785	0.64875	19.7739	0.197739
10	10.125	0.8438	25.718	0.25718	K	9.449	0.78742	24.0005	0.240005
					L	9.625	0.80208	24.4475	0.244475
					M	9.701	0.80842	24.6405	0.246405
12	12.125	1.0104	30.798	0.30798	K	11.315	0.94292	28.7401	0.287401
					L	11.565	0.96375	29.3751	0.293751
					M	11.617	0.96808	29.5072	0.295072

Source: The data were taken from *The Copper Tube Handbook*, Copper Development Association, Inc., New York, 2010.

Appendix E: Resistance Coefficients for Valves and Fittings

This appendix contains expressions for the calculation of resistance coefficients for various valves and fittings. The expressions shown are taken from Crane Technical Paper 410 (The Crane Company 2013). Page 2–11, which is referenced on the next page, is included at the end of this appendix. Permission has been granted by the Crane Company to use these pages.

CRANE

Representative Resistance Coefficient K for Valves and Fittings

(K is based on use of schedule pipe as listed on page 2-9.)

Pipe Friction Data for Schedule 40 Clean Commercial Steel Pipe with Flow in Zone of Complete Turbulence

Nominal Size	½"	¾"	1"	1¼"	1½"	2"	2½"	3"	4"	5, 6"	8"	10-14"	16-22"	24-36"
Friction Factor (f_T)	.026	.024	.022	.021	.020	.019	.018	.017	.016	.015	.014	.013	.012	.011

$$f_T = \frac{0.25}{\left[log\left(\frac{\varepsilon/D}{3.7}\right)\right]^2}$$

Formulas For Calculating K Factors* For Valves and Fittings with Reduced Port

(Refer to page 2-11)

Formula 1

$$K_2 = \frac{0.8 \, sin\frac{\theta}{2}(1-\beta^2)}{\beta^4} = \frac{K_1}{\beta^4}$$

Formula 2

$$K_2 = \frac{0.5\sqrt{sin\frac{\theta}{2}}\,(1-\beta^2)}{\beta^4} = \frac{K_1}{\beta^4}$$

Formula 3

$$K_2 = \frac{2.6 \, sin\frac{\theta}{2}\,(1-\beta^2)^2}{\beta^4} = \frac{K_1}{\beta^4}$$

Formula 4

$$K_2 = \frac{(1-\beta^2)^2}{\beta^4} = \frac{K_1}{\beta^4}$$

Formula 5

$$K_2 = \frac{K_1}{\beta^4} + Formula\ 1 + Formula\ 3$$

$$K_2 = \frac{K_1 + sin\frac{\theta}{2}[0.8\,(1-\beta^2) + 2.6\,(1-\beta^2)^2]}{\beta^4}$$

*Use K furnished by valve or fitting supplier when available.

Formula 6

$$K_2 = \frac{K_1}{\beta^4} + Formula\ 2 + Formula\ 4$$

$$K_2 = \frac{K_1 + 0.5\sqrt{sin\frac{\theta}{2}}\,(1-\beta^2) + (1-\beta^2)^2}{\beta^4}$$

Formula 7

$$K_2 = \frac{K_1}{\beta^4} + \beta\,(Formula\ 2 + Formula\ 4) \qquad When\ \theta = 180°$$

$$K_2 = \frac{K_1 + \beta\,[0.5\,(1-\beta^2) + (1-\beta^2)^2]}{\beta^4}$$

$$\beta = \frac{d_1}{d_2}$$

$$\beta^2 = \left(\frac{d_1}{d_2}\right)^2 = \frac{a_1}{a_2}$$

Subscript 1 defines dimensions and coefficients with reference to the smaller diameter. Subscript 2 refers to the larger diameter.

Sudden and Gradual Contraction	**Sudden and Gradual Enlargement**

If: $\theta \gtrless 45°$ . . . .K_2 = Formula 1	If: $\theta \gtrless 45°$ . . . .K_2 = Formula 3
$45° < \theta \gtrless 180°$ . . . . .K_2 = Formula 2	$45° < \theta \gtrless 180°$ . . . . .K_2 = Formula 4

CRANE

Representative Resistance Coefficient K for Valves and Fittings

For formulas and friction data, see page A-27. K is based on use of schedule pipe as listed on page 2-9.

GATE VALVES
Wedge Disc, Double Disc, or Plug Type

If: $\beta = 1$, $\theta = 0$ . . . .$K_1 = 8\,f_T$

$\beta < 1$ and $\theta \lesssim 45°$ . . . .$K_2 = $ Formula 5

$\beta < 1$ and $45° < \theta \lesssim 180°$ . . . .$K_2 = $ Formula 6

SWING CHECK VALVES

$K = 100\,f_T$ $K = 50\,f_T$

Minimum pipe velocity   Minimum pipe velocity
(fps) for full disc lift   (fps) for full disc lift

$= 35\sqrt{V}$ $= 60\sqrt{V}$ except

U/L listed $= 100\sqrt{V}$

GLOBE AND ANGLE VALVES

If: $\beta = 1$ . . . $K_1 = 340\,f_T$

If: $\beta = 1$ . . . $K_1 = 55\,f_T$

LIFT CHECK VALVES

If: $\beta = 1$. . .$K_1 = 600\,f_T$

$\beta < 1$. . .$K_2 = $ Formula 7

Minimum pipe velocity (fps) for full disc lift

$= 40\,\beta^2\sqrt{V}$

If: $\beta = 1$. . .$K_1 = 55\,f_T$

$\beta < 1$. . .$K_2 = $ Formula 7

Minimum pipe velocity (fps) for full disc lift

$= 140\,\beta^2\sqrt{V}$

TILTING DISC CHECK VALVES

	$\alpha = 5°$	$\alpha = 15°$
Sizes 2 to 8" . . .$K =$	$40\,f_T$	$120\,f_T$
Sizes 10 to 14" . . .$K =$	$30\,f_T$	$90\,f_T$
Sizes 16 to 48". . .$K =$	$20\,f_T$	$60\,f_T$
Minimum pipe velocity (fps) for full disc lift $=$	$80\sqrt{V}$	$30\sqrt{V}$

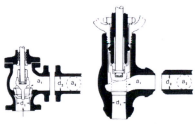

If: $\beta = 1$ . . . $K_1 = 150 f_T$ If: $\beta = 1$ . . . $K_1 = 55 f_T$

All globe and angle valves,
whether reduced seat or throttled,
If: $\beta < 1$ . . . $K_2 = $ Formula 7

Representative Resistance Coefficient K for Valves and Fittings

For formulas and friction data, see page A-27. K is based on use of schedule pipe as listed on page 2-9.

STOP-CHECK VALVES
(Globe and Angle Types)

If:
$\beta = 1 \ldots K_1 = 400\, f_T$
$\beta < 1 \ldots K_2 = $ Formula 7

Minimum pipe velocity
for full disc lift
$= 55\, \beta^2 \sqrt{\overline{V}}$

If:
$\beta = 1 \ldots K_1 = 200\, f_T$
$\beta < 1 \ldots K_2 = $ Formula 7

Minimum pipe velocity
for full disc lift
$= 75\, \beta^2 \sqrt{\overline{V}}$

If:
$\beta = 1 \ldots K_1 = 300\, f_T$
$\beta < 1 \ldots K_2 = $ Formula 7

If:
$\beta = 1 \ldots K_1 = 350\, f_T$
$\beta < 1 \ldots K_2 = $ Formula 7

Minimum pipe velocity (fps) for full disc lift
$= 60\, \beta^2 \sqrt{\overline{V}}$

If:
$\beta = 1 \ldots K_1 = 55\, f_T$
$\beta < 1 \ldots K_2 = $ Formula 7

If:
$\beta = 1 \ldots K_1 = 55\, f_T$
$\beta < 1 \ldots K_2 = $ Formula 7

Minimum pipe velocity (fps) for full disc lift
$= 140\, \beta^2 \sqrt{\overline{V}}$

FOOT VALVES WITH STRAINER

Poppet Disc　　　　　　　　　Hinged Disc

$K = 420\, f_T$　　　　　　　　$K = 75\, f_T$

Minimum pipe velocity
(fps) for full disc lift
$= 15 \sqrt{\overline{V}}$

Minimum pipe velocity
(fps) for full disc lift
$= 35 \sqrt{\overline{V}}$

BALL VALVES

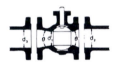

If: $\beta = 1,\ \theta = 0 \ldots\ldots\ldots\ldots\ldots\ldots\ldots K_1 = 3\, f_T$
$\beta < 1$ and $\theta \gtrless 45° \ldots\ldots\ldots\ldots K_2 = $ Formula 5
$\beta < 1$ and $45° < \theta \gtrless 180° \ldots\ldots K_2 = $ Formula 6

BUTTERFLY VALVES

SIZE RANGE	CENTRIC	DOUBLE OFFSET	TRIPLE OFFSET
2" - 8"	$K = 45\, f_T$	$K = 74\, f_T$	$K = 218\, f_T$
10" - 14"	$K = 35\, f_T$	$K = 52\, f_T$	$K = 96\, f_T$
18" - 24"	$K = 25\, f_T$	$K = 43\, f_T$	$K = 55\, f_T$

DIAPHRAGM VALVES

Weir　　　　　　　　　　　Straight
　　　　　　　　　　　　　Through
$\beta = 1 \ldots K = 149\, f_T$　　$\beta = 1 \ldots K = 39\, f_T$

Representative Resistance Coefficient K for Valves and Fittings

For formulas and friction data, see page A-27. K is based on use of schedule pipe as listed on page 2-9.

PLUG VALVES AND COCKS

Straight Way

3-Way

View X—X

If: $\beta = 1$,
$K_1 = 18 f_T$

If: $\beta = 1$,
$K_1 = 30 f_T$

If: $\beta = 1$,
$K_1 = 90 f_T$

If: $\beta < 1 \ldots K_2 =$ Formula 6

STANDARD ELBOWS

90°

45°

$K = 30 f_T$ $K = 16 f_T$

STANDARD TEES AND WYES

Refer to Chapter 2, pages 2-14 through 2-16

MITRE BENDS

α	K
0°	$2 f_T$
15°	$4 f_T$
30°	$8 f_T$
45°	$15 f_T$
60°	$25 f_T$
75°	$40 f_T$
90°	$60 f_T$

90° PIPE BENDS AND FLANGED OR BUTT-WELDING 90° ELBOWS

r/d	K	r/d	K
1	$20 f_T$	8	$24 f_T$
1.5	$14 f_T$	10	$30 f_T$
2	$12 f_T$	12	$34 f_T$
3	$12 f_T$	14	$38 f_T$
4	$14 f_T$	16	$42 f_T$
6	$17 f_T$	20	$50 f_T$

The resistance coefficient, K_B, for pipe bends other than 90° may be determined as follows:

$$K_B = (n - 1) \left(0.25 \, \pi \, f_i \, \frac{r}{d} + 0.5 \, K \right) + K$$

n = number of 90° bends
K = resistance coefficient for one 90° bend (per table)

PIPE ENTRANCE

Inward Projecting

$K = 0.78$

Flush

For K, see table

r/d	K
0.00*	0.5
0.02	0.28
0.04	0.24
0.06	0.15
0.10	0.09
0.15 & up	0.04

*Sharp-edged

CLOSE PATTERN RETURN BENDS

$K = 50 f_T$

PIPE EXIT

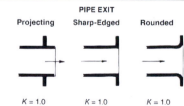

Projecting   **Sharp-Edged**   **Rounded**

$K = 1.0$ $K = 1.0$ $K = 1.0$

Contraction and Enlargement

The resistance to flow due to sudden enlargements may be expressed by,

$$K_1 = \left(1 - \frac{d_1^2}{d_2^2}\right)^2 \qquad \text{Equation 2-15}$$

and the resistance due to sudden contractions, by

$$K_1 = 0.5\left(1 - \frac{d_1^2}{d_2^2}\right) \qquad \text{Equation 2-16}$$

Subscripts 1 and 2 define the internal diameters of the small and large pipes respectively.

It is convenient to identify the ratio of diameters of the small to large pipes by the Greek letter β (beta). Using this notation, these equations may be written,

Sudden Enlargement

$$K_1 = (1 - \beta^2)^2 \qquad \text{Equation 2-17}$$

Sudden Contraction

$$K_1 = 0.5\,(1 - \beta^2) \qquad \text{Equation 2-18}$$

Equation 2-15 is derived from the momentum equation together with the Bernoulli equation. Equation 2-16 uses the derivation of Equation 2-15 together with the continuity equation and a close approximation of the contraction coefficients determined by Julius Weisbach.[20]

The value of the resistance coefficient in terms of the larger pipe is determined by dividing Equations 2-15 and 2-16 by β^4,

$$K_2 = \frac{K_1}{\beta^4} \qquad \text{Equation 2-19}$$

The losses due to gradual enlargements in pipes were investigated by A.H. Gibson,[21] and may be expressed as a coefficient, C_e, applied to Equation 2-15. Approximate averages of Gibson's coefficients for different included angles of divergence, θ, are defined by the equations:

$$if \ \theta \le 45° \qquad C_e = 2.6 \ sin\frac{\theta}{2} \qquad \text{Equation 2-20}$$

$$if \ 45° < \theta \le 180° \quad C_e = 1 \qquad \text{Equation 2-21}$$

The losses due to gradual contractions in pipes were established by the analysis of Crane test data, using the same basis as that of Gibson for gradual enlargements, to provide a contraction coefficient, C_c, to be applied to Equation 2-16.

The approximate averages of these coefficients for different included angles of convergence, θ, are defined by the equations:

$$if \ \theta \le 45° \qquad C_c = 1.6 \ sin\frac{\theta}{2} \qquad \text{Equation 2-22}$$

$$if \ 45° < \theta \le 180° \quad C_c = \sqrt{sin\frac{\theta}{2}} \qquad \text{Equation 2-23}$$

The resistance coefficient K for sudden and gradual enlargements and contractions, expressed in terms of the large pipe, is established by combining equations 2-15 to 2-23 inclusive.

Sudden and Gradual Enlargements

$$\text{Equation 2-24}$$

$$if \ \theta \le 45° \qquad K_2 = \frac{2.6 \ sin\frac{\theta}{2} \ (1 - \beta^2)^2}{\beta^4}$$

$$if \ 45° < \theta \le 180° \quad K_2 = \frac{(1 - \beta^2)^2}{\beta^4} \qquad \text{Equation 2-25}$$

Sudden and Gradual Contractions

$$\text{Equation 2-26}$$

$$if \ \theta \le 45° \qquad K_2 = \frac{0.8 \ sin\frac{\theta}{2}(1 - \beta^2)}{\beta^4}$$

$$\text{Equation 2-27}$$

$$if \ 45° < \theta \le 180° \quad K_2 = \frac{0.5\sqrt{sin\frac{\theta}{2}} \ (1 - \beta^2)}{\beta^4}$$

Appendix F: Optimization Using Engineering Equation Solver

There are many instances in thermal energy system design and analysis where optimization can be implemented. The purpose of this appendix is to show the reader how Engineering Equation Solver (EES) can be used for various optimization problems.

F.1 Univariate (Single Variable) Optimization

Consider the following two functions:

$$y_1 = \frac{x}{4} \quad \text{and} \quad y_2 = e^{-x} \tag{F.1}$$

A third function that represents the sum of these two functions is given by the following equation:

$$y_3 = y_1 + y_2 \tag{F.2}$$

In the range from $0 \leq x \leq 3$, these functions are shown graphically in Figure F.1.

As Figure F.1 shows, the function y_3 has a minimum value. To find this minimum using EES, the Min/Max solution is used. To begin the process, the following equations are entered into the Equations Window:

```
$IFNOT Parametric Study or Min/Max
  x = 2
$ENDIF
  y_1 = x/4
  y_2 = exp(-x)
  y_3 = y_1 + y_2
```

Notice the use of the EES directive $IFNOT. This directive *includes* the equation $x = 2$ when the Solve button is clicked. However, the directive *ignores* the equation $x = 2$ if a Parametric Study or a Min/Max solution is being conducted. The use of directives gives the user quite a bit of flexibility in building versatile code that can do several things without commenting out or deleting equations. For example, the plot shown in Figure F.1 was generated from a Parametric Table by varying x from 0 to 3.

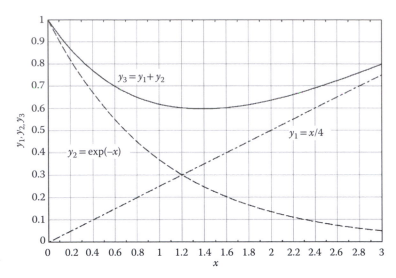

FIGURE F.1
Graphical representation of the functions defined in Equations F.1 and F.2.

To find the minimum of the y_3 function, a Min/Max study will be conducted in EES. When the Min/Max study is initiated, the dialog box shown in Figure F.2 becomes available.

Notice that EES allows for *minimization* or *maximization* of a value. In Figure F.2, EES is being instructed to *minimize* the variable y_3 while allowing x to change. A good question to ask at this point is, "How did EES know that there is only one independent variable in the problem?" When the Min/Max study is initiated, the equation $x = 2$ is ignored. Therefore, the remaining *three* equations defining y_1, y_2, and y_3 contain a total of *four* unknowns. This is a system of three equations with four unknowns. However, if we specify $4 - 3 = 1$ of the variables (say x), then the system can be solved. This is exactly how EES goes about a solution using sophisticated mathematical algorithms to constantly readjust x until the minimum of y_3 is found. For the univariate case, the user can select between the Golden Section or Quadratic Approximation algorithms (see Figure F.2).

The difference between the number of equations and number of unknowns in the problem is called the number of *degrees of freedom* in the optimization problem. In this example, there is $4 - 3 = 1$ degree of freedom. That is why EES is telling us we can select *one* independent variable to change as it seeks out the optimum value of y_3 using the selected search algorithm.

As discussed previously, EES uses sophisticated mathematically based algorithms to quickly find the solution to the optimization problem. However, the method requires an initial guess of the independent variable. EES defaults the initial guesses of unknown variables to a value of 1. In many cases, this may be sufficient. However, as problems become more complex, it is necessary to have initial guesses that are fairly close to the solution. Before EES

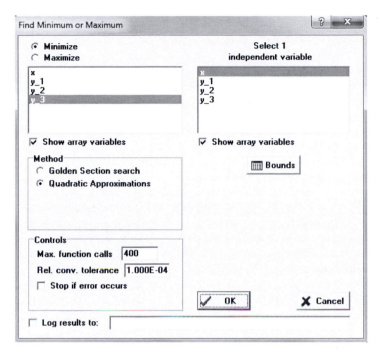

FIGURE F.2
Min/Max dialog box for univariate optimization.

can solve the optimization problem, it also needs to know the realistic bounds of the independent variable(s). The default bounds on all variables are ±∞. These bounds are not sufficient for optimization. Therefore, the user must set the bounds before a solution can be found. Both the initial guess(es) for the independent variable(s) and their bounds can be set by clicking the "Bounds" button in the Min/Max dialog box (see Figure F.2). This brings up the Variable Information window that allows the user to change the initial guess and bounds. For the univariate problem being considered here, the bounds were changed from 0 to 3, and the initial guess was left at the default value of 1. Figure F.3 shows the Variable Information window for this problem.

Once the initial guesses and bounds are set, the "OK" button can be clicked in the Min/Max dialog box and the optimization proceeds as the user instructed. For this problem, the solution is determined fairly quickly. The results are $x = 1.386$, $y_1 = 0.3466$, $y_2 = 0.25$, and $y_3 = 0.5966$. The solution indicates that the *minimum* value of y_3 is 0.5966, corresponding to an x value of 1.386. This numerical solution is easily verified by referring to Figure F.1.

It is always a good idea to run the optimization a few more times by updating the variables in the Variable Information window. This gives the optimization a new starting point. If the subsequent optimization produces the same result, then the user can feel confident that the optimum point has been found.

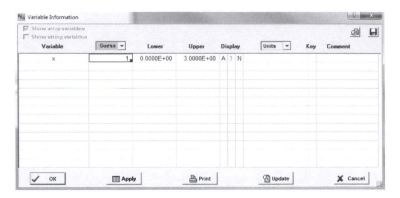

FIGURE F.3
Variable Information window used to specify initial guesses and bounds of independent variables in an optimization problem.

F.2 Multivariate Optimization

The example discussed in Section F.1 is a fairly simple one in that all the equations are two-dimensional (y vs. x) behavior. While simple, this example allows for visualization of the optimization process. Unfortunately, not many thermal energy system design and analysis problems are two-dimensional. Full system optimization problems may involve *hundreds* of variables. Problems like this are known as *multivariate* (multidimensional) problems. Even though we cannot easily visualize a multivariate problem, the procedure for solving the problem is exactly the same as the simple two-dimensional problem discussed in the previous section.

Multivariate optimization requires very sophisticated and complex algorithms to find the optimum condition. EES allows the user to select from several different optimization algorithms. In multivariate problems, it is a very good idea to run the optimization several times, updating variables between optimizations, to ensure that the true optimum is found.

Optimization problems with many degrees of freedom (many independent variables) may become increasingly difficult to solve. However, successful convergence on an optimum solution can be accomplished by very careful selection of initial guesses and bounds on the independent variables. As previously mentioned, EES defaults all initial guesses to a value of 1. In thermal system problems, this may be problematic. For example, a temperature of 1°F or 1 K may be completely unreasonable for the system being modeled. It may be quite difficult to bring up the Variable Information window and systematically go through each independent variable and simply set initial guesses that you think are accurate. Sometimes, preliminary calculations should be done to make sure the conservation laws are obeyed. For example, if a heat exchanger is involved in the problem, it may be possible to apply the conservation of energy to the heat exchanger and calculate unknown temperatures or flow rates based on other values that the user has specified for other independent variables. This may seem counterintuitive to the purpose of EES. After all, EES should be able to find these values as long as the equations are correct. However, as discussed previously, as the complexity of the optimization increases (more degrees of freedom), chances of successful convergence on an optimum solution require very good initial guesses of the unknown variables. Anything that the engineer can do to determine reasonable guesses will contribute toward a successful optimization.

Appendix G: Pump Curves

This appendix contains complete technical booklets for the Bell & Gossett Series 80-SC in-line centrifugal pump and Series 1531 end suction centrifugal pump. This information is included here for the purposes of learning how to use the pump curves. Permission has been granted by Xylem Inc. to include these pump curves in this textbook.

Series 80®-SC
Spacer-Coupled Vertical In-Line Centrifugal Pump
60 HZ PERFORMANCE CURVES

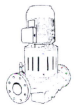

Bell & Gossett
a xylem brand

Bell & Gossett
a xylem brand

CURVES
B-360E

SERIES 1531
CLOSE COUPLED CENTRIFUGAL PUMP PERFORMANCE CURVES

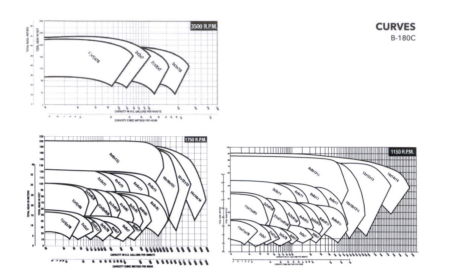

Series 80®-SC

Spacer-Coupled Vertical In-Line Centrifugal Pump

60 HZ PERFORMANCE CURVES

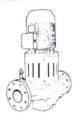

Bell & Gossett
a **xylem** brand

TABLE OF CONTENTS

USEFUL PUMP FORMULAS

$$\frac{\text{Pressure}}{\text{(PSI)}} = \frac{\text{Head (Feet) x Specific Gravity}}{2.31}$$

$$\frac{\text{Head}}{\text{(Feet)}} = \frac{\text{Pressure (PSI) x 2.31}}{\text{Specific Gravity}}$$

$$\frac{\text{Vacuum}}{\text{(Inches of Mercury)}} = \frac{\text{Dynamic Suction Lift (Feet) x .883}}{\text{x Specific Gravity}}$$

$$\frac{\text{Horsepower}}{\text{(Brake)}} = \frac{\text{GPM x Head (Feet) x Specific Gravity}}{3960 \times \text{Pump Efficiency}}$$

$$\frac{\text{Horsepower}}{\text{(Water)}} = \frac{\text{GPM x Head (Feet) x Specific Gravity}}{3960}$$

$$\frac{\text{Efficiency}}{\text{(Pump)}} = \frac{\text{Horsepower (Water)}}{\text{Horsepower (Brake)}} \times 100 \text{ Per Cent}$$

$$\frac{\text{NPSH}}{\text{(Available)}} = \text{Positive Factors} - \text{Negative Factors}$$

Affinity Laws: Effect of change of speed or impeller diameter on centrifugal pumps.

	GPM Capacity	Ft. Head	BHP
Impeller Diameter Change	$Q_2 = \frac{D_2}{D_1} Q_1$	$H_2 = \left(\frac{D_2}{D_1}\right)^2 H_1$	$P_2 = \left(\frac{D_2}{D_1}\right)^3 P_1$
Speed Change	$Q_2 = \frac{RPM_2}{RPM_1} Q_1$	$H_2 = \left(\frac{RPM_2}{RPM_1}\right)^2 H_1$	$P_2 = \left(\frac{RPM_2}{RPM_1}\right)^3 P_1$

Where Q = GPM, H = Head, P = BHP, D = Impeller Dia., RPM = Pump Speed

3500 RPM PUMP CURVES

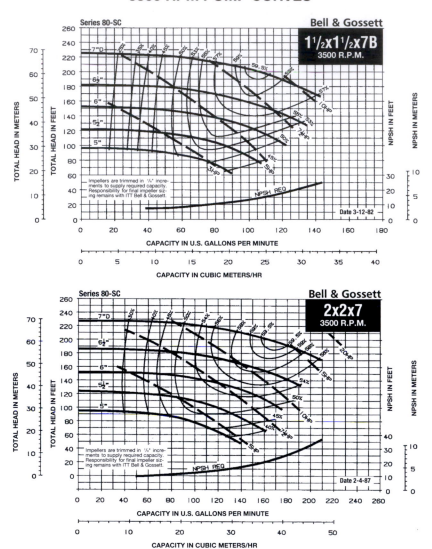

3500 RPM PUMP CURVES

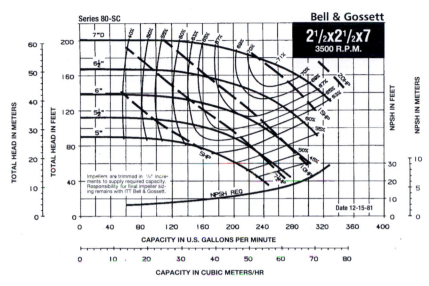

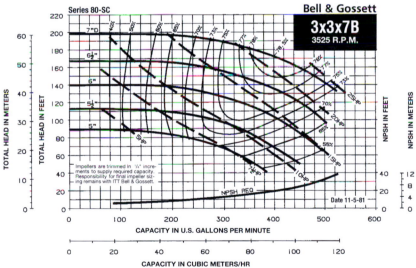

1750 RPM PUMP CURVES

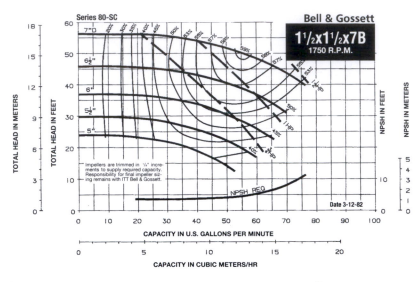

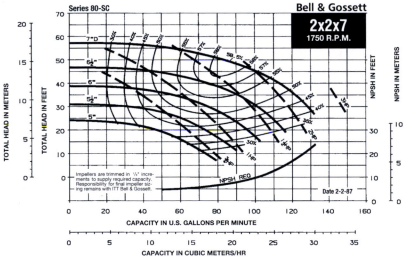

1750 RPM PUMP CURVES

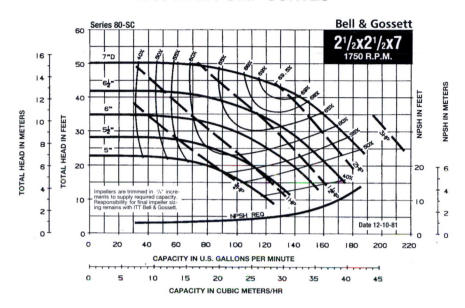

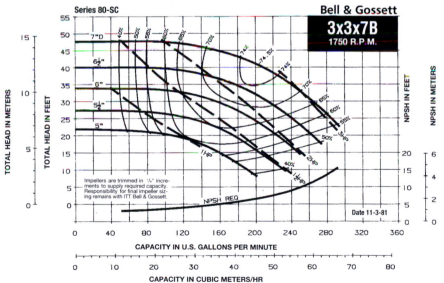

1750 RPM PUMP CURVES

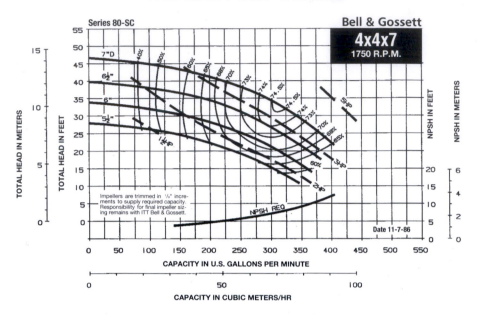

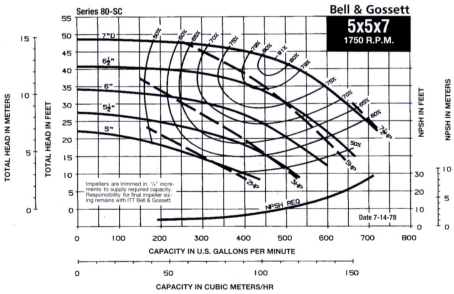

1750 RPM PUMP CURVES

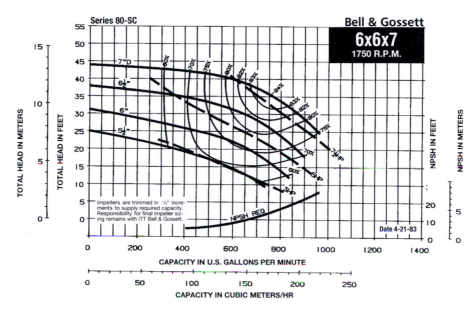

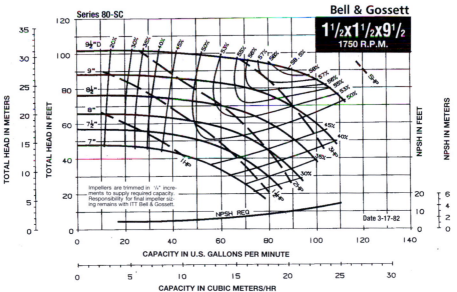

1750 RPM PUMP CURVES

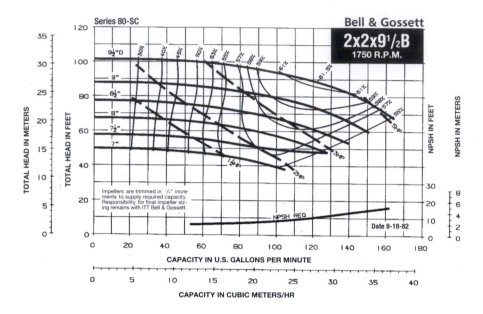

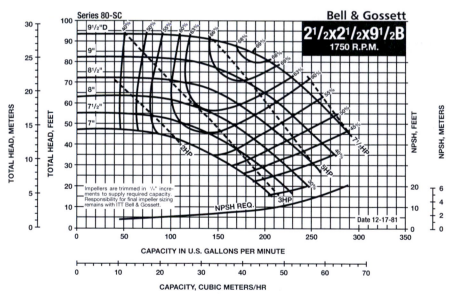

1750 RPM PUMP CURVES

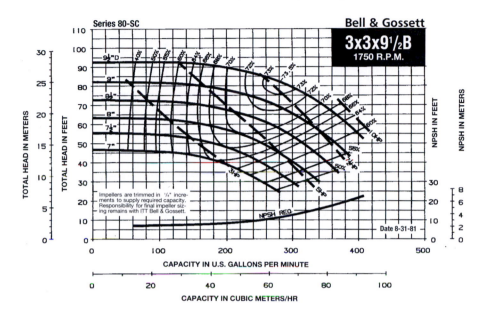

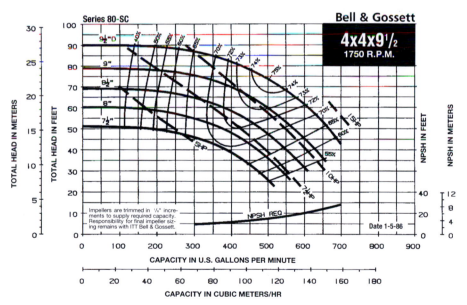

1750 RPM PUMP CURVES

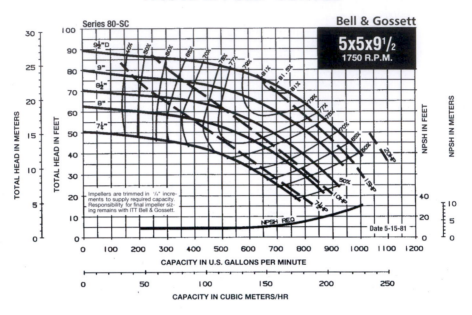

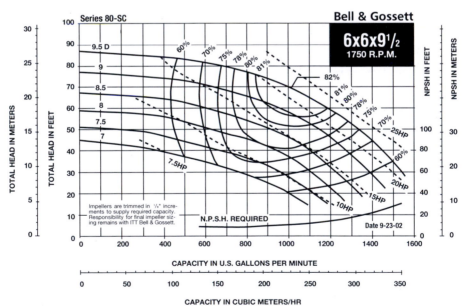

1750 RPM PUMP CURVES

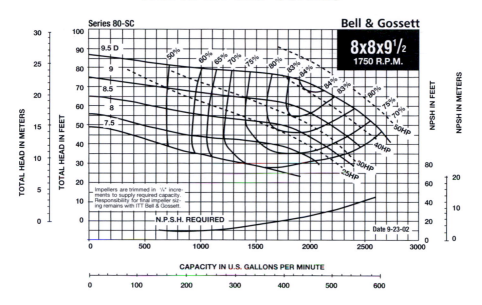

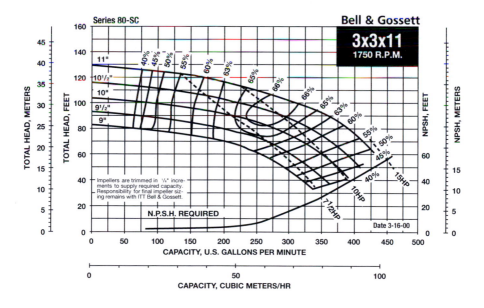

1750 RPM PUMP CURVES

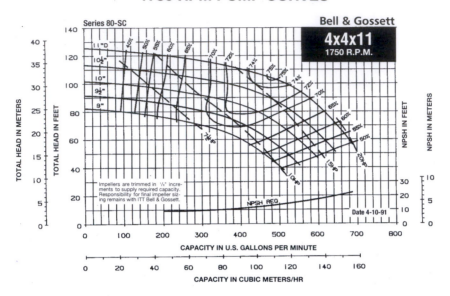

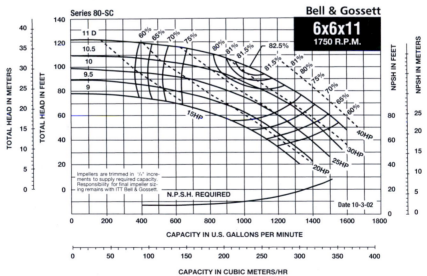

1770 RPM PUMP CURVES

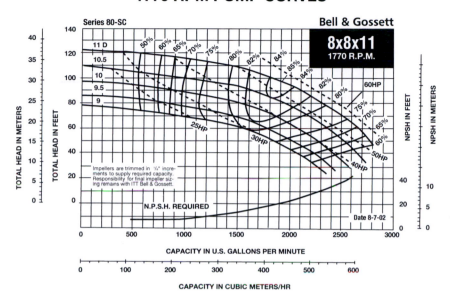

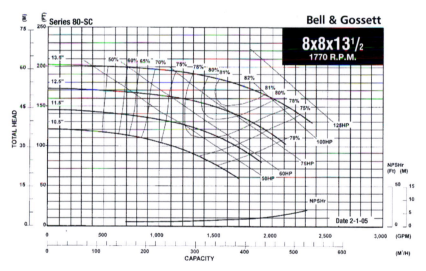

1770 RPM PUMP CURVES

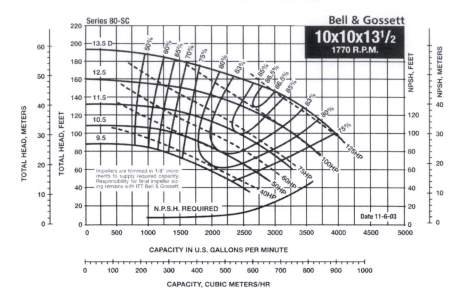

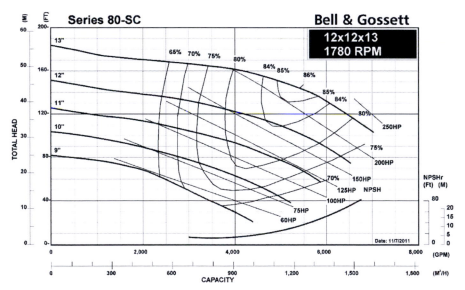

1770 RPM PUMP CURVES

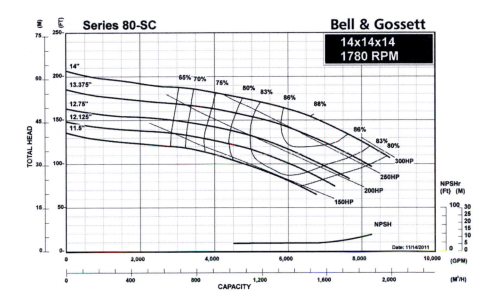

1150 RPM PUMP CURVES

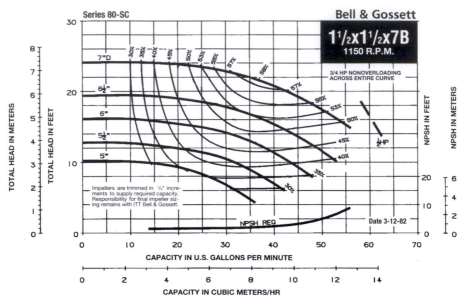

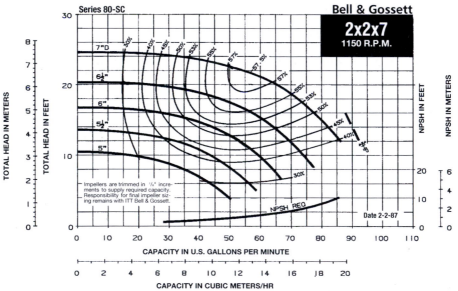

1150 RPM PUMP CURVES

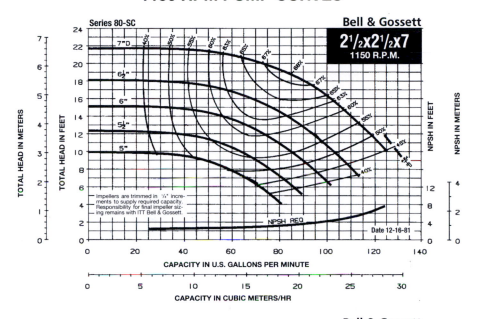

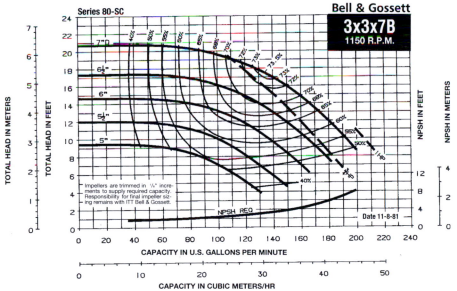

1150 RPM PUMP CURVES

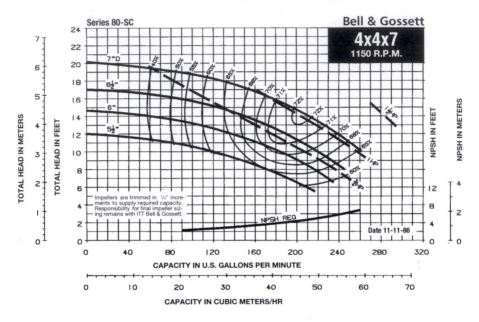

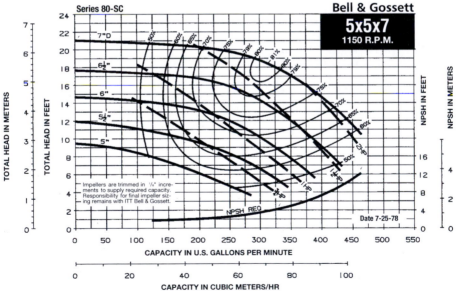

1150 RPM PUMP CURVES

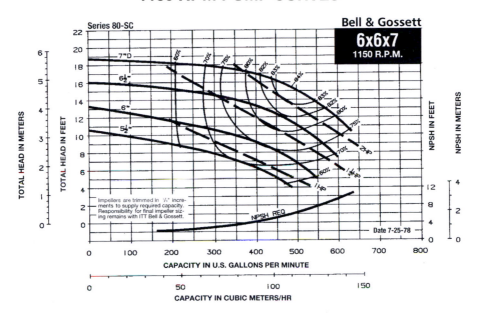

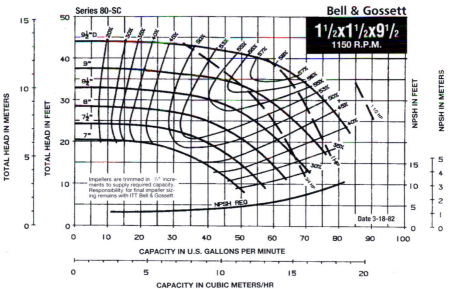

1150 RPM PUMP CURVES

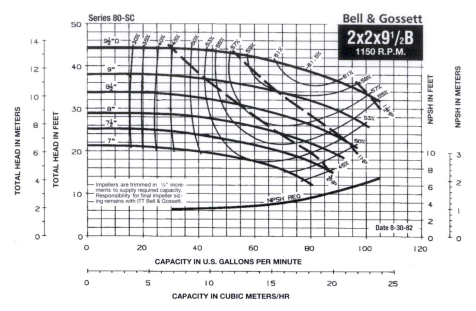

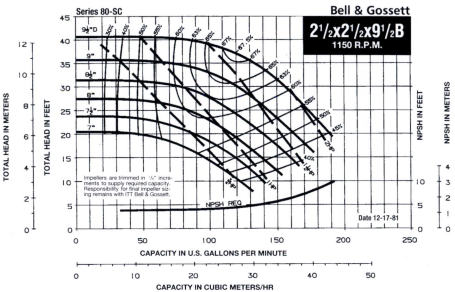

1150 RPM PUMP CURVES

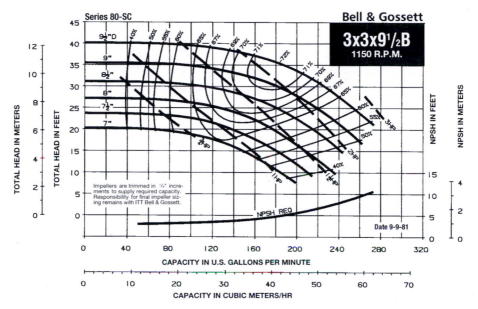

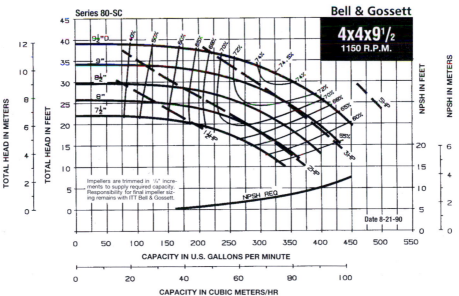

1150 RPM PUMP CURVES

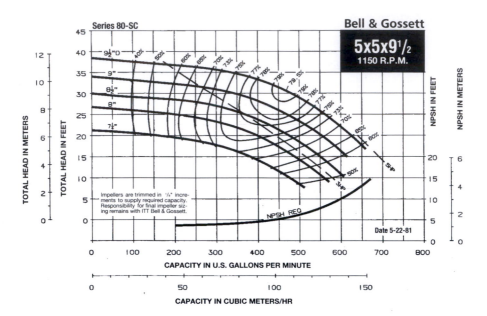

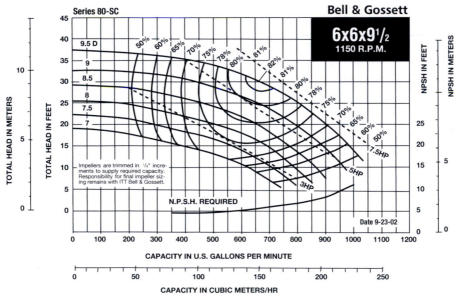

1150 RPM PUMP CURVES

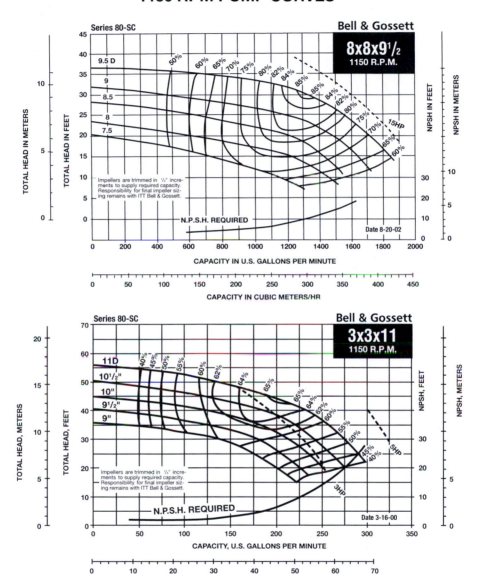

1150 RPM PUMP CURVES

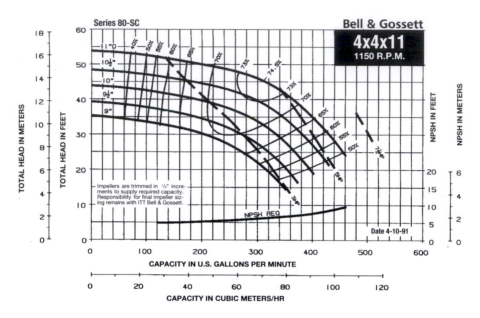

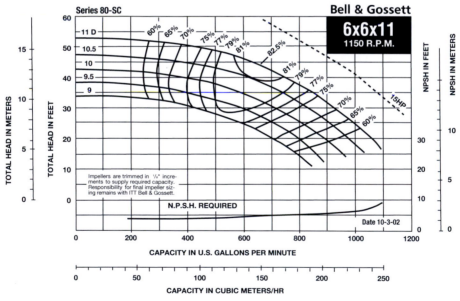

1150 RPM PUMP CURVES

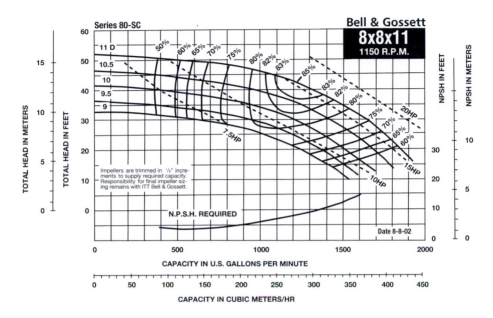

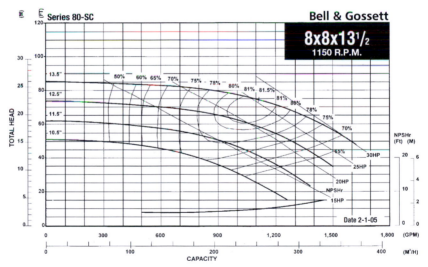

1150 RPM PUMP CURVES

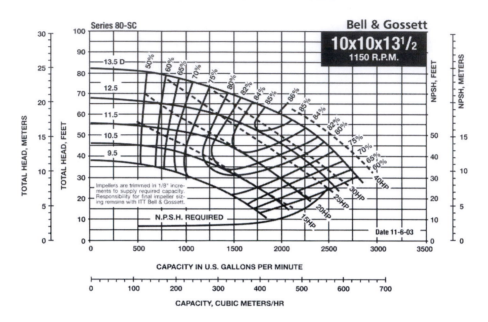

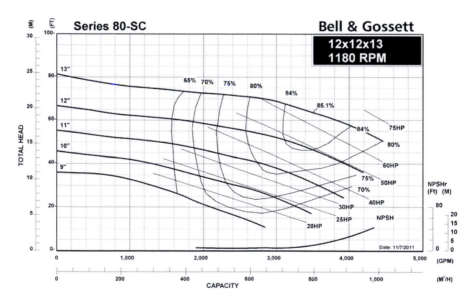

1150 RPM PUMP CURVES

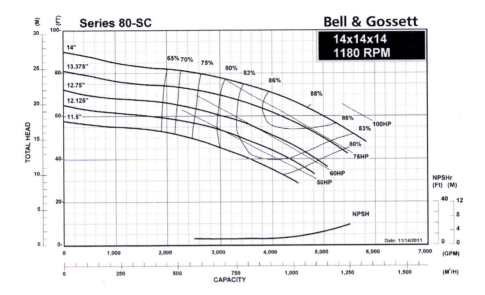

Xylem |'zīləm|

1) The tissue in plants that brings water upward from the roots;
2) a leading global water technology company.

We're 12,000 people unified in a common purpose: creating innovative solutions to meet our world's water needs. Developing new technologies that will improve the way water is used, conserved, and re-used in the future is central to our work. We move, treat, analyze, and return water to the environment, and we help people use water efficiently, in their homes, buildings, factories and farms. In more than 150 countries, we have strong, long-standing relationships with customers who know us for our powerful combination of leading product brands and applications expertise, backed by a legacy of innovation.

For more information on how Xylem can help you, go to www.xyleminc.com

Xylem Inc.
8200 N. Austin Avenue
Morton Grove, Illinois 60053
Phone: (847) 966-3700
Fax: (847) 965-8379
www.xyleminc.com/brands/bellgossett

Bell & Gossett is a trademark of Xylem Inc. or one of its subsidiaries.
© 2011 Xylem Inc. B-180C January 2012

Bell & Gossett
a **xylem** brand

SERIES 1531

CLOSE COUPLED CENTRIFUGAL PUMP PERFORMANCE CURVES

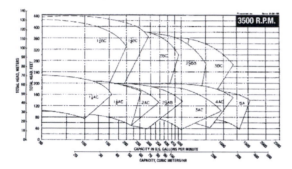

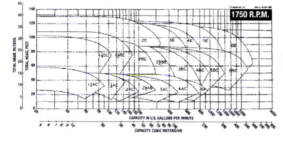

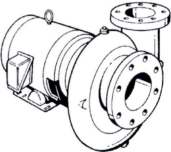

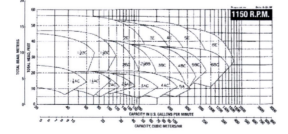

TABLE OF CONTENTS

USEFUL PUMP FORMULAS

$$\frac{\text{Pressure}}{\text{(PSI)}} = \frac{\text{Head (Feet) x Specific Gravity}}{2.31}$$

$$\frac{\text{Head}}{\text{(Feet)}} = \frac{\text{Pressure (PSI) x 2.31}}{\text{Specific Gravity}}$$

$$\frac{\text{Vacuum}}{\text{(Inches of Mercury)}} = \frac{\text{Dynamic Suction Lift (Feet) x .883}}{\text{x Specific Gravity}}$$

$$\frac{\text{Horsepower}}{\text{(Brake)}} = \frac{\text{GPM x Head (Feet) x Specific Gravity}}{3960 \text{ x Pump Efficiency}}$$

$$\frac{\text{Horsepower}}{\text{(Water)}} = \frac{\text{GPM x Head (Feet) x Specific Gravity}}{3960}$$

$$\frac{\text{Efficiency}}{\text{(Pump)}} = \frac{\text{Horsepower (Water)}}{\text{Horsepower (Brake)}} \text{ x 100 Per Cent}$$

$$\frac{\text{NPSH}}{\text{(Available)}} = \text{Positive Factors – Negative Factors}$$

Affinity Laws: Effect of change of speed or impeller diameter on centrifugal pumps.

	GPM Capacity	Ft. Head	BHP
Impeller Diameter Change	$Q_2 = \frac{D_2}{D_1} Q_1$	$H_2 = \left(\frac{D_2}{D_1}\right)^2 H_1$	$P_2 = \left(\frac{D_2}{D_1}\right)^3 P_1$
Speed Change	$Q_2 = \frac{RPM_2}{RPM_1} Q_1$	$H_2 = \left(\frac{RPM_2}{RPM_1}\right)^2 H_1$	$P_2 = \left(\frac{RPM_2}{RPM_1}\right)^3 P_1$

Where Q = GPM, H = Head, P = BHP, D = Impeller Dia., RPM = Pump Speed

3500 RPM PUMP CURVES

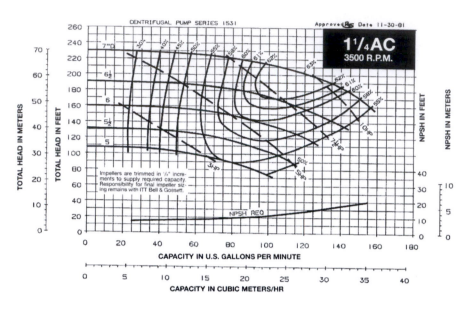

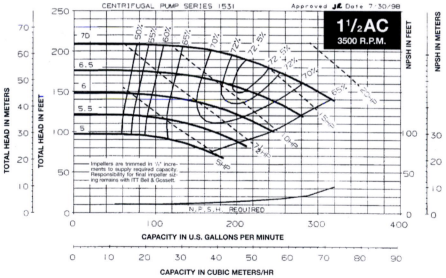

3500 RPM PUMP CURVES

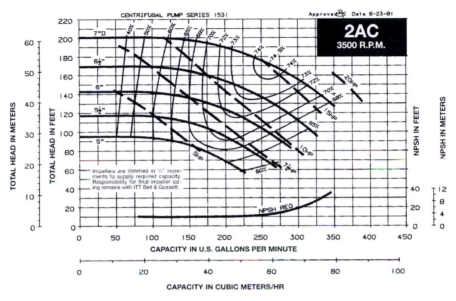

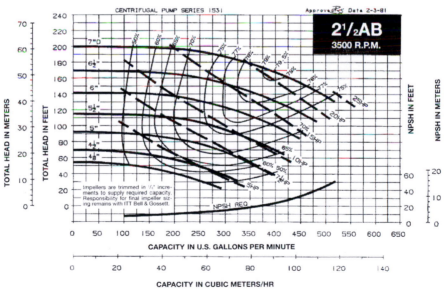

3500 RPM PUMP CURVES

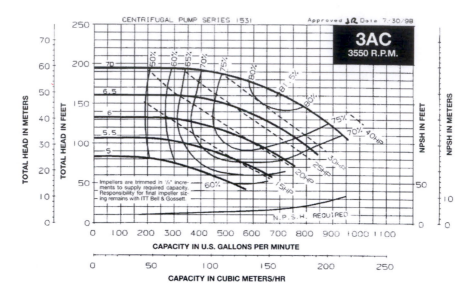

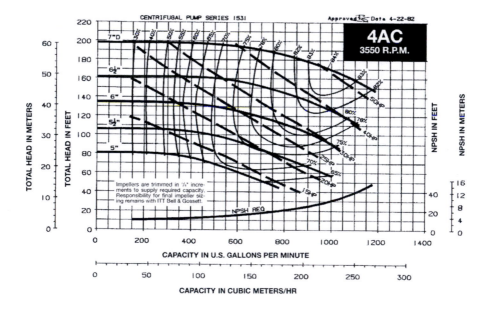

3500 RPM PUMP CURVES

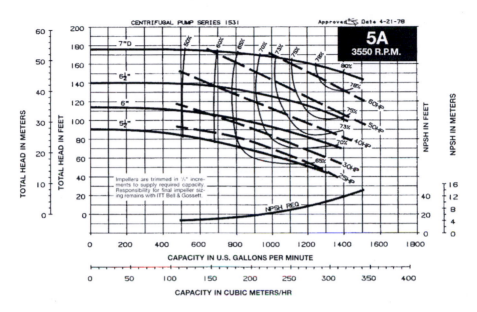

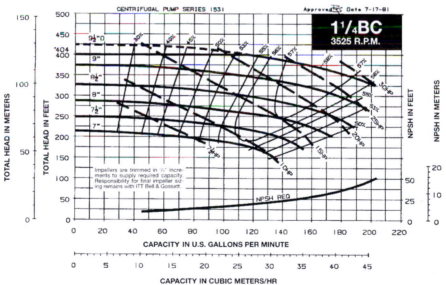

*Operation above 404' will exceed the maximum pressure rating of the pump.

3500 RPM PUMP CURVES

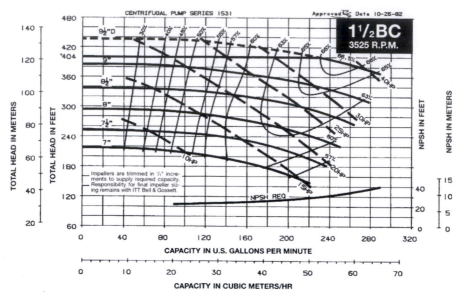

*Operation above 404' will exceed the maximum pressure rating of the pump.

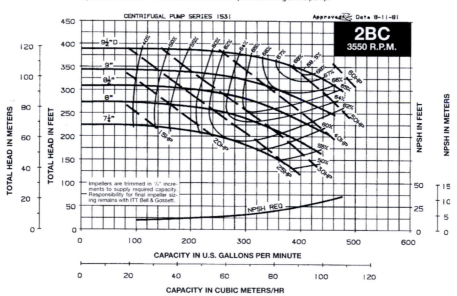

3500 RPM PUMP CURVES

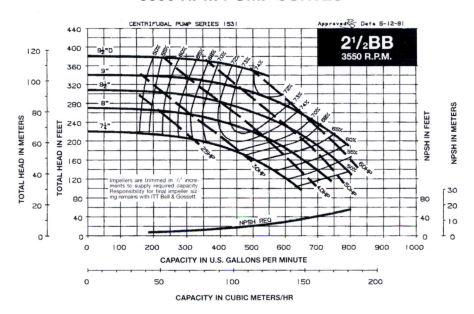

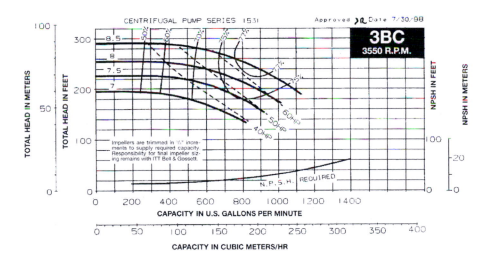

1750 RPM PUMP CURVES

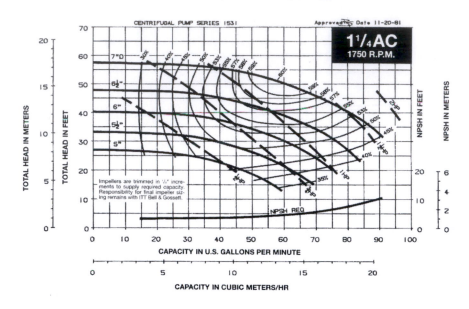

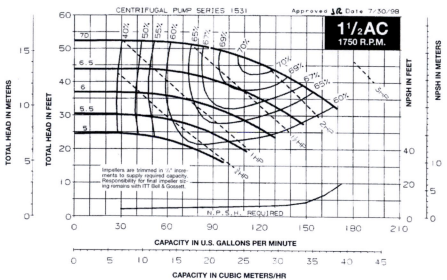

1750 RPM PUMP CURVES

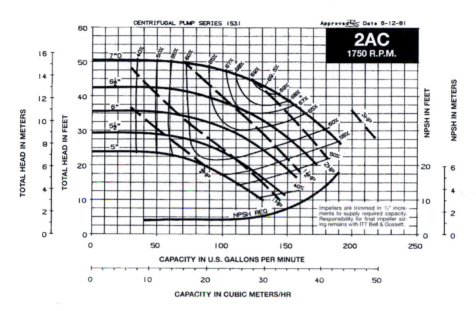

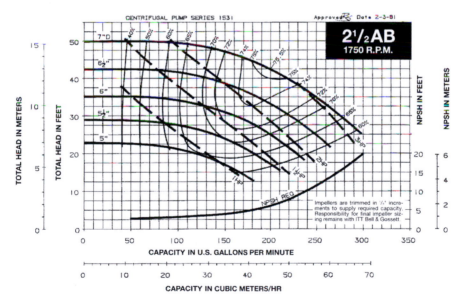

1750 RPM PUMP CURVES

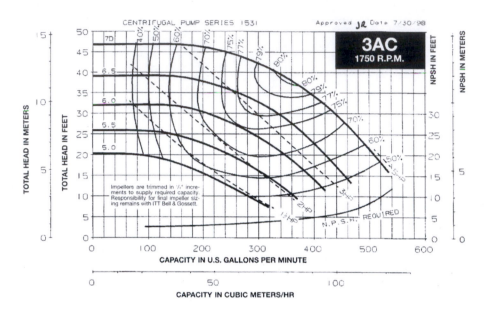

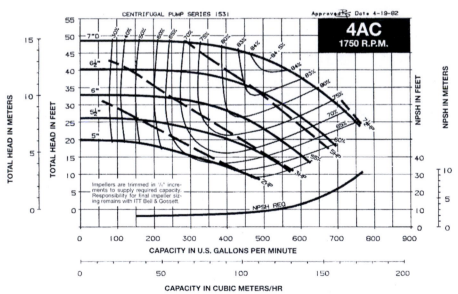

1750 RPM PUMP CURVES

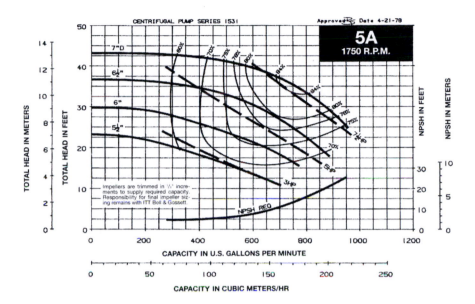

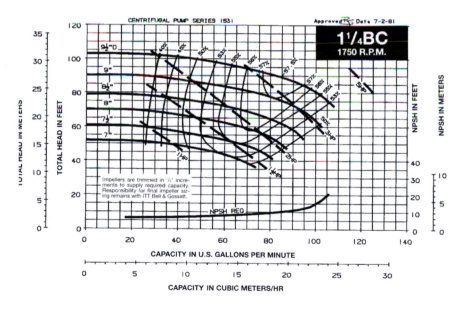

1750 RPM PUMP CURVES

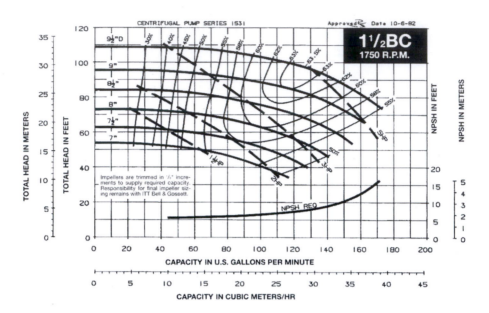

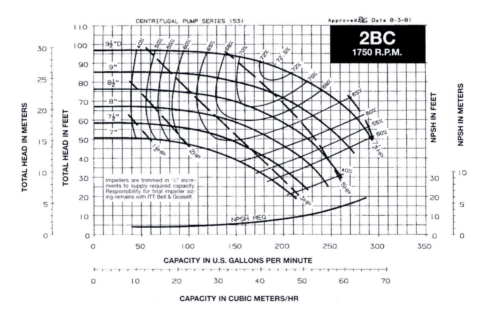

1750 RPM PUMP CURVES

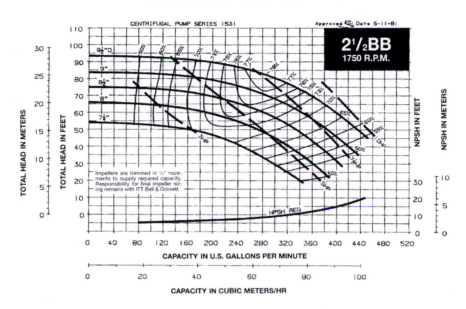

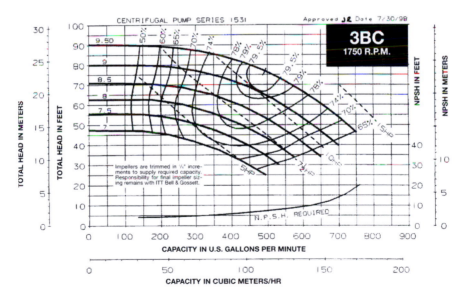

1750 RPM PUMP CURVES

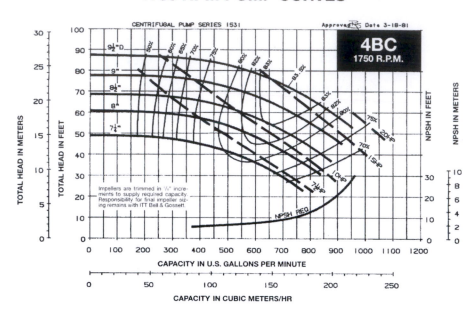

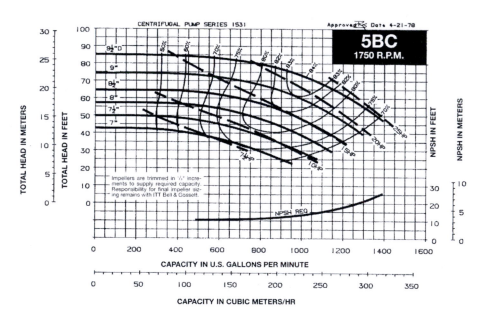

1750 RPM PUMP CURVES

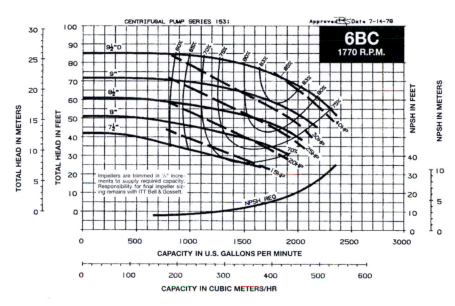

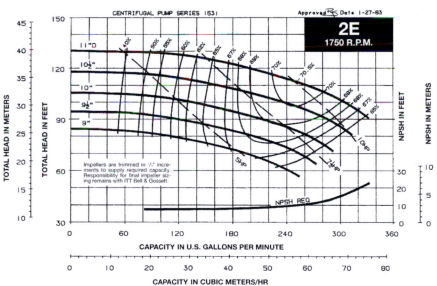

1750 RPM PUMP CURVES

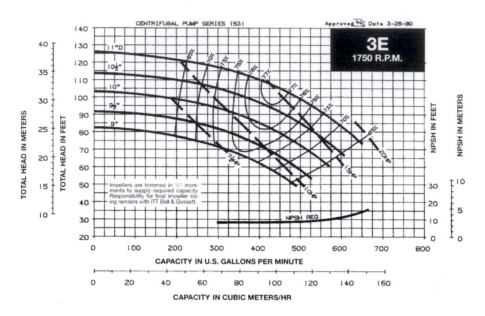

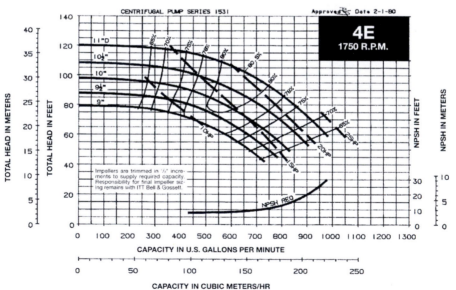

1750 RPM PUMP CURVES

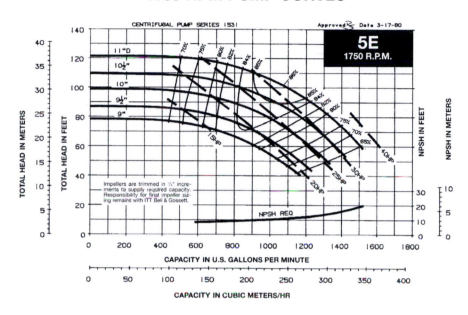

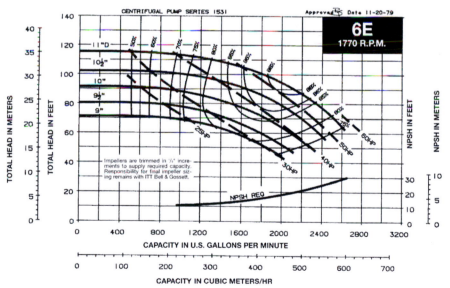

1150 RPM PUMP CURVES

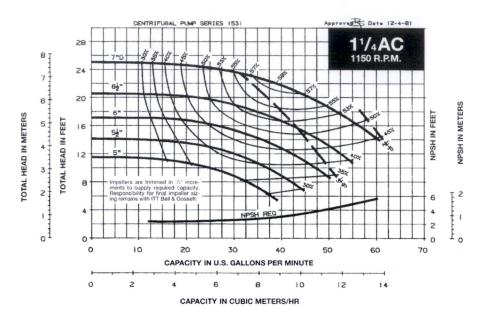

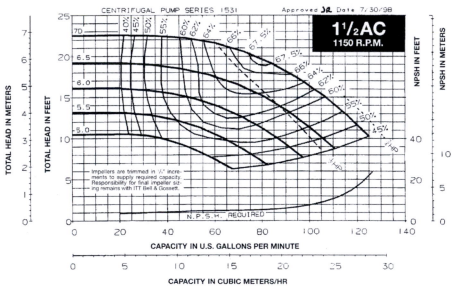

1150 RPM PUMP CURVES

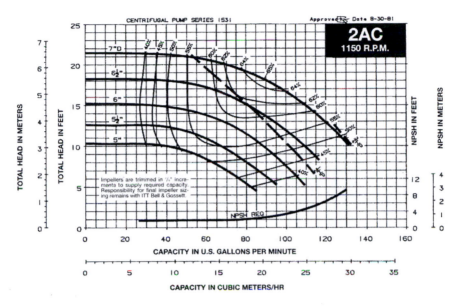

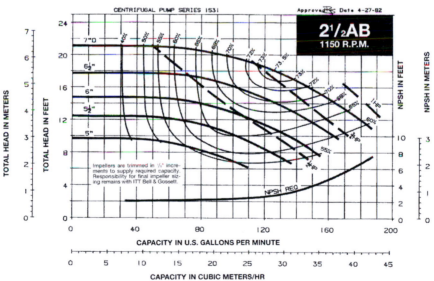

1150 RPM PUMP CURVES

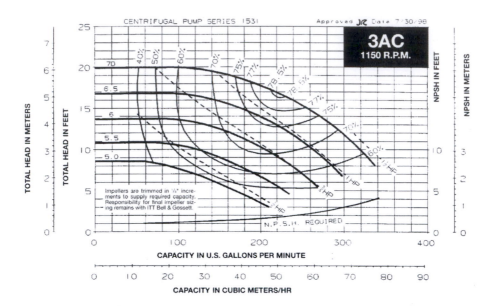

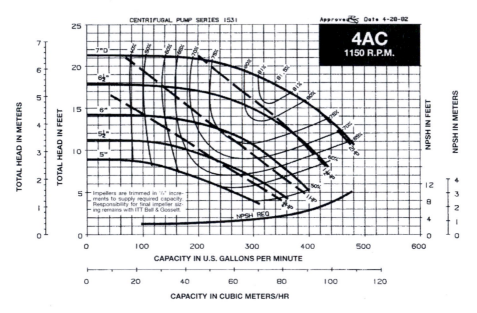

1150 RPM PUMP CURVES

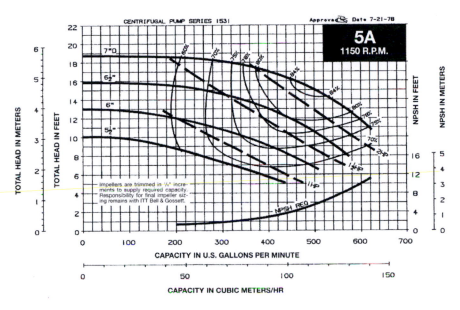

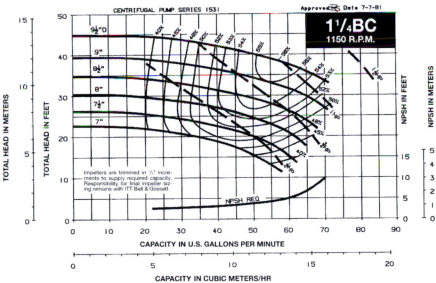

1150 RPM PUMP CURVES

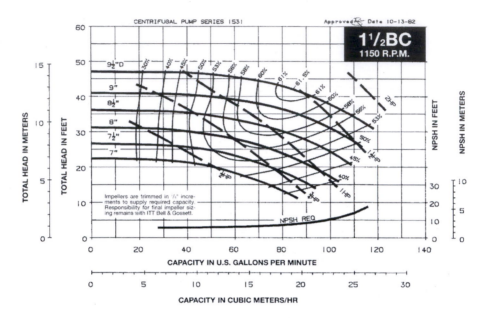

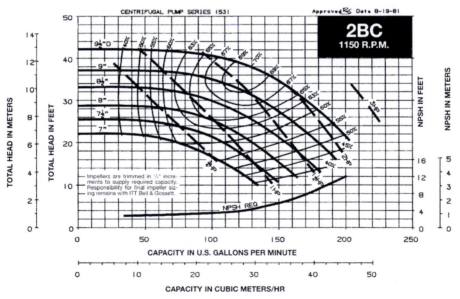

1150 RPM PUMP CURVES

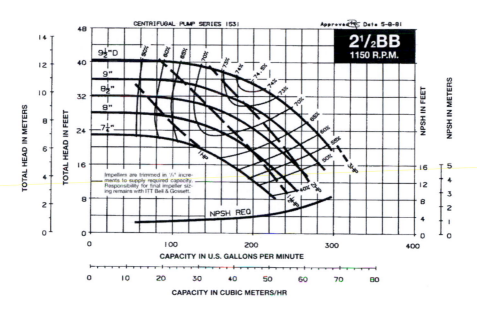

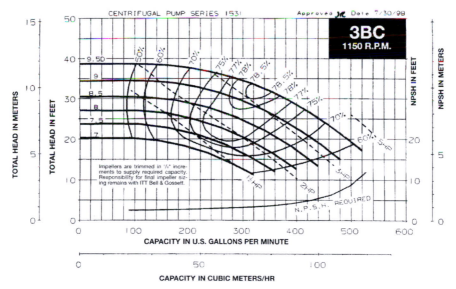

1150 RPM PUMP CURVES

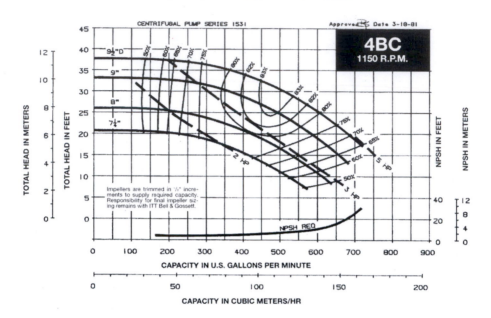

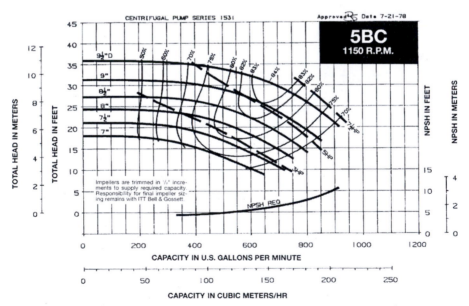

1150 RPM PUMP CURVES

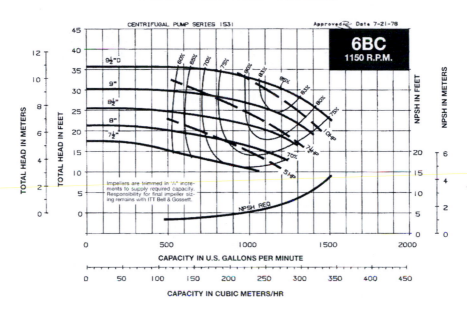

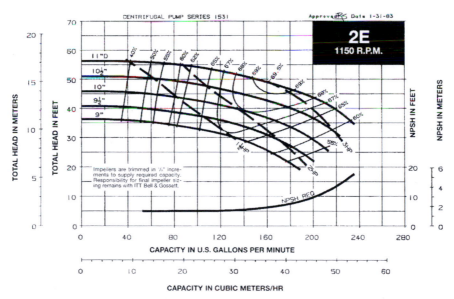

1150 RPM PUMP CURVES

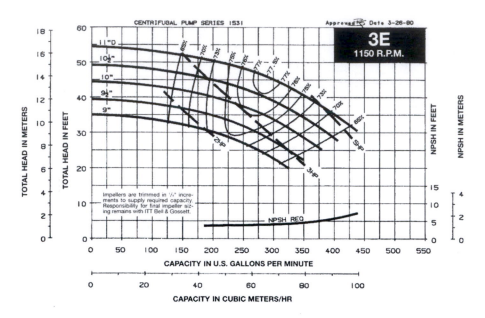

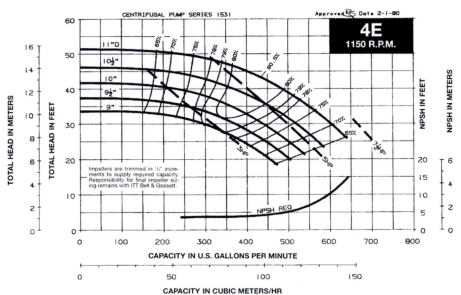

1150 RPM PUMP CURVES

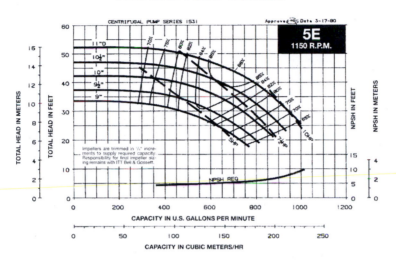

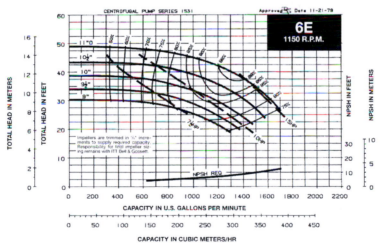

Xylem Inc.

8200 N. Austin Avenue, Morton Grove, Illinois 60053
Phone: (847) 966-3700 Fax: (847) 965-8379
www.xyleminc.com/brands/bellgossett

Bell & Gossett
a **xylem** brand

References

Baehr, H. D. and K. Stephan. *Heat Transfer*. Berlin, Germany: Springer, 2006.

Bejan, A. *Advanced Engineering Thermodynamics*. Hoboken, NJ: John Wiley & Sons, 2006.

Bejan, A. and A. D. Kraus. *Heat Transfer Handbook*. New York: Wiley, 2003.

Bell and Gossett. *GPX Plate and Frame Heat Exchangers*. Buffalo, NY: Xylem, Inc., 2011.

Bell and Gossett. *Series 80-SC Spacer-Coupled Vertical In-Line Centrifugal Pump: 60 Hz Performance Curves*. Morton Grove, IL: Xylem, Inc., 2012.

Bergman, T. L., A. S. Lavine, F. P. Incropera, and D. P. DeWitt. *Fundamentals of Heat and Mass Transfer*. Hoboken, NJ: Wiley, 2011.

Cengel, Y. and Boles, M.A. *Thermodynamics: An Engineering Approach*. New York: McGraw-Hill, 2011.

Churchill, S. W. and M. Bernstein. Correlating equation for forced-convection from gases and liquids to a circular-cylinder in cross-flow. *Journal of Heat Transfer—Transactions of the ASME* 99(2) (1977): 300–306.

Churchill, S. W. and H. Ozoe. Correlations for laminar forced convection in flow over an isothermal flat plate and in developing and fully developed flow in an isothermal tube. *Journal of Heat Transfer—Transactions of the ASME* 95(3) (1973): 416–419.

Darby, R. and J. D. Melson. Direct determination of optimum economic pipe diameter for non-Newtonian fluids. *Journal of Pipelines* 2 (1982): 11–21.

Fraas, A. P. *Heat Exchanger Design*. New York: Wiley, 1989.

Gnielinski, V. New equations for heat and mass-transfer in turbulent pipe and channel flow. *International Chemical Engineering* 16(2) (1976): 359–368.

Grundfos Research and Technology. *The Centrifugal Pump*. Bjerringbro, Denmark: Grundfos, 2014.

Janna, W. S. *Design of Fluid Thermal Systems*, pp. 163–184. Stamford, CT: Cengage Learning, 2015.

Kakac, S., H. Liu, and A. Pramuanjaroenkij. *Heat Exchangers: Selection, Rating, and Thermal Design*. Boca Raton, FL: CRC Press, 2012.

Karassik, I. J., J. P. Messina, P. Cooper, and C. C. Heald. *Pump Handbook*. New York: McGraw-Hill, 2008.

Kays, W. M. and A. L. London. *Compact Heat Exchangers*. New York: Krieger Publishing, 1998.

Kern, D. Q. *Process Heat Transfer*. New York: McGraw-Hill, 1950.

Klein, S. and Nellis, G. *Thermodynamics*. Cambridge: Cambridge University Press, 2012.

Mason, J. L. Heat transfer in cross-flow. Proceedings of the Second U.S. National Congress of Applied Mechanics, pp. 801–803, 1954.

Moody, L. F. Friction factors for pipe flow. *Transactions of the ASME* 68(8) (1944): 671–684.

Moran, M.J., Shapiro, H.N., Boettner, D.D., and Bailey, M.B. *Fundamentals of Engineering Thermodynamics, 8e*. Hoboken, NJ: John Wiley & Sons, 2014.

Rao, B. P., B. Sunden, and S. K. Das. An experimental and theoretical investigation of the effect of flow maldistribution on the thermal performance of plate heat exchangers. *Journal of Heat Transfer*, 127(3) (2005): 332–343.

Rohsenow, W. M., J. P. Hartnett, and Y. I. Cho. *Handbook of Heat Transfer*. New York: McGraw-Hill, 1998.

Smith, P. and R. Zappe. *Valve Selection Handbook: Engineering Fundamentals for Selecting the Right Valve Design for Every Industrial Flow Application*. Oxford, UK: Gulf Professional Publishing, 2004.

The Crane Company. *Flow of Fluids Through Valves, Fittings and Pipe*. Stamford, CT: The Crane Company, 2013.

Thompson, A. and B. N. Taylor. *Guide for the Use of the International System of Units (SI)*. Gaithersburg, MD: National Institute of Standards and Technology, 2008.

Waide, P. and C. U. Brunner. *Energy-Efficiency Policy Opportunities for Electric Motor-Driven Systems*. Paris, France: International Energy Agency, 2011.

Wang, C., Y. Chang, Y. Hsieh, and Y. Lin. Sensible heat and friction characteristics of plate fin-and-tube heat exchangers having plate fins. *International Journal of Refrigeration* 19(4) (1996): 223–230.

Index